STUDENT STUDY GUIDE WITH SELECTED SOLUTIONS

DAVID REID

Eastern Michigan University

PHYSICS

JAMES S. WALKER

Upper Saddle River, NJ 0745

Executive Editor: Alison Reeves
Assistant Editor: Christian Botting
Executive Managing Editor: Kathleen Schiaparelli
Assistant Managing Editor: Dinah Thong
Production Editor: Natasha Wolfe
Supplement Cover Manager: Paul Gourhan
Supplement Cover Designer: PM Workshop Inc.
Manufacturing Buyer: Lisa McDowell
Cover Photos: Jim Ross, NASA/Dryden Flight Research Center; Thomas Ives, The Stock Market; David Jacobs, Tony Stone Images; Roger Tully, Toney Stone Images; Wayne Lynch, DRK Photo; Stocktrek, PhotoDisc, Inc.; Jamie Squire, Allsport Photography (USA), Inc.

Upper Saddle River, NJ 07458

Printed in the United States of America

10 9 8 7 6 5 4 3

ISBN 0-13-027064-4

Prentice-Hall International (UK) Limited, London
Prentice-Hall of Australia Pty. Limited, Sydney
Prentice-Hall Canada, Inc., Toronto
Prentice-Hall Hispanoamericana, S.A., Mexico City
Prentice-Hall of India Private Limited, New Delhi
Pearson Education Asia Pte. Ltd., Singapore
Prentice-Hall of Japan, Inc., Tokyo
Editora Prentice-Hall do Brazil, Ltda., Rio de Janeiro

TABLE OF CONTENTS

PREFACE

This *Student Study Guide with Selected Solutions* is designed to assist you in your study of the fascinating and sometimes challenging world of physics using *Physics, First Edition,* by James S. Walker. To do this, I have provided a Chapter Review, which consists of a comprehensive (but brief) review of every section in the text. Numerous solved examples and exercises appear throughout each Chapter Review. The examples follow the two-column format of the text, while the solutions to the exercises have a more traditional layout. Together with the Chapter Review, each chapter of the Study Guide contains a list of objectives, a practice quiz, a glossary of key terms and phrases, a table of important formulas, and a table that reviews the units of the new quantities introduced.

In addition to the above materials that I have provided, you will also find Warm-Up and Puzzle questions by *Just in Time Teaching* innovators Gregor Novak and Andrew Gavrin (Indiana University-Purdue University, Indianapolis), Practice Problems by Carl Adler (East Carolina University) and Solutions to Select Problems from the *Instructor's Solutions Manual*. Taken together, the information in this Study Guide, when used in conjunction with the main text, should enhance your ability to master the many concepts and skills needed to understand physics, and, therefore, the world around you. Work hard, and most importantly, have fun doing it!

I am indebted to many for helping me to complete this work. Most directly, I thank Mr. Christian Botting of Prentice Hall (and Elizabeth Kell before him) for his continued work with me on this project. Thanks is also due to Ms. Nicole Blair whose occasional comic relief made the process of writing this study guide more enjoyable that it would have been otherwise. I most especially wish to acknowledge Dr. Anand P. Batra of Howard University. He provided an excellent review of the physics content and made countless valuable suggestions.

David D. Reid
Eastern Michigan University
July, 2001

To my wife, *Annie*
who waited patiently for me to complete this work

CHAPTER 1
INTRODUCTION

Chapter Objectives

After studying this chapter, you should:

1. know the three most common basics physical quantities in physics and their units.
2. know how to determine the dimension of a quantity and perform a dimensional check on any equation.
3. be familiar with the most common metric prefixes.
4. be able to perform calculations keeping proper account of the number of significant figures.
5. be able to convert quantities from one set of units to another.
6. be able to perform quick order-of-magnitude calculations.

Warm-Ups

1. Do you believe that the metric system is superior to the previous systems of measure, such as the everyday system used in the United States? Whatever your answer, what arguments would you use to persuade a person who has a different opinion?
2. Estimate the number of seconds in a human lifetime. We'll let you choose the definition of lifetime. Do all reasonable choices of lifetime give answers that have the same order of magnitude?
3. Which is a faster speed, 30 mi/h or 13 m/s? Describe in words how you obtained your answer.
4. Estimate how many 20-cm × 20-cm tiles it would take to tile the floor and three sides of a shower stall. The stall has a 16 ft^2 floor and 5-ft walls.

Chapter Review

1–1 – 1–2 Basic Physical Quantities and Standard Prefixes

The study of **physics** deals with the fundamental laws of nature and many of their applications. These laws govern the behavior of all physical phenomena. We describe the behavior of physical systems using various quantities that we create for this purpose; however, there are three quantities — **length**, **mass**, and **time** — that we take as fundamental quantities and we use these three to create other quantities.

We define a system of units for these quantities so that we can specify how much length, mass, or time we have. The system of units used in this book is the SI, which stands for Système International. In this system the unit of length is the **meter** (m), the unit of mass is the **kilogram** (kg), and the unit of time is the **second** (s). This system of units is still sometimes referred to by its former name, the mks system.

As you probably noticed, SI units are based on the metric system. An important aspect of this system is its hierarchy of prefixes used for quantities of different magnitudes. Certain of these prefixes are used very frequently in physics, so you should become very familiar with them. Some of the more common ones are listed here:

Power	Prefix	Symbol
10^{-15}	femto	f
10^{-12}	pico	p
10^{-9}	nano	n
10^{-6}	micro	μ
10^{-3}	milli	m
10^{-2}	centi	c
10^{3}	kilo	k
10^{6}	mega	M

Exercise 1–1 Metric Prefixes Write the following quantities using a convenient metric prefix.
(a) 0.00025 m, **(b)** 25,000 m **(c)** 250 m **(d)** 250,000,000 m **(e)** 0.0000025 m

Solution **(a)** 0.25 mm **(b)** 25 km **(c)** 0.25 km **(d)** 250 Mm **(e)** 2.5 μm

Practice Quiz

1. Which of the following quantities is not one of the fundamental quantities?
   (a) length (b) speed (c) time (d) mass

1–3 Dimensional Analysis

As stated previously, in physics we derive the physical quantities of interest from the set of fundamental quantities of length, mass, and time. The **dimension** of a quantity tells us what *type* of quantity it is.

When indicating the dimension of a quantity only, we use capital letters and enclose them in brackets. Thus, the dimension of length is represented by [L], mass by [M], and time by [T].

We use many equations in physics, and these equations must be dimensionally consistent. It is extremely useful to perform a dimensional analysis on any equation about which you are unsure. If the equation is not dimensionally consistent, it cannot be a correct equation. The rules are simple:

* Two quantities can be added or subtracted only if they are of the same dimension.
* Two quantities can be equal only if they are of the same dimension.

Notice that only the dimension needs to be the same, not the units. It is perfectly valid to write 12 inches = 1 foot because both quantities are lengths, [L] = [L], even though their units are different. However, it is not valid to write x inches = t seconds because the quantities have different dimensions; [L] ≠ [T].

Example 1–2 Checking the Dimensions Given that the quantities x (m), v (m/s), a (m/s^2), and t (s) are measured in the units shown in parentheses, perform a dimensional analysis on the following equations.

(a) $x = t$ **(b)** $x = 2vt$ **(c)** $v = at + t/x$ **(d)** $x = vt + 3at^2$

Picture the Problem There is no picture.

Strategy Write each equation in terms of its dimensions and check if the equation obeys the preceding rules.

Solution

Part (a)

1. Write the equation with dimensions only: $[L] = [T]$

Since these dimensions are not the same, the equation is not valid.

Part (b)

2. Write out the dimensions of this equation: $[L] = \frac{[L]}{[T]} \times [T] = [L]$

The right-hand-side dimension is equal to the dimension on the left, so the equation is dimensionally correct.

Part (c)

3. Write out the dimension of this equation: $\frac{[L]}{[T]} = \frac{[L]}{[T^2]} \times [T] + \frac{[T]}{[L]} = \frac{[L]}{[T]} + \frac{[T]}{[L]}$

The first and second terms on the right-hand-side are not of equal dimension and cannot be added. This not a valid equation.

Part (d)

4. Write out the dimension of this equation: $[L] = \frac{[L]}{[T]} \times [T] + \frac{[L]}{[T^2]} \times [T^2] = [L] + [L]$

Here, both terms on the right have the same dimension, which is also equal to the dimension on the left. This equation is dimensionally correct.

Insight Notice that in dimensional analysis purely numerical factors are ignored because they are **dimensionless**. Since there are dimensionless quantities, dimensional consistency does not guarantee that the equation is physically correct, but it makes for a quick and easy first check.

Practice Quiz

2. Which of the following expressions is dimensionally correct?

(a) [L] = [M] x [T] **(b)** [T] = [L]/[T] **(c)** $[L] = \frac{[L]}{[T]} \times [T]$ **(d)** $[M] = \frac{[L^2]}{[T]}$

3. If speed v has units of m/s, distance d has units of m, and time t has units of s, which of the following expressions in dimensionally correct?

(a) v = t/d **(b)** t = vd **(c)** d = v/t **(d)** t = d/v

1–4 Significant Figures

All measured quantities carry some uncertainty in their values. When working with the values of quantities it is important to keep proper account of the digits that are reliably known. Such digits are called **significant figures**. The rules for working with significant figures are as follows:

* *Multiplication and Division*: The number of significant figures in the result of a multiplication or division equals the number of significant figures in the factor containing the fewest significant figures.

* *Addition and Subtraction*: The significant figures in the result of an addition or subtraction are located only in the *places* (hundreds, ones, tenths, etc.) that are reliably known for *every* value in the sum.

Example 1–3 Driving in a Residential Zone: On most residential streets in the United States the speed limit is 25 mi/h (= 11 m/s). If a car drives down a neighborhood side street at the legal speed limit for 120.46 s, how much distance does the car cover?

Picture the Problem Our sketch shows the car moving along a straight road.

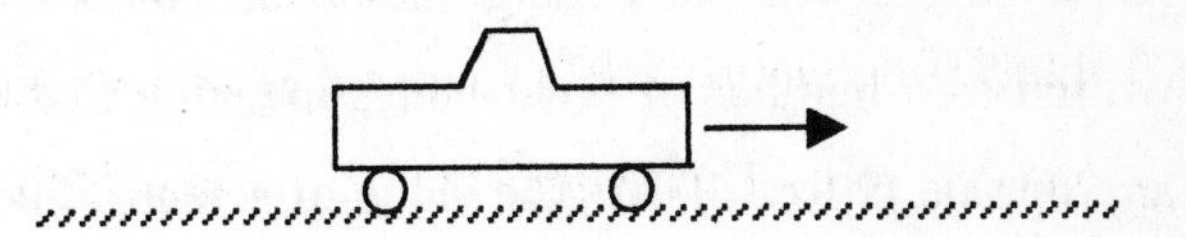

Strategy The distance traveled is the speed multiplied by the time of travel.

Solution

Multiply the speed and the time to get distance: $d = 11\ \text{m/s} \times 120.46\ \text{s} = 1300\ \text{m}$

Insight There are two important things to notice about the result. First, despite the fact that the time is known to five significant figures, the speed is known only to two, and so the result has only two significant figures. Second, the final two zeros in the value 1300 are not significant. They must be written, however, to give the proper magnitude of the value. It can often be unclear whether such zeros are significant. This problem can be avoided by using **scientific notation** (discussed next).

Exercise 1–4 Significant Figures A calculation involves the addition of two measured distances d_1 = 1250 m, and d_2 = 336 m. If each measurement is given to three significant figures, what is the result of the calculation?

Solution Adding the two distances we get $d_1 + d_2 = 1250\ \text{m} + 336\ \text{m} = 1590\ \text{m}$.

The answer is not 1586 because even though the 6 is significant in 336 m, the one's place of 1250 m (the 0) is not significant, so the one's place of the result cannot be significant. You may wonder about the fact that there is no significant figure in the thousand's place of 336 m because the value requires no digit there: however, because there is no digit there we know that place with certainty.

Be aware that to avoid excessive round off error you should round only to the proper number of significant figures at the very end of a calculation. In Example 1.4, if the distance calculated is only an intermediate step in a longer calculation, then the value 1586 m should be used in the subsequent steps. In general, keep at least one additional digit for values calculated in intermediate steps. Another, even better, approach is not to calculate intermediate values numerically but just carry through the formulas inserting numerical values only once at the end. This will be illustrated in the "Tips" section below.

Scientific Notation

A very useful way of writing numerical values is to use scientific notation. In this notation a value is written as a number of order unity (meaning that only one digit is left of the decimal point) times the appropriate power of 10. The value of scientific notation is that it allows for quick identification of the **order-of-magnitude** (power of 10) of a quantity, calculations are often easier to perform when the values are listed this way, and it removes any ambiguity in the number of significant figures. For example, if the number 3500 has only one significant figure, we write it as 4×10^3; if it has two, we write 3.5×10^3; if it has three, we write 3.50×10^3; and if it has four significant figures we write 3.500×10^3.

Exercise 1–5 Scientific Notation Write the following quantities using scientific notation assuming three significant figures. **(a)** 0.00250 m **(b)** 12,060 m **(c)** 451 m **(d)** 8.00 m **(e)** 0.00003593 m

Answer: (a) 2.50×10^{-3} m **(b)** 1.21×10^4 m **(c)** 4.51×10^2 m **(d)** 8.00×10^0 m **(e)** 3.59×10^{-5} m

When the value is already of order unity, as with part (d), the power of 10 is often dropped. In such cases, the fact that the two zeros are written after the decimal point indicates that they are significant figures.

Practice Quiz

4. Assuming that every nonzero digit is significant, consider the following product of numbers: $1.34 \times 10.75 \times 0.042$. Which answer is correct to the proper number of significant figures?

   (a) 6 **(b)** 0.61 **(c)** 6.05 **(d)** 6.0501

5. Assuming that only nonzero digits are significant, consider the following sum of numbers: 1700 + 338 + 13. Which answer is correct to the proper number of significant figures?

   (a) 2051 **(b)** 2050 **(c)** 2100 **(d)** 2000

6. Consider the following expression: $(5.93) \times (8.762) + (2.116) \times (3.70)$. Which answer is correct to the proper number of significant figures?

(a) 59.78786 (b) 59.79 (c) 59.83 (d) 59.8

7. Which of the following numbers is proper scientific notation for 25,300?

(a) 2.53×10^4 (b) 25.3×10^3 (c) 2.53×10^3 (d) 0.253×10^5

8. The number 7.4×10^5 is equivalent to which of the following?

(a) 7.4 (b) 740 (c) 7,400 (d) 740,000

1–5 Converting Units

Even though we predominantly use SI units, it will often be necessary to convert between SI and other units. A conversions can be accomplished using a **conversion factor** that is constructed by knowing how much of a quantity in one unit equals that same quantity in another unit. A conversion factor is a ratio of equal quantities written such that, when multiplied by a quantity, the undesired unit algebraically cancels leaving only the desired unit. This concept is best illustrated by example.

Example 1–6 Volume of a Box A typical cardboard box provided by moving companies measures 1.50 ft × 1.50 ft × 1.33 ft. Determine the volume (V) of clothes that you can pack into this box in cubic meters.

Picture the Problem The diagram represents the box whose volume we wish to determine.

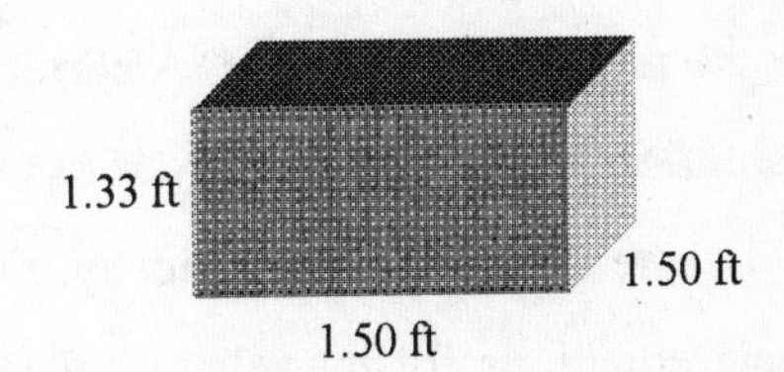

Strategy We first calculate the volume in the given units, determine the conversion factor, then convert the volume to cubic meters.

Solution

1. Calculate the volume as given: $V = 1.50 \text{ ft} \times 1.50 \text{ ft} \times 1.33 \text{ ft} = 2.993 \text{ ft}^3$

2. Write the number of meters in a foot: $1 \text{ m} = 3.281 \text{ ft}$

3. Write the conversion factor from feet to meters: $\dfrac{1 \text{ m}}{3.281 \text{ ft}}$

4. The conversion factor from ft^3 to m^3 is: $\left(\frac{1\text{ m}}{3.281\text{ ft}}\right)^3 = \frac{1\text{ m}^3}{35.32\text{ ft}^3}$

5. Multiply the volume by the conversion factor: $V = 2.993\ \text{ft}^3\left(\frac{1\text{ m}^3}{35.32\ \text{ft}^3}\right) = 0.0847\ \text{m}^3$

Insight In the final step, the unit ft^3 cancels just as numbers would. Setting up this cancellation is the crucial step in unit conversion. You will get plenty of practice converting units in your study of physics.

Practice Quiz

9. Given that 1 in. = 2.54 cm, convert 250.0 cm to inches.

(a) 635 in. **(b)** 0.394 in. **(c)** 98.4 in. **(d)** 150.0 in.

10. Convert the speed 1.00×10^2 m/s to km/h.

(a) 36 km/h **(b)** 360 km/h **(c)** 3.60×10^8 km/h **(d)** 27.8 km/h

1–7 Problem Solving in Physics

Solving physics problems is a logical and creative endeavor for which there is no set prescription; however, there are certain practices that help this creative process to flourish. First, a *careful reading* of the problem is necessary to fully grasp the question being posed and the information being given. It is often useful to separately write out all the given and required information; several of the solved examples in this study guide illustrate that approach. It is also good practice to make a *sketch* of the problem and *visualize* the physics taking place. A correct mental picture of the problem takes you a long way toward a correct solution. Next, map out your *strategy* for the solution. Here, you basically solve the problem logically before doing it mathematically. For the mathematical solution you need to *identify and solve* the appropriate equations for the relevant physics. Finally, you should *check and explore* your result to be sure that the answer makes sense in the context of the problem.

Reference Tools and Resources

I. Key Terms and Phrases

physics the study of the fundamental laws of nature and many of their applications

SI units the internationally adopted standard system of units (based on meters, kilograms, and seconds) for quantitatively measuring quantities

dimension of a quantity the fundamental type of a quantity such as length, mass, or time

dimensional analysis a type of calculation that checks the dimensional consistency of an equation

significant figures the digits in the numerical value of a quantity that are known with certainty

scientific notation a method of writing numbers that consists of a number of order unity times the appropriate power of 10

conversion factor a factor (equal to 1) that multiplies a quantity to convert its value to another unit

order-of-magnitude the power of ten characterizing the size of a quantity

II. Tips

Dimensional Analysis

You should be aware that, typically, arguments of mathematical functions are dimensionless. Angles, for example, are dimensionless, as can be seen by the equation for the length of a circular arc, $s = r\theta$, where θ is in radians; hence, angular measures like radians and degrees signify only how we choose to measure the angle. The trigonometric functions, therefore, such as sine, cosine, and tangent are applied to dimensionless quantities. Other examples of dimensionless functions are log(x), ln(x), and their inverse functions 10^x and exp(x).

Round-off Error

It was stated previously that excessive round-off error can often be avoided by keeping at least one additional figure in intermediate calculations. An often better approach is to avoid calculating numerical values in intermediate steps. Instead, just carry through the formulas from the intermediate steps and only plug in numerical values once at the end. You will see examples of both approaches throughout this study guide.

Example 1–7 Don't Round-off too Soon A cardboard box has measurements of L = 1.92 m, and W = 0.725 m. Its height is H = 1.88 m. **(a)** Calculate the area (A) of the base of the box. **(b)** Calculate its volume (V) using the result of (a). **(c)** Calculate its volume using the formula for volume.

Picture the Problem The diagram shows a box representing the box whose base area and volume we wish to determine.

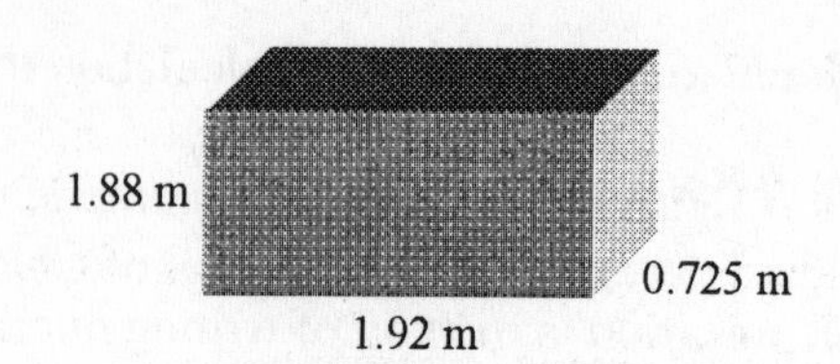

Strategy First calculate the area of the base.

Solution

1. Calculate the area of the base: $A = LW = 1.92 \text{ m} \times 0.725 \text{ m} = 1.39 \text{ m}^2$

2. The volume of the box is area × height: $V = A \times H = 1.39 \text{ m}^2 \times 1.88 \text{ m} = 2.61 \text{ m}^3$

3. The volume is length × width × height: $V = A \times H = LWH = 1.92 \text{ m} \times 0.725 \text{ m} \times 1.88 \text{ m}$
$= 2.62 \text{ m}^3$

Insight The answers to parts (b) and (c) differ in the final digit. Which one is correct? Part (c) is correct because the full values were used. The round-off to three significant figures in part (a) is the reason for the difference.

Practice Problems

1. What is the decimal equivalent of 3.14×10^7?
2. What is the decimal equivalent of 2.9e–4?
   (In the JavaScript web language, scientific notation normally written as 2×10^3 is written as 2e3.)
3. 10.2 * 8 =
4. 27.1/5.08 =
5. 2.712 + 10.8 =
6. What is the volume in cubic centimeters of a sphere with a radius of 2 cm?
7. A sphere has a volume of 106 cm^3, what is its radius (in cm)?
8. What is the area in square meters of a triangle with a base 2.9 m and a height of 10.4 m?
9. What is the volume in cubic centimeters of a cylinder of diameter 1.2 cm with a height of 2.6 cm?
10. 85 miles (exactly) is how many meters (exactly)?

Puzzle

WHERE ARE YOU?

The standard geographical coordinates of Chicago are as follows:

Latitude: 41 degrees 50 minutes

Longitude: 87 degrees 45 minutes.

What are the x, y, z coordinates of Chicago in a coordinate system centered at the center of Earth with the z-axis pointing from the South Pole to the North Pole, and the x-axis passing through the zero longitude meridian pointing away from Europe into space? Answer this question in words, not equations, briefly explaining how you obtained your answer.

Selected Solutions

9. We first solve the equation for k:

$$T = 2\pi\sqrt{\frac{m}{k}} \quad\Rightarrow\quad T^2 = 4\pi^2\frac{m}{k} \quad\Rightarrow\quad k = 4\pi^2\frac{m}{T^2}.$$

The factor $4\pi^2$ is dimensionless, so dimensional analysis gives

$$\boxed{[k] = \frac{[M]}{[T^2]}}$$

13. The total weight is given by the sum

$$Wt_{tot} = Wt_{bass} + Wt_{cod} + Wt_{salmon}$$

Taking the direct sum we obtain

$$Wt_{tot} = 2.65\text{ lb} + 10.1\text{ lb} + 17.23\text{ lb} = 29.98\text{ lb}$$

This result must be rounded to the proper number of significant figures. The 10.1-lb cod gives us only one decimal place. Therefore, the result must be rounded to one decimal place:

$$Wt_{tot} = \boxed{30.0\text{ lb}}$$

15. The area of a circle is given by the expression $A = \pi r^2$. The value of π is known to many significant digits. The number of significant digits in the results is limited only by the number in the radius r. In each case, we need to use only one more significant digits in π than in r to obtain the most accurate result available.

(a) $A = (3.1416)(5.342\text{ m})^2 = \boxed{89.65\text{ m}^2}$

(b) $A = (3.14)(2.7\text{ m})^2 = \boxed{23\text{ m}^2}$

29. **(a)** For this part we need only to convert meters to feet:

$$\left(\frac{20.0\text{ m}}{\text{s}}\right)\left(\frac{3.28\text{ ft}}{\text{m}}\right)=\boxed{65.6\text{ft/s}}$$

(b) For this part we must convert both the distance and time units.

$$\left(\frac{65.6\text{ ft}}{\text{s}}\right)\left(\frac{1\text{ mi}}{5280\text{ ft}}\right)\left(\frac{3600\text{ s}}{\text{h}}\right)=\boxed{44.7\text{mi/h}}$$

33. **(a)** From the given information, the surface of Earth must move through the 3000 miles in the 3-h time difference. Thus, we have for the rotational speed,

$$\text{v}=\frac{3000\text{ mi}}{3\text{ h}}=\boxed{1000\text{ mi/h}}$$

(b) Earth completes one rotation in 24 hours. Given the above rotational speed, we can say

$$\frac{1000\text{ mi}}{1\text{ h}}=\frac{\text{Circum.}}{24\text{ h}}\quad\therefore\quad\text{Circum.}=24\text{ h}\times\frac{1000\text{ mi}}{\text{h}}=\boxed{24{,}000\text{ mi}}\quad\text{(approx.)}$$

(c) The circumference of a circle is given by Circum. = $2\pi r$. Therefore,

$$\text{r}=\frac{\text{Circum.}}{2\pi}=\frac{24{,}000\text{ mi}}{2\pi}=\boxed{4000\text{ mi}}$$

Answers to Practice Quiz

1. (b) **2.** (c) **3.** (d) **4.** (b) **5.** (c) **6.** (d) **7.** (a) **8.** (d) **9.** **(c)** **10.** (b)

Answers to Practice Problems

1. 31,400,000 — 2. 0.00029

3. 81.6 — 4. 5.33

5. 13.5 — 6. 30 cm^3

7. 2.94 cm — 8. 15 m^2

9. 2.9 cm^3 — 10. 136794.24 m

CHAPTER 2

ONE-DIMENSIONAL KINEMATICS

Chapter Objectives

After studying this chapter, you should:

1. know the difference between distance and displacement.
2. know the difference between speed and velocity.
3. know the difference between velocity and acceleration.
4. be able to define acceleration and give examples of both positive and negative acceleration.
5. be able to calculate displacements, velocities, and accelerations using the equations of one-dimensional motion.
6. be able to interpret x-versus-t and v-versus-t plots for both motion with constant velocity and constant acceleration.
7. be able to describe the motion of freely falling objects.

Warm-Ups

1. During aerobic exercise, people often suffer injuries to knees and other joints due to high accelerations. When do these high accelerations occur?
2. Estimate the acceleration you subject yourself to if you walk into a brick wall at normal walking speed. (Make a reasonable estimate of your speed and of the time it takes you to come to a stop.)
3. A man drops a baseball from the edge of a roof of a building. At exactly the same time another man shoots a baseball vertically up toward the man on the roof in such a way that the ball just barely reaches the roof. Does the ball from the roof reach the ground before the ball from the ground reaches the roof, or vice versa?
4. Estimate the time it takes for a free-fall drop from a height of 10 m. Also estimate the time a 10-m platform diver is in the air if he takes off straight up with a vertical speed of 2 m/s (and clears the platform of course!)

Chapter Review

2–1 Position, Distance, and Displacement

Any description of motion takes place in a **coordinate system** that allows us to track the **position** of an object. **One-dimensional motion** means that objects are free to move back and forth only along a single line. As a coordinate system for one-dimensional motion, we choose this line to be an x-axis together with a specified origin and positive and negative directions. The location of the origin and which way is called positive or negative may be chosen according to convenience.

The **distance** traveled by an object that moves from one position to another is just the total length of travel during the trip. This total length of travel depends on the path taken as the object moves from its initial position x_i to its final position x_f. Distance should be distinguished from **displacement,** $\Delta x = x_f - x_i$, which is the change in position of the object regardless of the path taken. The primary difference between the two is that the distance an object travels tells us nothing about the direction of travel, whereas displacement tells us precisely how far, and in what direction, from its initial position an object is located. To put it succinctly, distance is the *total* length of travel, and displacement is the *net* length of travel accounting for direction.

Example 2–1 Parking in the Same Spot In your car, you leave your favorite parking spot and drive 4.83 km east on Main Street to go to the grocery store. After shopping, you go back home by traveling west on Main Street and find that your favorite parking spot is still available. **(a)** What distance do you travel during this trip? **(b)** What is your displacement?

Picture the Problem We choose the x-axis to represent Main Street. The origin is placed at the initial parking spot.

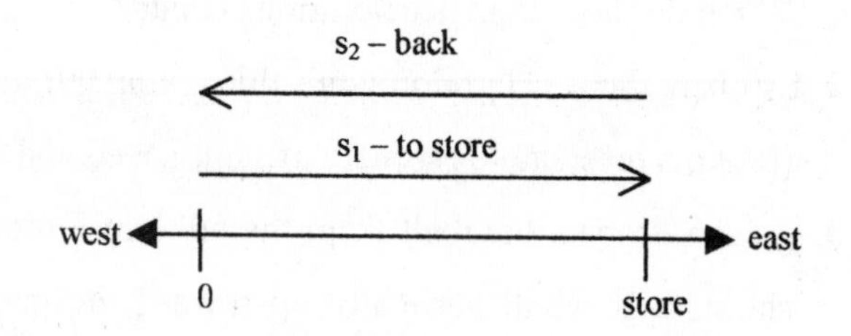

Strategy For this problem we must remember the distinction between distance and displacement. For the distance we must consider the entire length of the path taken. For displacement we ignore the path and focus on the initial and final positions only.

Solution

Part (a)

The trip has two length segments: s_1, going to the store, and s_2, coming back home. The total length of travel, s, must be the sum of these two segments:

$$s = s_1 + s_2 = 4.83\,\text{km} + 4.83\,\text{km} = 9.66\,\text{km}$$

Part (b)
Here we notice that the initial and final positions are the same. Subtract the initial position from the final position to get the displacement:

$$\Delta x = x_f - x_i = 0\,\text{m} - 0\,\text{m} = 0\,\text{m}$$

Insight This example clearly shows how different the distance and displacement can be. It doesn't matter where the origin is placed; the results for both parts (a) and (b) would be the same. As an exercise, you should convince yourself of this fact.

Practice Quiz

1. If you walk exactly four times around a quarter-mile track, what is your displacement?

   (a) one mile **(b)** half a mile **(c)** one-quarter mile **(d)** zero

2–2 – 2–3 Speed and Velocity

The concepts of distance and displacement relate to the fact that an object moves from one position to another. Additionally, an important part of describing motion is to specify how rapidly an object moves. One way to do this is to quote an **average speed**,

$$\text{average speed} = \frac{\text{distance}}{\text{elapsed time}}$$

Thus, average speed is the distance traveled divided by the amount of time it took to travel that distance.

Another, sometimes more appropriate, way to describe the rate of motion is to quote an **average velocity**,

$$v_{av} = \frac{\text{displacement}}{\text{elapsed time}} = \frac{\Delta x}{\Delta t}$$

Thus, average velocity is the displacement divided by the amount of time it took to undergo that displacement. The difference between average speed and average velocity is that average speed relates to the distance traveled, whereas average velocity relates to the displacement. Therefore, average speed tells us about the average rate of motion over the entire path taken and contains no directional information. Average velocity, in contrast, relates only to the rate at which an object goes from x_i to x_f, regardless of the path taken, and thereby specifies direction.

Example 2–2 Average Speed versus Average Velocity For the trip described in Example 2–1, if it took $\Delta t_1 = 10.0$ min to drive to the store and $\Delta t_2 = 12.0$ min to drive back home, calculate the average speed and average velocity for the trip.

Picture the Problem The same picture from example 2.1 applies here.

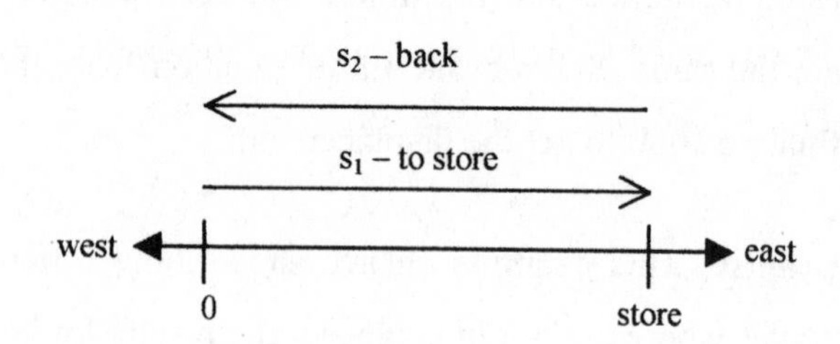

Strategy Knowing the definitions of average speed and average velocity, we can apply them directly using the results of Example 2–1.

Solution

1. To get the answers in SI units let us first determine the elapsed time for the entire trip and convert it to seconds:

$$\Delta t = \Delta t_1 + \Delta t_2 = 10.0\text{ min} + 12.0\text{ min} = 22.0\text{ min}$$
$$= 22.0\text{ min}\left(\frac{60\text{ s}}{\text{min}}\right) = 1320\text{ s}$$

2. Convert the distance traveled to meters:

$$9.66\text{ km} = 9.66 \times 10^3\text{ m}$$

3. Use the definition of average speed:

$$\text{ave speed} = \frac{\text{distance}}{\Delta t} = \frac{9.66 \times 10^3\text{m}}{1320\text{ s}} = 7.32\text{ m/s}$$

4. Similarly, use the definition of average velocity to calculate its value:

$$v_{av} = \frac{\Delta x}{\Delta t} = \frac{0\text{ m}}{1320\text{ s}} = 0\text{ m/s}$$

Insight Notice here that just as distance and displacement can be very different, so can average speed and average velocity. Don't confuse the terms.

Some situations require more than just the average rate of motion. Often, we require the velocity that an object has at a specific instant in time; this velocity is called the **instantaneous velocity**. The instantaneous velocity, v, can be defined in terms of the average velocity measured over an infinitesimally small elapsed time,

$$v = \lim_{\Delta t \to 0} \frac{\Delta x}{\Delta t}$$

Thus, the instantaneous velocity would be the velocity that an object has right at $t = 2.0$ s, for example, instead of the average velocity over a time period $\Delta t = 2.0$ s. The magnitude of the instantaneous velocity of an object (how fast it is going) is its **instantaneous speed.**

In general, average and instantaneous velocities will have very different values; however, the case of **constant-velocity motion** is special in that the average velocity over any time interval equals the instantaneous velocity at any time. Thus, the definition of average velocity also serves as an equation that describes constant velocity motion. For this special case, the equation is often written as

$$\Delta x = v\,\Delta t$$

where v no longer needs to be called v_{av}.

Example 2–3 Time the Moving Ball How much time is required for a ball that rolls with a constant velocity of 0.64 m/s to roll across a 2.3-m-long table?

Picture the Problem For this problem we take the length along the tabletop as the x-axis with one end as the origin. The ball is taken to roll in the positive direction.

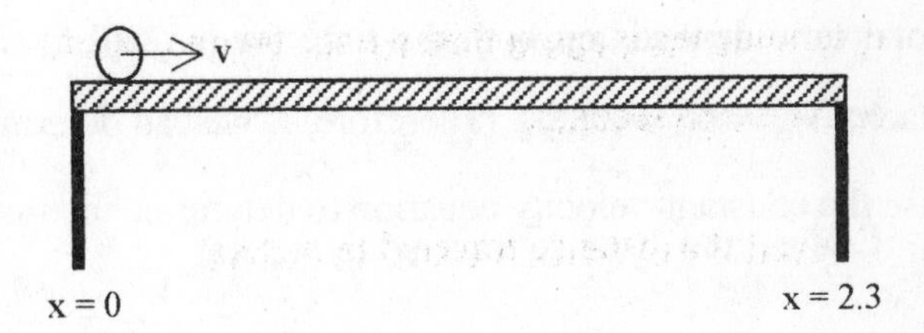

Strategy From the coordinate system established in the picture and the information given in the problem, it is clear that we know both the displacement and the velocity. Since we can take both x_i and t_i to be zero, we can simply write $\Delta x = x = 2.3$ m and $\Delta t = t$.

Solution

1. Solve the constant-velocity equation for t: $x = vt \;\Rightarrow\; t = \dfrac{x}{v}$

2. Insert the values to get the numerical solution: $t = \dfrac{2.3\text{ m}}{0.64\text{ m/s}} = 3.6\text{ s}$

Insight Although this problem may seem short, it gets to the core of good problem solving. You should identify the information given to you, as specifically as possible, and relate it to the physics you've been learning. This is especially true for longer problems.

Exercise 2–4 Playing Catch A college athlete at softball practice stands 13.2 m away from her friend and throws a ball to him at 26.8 m/s (assumed constant while in the air). If the total time, from when she throws the ball to when she hears the sound of her friend catching it, is 0.531 s, what is the speed at which the sound travels?

Solution Try to sketch a picture of the problem. It has two different parts, the athlete throwing the ball, with constant velocity, to her friend, followed by the sound traveling, with a different constant velocity, from the friend to the softball player. The information supplied by the problem can be listed as:

Given: $x = 13.2$ m, $v_{ball} = 26.8$ m/s, $T_{tot} = 0.531$ s; **Find**: v_{sound}

We first address the quantity of interest and let it guide us to the next step in the solution. We want to calculate the constant velocity of sound. From the equation for constant velocity we know that $v_{sound} = x/t_{sound}$. Since we already know x, we must now find t_{sound}. What else depends on t_{sound}? The total time depends on both the time for the ball to travel to the friend and the time for sound to travel back: $T_{tot} = t_{ball} + t_{sound}$. Therefore, if we can determine t_{ball}, we'll be able to determine t_{sound}. We can use the constant-velocity equation to determine the time for the ball to reach the friend:

$$t_{ball} = \frac{x}{v_{ball}} = \frac{13.2\text{ m}}{26.8\text{ m/s}} = 0.4925\text{ s}$$

With this result, we now use the equations involving time to solve for t_{sound}:

$$t_{sound} = T_{tot} - t_{ball} = 0.531\text{ s} - 0.4925\text{ s} = 0.0385\text{ s}$$

Finally, use the constant-velocity equation to solve for v_{sound}:

$$v_{sound} = \frac{x}{t_{sound}} = \frac{13.2\text{ m}}{0.0385\text{ s}} = 343\text{ m/s}$$

There are several things to notice about this problem. At first glance it may not appear as a problem that can be solved using the constant-velocity equation, since there is clearly more than one velocity involved. However, by dividing the problem into two segments we were able to apply constant-velocity motion to each. Also, be sure you understand why the same value of x was used for both the ball and the sound. Here x actually represents *distance*, the distance between the athlete and her friend; the *displacements* of the ball and the sound in this problem are of opposite signs. Finally, you'll notice that the answer for t_{ball} contains four digits instead of three. This is because it is only an intermediate step. As discussed in Chapter 1 of this study guide, you should retain an extra digit in intermediate calculations to help avoid excessive round-off error.

Practice Quiz

2. Which of the following is the correct SI unit for the product of velocity and time?

(a) m **(b)** m/s **(c)** s **(d)** s/m **(e)** m/s^2

3. In a coordinate system in which east is the positive direction and west is the negative direction, you take a total time of 105 seconds to walk 20 m west, then 10 m east, followed by 15 m west. With what average speed have you walked?

(a) 0.24 m/s **(b)** 0.43 m/s **(c)** – 0.24 m/s **(d)** – 0.43 m/s **(e)** 0 m/s

4. For the information given in question 3, with what average velocity have you walked?

(a) 0.24 m/s **(b)** 0.43 m/s **(c)** – 0.24 m/s **(d)** – 0.43 m/s **(e)** 0 m/s

5. How long does it take a person on a bicycle to travel exactly 1 km if she rides at a constant velocity of 20 m/s?

(a) 20000 s **(b)** 0.020 s **(c)** 50 s **(d)** 20 s **(e)** 5.0 s

2–4 Acceleration

A very important concept in physics is the **acceleration** of an object. Acceleration is the rate at which an object's velocity is changing. This velocity can be changing because the object is slowing down, speeding up, turning around, or any combination thereof. If there is any change in the speed and/or direction of motion of an object, it is accelerating. Sometimes we need only the average rate of change of velocity, or **average acceleration,**

$$a_{av} = \frac{v_f - v_i}{t_f - t_i} = \frac{\Delta v}{\Delta t}$$

At other times we may need the acceleration at a specific instant in time, or **instantaneous acceleration,** which is defined in terms of the average acceleration measured over an infinitesimally small elapsed time,

$$a = \lim_{\Delta t \to 0} \frac{\Delta v}{\Delta t}$$

As an object accelerates from an initial velocity v_i to a final velocity v_f it may have many different values of instantaneous acceleration along the way. Basically, the average acceleration tells us what constant acceleration would produce the same velocity change Δv in the same amount of elapsed time Δt.

The change in velocity that results from a particular acceleration depends on how the velocity and acceleration relate to each other. In general, when the velocity and acceleration have opposite signs, the object in question will slow down. When the velocity and acceleration have the same sign (whether both are negative or positive), the object will speed up.

Example 2–5 Leaving a Stop Sign After stopping at a stop sign, you begin to accelerate with an average acceleration of $-1.4\ \text{m/s}^2$. What is your speed after 2.0 s have passed?

Picture the Problem Here we picture your automobile accelerating along the negative x-axis.

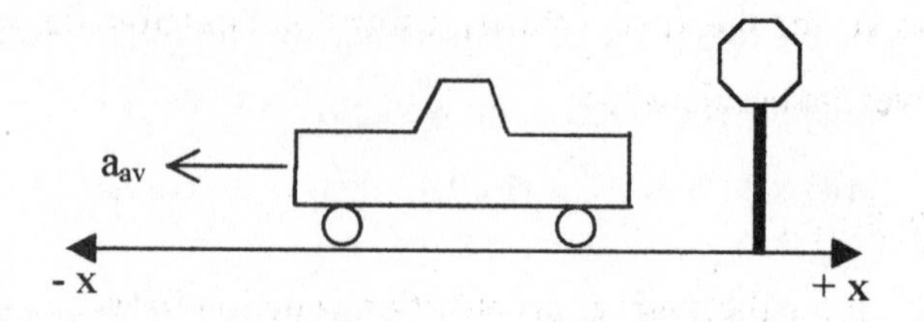

Strategy We first take note of the information given in the problem. Since you were at a stop sign we know your car starts from rest, so that $v_i = 0$. We are also directly given the elapsed time and the average acceleration. Comparing these quantities with the equation for average acceleration we see that we know everything except v_f which is closely related to the final speed.

Solution

1. Solve the average acceleration equation for v_f: $a_{av} = \dfrac{v_f - v_i}{\Delta t} \Rightarrow v_f = a_{av}\,\Delta t$

2. Insert the values to get the final velocity: $v_f = -1.4\ \text{m/s}^2(2.0\ \text{s}) = -2.8\ \text{m/s}$

3. Obtain the final speed as the magnitude of the final velocity: $\text{speed} = |v_f| = 2.8\ \text{m/s}$

Insight Notice that your car sped up even though it had a negative acceleration; it simply went faster in the negative direction. Also note that the problem asked for the *speed* after 2.0 s, not the velocity. Always pay careful attention to precisely what is being asked in a problem; it would have been very easy to just stop at $v_f = -2.8\ \text{m/s}$, but this answer cannot be correct because speed is never negative.

Example 2–6 Average Acceleration At a particular instant in time a particle has a velocity of 6.00 m/s. Over the next 4.00 s it experiences an average acceleration of $-3.00\ \text{m/s}^2$. Determine its velocity at the end of this 3-s time interval.

Picture the Problem Our sketch shows the initial situation in which the particle has velocity in the positive direction while accelerating in the negative direction.

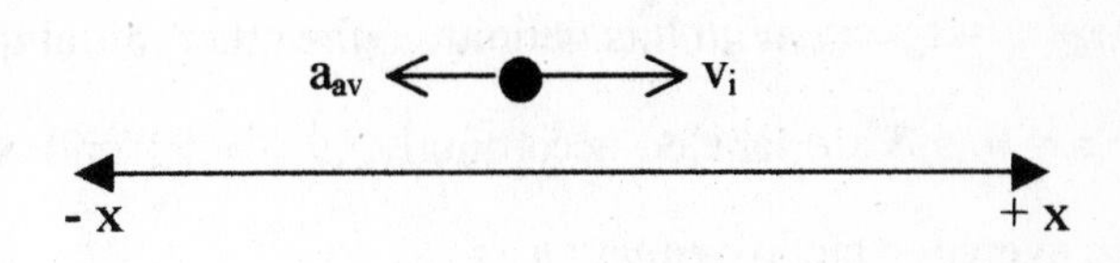

Strategy As in the previous example, we need to solve for the final velocity using the equation for average acceleration.

Solution

1. Solve the average acceleration equation for v_f: $v_f = v_i + a_{av}\,\Delta t$

2. Substitute in the values given in the problem: $v_f = 6.00\text{ m/s} + \left(-3.00\text{ m/s}^2\right)\left(4.00\text{ s}\right) = -6.00\text{ m/s}$

Insight Getting the numerical solution to this problem was easy; however, make sure you understand the physics just described. Since the particle's acceleration was opposite its initial velocity, it slowed down until it finally stopped moving in the positive direction and starting moving in the negative direction. Once moving in the negative direction, the particle sped up and, at the time of interest, reached the same speed that it had initially.

Practice Quiz

6. An object is moving east with a speed of 3.0 m/s. If its acceleration is westward, the object is
   (a) speeding up **(b)** slowing down **(c)** maintaining constant speed **(d)** none of these

2–5 – 2–6 Motion with Constant Acceleration

An important special case of accelerated motion occurs when the acceleration is constant. Conceptually, this simply means that the rate at which the velocity increases or decreases is the same at every instant in time during the motion. Therefore, an object undergoing constant acceleration will take the same amount of time to increase its velocity from 5 m/s to 10 m/s as it will to increase its velocity from 50 m/s to 55 m/s because in each case it has an equal change in velocity of $\Delta v = 5$ m/s.

As was the case with velocity, when the acceleration of an object is constant, the average acceleration equals the instantaneous acceleration. This means that the definition of average acceleration provides an equation that can be used to describe motion with constant acceleration, $a = \Delta v/\Delta t$. In this context it is customary to simply call the final time t ($t_f = t$) and to define the initial time to be $t_i = 0$; hence,

$\Delta t = t_f - t_i = t$. With this definition, the other initial quantities, x_i and v_i, are the values that correspond to $t = 0$ and are labeled accordingly: $v_i \rightarrow v_0$ and $x_i \rightarrow x_0$. With these modifications, we can rewrite the average velocity equation as

$$v = v_0 + at$$

Notice, however, that the preceding equation does not involve position. This suggests that to completely describe motion with constant acceleration we will need more than just this one equation. In fact, we use four equations, relating the different quantities of interest, to mathematically describe this type of motion. The other three equations are

$$x = x_0 + \tfrac{1}{2}(v_0 + v)t;\; x = x_0 + v_0 t + \tfrac{1}{2}at^2;\; \text{and}\; v^2 = v_0^2 + 2a(x - x_0).$$

These four equations contain all the information needed to describe motion with constant acceleration.

Example 2–7 Training a Sprinter In 1999 Maurice Greene set the world record in the 100-m dash with a time of 9.79 s. Coming off the blocks, top sprinters have an acceleration of about 4.00 m/s². If a sprinter could train himself to maintain this acceleration for an entire 100-m dash, what would be his time?

Picture the Problem Our sketch shows the situation in which the sprinter has a constant forward acceleration on a 100-m track.

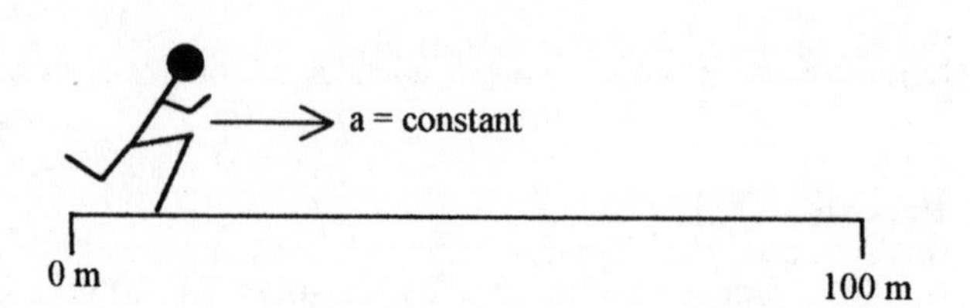

Strategy To solve this problem we focus on the three equations that contain the quantity we seek, time. We also note that $v_0 = 0$, and $x_0 = 0$; so we must select the most appropriate equation for the given information.

Solution

1. The best equation to use contains t with all other quantities known: $x = \tfrac{1}{2}at^2$

2. Solve for t and insert numerical values: $t = \sqrt{\dfrac{2x}{a}} = \sqrt{\dfrac{2(100\text{ m})}{4.00\text{ m/s}^2}} = 7.07\text{ s}$

Insight This problem shows a typical approach with constant acceleration. Find an equation containing the quantity you want in which you already know all the other quantities. Also, you should always examine your solutions for its physical meaning. Considering the result of this example, do you think it is feasible to train an athlete to have a constant acceleration of 4.00 m/s^2 throughout a 100-m dash?

Exercise 2–8 Constant Acceleration An object moves with a constant acceleration of 1.75 m/s^2 and reaches a velocity of 7.80 m/s after 3.20 s. How far does it travel during those 3.20 s?

Solution Try to sketch a picture of the problem. The information breakdown in this problem follows:

Given: a = 1.75 m/s^2, v = 7.80 m/s, t = 3.20 s; **Find**: Δx

We are free to take $x_0 = 0$ and to seek only the final position x. A comparison of the given information with the equations for constant acceleration shows that none of them allows us to immediately solve for x because we are not given v_0; hence, we must first solve for the initial velocity. To solve for v_0 we start the process all over again. Looking at the equations that contain v_0, we can see that the given information allows us to obtain v_0 from the equation

$$v = v_0 + at \quad \Rightarrow \quad v_0 = v - at = 7.80 \text{ m/s} - \left(1.75 \text{ m/s}^2\right)\left(3.20 \text{ s}\right) = 2.20 \text{ m/s}$$

Now that we have v_0, we can use any of the equations involving x to calculate it.

$$x = \tfrac{1}{2}\left(v_0 + v\right)t = \tfrac{1}{2}\left(2.20 \text{ m/s} + 7.80 \text{ m/s}\right)\left(3.20 \text{ s}\right) = 16.0 \text{ m}$$

In this case, the "typical" approach of finding an equation that contains the desired unknown with all other variables known didn't work immediately. We needed an intermediate step because all the equations contain the initial velocity. See the section "Tips" for a quicker approach to this problem.

Practice Quiz

7. Starting from rest, an object accelerates at 3.32 m/s^2. What will its instantaneous velocity be 3.00 s later?

   (a) 14.9 m/s **(b)** 1.11 m/s **(c)** 0.904 m/s **(d)** 9.96 m/s

8. The driver of a car that is initially moving at 25.0 m/s west, applies the brakes until he is going 15.0 m/s west. If the car travels 13.5 m while slowing at a constant rate, what is its acceleration and in what direction is it?

   (a) 0.675 m/s^2, east **(b)** 20.0 m/s^2, west **(c)** 14.8 m/s^2, east **(d)** 22.2 m/s^2, west

9. If a car cruises at 11.3 m/s for 75.0 s, then uniformly speeds up until, after 45.0 s, it reaches a speed of 18.5 m/s. What is the car's displacement during this motion?

(a) 671 m **(b)** 1.52 km **(c)** 3.58 km **(d)** 848 m **(e)** 0.160 km

2–7 Freely Falling Objects

Constant acceleration can be applied to objects falling near Earth's surface. It is an experimental fact that when air resistance is negligible, objects near Earth's surface fall with a constant acceleration g. The symbol g represents the magnitude of this acceleration, which has an average value of

$$g = 9.81\ \text{m/s}^2$$

Objects undergoing this type of motion, when gravity is the only important influence, are said to be in **free fall.** The direction of this acceleration is downward. This direction is commonly taken to be the negative direction along the axis defining the coordinate system. In this case the acceleration, a, is given by $a = -g$, but notice that the value of g is always positive. In some cases it may be more convenient to choose downward as the positive direction, in which case $a = +g$.

Example 2–9 Learning to Juggle An entertainer is learning to juggle balls thrown very high. One of the balls is thrown vertically upward from 1.80 m above the ground with an initial velocity of 4.92 m/s. If he fails to catch the ball and it hits the ground, how long is it in the air?

Picture the Problem We choose up as the positive direction in this picture. The sketch shows the ball's upward trip on the left and downward trip on the right.

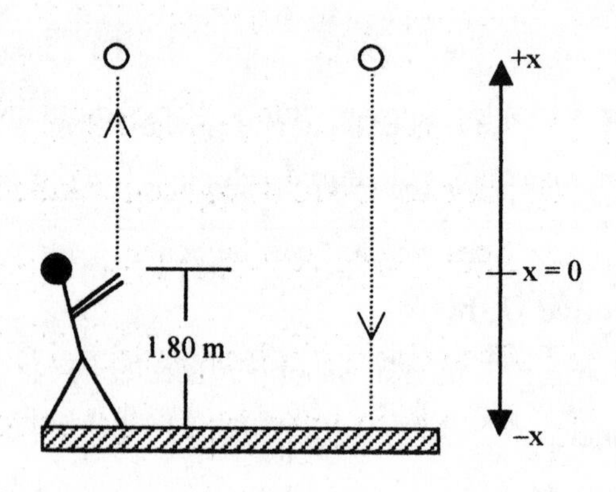

Strategy To solve this problem we need an expression that relates time to position, initial velocity, and acceleration. Because of how the coordinate system is chosen, $x_0 = 0$, $a = -g$, and when the ball hits the ground $x = -1.80\ \text{m}$.

Solution

1. Choose the equation relating t with v_0, a, and x. Notice that –g has been used for a: $x = v_0 t - \frac{1}{2} g t^2$

2. Put the quadratic equation for t in standard form: $\left(-\frac{g}{2}\right)t^2 + v_0 t - x = 0$

3. Apply the quadratic formula to get the solutions: $t = \frac{-v_0 \pm \sqrt{v_0^2 - 2gx}}{-g} = 0.5015\text{ s} \pm 0.7865\text{ s}$

4. The solution that gives a positive time is the correct answer: $t = 0.5015\text{ s} + 0.7865\text{ s} = 1.29\text{ s}$

Insight The above solution is only one approach to solving this problem. Another way might be to separately determine the times for the ball to travel upward and downward, then add them (try it). The above approach combines both motions. This can be done because the acceleration is the same going up and coming down. Make sure you can reproduce the numerical values in the solution. Be careful to properly account for all the minus signs. To what does the negative solution correspond?

There are a few facts concerning free-fall motion that you can utilize in analyzing situations. These facts can be deduced from the four equations for motion with constant acceleration.

* When an object launched vertically upward reaches the top of its path (its maximum height), its instantaneous velocity is zero, even though its acceleration continues to be 9.81 m/s^2 downward.

* An object launched upward from a given height takes an equal amount of time to reach the top of its path as it takes to fall from the top of its path back to the height from which it was launched.

* The velocity an object has at a given height, on its way up, is equal and opposite to the velocity it will have at that same height on its way back down.

Example 2–10 A Childs Toy A toy rocket is launched vertically upward from the ground. If its initial speed is 8.93 m/s, with what speed will it strike the ground?

Picture the Problem We choose up as the positive direction. The diagram shows the rocket's upward trip on the left and downward trip on the right.

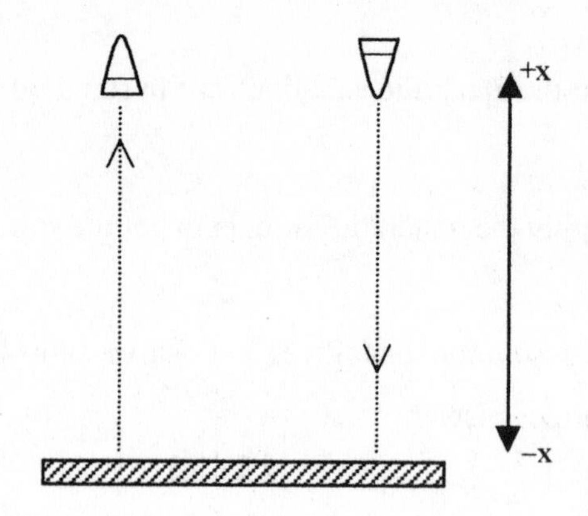

Strategy/Solution Instead of using equations, we can get the solution by using the preceding symmetry results. Since the rocket is launched upward from the ground with a certain speed, we know that when it comes back to the ground it will have the same speed (in the opposite direction). Therefore, it strikes the ground with a speed of 8.93 m/s.

Insight Sometimes we can use symmetry to get the answer with little or no calculation. Convince yourself of the above result by using the equation $v^2 = v_0^2 + 2a(x - x_0)$ to solve the problem.

Practice Quiz

10. A ball is dropped from a height of 2.89 m above the ground. How long does it take to reach the ground?

(a) 0.768 s **(b)** 0.589 s **(c)** 9.81 s **(d)** 28.4 s

11. A ball is thrown vertically upward from a height of 2.00 m above the ground with a speed of 17.3 m/s. If the ball is caught by the same person at the same height, how long is the ball in the air?

(a) 1.76 s **(b)** 4.00 s **(c)** 3.53 s **(d)** 8.65 s

12. Can an object that is moving upward be in free fall?

(a) No, because an object cannot be falling if it is moving upward

(b) Yes, because Earth's gravity sometimes pushes objects upward

(c) No, because Earth's gravity never pushes objects upward

(d) Yes, as long as gravity is the only force acting on it

13. Stone A is thrown upward from the top of a high bridge while stone B is dropped from the same height. Both stones fall and strike the water below. Which stone strikes the water with greater speed?

(a) stone A **(b)** stone B **(c)** They strike with equal speeds.

Graphical Analysis of Motion

The motion of an object is often analyzed graphically. Graphical analysis is useful for many things and can be used to determine what kind of motion is being observed. In order to do that we must first know what kinds of graphs the different types of motion produces and how to obtain information from them. Specifically, we focus on graphs of position as a function of time, **x-versus-t**, and velocity as a function of time, **v-versus-t**.

Position-versus-Time

In general, plots of position-versus-time will be curved. Regardless of the shape of the curve, however, information about the velocity of the motion can be determined from the graph. Any two points on the graph can be connected by a straight line. The slope of this connecting line equals the average velocity of the object over the corresponding time interval. Also, for any given point on the curve there is a line, called the **tangent line**, that intersects the curve at that point. The slope of the tangent line at a point equals the instantaneous velocity of the object at the corresponding time.

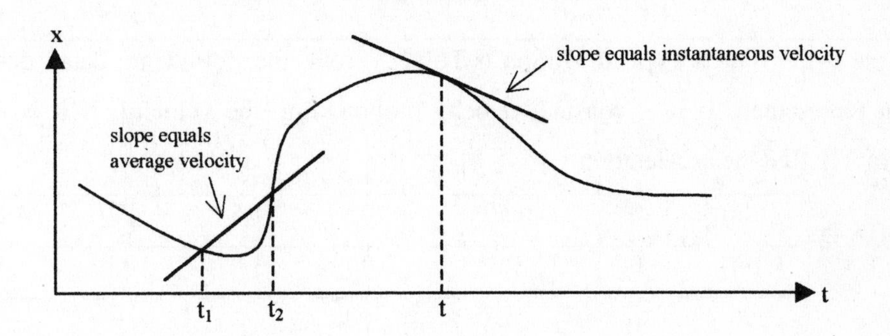

For the special case of constant-velocity motion, the equation $x = vt$ shows that we expect the x-versus-t graph to be linear. The slope of the line will give us the velocity of the motion.

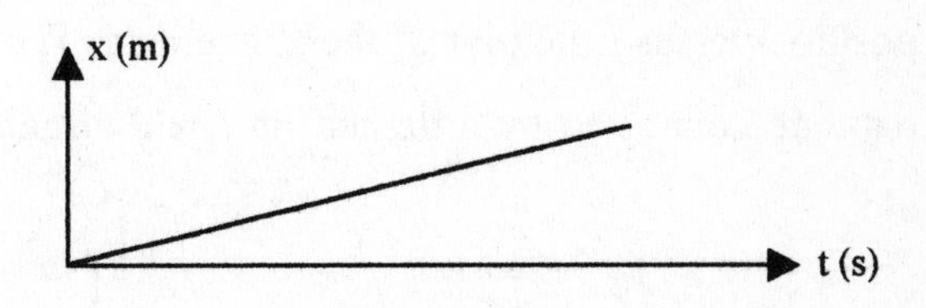

For the special case of constant acceleration, the equation $x = (1/2)at^2$ (with $v_0 = 0$) shows that we expect the x-versus-t graph to be a parabola. From this curve we can determine average and instantaneous velocities.

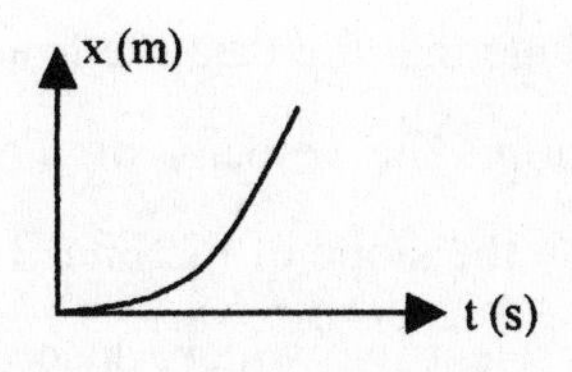

Velocity-versus-Time

For motion with constant acceleration, which includes constant-velocity motion ($a = 0$), the graph of velocity as a function of time is linear, as can be seen by the equation $v = v_0 + at$. The slope of the straight line equals the acceleration of the motion.

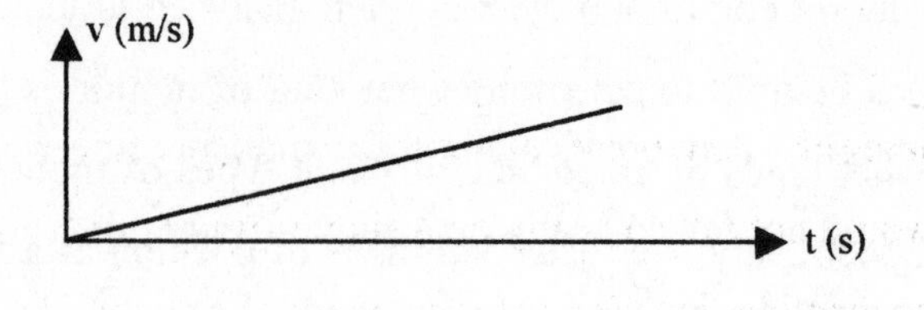

Whether the v-versus-t curve is linear or nonlinear ($a \neq \text{constant}$), the distance traveled by an object from one time to another equals the area under the curve between those two times. For the cases of constant velocity and constant acceleration, these areas are rectangles and triangles, respectively.

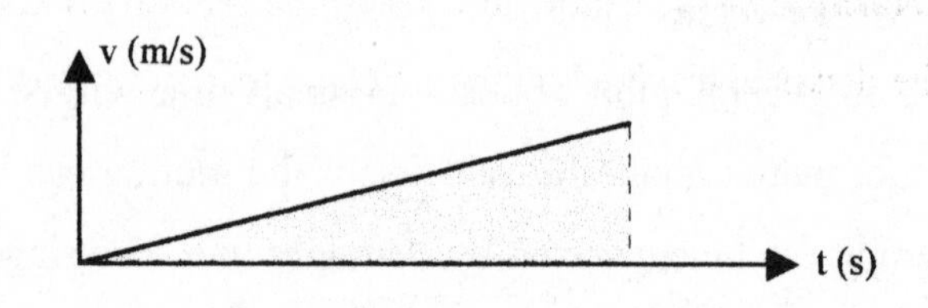

Example 2–11 Which Type of Motion Is This? From the following data, determine the type of motion represented. If it is constant-velocity motion, find the velocity. If it is motion with constant acceleration, find the acceleration.

t (s):	0	0.50	1.0	1.5	2.0	2.5	3.0
x (m):	0	1.0	2.0	3.0	4.0	5.0	6.0

Picture the Problem For the picture we make a position-versus-time plot of the above data. The data are connected by a dashed line as a visual aid.

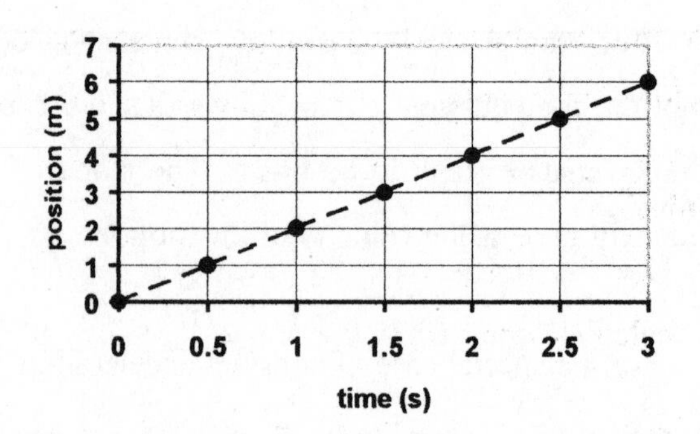

Strategy/Solution We examine the features of the plot. It clearly shows a linear relationship so we can conclude that the data are from constant-velocity motion. The velocity of the motion is determined by the slope of the line. To calculate the slope we select two widely spaced points on the line (1.0 m, 0.50 s) and (5.0 m, 2.5 s).

Slope equals rise divided by run:

$$v = \frac{\text{rise}}{\text{run}} = \frac{5.0\text{ m} - 1.0\text{ m}}{2.5\text{ s} - 0.50\text{ s}} = 2.0\text{ m/s}$$

Insight Any points along the connecting line would have given the same slope. Real-world data points would not fall so neatly on a single straight line. In such cases you would draw a best-fit line instead of a connecting line. The velocity would then equal the slope of this best-fit line.

Example 2–12 Determine Distance from a Graph Use the following v-versus-t graph below to find the distance traveled by the object whose motion is represented.

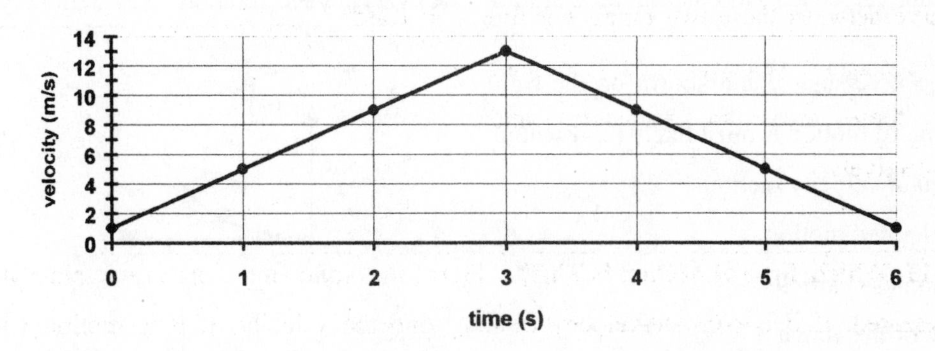

Picture the Problem The figure shows the graph of v-versus-t referred to in the problem.

Strategy To get the distance we need the area under the curve. The diagram shows that this area can be divided into the area of the triangle bounded by the two lines plus the area of the rectangle below this triangle.

Solution

1. Use the formula for the area of a triangle to calculate one term in the sum x_1:

$$\text{area of triangle} = \tfrac{1}{2} \times \text{base} \times \text{height}$$

$$\therefore \quad x_1 = \tfrac{1}{2} \times (6.0\text{ s}) \times (12\text{ m/s}) = 36\text{ m}$$

2. Calculate the area of the rectangle beneath the triangle for the second term in the sum x_2:

$$\text{area of rectangle} = \text{width} \times \text{height}$$

$$\therefore \quad x_2 = (6.0\text{ s}) \times (1.0\text{ m/s}) = 6.0\text{ m}$$

3. The distance traveled is the sum of the two:

$$\text{distance} = x_1 + x_2 = 42\text{ m}$$

Insight Notice that the height of the triangle is 12 m/s, not 13 m/s. This is because the motion of interest both started and ended with a velocity of 1.0 m/s.

Practice Quiz

14. On an x-versus-t plot, the slope of a tangent line equals

    (a) average velocity **(b)** average acceleration

    (c) instantaneous velocity **(d)** instantaneous acceleration

15. If an object's motion is described by a v-versus-t plot that is parabolic in shape, which kind of motion is it?

    (a) constant velocity **(b)** constant acceleration **(c)** varying acceleration **(d)** stationary object

16. Given the x-versus-t graph shown on the right, which type of motion is most likely represented

    (a) constant velocity motion

    (b) accelerated motion

    (c) motionless particle

    (d) none of the above

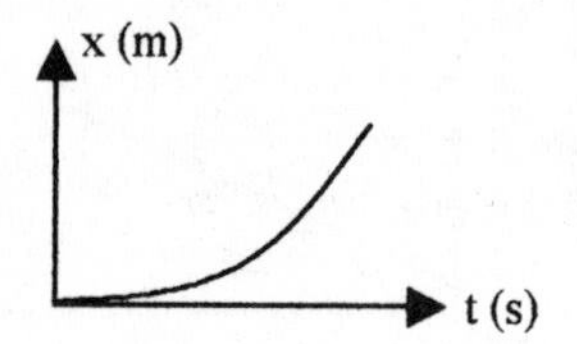

Reference Tools and Resources

I. Key Terms and Phrases

mechanics the study of how objects move and the forces that cause motion

kinematics the branch of physics that describes motion

distance the total length of travel

displacement the change in position of an object

average speed distance divided by elapsed time

velocity the rate of change of displacement with time

tangent line the straight line that intersects a curve at a point P as the result of a limiting process of secant lines through points surrounding P

acceleration the rate of change of velocity with time

free fall the motion of an object subject only to the influence of gravity

the acceleration of gravity the acceleration that results from Earth's gravitational pull

II. Important Equations

Name/Topic	Equation	Explanation
displacement	$\Delta x = x_f - x_i$	Displacement is the change in position of an object.
average velocity	$v_{av} = \frac{\Delta x}{\Delta t}$	Average velocity is the displacement divided by the elapsed time.
constant velocity motion	$\Delta x = v\,\Delta t$	When velocity is constant, the displacement equals velocity times elapsed time. For this case $v = v_{av}$.
average acceleration	$a_{av} = \frac{\Delta v}{\Delta t}$	Average acceleration is the velocity change divided by the elapsed time.
motion with constant acceleration	$v = v_0 + at$	velocity changes linearly with time
	$x = x_0 + \frac{1}{2}(v_0 + v)t$	position in terms of average velocity
	$x = x_0 + v_0 t + \frac{1}{2}at^2$	position in terms of acceleration and time
	$v^2 = v_0^2 + 2a(x - x_0)$	velocity squared in terms of displacement

III. Know Your Units

Quantity(ies)	Dimension	SI Unit
displacement, distance	[L]	m
velocity, speed	[L]/[T]	m/s
acceleration	$[L]/[T^2]$	m/s^2

IV. Tips

Motion with Constant Acceleration

The four equations for constant acceleration form a set in which each equation has a key quantity missing. Taking them in the order listed in the table Important Equations, the first equation is missing displacement ($x - x_0$), the second is missing acceleration a, the third is missing final velocity v, and the fourth is missing time t. From this point of view, if an important quantity is unknown, you have an equation that does not require it. However, you may have noticed that each equation contains the initial velocity v_0. Show that a fifth equation missing v_0 is

$$x = x_0 + vt - \tfrac{1}{2}at^2$$

Use of this equation would have considerably simplified the solution to Exercise 2–8.

Graphical Analysis

In the section on graphical analysis of accelerated motion it was pointed out that for constant acceleration the velocity-versus-time plot is linear, and its slope equals the acceleration. Suppose, however, that you only have position-versus-time data. This plot is parabolic; is there any good way to get the acceleration from this set of data? The answer is yes if the motion starts from rest. For this special case of starting from rest the equation for x as a function of t reduces to

$$x = \tfrac{1}{2}at^2$$

Notice that although x is quadratic in t, it is linear in t^2. So if you treat t^2 as a single variable and plot x-versus-t^2, you will get a linear curve. The slope of this line equals half of the acceleration. Just double the slope and you're done.

Practice Problems

1. A runner dashes from the starting line ($x = 0$) to a point 94 m away and then turns around and runs to a point 13 m away from the starting point in 20 s. To the nearest tenth of a m/s what is the runner's average speed?
2. What is the runner's average velocity in the previous problem?
3. A car accelerates at a constant rate from zero to 29.6 m/s in 10 s and then slows to 18.9 m/s in 5 s. What is its average acceleration to the nearest tenth of a m/s^2 during the 15 s?
4. What was the acceleration during the first 10 s in the previous problem?
5. A car traveling at 16 m/s accelerates at 1 m/s^2 for 10 s. To the nearest meter how far does it travel?
6. To the nearest tenth of a m/s, what is the final velocity of the car in the previous problem?

7. A passenger in a helicopter traveling upward at 23 m/s accidentlly drops a package out the window. If it takes the package 15 s to reach the ground, how high to the nearest meter was the helicopter when the package was dropped?
8. To the nearest meter what was the maximum height of the package above the ground in the previous problem?
9. A speeder traveling at 33 m/s passes a motorcycle policeman at rest at the side of the road. The policeman accelerates at 3.42 m/s^2. To the nearest tenth of a second how long does it take the policeman to catch the speeder?
10. How far to the nearest tenth of a meter can a runner running at 10 m/s run in the time it takes a rock to fall 67 m from rest?

Puzzle

ROUND TRIP

An airplane flying at constant air speed due east from My City to Your City in calm weather (no wind of any kind) would log the same flying time for both legs of the trip. Suppose the same trip is taken when there is wind from the west. How would the total (round trip) time in windy weather compare with the total time in calm weather? Answer this question in words, not equations, briefly explaining how you obtained your answer.

Selected Solutions

13. An arm is approximately 1 m long so take $\Delta x = 1$ m.

$$v_{av} = \frac{\Delta x}{t} \quad \therefore \quad t = \frac{\Delta x}{v_{av}} = \frac{1\text{ m}}{10^2\text{ m/s}} = 10^{-2}\text{ s} = \boxed{10\text{ ms}}$$

29. (a) The runner starts from rest, and at t = 2.0 s the runner is still accelerating; thus,

$$v = v_0 + at = 0 + \left(1.9\text{ m/s}^2\right)(2.0\text{ s}) = \boxed{3.8\text{ m/s}}$$

(b) The runner's speed at the end of the race is the same as that at t = 2.2 s.

$$v = v_0 + at = \left(1.9\text{ m/s}^2\right)(2.2\text{ s}) = \boxed{4.2\text{ m/s}}$$

35. (a) Since the car has constant deceleration, the equations for constant acceleration apply: $\Delta t = \Delta v/a$.

Therefore, doubling the driving speed (Δv) will increase the stopping time by a factor of 2.

(b) $\Delta t_{16} = \frac{\Delta v}{a} = \frac{0\text{ m/s} - 16\text{ m/s}}{-4.2\text{ m/s}^2} = 3.81\text{ s}$

$\Delta t_{32} = \frac{\Delta v}{a} = \frac{0\text{ m/s} - 32\text{ m/s}}{-4.2\text{ m/s}^2} = 7.62\text{ s}$

$\frac{\Delta t_{32}}{\Delta t_{16}} = \frac{7.62\text{ s}}{3.81\text{ s}} = \boxed{2}$

The time required to stop is doubled.

85. **(a)** The youngster's time in the air is given by $t_{air} = 2v_0/g$. Therefore, $\boxed{\text{if } v_0 \text{ doubles, so does } t_{air}}$.

(b) The youngster's maximum height is given by $y_{max} = v_0^2/2g$. Therefore, if v_0 doubles, then y_{max} increases by 2^2, so $\boxed{\text{it quadruples}}$.

(c) For part (a), $t_{air} = 2v_0/g$. Therefore,

$t_{2,air} = \frac{2v_0}{g} = \frac{2(2.0\text{ m/s})}{9.81\text{ m/s}^2} = 0.408\text{ s}$ and $t_{4,air} = \frac{2v_0}{g} = \frac{2(4.0\text{ m/s})}{9.81\text{ m/s}^2} = 0.815\text{ s}$

$\frac{t_{4,air}}{t_{2,air}} = \frac{0.815\text{ s}}{0.408\text{ s}} = \boxed{2.0}$, so it doubles.

For part (b), $y_{max} = v_0^2/2g$. Therefore,

$y_{2,max} = \frac{v_0^2}{2g} = \frac{(2.0\text{ m/s})^2}{2(9.81\text{ m/s}^2)} = 0.204\text{ m}$

$y_{4,max} = \frac{v_0^2}{2g} = \frac{(4.0\text{ m/s})^2}{2(9.81\text{ m/s}^2)} = 0.815\text{ m}$

$\frac{y_{4,max}}{y_{2,max}} = \frac{0.815\text{ m}}{0.204\text{ m}} = \boxed{4.0}$, so it quadruples

91. **(a)** Just after release, the only acceleration is due to gravity. Therefore, $a = \boxed{9.81\text{ m/s}^2}$, $\boxed{\text{downward}}$.

(b) At the maximum height, $v = 0$. Since we know v, v_0, and g, the appropriate expression to use is

$$v^2 = v_0^2 - 2g(y_{max} - y_0) = 0 \Rightarrow y_{max} = y_0 + \frac{v_0^2}{2g}$$

Using this result, we have

$$y_{max} = 14\text{ m} + \frac{(5.4\text{ m/s})^2}{2(9.81\text{ m/s}^2)} = \boxed{15\text{ m}}$$

(c) From the above result we can set $y_0 = 15.5$ m, $y = 14$ m, and $v_0 = 0$, then use the equations for free-fall to determine the time it takes for the shell to move between those two heights.

$$y = y_0 + v_0 t - \tfrac{1}{2}gt^2 = y_0 - \tfrac{1}{2}gt^2$$

$$\therefore \quad t^2 = \frac{2(y_0 - y)}{g} \quad \Rightarrow \quad t = \sqrt{\frac{2(y_0 - y)}{g}}$$

$$\text{so} \quad t = \sqrt{\frac{2(15.5\text{ m} - 14\text{ m})}{9.81\text{ m/s}^2}} = 0.55\text{ s}$$

Since the shell goes both up to the maximum height and then back down to 14 m, we must double this result:

$$t_{tot} = 2t = 2(0.55\text{ s}) = \boxed{1.1\text{ s}}$$

(d) Since, at a given level above the original height, the speed of an object in free-fall is the same coming down as it was going up, the speed when it returns to the 14-m height is $\boxed{5.4\text{ m/s}}$.

Answers to Practice Quiz

1. (d) **2.** (a) **3.** (b) **4.** (c) **5.** (c) **6.** (b) **7.** (d) **8.** (c) **9.** (b) **10.** (a) **11.** (c) **12.** (d) **13.** (a) **14.** (c) **15.** (c) **16.** (b)

Answers to Practice Problems

1.	8.8 m/s	2.	0.7 m/s
3.	1.3 m/s²	4.	3 m/s²
5.	210 m	6.	26 m/s
7.	759 m	8.	786 m
9.	19.3 s	10.	37 m

CHAPTER 3

VECTORS IN PHYSICS

Chapter Objectives

After studying this chapter, you should:

1. know how to represent vectors both graphically and mathematically.
2. know the difference between scalars and vectors.
3. be able to determine the magnitude and direction of a vector.
4. be able to determine the components of a vector.
5. be able to write a vector in unit vector notation.
6. know how to add and subtract vectors both graphically and algebraically.
7. be able to represent position, displacement, velocity, and acceleration as two-dimensional vectors.
8. be able to use velocity vectors to analyze constant-velocity relative motion.

Warm-Ups

1. Which of the following are vector quantities?
   - tension in a cable
   - weight of a rock
   - volume of a barrel
   - temperature of water in a pool
   - drift of an ocean current
2. Is there a place on Earth where you can walk due south, then due east, and finally due north and end up at the spot where you started?
3. Is it possible for the magnitude of the vector difference of two vectors to exceed the magnitude of the vector sum?
4. Reversing the algebraic sign on the velocity vector reverses the direction of motion. Is the same statement true for the acceleration vector?

Chapter Review

In Chapter 2 you studied the one-dimensional forms of several quantities, such as displacement and velocity, that are associated with directions. In this chapter we extend that study to two dimensions.

3–1 Scalars versus Vectors

For many quantities used in physics, such as mass and speed, a simple number together with its units suffices to specify the quantity. Such quantities are called **scalar** quantities. Other quantities, however, require a directional specification in addition to a numerical value. Such quantities are called **vector** quantities. The numerical value of a vector quantity is called its **magnitude**. A two-dimensional vector can be represented graphically by an arrow in a coordinate system.

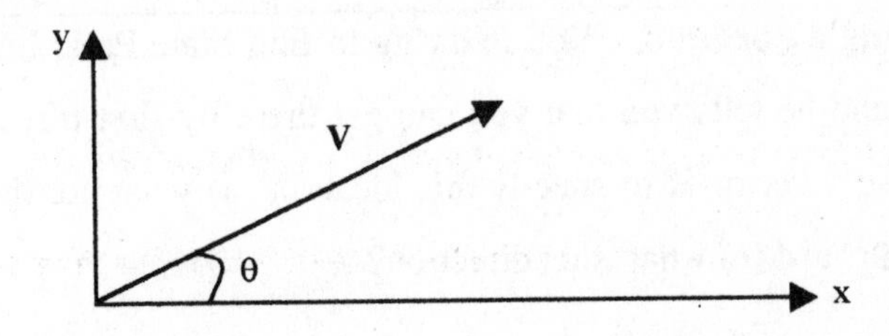

The vector quantity is represented in boldface, the **V** in the above figure. Its direction is specified by its orientation with respect to the axes of the coordinate system. In the figure, the x-axis is used. The magnitude of the vector is represented by the length of the arrow.

3–2 Components of a Vector

Working with vector quantities can often be simplified by resolving them into **components**. In two dimensions a vector has two components, one corresponding to its extent along the x-axis and the other corresponding to its extent along the y-axis. In the figure below the vector **V** is resolved into two *scalar components*, V_x and V_y.

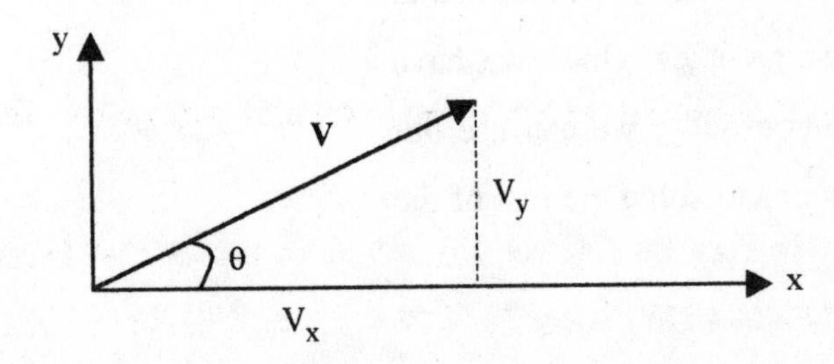

Notice that the magnitude of the vector V, (not boldface), and the two components form a right triangle; hence, we can use the trigonometric functions to relate all the relevant quantities

$$V_x = V\cos\theta \quad V_y = V\sin\theta \quad \theta = \tan^{-1}\left(\frac{V_y}{V_x}\right)$$

As you can see from these relations, the components can be positive, negative, or zero.

Be careful to note that the third of the preceding equations, for the angle, provides only a *reference angle* for the vector with respect to one of its axes. To know precisely what the direction is, you must also account for the signs of the components or, equivalently, the quadrant of the coordinate system in which the vector lies. This last point will be illustrated in solved examples. If you know the components and want the magnitude, then you can use the Pythagorean theorem

$$V = \sqrt{V_x^2 + V_y^2}$$

Example 3–1 Specifying a Location You're trying to find State Park, but you're lost. You ask a kind stranger for directions and he tells you that you can get there by first traveling 250 m west then 310 m north. If you wish to use a vector **R** to specify this location **(a)** what are the components of this vector? **(b)** what is its magnitude? and **(c)** what is its direction?

Picture the Problem The diagram shows a coordinate system with the various directions labeled. The location is indicated by the open circle, and the arrow is the vector we wish to use to specify this location. The angle θ is used to specify the direction. Try to draw R_x and R_y on this diagram yourself.

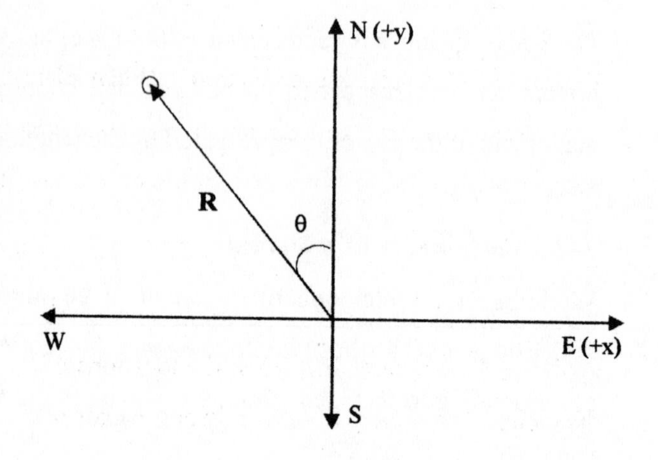

Strategy A careful reading of the problem reveals that the components are given in terms of geographic directions. We need to translate that information to the x- and y-axes. Once we have identified the components exactly, we can use our knowledge of right triangles to solve the rest of the problem.

Solution

Part (a): In our coordinate system north

corresponds to the +y-axis, and west to the −x-axis.

1. Write the x component: $R_x = -250\text{ m}$
2. Write the y component: $R_y = 310\text{ m}$

Part (b): Use the Pythagorean theorem to get the magnitude of R: $R = \sqrt{R_x^2 + R_y^2} = \sqrt{(-250\text{ m})^2 + (310\text{ m})^2} = 400\text{ m}$

Part (c): To determine θ use the tangent function, which shows that the direction of R is 39° west of north: $\theta = \tan^{-1}\left(\frac{|R_x|}{R_y}\right) = \tan^{-1}\left(\frac{250\text{ m}}{310\text{ m}}\right) = 39^\circ$

Insight There are several points to notice here. First, even though the problem gives only positive values, because of the coordinate system chosen, R_x must be given a negative value or you will arrive at the wrong location. Also, in determining the direction, we did not just naively apply the formula $\tan\theta = R_y/R_x$. You must pay close attention to where the vector lies in the coordinate system. Thus, in part (c) the absolute value of R_x is used to give a positive result for the angle.

Example 3–2 Designing a Garden When planting your garden you notice that one plant is 0.75 m away from another at approximately 60° above the horizontal. If you want to reproduce this pattern with another pair of plants, how far should you measure horizontally and vertically to determine where to dig the hole?

Picture the Problem The diagram shows the garden and two points that mark the locations of the plants. The vector **R** is the position vector of one plant relative to the other. The dashed lines represent its horizontal and vertical components.

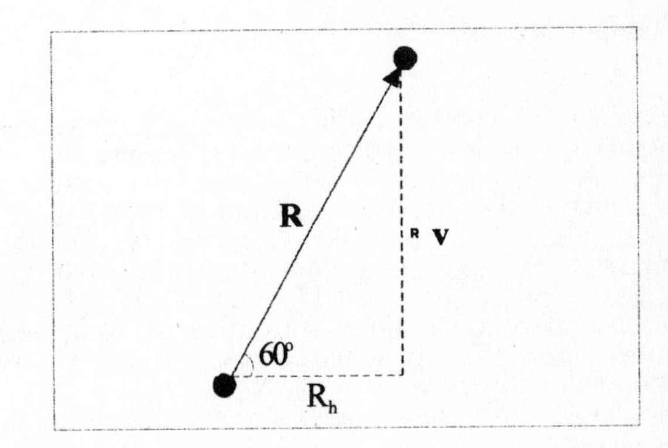

Strategy Comparing the information given with the picture, we see that the problem gives us the magnitude and direction of the position vector. Therefore, the distances we seek correspond to the horizontal and vertical components of this vector.

Solution

1. Calculate the horizontal distance as the horizontal component of the vector **R**: $R_h = R\cos\theta = (0.75\text{ m})\cos(60°) = 0.38\text{ m}$
2. Calculate the corresponding vertical distance: $R_v = R\sin\theta = (0.75\text{ m})\sin(60°) = 0.65\text{ m}$

Insight In this example, unlike in Example 3–1, we started out knowing the magnitude and direction of the vector and used these to determine its components.

Practice Quiz

1. Taking north to be the +y direction and east to be +x, calculate the x and y components (x, y) of a vector whose magnitude is 15 m and is directed 40° south of west.

   (a) (11 m, 9.6 m) **(b)** (-9.6 m, 11m) **(c)** (-13 m, -11 m)
   (d) (9.6 m, 13 m) **(e)** (-11 m , -9.6 m)

2. A certain vector has a y component of 17 in arbitrary units. If its direction is 153° counterclockwise from the +x-direction, what is its magnitude?

   (a) 44 **(b)** 37 **(c)** 16 **(d)** -44 **(e)** 40

3–3 Adding and Subtracting Vectors

There are two approaches to vector addition and subtraction; a graphical method based on geometry, and a component method based on algebra. For precise calculations you will predominantly use the component method, but the graphical approach can be invaluable for picturing the physical situation being described.

Adding and Subtracting Vectors Graphically

To picture the graphical addition of two vectors **A** and **B** to form a third vector **C**, i.e., **C** = **A** + **B**, imagine traveling along the two vectors. First you travel along **A** from its tail to its head, and then you immediately travel along **B** from its tail to its head. Your net trip, from where you started to where you stopped, will be **C**.

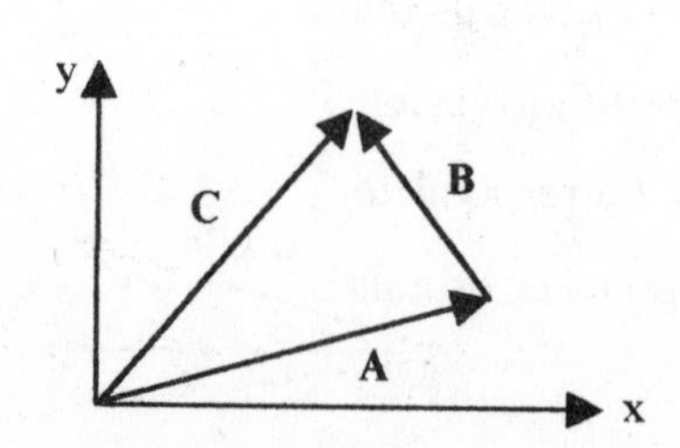

We can summarize this procedure by the following rule:

> *To add two vectors, place the tail of the second vector at the head of the first. The sum of the two then is the vector extending from the tail of the first vector to the head of the second.*

The act of moving vector **B** and placing it at the head of **A** is perfectly correct because vectors are characterized only by their length (magnitude) and orientation (direction), not by their location in the coordinate system.

The order in which two vectors are added does not matter. To see this graphically compare the following figure, which represents the addition **C** = **B** + **A**, with the diagram above for **C** = **A** + **B**.

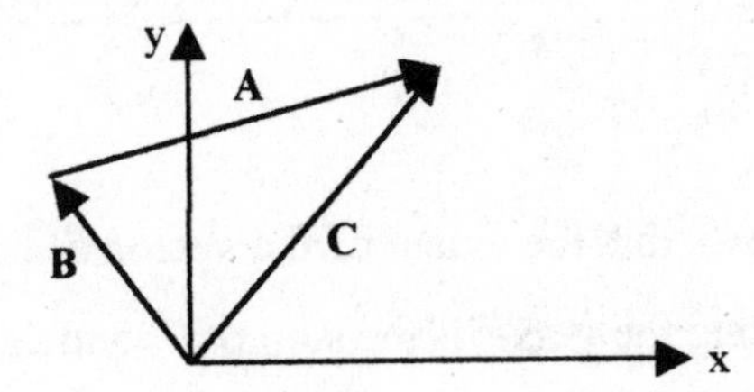

Observe that these two additions produce the same result.

We can approach vector subtraction in precisely the same way as addition by recalling that subtraction is the addition of the negative of a quantity; that is, the expression **D** = **A** – **B** is equivalent to the expression **D** = **A** + (–**B**). Therefore, once we form the vector –**B** we can proceed with vector addition as already described. The negative of a vector is formed by rotating the vector 180°, so that you end up with a vector of equal length pointing in the opposite direction as shown below.

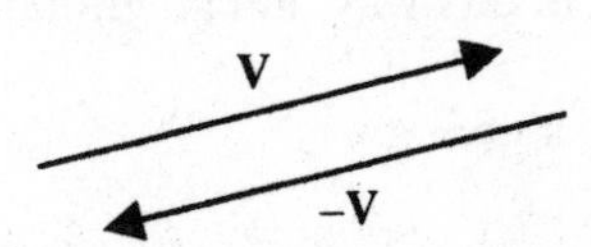

If we consider the same two vectors **A** and **B** from the addition example, the difference **A** – **B** would be drawn as follows:

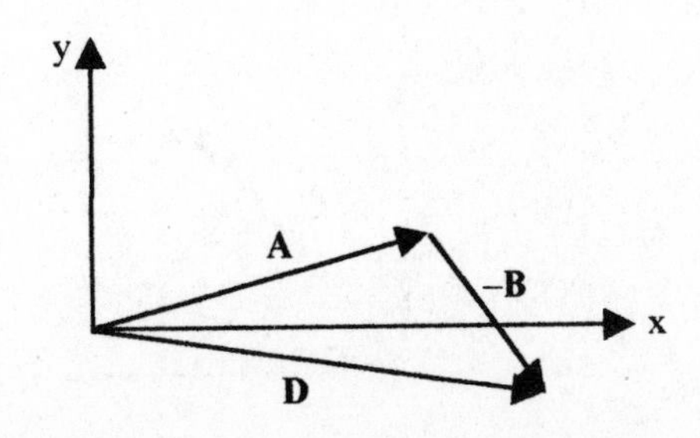

Adding and Subtracting Vectors Using Components

The graphical method of vector addition can also tell us how to add vectors using components if we resolve each of the vectors being added into its components. Consider the following figure showing the vector addition **E** = **F** + **G**.

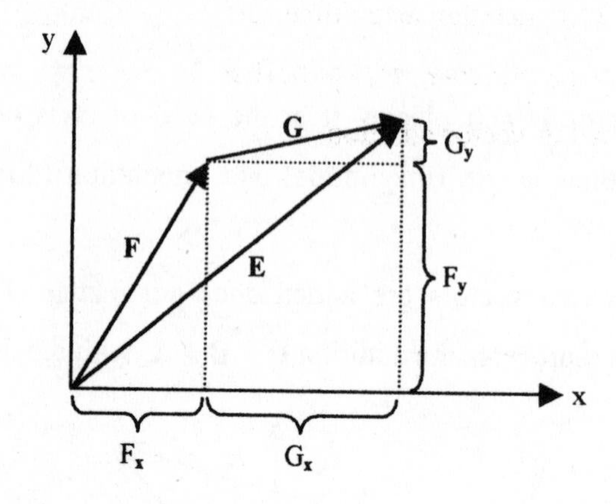

Inspection of this diagram shows that the extent of the vector **E** along the x-axis, i.e., E_x, is the sum of F_x and G_x, and the extent of **E** along the y-axis is the sum of F_y and G_y. Thus,

$$E_x = F_x + G_x \quad \text{and} \quad E_y = F_y + G_y$$

For vector subtraction we follow the same procedure as we did with the graphical method by using the fact that subtraction is the addition of the negative of a vector. Algebraically, the negative of a vector is obtained by changing the sign of each scalar component. Therefore, if vector **A** has components A_x and A_y, then vector −**A** will have components $-A_x$ and $-A_y$. Thus, for the vector difference **H** = **F** − **G**,

$$H_x = F_x - G_x \quad \text{and} \quad H_y = F_y - G_y$$

Once the components are determined in this way, the magnitude and direction can be calculated using the trigonometric functions.

Example 3–3 Vector Addition A vector **r** has components $r_x = -12.0$ and $r_y = 15.0$. Another vector **s** has components $s_x = 9.00$ and $s_y = -4.00$. Determine the magnitude and direction of the sum **r** + **s**.

Picture the Problem The sketch shows the graphical addition of **r** and **s**; the dashed vector is the sum **r** + **s**, which has been called **d**. The angle ϕ is used for the direction of **d**.

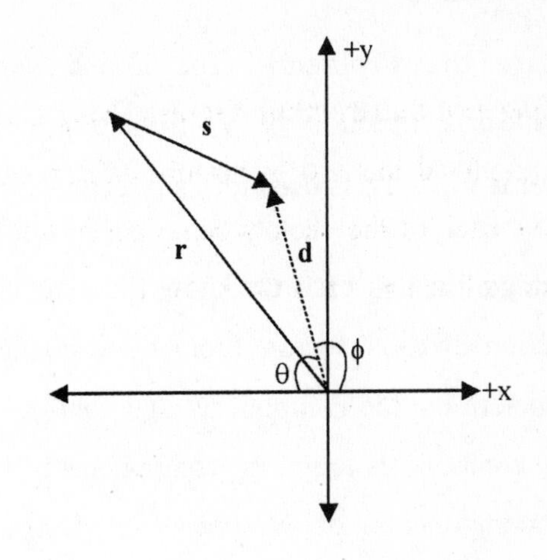

Strategy We know how to get the magnitude and direction of **d** from its components, so we'll first find the components by using vector addition.

Solution

1. Obtain the x component of **d**: $d_x = r_x + s_x = -12.0 + 9.00 = -3.00$

2. Obtain the y component of **d**: $d_y = r_y + s_y = 15.0 + (-4.00) = 11.0$

3. Obtain the magnitude of **d** from its components: $d = \left(d_x^2 + d_y^2\right)^{1/2} = \left[(-3.00)^2 + (11.0)^2\right]^{1/2} = 11.4$

4. Obtain the reference angle θ for the direction. This is the angle **d** makes with the negative x-axis: $\theta = \tan^{-1}\left(\frac{d_y}{|d_x|}\right) = \tan^{-1}\left(\frac{11.0}{3.00}\right) = 74.7°$

5. Use θ to determine ϕ: $\phi = 180° - \theta = 180° - 74.7° = 105°$

Insight Notice that the sketch alone tells us that **d** should have a negative x component and a positive y component before any calculations are done. Always try to check for these kinds of consistencies.

Example 3–4 Vector Subtraction A vector $\mathbf{v}_1$ has a magnitude of 25.0 and makes an angle of – 37.0° with the positive x direction. Another vector, $\mathbf{v}_2$, has a magnitude of 15.0 and makes an angle of 70.0° with the positive x direction. Determine the magnitude and direction of the vector $\mathbf{v}_3$ if $\mathbf{v}_3 = \mathbf{v}_1 - \mathbf{v}_2$.

Picture the Problem The sketch shows the graphical subtraction of $\mathbf{v}_2$ from $\mathbf{v}_1$; the dashed vector is $\mathbf{v}_3$, and ϕ is used for its direction.

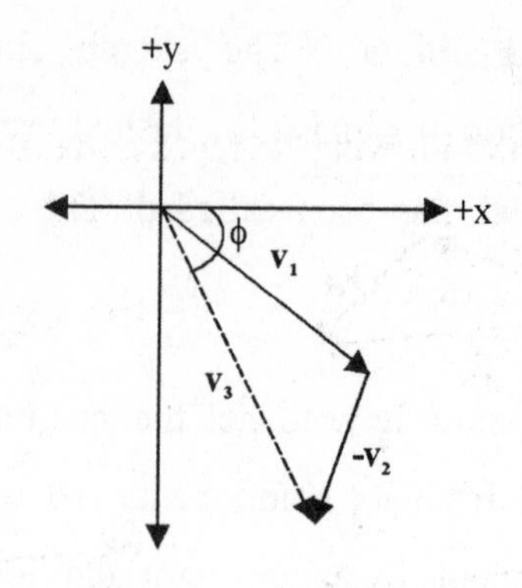

Strategy In this case we know the magnitude and direction of each vector. From these quantities we can determine the components of $\mathbf{v}_1$ and $\mathbf{v}_2$ and use these components to get the components of $\mathbf{v}_3$. From the components of $\mathbf{v}_3$ we can determine its magnitude and direction.

Solution

1. Obtain the x component of $\mathbf{v}_1$: $v_{1x} = v_1 \cos\theta_1 = (25.0)\cos(-37.0^\circ) = 19.97$

2. Obtain the y component of $\mathbf{v}_1$: $v_{1y} = v_1 \sin\theta_1 = (25.0)\sin(-37.0^\circ) = -15.05$

3. Obtain the x component of $\mathbf{v}_2$: $v_{2x} = v_2 \cos\theta_2 = (15.0)\cos(70.0^\circ) = 5.13$

4. Obtain the y component of $\mathbf{v}_2$: $v_{2y} = v_2 \sin\theta_2 = (15.0)\sin(70.0^\circ) = 14.10$

5. Obtain the x component of $\mathbf{v}_3$: $v_{3x} = v_{1x} - v_{2x} = 19.97 - 5.13 = 14.84$

6. Obtain the y component of $\mathbf{v}_3$: $v_{3y} = v_{1y} - v_{2y} = -15.05 - 14.10 = -29.15$

7. Obtain the magnitude of $\mathbf{v}_3$ from its components: $v_3 = \left(v_{3x}^2 + v_{3y}^2\right)^{1/2} = \left[(14.84)^2 + (-29.15)^2\right]^{1/2} = 32.7$

8. Obtain the reference angle for the direction of $\mathbf{v}_3$. Note that this is the angle ϕ that $\mathbf{v}_3$ makes with the positive x-axis: $\phi = \tan^{-1}\left(\frac{v_{3y}}{v_{3x}}\right) = \tan^{-1}\left(\frac{-29.15}{14.84}\right) = -63.0^\circ$

Insight Notice that the solution to this problem followed a straightforward procedure with which you should become very familiar. In the final calculation of ϕ we don't need to use absolute values because this angle gives the final direction of the vector. We then rely on our sketch, or knowledge of the signs of the components, to accurately place the vector in the coordinate system.

Practice Quiz

3. Which of the following sketches correctly represents the vector addition $\mathbf{R}_3 = \mathbf{R}_1 + \mathbf{R}_2$?

(a) 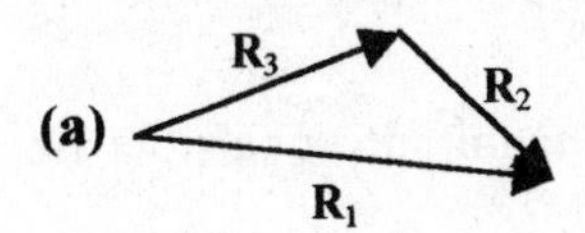

(b) 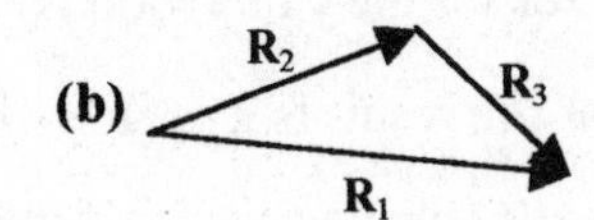

(c) 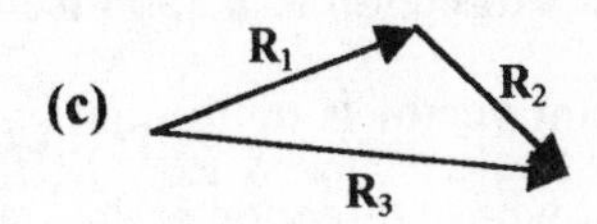

(d) 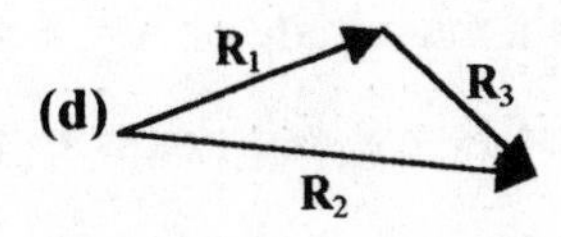

(e) 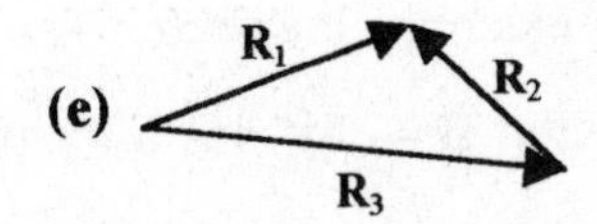

4. Which of the following sketches correctly represents the vector subtraction $\mathbf{R}_3 = \mathbf{R}_1 - \mathbf{R}_2$?

(a) 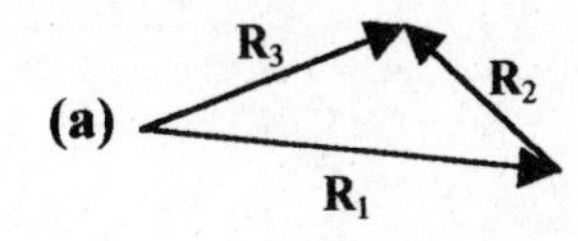

(b) 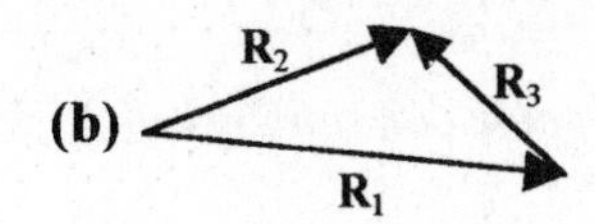

(c) 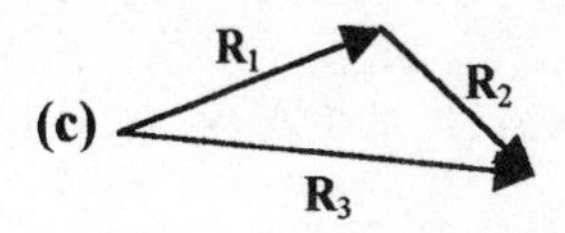

(d) 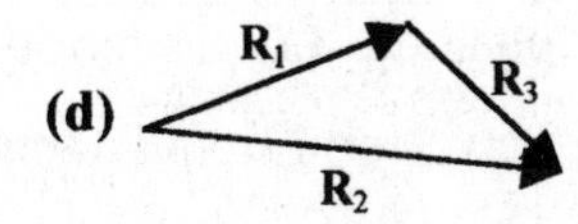

(e) 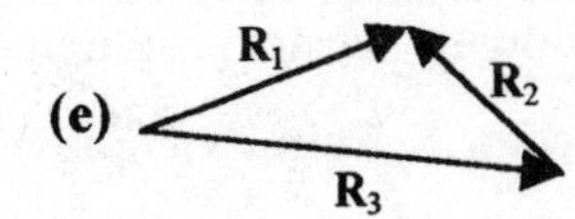

5. If a vector **Q** has x and y components of –3.0 and 5.5 respectively, and vector **R** has x and y components of 9.2 and 4.4, respectively, what are the x and y components (x, y) of the sum **Q** + **R**?

(a) (14.7, 1.4) **(b)** (6.2, 9.9) **(c)** (3.7, 7.4) **(d)** (2.5, –4.8) **(e)** (–3.0, 4.4)

6. Given vectors **Q** and **R** from question 5, what are the x and y components (x, y) of the vector **R** – **Q**?

(a) (–7.4, –3.7) **(b)** (6.2, 9.9) **(c)** (6.2, 1.1) **(d)** (12.2, –1.1) **(e)** (9.2, –4.4)

7. Suppose a vector **A** has x and y components of 41 and 28, respectively, and vector **B** has x and y components of 12 and –18, respectively. Determine the magnitude of the vector **C** if **C** = **B** + **A**.

(a) 50 **(b)** 22 **(c)** 54 **(d)** 28 **(e)** 72

8. Given the vectors **A**, **B**, and **C** in question 7, what is the direction of **C** in a coordinate system in which the +x-axis points horizontally to the right and the +y-axis points vertically upward?

(a) 11° **(b)** 34° **(c)** –22° **(d)** 56° **(e)** 91°

3–4 Unit Vectors

A very convenient way to completely specify a vector is by using **unit vectors**. A unit vector is a dimensionless vector whose magnitude equals 1. Generally, unit vectors are used to indicate specific directions. Our most common application of unit vectors will be those for specifying the x- and y-directions. A unit vector will be distinguished from other vectors by having a ^ over it. Therefore, the unit vector for the x direction is $\hat{\mathbf{x}}$, and the unit vector for the y direction is $\hat{\mathbf{y}}$.

When a unit vector is multiplied by a scalar the result is a vector whose magnitude equals the size of the scalar and whose direction is the same as that of the unit vector if the scalar is positive, or the opposite of the unit vector if the scalar is negative. For example, the vector $\mathbf{A} = 5\hat{\mathbf{x}}$ has a magnitude of A = 5 and points in the positive x direction, and the vector $\mathbf{B} = -8\hat{\mathbf{y}}$ has a magnitude of B = 8 and points in the negative y direction. Using this fact we can write any arbitrary vector in terms of unit vectors by multiplying its scalar components by the corresponding unit vectors and summing them:

$$\mathbf{A} = A_x\hat{\mathbf{x}} + A_y\hat{\mathbf{y}}$$

The quantities $A_x\hat{\mathbf{x}}$ and $A_y\hat{\mathbf{y}}$ are called the *vector components* of **A**.

Exercise 3–5 Unit Vectors Express the following vectors in unit-vector notation: **(a)** the position vector **R** from Example 3–1, **(b)** the position vector **R** from Example 3–2, **(c)** the vector **d** from Example 3–3, and **(d)** the vector $\mathbf{v}_3$ from Example 3–4.

Solution Following the preceding discussion we construct the vectors by multiplying the scalar components determined in the relevant examples by the appropriate unit vectors and summing them.

(a) Since in Example 3–1 we determined that $R_x = -250$ m, and $R_y = 310$ m we can write for vector **R**

$$\mathbf{R} = -(250\text{m})\hat{\mathbf{x}} + (310\text{m})\hat{\mathbf{y}}$$

(b) Since in Example 3–2 we determined that $R_x = 0.38$ m and $R_y = 0.65$ m, we can write for vector **R**

$$\mathbf{R} = (0.38\text{m})\hat{\mathbf{x}} + (0.65\text{m})\hat{\mathbf{y}}$$

(c) In Example 3–3 we found $d_x = -3.00$ and $d_y = 11.0$, so that

$$\mathbf{d} = -3.00\hat{\mathbf{x}} + 11.0\hat{\mathbf{y}}$$

(d) In Example 3–4 we calculated $v_{3x} = 14.84$ and $v_{3y} = -29.15$, which gives

$$\mathbf{v}_3 = 14.84\hat{\mathbf{x}} - 29.15\hat{\mathbf{y}}$$

3–5 Position, Displacement, Velocity, and Acceleration Vectors

If you recall the discussions of position, displacement, velocity, and acceleration from Chapter 2, you may remember that associated with each of these quantities was a size (magnitude) and a direction (positive or negative). Therefore, these are all vector quantities and in more than one dimension we wish to represent these quantities in the more general vector notation discussed previously. The SI units of all these quantities are given in Chapter 2.

The position vector **r** for an object in a coordinate system is the arrow from the origin of the coordinate system to the location of the object. The magnitude of this vector is the distance of the object from the origin. The x and y scalar components of **r** are the x and y coordinates of the object. Therefore, in unit vector notation the two-dimensional position vector is written as

$$\mathbf{r} = x\hat{\mathbf{x}} + y\hat{\mathbf{y}}$$

Examples 3–1 and 3–2 involved two-dimensional position vectors.

The displacement of an object is its change in position, so the displacement vector $\Delta\mathbf{r}$ is the difference between its final and initial positions, $\Delta\mathbf{r} = \mathbf{r}_f - \mathbf{r}_i$. Given what we learned in the Section 3–3 about vector addition and subtraction, the unit vector notation for displacement in two dimensions is

$$\Delta\mathbf{r} = (x_f - x_i)\hat{\mathbf{x}} + (y_f - y_i)\hat{\mathbf{y}}$$

The average velocity is the displacement divided by the elapsed time Δt. Dividing the vector $\Delta\mathbf{r}$ by the scalar Δt, we divide each component of $\Delta\mathbf{r}$ by Δt; hence,

$$\mathbf{v}_{av} = \frac{\Delta\mathbf{r}}{\Delta t} = \frac{\Delta x}{\Delta t}\hat{\mathbf{x}} + \frac{\Delta y}{\Delta t}\hat{\mathbf{y}}$$

where $\Delta x = x_f - x_i$ and $\Delta y = y_f - y_i$. The instantaneous velocity, once again, is given by the limit of the average velocity as the time interval approaches zero:

$$\mathbf{v} = \lim_{\Delta t \to 0} \frac{\Delta\mathbf{r}}{\Delta t}$$

The average acceleration is the change in velocity, $\Delta\mathbf{v}$, divided by the elapsed time, Δt,

$$\mathbf{a}_{av} = \frac{\Delta\mathbf{v}}{\Delta t}$$

The change in velocity is the difference between the final and initial velocities, $\Delta\mathbf{v} = \mathbf{v}_f - \mathbf{v}_i$. The instantaneous acceleration is given by the limit of the average acceleration as the time interval approaches zero:

$$\mathbf{a} = \lim_{\Delta t \to 0} \frac{\Delta\mathbf{v}}{\Delta t}$$

Example 3–6 Out of Gas Suppose you are driving northwest at 13.5 m/s for exactly half an hour when you run out of gas. Frustrated, you walk for 40.0 min to the nearest gas station, which is 2.40 km away at 30° north of west from where your car stopped. Determine your average velocity for this trip. State the answer both as a magnitude and a direction and using unit vectors.

Picture the Problem The diagram shows the three relevant displacement vectors for the trip. The vector $\mathbf{d}_c$ is your displacement while driving the car, $\mathbf{d}_w$ is your displacement while walking, and $\mathbf{r}$ is your total displacement.

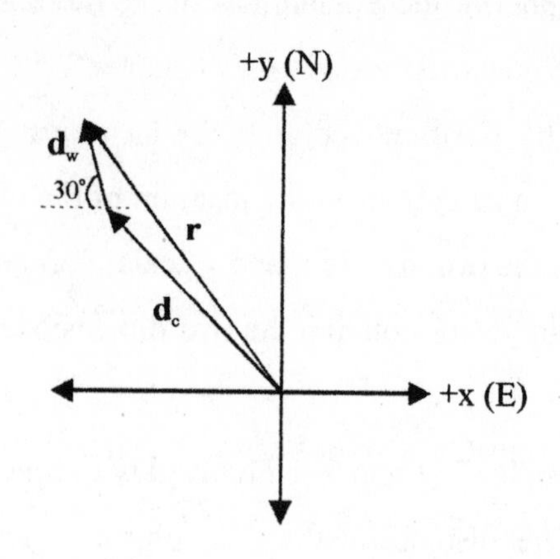

Strategy Since average velocity is displacement divided by elapsed time, we will first calculate the displacement vectors, then determine the average velocity.

Solution

Step 1: Determine the displacement $\mathbf{d}_c$.

1. Calculate the magnitude of $\mathbf{d}_c$ from the speed and time while in the car:

$$d_c = v_c t_c = (13.5\ \text{m/s})(30\ \text{min})\left(\frac{60\ \text{s}}{\text{min}}\right) = 24{,}300\ \text{m}$$

2. Going northwest implies a direction for $\mathbf{d}_c$ of:

$$\phi_c = 90^\circ + 45^\circ = 135^\circ$$

3. Determine the x and y components of $\mathbf{d}_c$:

$$d_{cx} = d_c \cos\phi_c = (24{,}300\ \text{m})\cos(135^\circ) = -17{,}183\ \text{m}$$
$$d_{cy} = d_c \sin\phi_c = (24{,}300\ \text{m})\sin(135^\circ) = 17{,}183\ \text{m}$$

Step 2: Determine the displacement $\mathbf{d}_w$.

4. 30° north of west implies a direction for $\mathbf{d}_w$ of:

$$\phi_w = 180^\circ - 30^\circ = 150^\circ$$

5. Determine the x and y components of $\mathbf{d}_w$:

$$d_{wx} = d_w \cos\phi_w = (2400\ \text{m})\cos(150^\circ) = -2{,}078\ \text{m}$$
$$d_{wy} = d_w \sin\phi_w = (2400\ \text{m})\sin(150^\circ) = 1{,}200\ \text{m}$$

Step 3: The total displacement $\mathbf{r}$ is $\mathbf{d}_c + \mathbf{d}_w$

6. Determine the x and y components of $\mathbf{r}$:

$$r_x = d_{cx} + d_{wx} = -17{,}183\ \text{m} - 2{,}078\ \text{m} = -19{,}261\ \text{m}$$
$$r_y = d_{cy} + d_{wy} = 17{,}183\ \text{m} + 1{,}200\ \text{m} = 18{,}383\ \text{m}$$

7. Determine the total elapsed time, t:

$$t = t_c + t_w = 30\ \text{min} + 40\ \text{min} = 70\ \text{min}\left(\frac{60\ \text{s}}{\text{min}}\right)$$
$$= 4200\ \text{s}$$

8. Determine the x and y components of $\mathbf{v}_{av}$:

$$v_{av,x} = \frac{r_x}{t} = \frac{-19{,}261\ \text{m}}{4200\ \text{s}} = -4.586\ \text{m/s}$$
$$v_{av,y} = \frac{r_y}{t} = \frac{18{,}383\ \text{m}}{4200\ \text{s}} = 4.377\ \text{m/s}$$

9. State the average velocity using unit vectors: $\mathbf{v}_{av} = -(4.59\,\text{m/s})\hat{\mathbf{x}} + (4.38\,\text{m/s})\hat{\mathbf{y}}$

10. Find the magnitude of $\mathbf{v}_{av}$:

$$v_{av} = \left(v_{av,x}^2 + v_{av,y}^2\right)^{1/2}$$

$$= \left[(-4.586\,\text{m/s})^2 + (4.377\text{m/s})^2\right]^{1/2} = 6.34\,\text{m/s}$$

11. Calculate a reference angle θ for $\mathbf{v}_{av}$:

$$\theta = \tan^{-1}\left(\frac{v_{av,y}}{v_{av,x}}\right) = \tan^{-1}\left(\frac{4.377\,\text{m/s}}{-4.586\,\text{m/s}}\right) = -43.7^\circ$$

12. From the signs of the components we can see that θ is measured above the –x-axis; therefore, the magnitude and direction of $\mathbf{v}_{av}$ is: 6.34 m/s at 43.7° north of west.

Insight In the final calculation of θ the negative signs comes strictly as a result of the negative x component. Without any further statement it would be assumed that the angle is measured clockwise from the +x-axis. This potential mistake is why the interpretation in the final step is needed.

Exercise 3–7 Turn Left at the Light Suppose that you are driving at 35 mi/h down Main Sreet while approaching a green light at the intersection with State Street. In order to turn left onto State St. you slow down, make the turn, then speed up to 35 mi/h on State Street. If it takes you a total time of 1.75 minutes from the time you begin to slow down on Main Sreet until you reach 35 mi/h on State Sreet, what is the magnitude and direction of your average acceleration?

Solution Try to sketch the velocity and average acceleration vectors for this problem. Let us choose the direction of the initial velocity on Main Sreet to be the +x direction, and the direction of the final velocity on State Sreet to be the +y direction.

Given: $v_i = 35$ mi/h, $\theta_i = 0.0°$, $v_f = 35$ mi/h, $\theta_f = 90°$, $t = 1.75$ min **Find:** a_{av}, θ_{av}

Since the average acceleration is determined by the change in velocity Δ**v**, lets us first calculate Δ**v** by getting the components of the final and initial velocities. For the initial velocity we have

$$v_{ix} = v_i \cos\theta_i = 35\text{ mi/h}\ \cos(0.0^\circ) = 35\text{ mi/h}\left(\frac{0.447\text{ m/s}}{1\text{ mi/h}}\right) = 15.6\text{ m/s}$$

$$v_{iy} = v_i \sin\theta_i = 35\text{ mi/h}\ \sin(0.0^\circ) = 0\text{ m/s}$$

For the final velocity we have

$$v_{fx} = v_f \cos\theta_f = 35 \text{ mi/h} \cos(90^\circ) = 0 \text{ mi/h} = 0 \text{ m/s}$$
$$v_{fy} = v_f \sin\theta_f = 35 \text{ mi/h} \sin(90^\circ) = 35 \text{ mi/h} = 15.6 \text{ m/s}$$

Now that we have the components we can determine the components of Δv.

$$\Delta v_x = v_{fx} - v_{ix} = 0 \text{ m/s} - 15.6 \text{ m/s} = -15.6 \text{ m/s}$$
$$\Delta v_y = v_{fy} - v_{iy} = 15.6 \text{ m/s} - 0 \text{ m/s} = 15.6 \text{ m/s}.$$

From these components we can determine the magnitude:

$$\Delta v = \left[\Delta v_x^2 + \Delta v_y^2\right]^{1/2} = \left[(-15.6 \text{ m/s})^2 + (15.6 \text{ m/s})^2\right]^{1/2} = 22.06 \text{ m/s}$$

The total elapsed time is 1.75×60 s = 105 s, so the magnitude of the average acceleration is

$$a_{av} = \frac{\Delta v}{t} = \frac{22.06 \text{ m/s}}{105 \text{ s}} = 0.21 \text{ m/s}^2$$

To determine the direction, notice that the $a_{av,x}$ will be negative, because Δv_x is negative, and $a_{av,y}$ will be positive for a similar reason. These facts mean that $\mathbf{a}_{av}$ lies in the second quadrant of the coordinate system. (If you couldn't sketch $\mathbf{a}_{av}$ before, then do so now.) Also notice that since the x and y components of Δv have the same size, 15.6 m/s, then Δv, and therefore $\mathbf{a}_{av}$, must make a 45 degree angle in that quadrant. Finally, we can say that the average acceleration is 0.21 m/s^2 at 135 degrees from the positive x-axis.

Notice that even though the magnitudes of the initial and final velocities are the same, the acceleration is not zero. This is because the directions of the velocities are different, and this difference is just as important for determining acceleration as a difference in speed.

Practice Quiz

9. In unit-vector notation determine the displacement of an object that goes from (x, y) coordinates of (–3.6, 2.1) to (4.2, –9.8) in arbitrary units.

   (a) $0.6\,\hat{x} - 7.7\,\hat{y}$ **(b)** $-0.6\,\hat{x} + 7.7\,\hat{y}$ **(c)** $-7.8\,\hat{x} + 11.9\,\hat{y}$ **(d)** $7.8\,\hat{x} - 11.9\,\hat{y}$ **(e)** $-0.6\,\hat{x} - 7.7\,\hat{y}$

10. Which of the following (in m/s) gives the correct average velocity of a particle that moves from position $\mathbf{r}_1 = 2.0\hat{x} + 5.0\hat{y}$ to position $\mathbf{r}_2 = 8.0\hat{x} - 3.0\hat{y}$, both in meters, in 5.0 s?

    (a) $1.2\,\hat{x} - 1.6\,\hat{y}$ **(b)** $6.0\,\hat{x} - 8.0\,\hat{y}$ **(c)** $2.0\,\hat{x} + 0.4\,\hat{y}$ **(d)** $30\,\hat{x} - 40\,\hat{y}$ **(e)** $1.6\,\hat{x} - 0.6\,\hat{y}$

11. An object has an average acceleration of $\mathbf{a}_{av} = 8.21\hat{x} + 1.71\hat{y}$ in m/s^2 after accelerating for 6.75 s. What was its change in velocity? (Answer are in m/s.)

    (a) $187\,\hat{x} + 39.0\,\hat{y}$ **(b)** $1.22\,\hat{x} + 0.253\,\hat{y}$ **(c)** $55.4\,\hat{x} + 11.5\,\hat{y}$ **(d)** $4.11\,\hat{x} - 0.855\,\hat{y}$ **(e)** 0

3–6 Relative Motion

A good and relevant application of vector addition and subtraction is relative motion. Specifically, we will focus on the velocity of an object as measured by two observers who are moving with respect to each other at a constant relative velocity. To identify who is measuring what, we use a system of subscripts in which the first subscript refers to the object whose velocity is being measured, and the second subscript refers to the observer (or coordinate system) making the measurement. [Note that the observer can be fictitious; we really need only the coordinate system with respect to which the velocity has the value in question]. For example, a velocity labeled $\mathbf{v}_{12}$ refers to the velocity *of* object 1 *relative to* (or as measured by) object 2. If you reverse the subscripts to get the velocity of 2 relative to 1, this velocity $\mathbf{v}_{21}$ relates to $\mathbf{v}_{12}$ by a minus sign: $\mathbf{v}_{21} = -\mathbf{v}_{12}$.

To state how relative velocity motion is handled in more general terms, consider the accompanying diagram which shows two coordinate systems, A and B, and a point P that is in motion relative to each coordinate system. Observers in system A identify the velocity of P as $\mathbf{v}_{PA}$; whereas observers in system B identify P's velocity as $\mathbf{v}_{PB}$. Let us assume that system B moves relative to system A with a velocity $\mathbf{v}_{BA}$. The relationship between these velocities can be written as

$$\mathbf{v}_{PA} = \mathbf{v}_{PB} + \mathbf{v}_{BA}$$

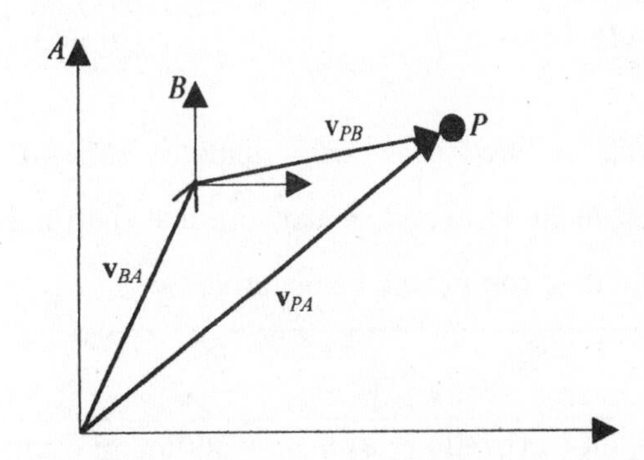

It is very helpful to notice how the subscript convention works in this velocity addition equation. The resultant velocity $\mathbf{v}_{PA}$ has P as the leftmost subscript and A as the rightmost subscript. Now, the subscripts on the two velocities whose addition produces the resultant are ordered $PBBA$. In this ordering P is the leftmost subscript and A is the rightmost, just as with the resultant. The subscript for the system not referenced by the resultant (B in this case) is repeated in the middle. This mnemonic can be used to help analyze almost any relative velocity situation as long as you stay consistent with how you label the velocities. You should be careful, however, not to fall into the trap of using this mnemonic as a substitute for understanding the physical situation being described.

Example 3–8 Crossing the Street Tom and Jan are standing at the side of the street waiting to cross. A car is coming at 10.0 m/s. Jan decides to go ahead and run directly across the street at 2.80 m/s whereas Tom chooses to wait. While Jan is running across, what is her velocity as measured by the driver of the car?

Picture the Problem The diagram indicates the motion of both Jan and the car. The velocity vectors are labeled such that g means ground.

Strategy We need to determine Jan's velocity with respect to the car. In our system of labels this velocity is $\mathbf{v}_{Jc}$.

Solution

1. The vector addition we need is: $\mathbf{v}_{Jc} = \mathbf{v}_{Jg} + \mathbf{v}_{gc}$

2. Write out $\mathbf{v}_{Jg}$ based on given information: $\mathbf{v}_{Jg} = 2.80\text{ m/s}\,\hat{\mathbf{y}}$

3. Write out $\mathbf{v}_{gc}$ based on given information: $\mathbf{v}_{gc} = -\mathbf{v}_{cg} = -10.0\text{ m/s}\,\hat{\mathbf{x}}$

4. Perform the vector addition: $\mathbf{v}_{Jc} = -(10.0\text{ m/s})\,\hat{\mathbf{x}} + (2.80\text{ m/s})\,\hat{\mathbf{y}}$

Insight The mnemonic outlined previously made it straightforward to identify the correct vector addition to solve this problem. However, make sure that the final answer for $\mathbf{v}_{Jc}$ makes intuitive sense to you as well. Why should its x component be negative?

Example 3–9 Holding an Umbrella You are walking on campus at 1.3 m/s directly against a 20 mi/h, horizontal wind in a light snowstorm. If the snowflakes fall vertically downward with a speed of 3.9 m/s with respect to still air, at what angle, with respect to the vertical, should you hold your umbrella to best protect yourself from the snow?

Picture the Problem The sketch shows you, relative to the ground, walking in the storm holding the umbrella at an angle. The motion of the snow relative to the ground is also indicated by the lines

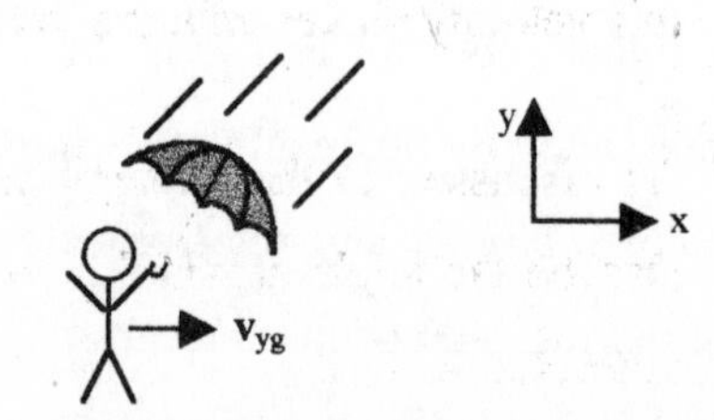

about to strike the umbrella.

Strategy The umbrella protects best if it is oriented so that the snow strikes the greatest amount of its surface. For this to happen the shaft of the umbrella should be oriented along the line of motion of the snow (as viewed by you). Therefore, the question will be answered if we determine the angle that the velocity of the snow makes with the vertical as measured by you, so we need $\mathbf{v}_{sy}$.

Solution

The various reference frames here are *you* (y), the *ground* (g), the *snow* (s), and the *air* (a).

1. The vector addition that we need is: $\mathbf{v}_{sy} = \mathbf{v}_{sg} + \mathbf{v}_{gy}$

2. We can immediately write down $\mathbf{v}_{gy}$: $\mathbf{v}_{gy} = -\mathbf{v}_{yg} = -1.3\text{ m/s}\,\hat{\mathbf{x}}$

3. The vector addition to get $\mathbf{v}_{sg}$ is: $\mathbf{v}_{sg} = \mathbf{v}_{sa} + \mathbf{v}_{ag}$

4. $\mathbf{v}_{sa}$ is given in the problem: $\mathbf{v}_{sa} = -3.9\text{ m/s}\,\hat{\mathbf{y}}$

5. From the given information we also have $\mathbf{v}_{ag}$: $\mathbf{v}_{ag} = -20\text{ mi/h}\left(\dfrac{0.447\text{ m/s}}{1\text{ mi/h}}\right)\hat{\mathbf{x}} = -8.94\text{ m/s}\,\hat{\mathbf{x}}$

6. The velocity of the snow relative to the ground: $\mathbf{v}_{sg} = -8.94\text{ m/s}\,\hat{\mathbf{x}} - 3.9\text{ m/s}\,\hat{\mathbf{y}}$

7. From step 1 the velocity of the snow relative to you is:

$$\mathbf{v}_{sy} = (-8.94\text{ m/s} - 1.3\text{ m/s})\hat{\mathbf{x}} - 3.9\text{ m/s}\,\hat{\mathbf{y}} = -10.24\text{ m/s}\,\hat{\mathbf{x}} - 3.9\text{ m/s}\,\hat{\mathbf{y}}$$

8. As shown in the following sketch, the angle θ this velocity makes with the vertical is:

$$\theta = \tan^{-1}\left(\frac{v_{sy,x}}{v_{sy,y}}\right) = \tan^{-1}\left(\frac{-10.24\text{ m/s}}{-3.9\text{ m/s}}\right) = 69°$$

9. As discussed in the strategy section, this must also be the angle at which you should hold the umbrella.

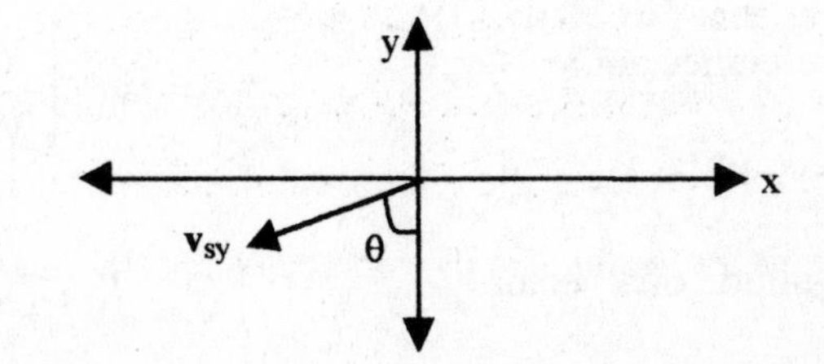

Insight It is easy to forget that the angle of the umbrella is determined relative to you. A common mistake here is to calculate only the angle of the snow relative to the ground. When dealing with relative motion always remember to ask yourself *with respect to whom* is this question being asked.

Practice Quiz

12. If $\mathbf{v}_{PA}$ is the velocity of object P with respect to frame A, and $\mathbf{v}_{PB}$ is the velocity of object P with respect to frame B, which of the following equals the velocity of frame B with respect to frame A?

(a) $\mathbf{v}_{AP} + \mathbf{v}_{BP}$ **(b)** $\mathbf{v}_{PA} + \mathbf{v}_{PB}$ **(c)** $\mathbf{v}_{AP} - \mathbf{v}_{BP}$ **(d)** $\mathbf{v}_{BP} - \mathbf{v}_{PA}$ **(e)** $\mathbf{v}_{BP} + \mathbf{v}_{PA}$

13. If $\mathbf{v}_{BA} = 12\hat{\mathbf{x}} + 22\hat{\mathbf{y}}$ in m/s and $\mathbf{v}_{PA} = -8\hat{\mathbf{x}} - 8\hat{\mathbf{y}}$ in m/s, then what is $\mathbf{v}_{PB}$ in SI units?

(a) $4\hat{\mathbf{x}} + 14\hat{\mathbf{y}}$ **(b)** $-20\hat{\mathbf{x}} - 30\hat{\mathbf{y}}$ **(c)** $-20\hat{\mathbf{x}} + 14\hat{\mathbf{y}}$ **(d)** $20\hat{\mathbf{x}} + 30\hat{\mathbf{y}}$ **(e)** $-4\hat{\mathbf{x}} - 14\hat{\mathbf{y}}$

Reference Tools and Resources

I. Key Terms and Phrases

vector a mathematical quantity having both magnitude and direction with appropriate units

scalar a numerical value with appropriate units

magnitude of a vector the full numerical value of the quantity being represented

direction of a vector the orientation within a coordinate system of the quantity being represented

component of a vector the part of a vector associated with a specific direction

unit vector a dimensionless vector of unit magnitude

II. Important Equations

Name/Topic	Equation	Explanation
scalar components of a vector **V**	$V_x = V\cos\theta,\ V_y = V\sin\theta$	Scalar components of a vector from its magnitude and direction. The angle θ is measured from the +x direction.
reference angle	$\theta = \tan^{-1}\left(\frac{V_y}{V_x}\right)$	The reference angle gives the direction of a vector with respect to the coordinate axes.
magnitude of a vector	$V = \sqrt{V_x^2 + V_y^2}$	The magnitude of a vector from its x- and y-components.

unit vectors	$\mathbf{V} = V_x\,\hat{\mathbf{x}} + V_y\,\hat{\mathbf{y}}$	A two-dimensional vector written in terms of unit vectors.
vector addition/subtraction	$\mathbf{A} \pm \mathbf{B} = (A_x \pm B_x)\hat{\mathbf{x}} + (A_y \pm B_y)\hat{\mathbf{y}}$	Vector addition and subtraction with components.
relative motion	$\mathbf{v}_{PA} = \mathbf{v}_{PB} + \mathbf{v}_{BA}$	Velocity addition for relative motion using the subscripting mnemonic.

III. Know Your Units

Quantity	Dimension	SI Unit
unit vector	—	dimensionless

IV. Tips

The component method of vector addition described in this chapter is very powerful and always works. However, the method can sometimes be cumbersome, and increasing the number of calculations provides greater opportunities for some sort of mistake to slip in. For cases when you know the magnitudes and direction of the two vectors you need to add, the law of cosines can provide a more direct route to the result than the component method. The law of cosines is illustrated below.

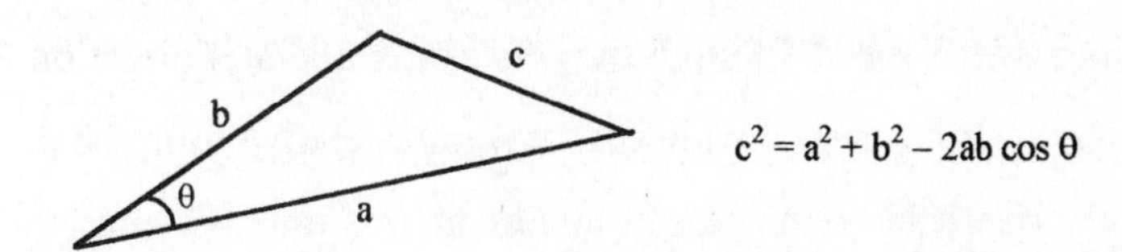

Consider the many calculations in the solution of example 3.4. If we only needed the magnitude of $\mathbf{v}_3$, having been given the magnitude and direction of both $\mathbf{v}_1$ and $\mathbf{v}_2$ would have allowed us to immediately draw the following triangle:

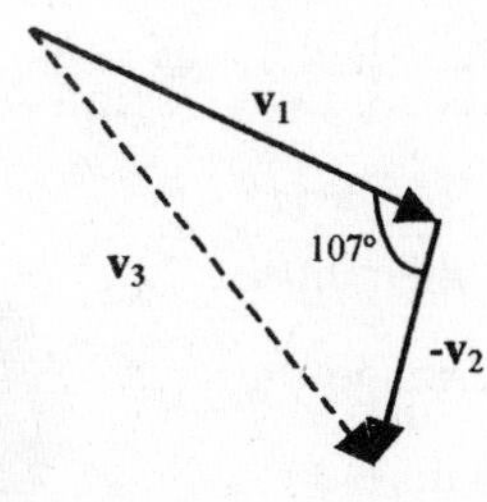

The law of cosines would then yield the result with just one additional step

$$v_3 = \left[25.0^2 + 15.0^2 - 2(25.0)(15.0)\cos(107°)\right]^{1/2} = 32.7.$$

Practice Problems

1. Vector A has a magnitude of 76 units and a direction counterclockwise from the east of 4.6 degrees. What is the value of its x component to one decimal place?
2. In problem 1 what is the vector's y component?
3. A vector has an x component of 9.3 and a y component of 2. What is its magnitude to the nearest tenth of a unit?
4. What is the direction of the vector in problem 3 to the nearest tenth of a degree measured counterclockwise from the east.
5. Vector **A** is 10 units long and is at angle of 65° counterclockwise from the east. Vector **B** is 10 units long and is at an angle of –30°. What is the direction measured counterclockwise from east of **A**–2**B** to the nearest tenth of a degree?
6. In problem 5 what is the magnitude of the resultant to the nearest tenth of a unit?
7. A car traveling at 8.2 m/s northwest reaches a curve, and 11.7 s later it is heading north at 15 m/s. What is the magnitude of its average acceleration to two decimal places?
8. In problem 7 what is the direction of the average acceleration to the nearest tenth of a degree measured counterclockwise from the east?
9. A boat is traveling with a speed 3.8 mi/h over the water and at a direction of 147° counterclockwise from the east. The river itself is flowing east at 4 mi/h relative to its bank. What is the speed of the boat relative to the riverbank to the nearest hundredth of a mile per hour?
10. In problem 9 what is the angle of progress to the nearest tenth of a degree of the boat with respect to the riverbank?

Puzzle

SHARING THE BURDEN

The helpful twin brothers are swinging their little 30-lb sister on a rope. Is it possible to arrange the rope so that each of the twins holds up 30 lbs?

Selected Solutions

15. (a)

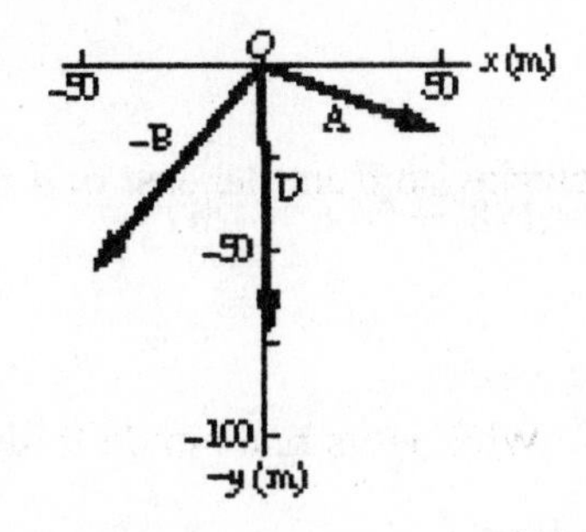

(b) From the information given in problem 14, we can determine vectors **A** and **B** by calculating the numerical value of their components.

$$\mathbf{A} = (50\text{ m})\cos(-20°)\hat{\mathbf{x}} + (50\text{ m})\sin(-20°)\hat{\mathbf{y}}$$
$$= (47.0\text{ m})\hat{\mathbf{x}} - (17.1\text{ m})\hat{\mathbf{y}}$$

$$\mathbf{B} = (70\text{ m})\cos(50°)\hat{\mathbf{x}} + (70\text{ m})\sin(50°)\hat{\mathbf{y}}$$
$$= (45.0\text{ m})\hat{\mathbf{x}} + (53.6\text{ m})\hat{\mathbf{y}}$$

We can now determine **D** using the component method:

$$\mathbf{D} = \mathbf{A} - \mathbf{B}$$
$$= (47.0\text{ m} - 45.0\text{ m})\hat{\mathbf{x}} + (-17.1\text{ m} - 53.6\text{ m})\hat{\mathbf{y}}$$
$$= (2.0\text{ m})\hat{\mathbf{x}} - (70.7\text{ m})\hat{\mathbf{y}}$$

Now that we know the components of **D** we can find its magnitude D and its direction θ:

$$D = \sqrt{(2.0\text{ m})^2 + (-70.7\text{ m})^2} = \boxed{71\text{ m}}$$

$$\theta = \tan^{-1}\left(\frac{\text{y component}}{\text{x component}}\right) = \tan^{-1}\left(\frac{-70.7\text{ m}}{2.0\text{ m}}\right) = \boxed{-88°}$$

19. From the given information we can write vector **A** as $\mathbf{A} = -5\hat{\mathbf{y}}$, and vector B as $\mathbf{B} = 10\hat{\mathbf{x}}$

(a) $\mathbf{A} + \mathbf{B} = (0 + 10)\hat{\mathbf{x}} + (-5 + 0)\hat{\mathbf{y}} = 10\hat{\mathbf{x}} - 5\hat{\mathbf{y}}$

From the above components we can determine the magnitude:

$$|\mathbf{A} + \mathbf{B}| = \sqrt{10^2 + (-5)^2} = \sqrt{125} = \boxed{5\sqrt{5}}$$

The direction is

$$\theta = \tan^{-1}\left(\frac{-5}{10}\right) = \boxed{-26.6°} \text{ relative to the +x axis}$$

(b) $\mathbf{A} - \mathbf{B} = (0 - 10)\hat{\mathbf{x}} + (-5 - 0)\hat{\mathbf{y}} = -10\hat{\mathbf{x}} - 5\hat{\mathbf{y}}$

From the above components we can determine the magnitude:

$$|\mathbf{A} - \mathbf{B}| = \sqrt{(-10)^2 + (-5)^2} = \boxed{5\sqrt{5}}$$

The direction is

$$\theta_{rel} = \tan^{-1}\left(\frac{-5}{-10}\right) = 26.6^\circ$$ below the −x-axis. Therefore, $\theta = 180° + 26.6° = \boxed{207°}$

This answer is also equivalent to −153°.

(c) $\mathbf{B} - \mathbf{A} = -(\mathbf{A} - \mathbf{B}) = 10\hat{\mathbf{x}} + 5\hat{\mathbf{y}}$

From the above components we can determine the magnitude:

$$|\mathbf{B} - \mathbf{A}| = \sqrt{10^2 + 5^2} = \boxed{5\sqrt{5}}$$

The direction is

$$\theta = \tan^{-1}\left(\frac{5}{10}\right) = \boxed{26.6°}$$ relative to the +x axis

23. (a) By the component method the direction and magnitude of **A** are

$$\theta = \tan^{-1}\left(\frac{A_y}{A_x}\right) = \tan^{-1}\left(\frac{-2.0\text{ m}}{5.0\text{ m}}\right) = \boxed{-22°}$$

$$A = \sqrt{(5.0\text{ m})^2 + (-2.0\text{ m})^2} = \sqrt{25\text{ m}^2 + 4.0\text{ m}^2} = \sqrt{29}\text{ m} = \boxed{5.4\text{ m}}$$

(b) The direction and magnitude of **B** are

$$\theta_{rel} = \tan^{-1}\left(\frac{B_y}{B_x}\right) = \tan^{-1}\left(\frac{5.0\text{ m}}{-2.0\text{ m}}\right) = -68.2^\circ$$, or 68.2 degrees above the −y axis

This result is equivalent to $\theta = 180° - 68.2° = \boxed{110°}$ to 2 significant figures relative to the +x axis.

$$B = \sqrt{(-2.0\text{ m})^2 + (5.0\text{ m})^2} = \sqrt{4.0\text{ m}^2 + 25\text{ m}^2} = \sqrt{29}\text{ m} = \boxed{5.4\text{ m}}$$

(c) $\mathbf{A} + \mathbf{B} = (5.0\text{ m} - 2.0\text{ m})\hat{\mathbf{x}} + (-2.0\text{ m} + 5.0\text{ m})\hat{\mathbf{y}} = (3.0\text{ m})\hat{\mathbf{x}} + (3.0\text{ m})\hat{\mathbf{y}}$

$$\theta = \tan^{-1}\left(\frac{3.0\text{ m}}{3.0\text{ m}}\right) = \tan^{-1}(1.0) = \boxed{45°}$$

$$|\mathbf{A} + \mathbf{B}| = \sqrt{(3.0\text{ m})^2 + (3.0\text{ m})^2} = \sqrt{2(3.0\text{ m})^2} = 3\sqrt{2}\text{ m} = \boxed{4.2\text{ m}}$$

33. (a) $v_x = (3.5\,\tfrac{\text{m}}{\text{s}})\cos(20.0°) = \boxed{3.3\text{ m/s}}$

$v_y = (3.5\,\tfrac{\text{m}}{\text{s}})\sin(20.0°) = \boxed{1.2\text{ m/s}}$

(b) The components will be halved.

41. The solution will be easier to follow if we make the following definitions:

$\mathbf{v}_{pg}$ = velocity of the plane relative to the ground

$\mathbf{v}_{pa}$ = velocity of the plane relative to the air

$\mathbf{v}_{ag}$ = velocity of the air relative to the ground

(a) We are interested in the direction the plane should be headed, which is different from the direction in which it will actually travel. The reason for this difference is the air, and so we need the direction of $\mathbf{v}_{pa}$. A velocity addition equation involving $\mathbf{v}_{pa}$ is

$$\mathbf{v}_{pg} = \mathbf{v}_{pa} + \mathbf{v}_{ag}$$

The magnitude of $\mathbf{v}_{pa}$ is given to be 310 km/h, and so we can write the velocity vector as

$$\mathbf{v}_{pa} = \left(310\ \frac{\text{km}}{\text{h}}\right)(\cos\theta)\hat{\mathbf{x}} + \left(310\ \frac{\text{km}}{\text{h}}\right)(\sin\theta)\hat{\mathbf{y}},$$

where θ is the direction we seek to determine and we take east to be along the +x-axis and north to be along the +y-axis.

Since east is along the +x-axis, we also have $\mathbf{v}_{ag} = (75\ \text{km/h})\hat{\mathbf{x}}$. The velocity addition equation then becomes

$$\mathbf{v}_{pg} = \left(310\ \frac{\text{km}}{\text{h}}\right)(\cos\theta)\hat{\mathbf{x}} + \left(310\ \frac{\text{km}}{\text{h}}\right)(\sin\theta)\hat{\mathbf{y}} + \left(75\ \frac{\text{km}}{\text{h}}\right)\hat{\mathbf{x}}$$

For the plane to travel due north, the net velocity in the x direction relative to the ground must equal zero.

$$v_{pg,x} = 0 = \left(310\ \frac{\text{km}}{\text{h}}\right)\cos\theta + \left(75\ \frac{\text{km}}{\text{h}}\right)$$

$$\therefore\ \cos\theta = \frac{-75\ \text{km/h}}{310\ \text{km/h}} = \frac{-15}{62} \Rightarrow \theta = \cos^{-1}\left(\frac{-15}{62}\right) = 104° = \boxed{14° \text{ west of north}}$$

(b)

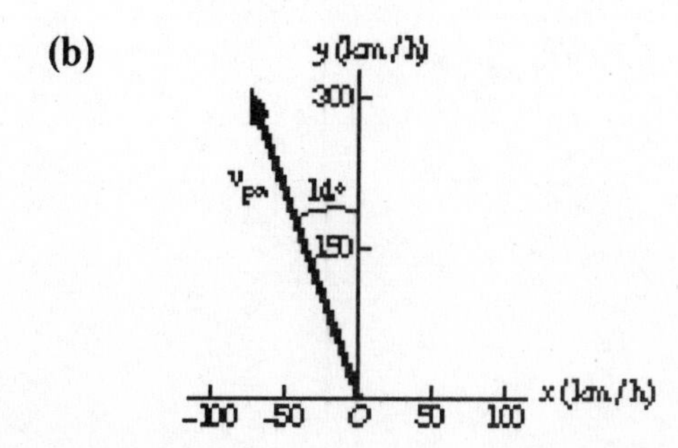

For the purposes of the sketch we calculate the components:

$$v_{pa,x} = \left(310\ \frac{\text{km}}{\text{h}}\right)\cos 104° = \boxed{-75\ \frac{\text{km}}{\text{h}}}$$

$$v_{pa,y} = \left(310\ \frac{\text{km}}{\text{h}}\right)\sin 104° = \boxed{301\ \frac{\text{km}}{\text{h}}}$$

(c) If the plane's speed is decreased from 310 km/h, then the factor cos θ will need to increase so that the product of the two will equal 75 km/h. The factor cos θ will increase if θ increases.

Answers to Practice Quiz

1. (e) **2.** (b) **3.** (c) **4.** (e) **5.** (b) **6.** (d) **7.** (c) **8.** (a) **9.** (d) **10.** (a) **11.** (c) **12.** (e) **13.** (b)

Answers to Practice Problems

1. 75.8	**2.** 6.1
3. 9.5	**4.** 12.1°
5. 124.4°	**6.** 23.1
7. 0.93 m/s^2	**8.** 57.8°
9. 2.22 mi/h	**10.** 68.6°

CHAPTER 4

TWO-DIMENSIONAL KINEMATICS

Chapter Objectives

After studying this chapter, you should:

1. know how to treat motion with constant velocity in two dimensions.
2. know how to treat motion with constant acceleration in two dimensions.
3. be able to apply the equations for two-dimensional motion to a projectile.
4. be able to calculate positions, velocities, and times for various types of projectile motion.

Warm-Ups

1. To attain the maximum range a projectile has to be launched at 45 degrees if the landing spot and the launch spot are at the same height (neglecting air resistance effects.) Explain in a few sentences how the relation between the vertical and the horizontal components of the initial velocity affects the projectile range.
2. On the Moon the acceleration due to gravity is about one-sixth that on Earth. If a golfer on the Moon imparts the same initial velocity to the ball as she does on Earth, how much farther will the ball go?
3. Three swimmers start across the river at the same time. They all swim with the same speed relative to the water. Peter swims in a direction perpendicular to the current. Albert swims slightly upstream along a line that makes an 80° angle with the shore. Dawn swims slightly downstream along a line that makes an 80° angle with the shore. Which of the three swimmers will reach the opposite shore first?
4. The pilot of a small plane is trying to maintain a course due north. His air speed is 120 mi/h. There is a 10 mi/h wind from the East. Estimate the direction in which the plane should be pointed to maintain the correct course.

Chapter Review

In Chapter 2 you studied motion in one dimension. Everything you learned there applies to two-dimensional motion. In fact, two-dimensional motion is really just two completely independent cases of motion in one dimension. We call these two dimensions the horizontal (x) and vertical (y) directions. The running theme of this chapter is the following: **horizontal and vertical motions are independent**.

4–1 Motion in Two Dimensions

Constant Velocity in Two Dimensions

Using the fact that two-dimensional motion is really two separate cases of one-dimensional motion makes it easy to find the equations for describing constant velocity in two dimensions: we simply take the one-dimensional equation, $\Delta x = vt$, and apply it to both the x and y directions. The key difference is that (a) *only the horizontal component of velocity,* v_{0x}, *applies to the horizontal motion,* and (b) *only the vertical component of velocity,* v_{0y}, *applies to the vertical motion.* Therefore, the equations are

$$x = x_0 + v_{0x}t$$
$$y = y_0 + v_{0y}t$$

Notice that since the velocity is constant, $v = v_0$ at all times during the motion.

Example 4–1 Pocket Billiards A billiard ball is struck so that it rolls across the pool table with a velocity of 25.0 cm/s at an angle of 52.0° above the horizontal. The horizontal direction is the short side of a standard 122-cm x 244-cm table. If the ball was initially located at a point $(x_0, y_0) = (11.0\text{ cm}, 17.0\text{ cm})$ from the leftmost corner on the short side, **(a)** how much time will pass before it strikes a side of the table? **(b)** Will it go into a pocket?

Picture the Problem The rough sketch shows the table and the ball. The lower left corner of the pool table is the origin of the coordinate system.

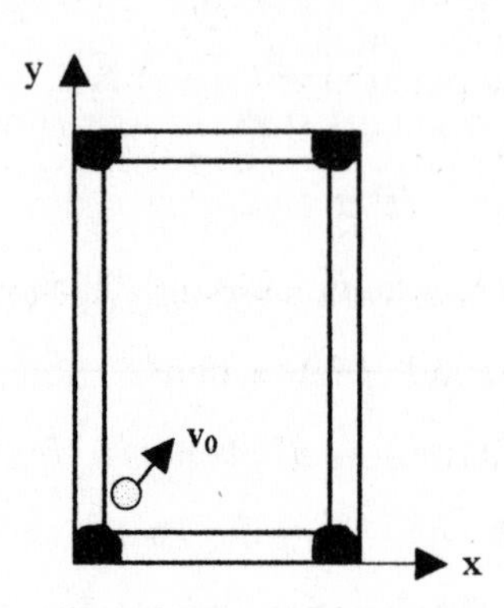

Strategy Treating the vertical and horizontal motions as independent, we need the x and y components of the velocity. We can then solve this problem by finding out where the ball will be at certain times.

Solution

Determine the components of the velocity:

$$v_{0x} = v_0 \cos\theta = (25.0\text{ cm/s})\cos(52.0°) = 15.39\text{ cm/s}$$
$$v_{0y} = v_0 \sin\theta = (25.0\text{ cm/s})\sin(52.0°) = 19.70\text{ cm/s}$$

Part (a)

1. Solve for the time it takes the ball to reach the right edge along the x-direction:

$$t = \frac{x - x_0}{v_{0x}} = \frac{122\text{ cm} - 11.0\text{ cm}}{15.39\text{ cm/s}} = 7.212\text{ s}$$

2. Determine the vertical position of the ball: $y = y_0 + v_{0y}t = 17.0\text{ cm} + (19.70\ \tfrac{\text{cm}}{\text{s}})(7.212\text{ s}) = 159\text{ cm}$

Since the upper edge is 244 cm from the lower edge, the ball has not reached the upper edge by the time it reaches the right edge. It strikes the right edge first.

Part (b)

3. Determine the ball's distance from the corner when it hits the right edge: $\Delta y = 244\text{ cm} - 159\text{ cm} = 85\text{ cm}$

The ball is 85 cm from the corner, which is too far for it to fall into the pocket.

Insight Clearly, we were able to treat the horizontal and vertical motions as simultaneous yet independent. Notice also that an additional digit was retained for the values of t, v_{0x}, and v_{0y} because they were used in intermediate steps.

Practice Quiz

1. If you walk in a direction 50.0° north of east at a pace of 2.30 m/s for one hour, how far north of your starting position will you be?

(a) 8.28 km **(b)** 138 m **(c)** 6.34 km **(d)** 5.32 km

2. A bird flies with a westerly speed of 20 m/s and a northerly speed of 15 m/s. How far will it fly over a time period of 32 seconds?

(a) 800 m **(b)** 640 m **(c)** 480 m **(d)** 160 m **(e)** 1100 m

Constant Acceleration in Two Dimensions

The same strategy used with constant velocity applies to motion with constant acceleration: we just apply the one-dimensional equations from Chapter 2 to both directions using only x components of velocity and acceleration for the horizontal motion and only y components for the vertical motion. Thus, the four equations become eight:

Horizontal Motion	Vertical Motion
$v_x = v_{0x} + a_x t$	$v_y = v_{0y} + a_y t$
$x = x_0 + \frac{1}{2}(v_{0x} + v_x)\ t$	$y = y_0 + \frac{1}{2}(v_{0y} + v_y)\ t$
$v_x^2 = v_{0x}^2 + 2a_x(x - x_0)$	$v_y^2 = v_{0y}^2 + 2a_y(y - y_0)$
$x = x_0 + v_{0x}t + \frac{1}{2}a_x t^2$	$y = y_0 + v_{0y}t + \frac{1}{2}a_y t^2$

These equations can be applied precisely as in Chapter 2. Here, they represent two independent yet simultaneous cases of one-dimensional motion.

Example 4–2 Horizontal Rocket Launch A toy rocket is launched horizontally from rest off the edge of a 0.850-m-high table. While in flight the rocket manages to remain pointed horizontally, and its thrust causes it to accelerate forward at 7.48 m/s^2 as gravity pulls it down. **(a)** How far from the edge of the table will it be when it hits the floor? **(b)** With what speed will it hit the floor?

Picture the Problem The diagram shows the rocket leaving the edge of the table and the path it takes. The positive coordinate directions are also indicated.

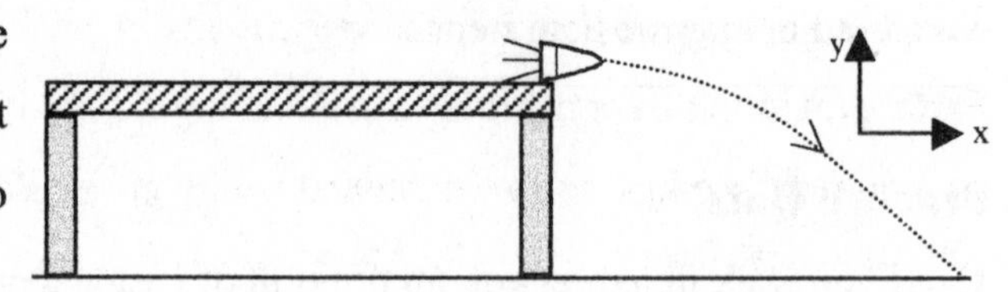

Strategy The rocket has constant acceleration in both directions, so the full set of eight equations applies.

Solution

Part (a): We need to determine when it hits the floor and get its horizontal position. Let's choose (x_0, y_0) = (0, 0). Also notice that the initial velocities are zero.

1. Choose a vertical equation that contains t with all other quantities known.: $y = \frac{1}{2}a_y t^2$

2. Solve this equation for the time to hit the floor: $t = \sqrt{\frac{2y}{a_y}} = \sqrt{\frac{2(-0.850\text{ m})}{-9.81\text{ m/s}^2}} = 0.4163\text{ s}$

3. Choose a horizontal equation involving x with all other quantities known: $x = \frac{1}{2}a_x t^2$

4. Solve to get the horizontal position: $x = \frac{1}{2}(7.48\ \text{m/s}^2)(0.4163\ \text{s})^2 = 0.648\ \text{m}$

Part (b):

5. Get the horizontal velocity when it hits the floor: $v_x = a_x t = (7.48\ \text{m/s}^2)(0.4163\ \text{s}) = 3.114\ \text{m/s}$

6. Get the vertical velocity when it hits the floor: $v_y = a_y t = (-9.81\ \text{m/s}^2)(0.4163\ \text{s}) = -4.084\ \text{m/s}$

7. Determine the magnitude of v to get the speed: $v = \sqrt{v_x^2 + v_y^2} = \sqrt{(3.114\ \text{m/s})^2 + (-4.084\ \text{m/s})^2}$
$= 5.14\ \text{m/s}$

Insight Be sure you understand the minus signs in part (a)-2.

4–2 – 4–4 Projectile Motion

The main application that is considered in this chapter is **projectile motion**. This motion is that of an object projected with some initial velocity and then allowed to free fall. Effects such as air resistance and Earth's rotation that sometimes causes an object's motion to differ significantly from that of pure free fall are ignored. The key point for understanding projectile motion is that the acceleration due to gravity acts only vertically, and since gravity is the only influence, there is no acceleration in the horizontal direction. This fact means that **although there is gravitational acceleration vertically, the horizontal motion is that of constant velocity**.

It is convenient to adopt a standard coordinate system for projectile motion. This standard takes the positive direction for vertical motion to be upward, making downward the negative direction. This means that the acceleration of gravity is given a negative value. For horizontal motion, the positive direction is to the right, and the negative direction to the left. Given this coordinate system we can rewrite the equations for two-dimensional motion in a form specific to projectile motion. To accomplish this rewriting we make use of several facts:

$$a_x = 0, \quad v_x = v_{0x} = v_0 \cos\theta$$
$$a_y = -g, \quad v_{0y} = v_0 \sin\theta$$

where θ is the **launch angle**, which is the angle of the initial velocity as measured from the horizontal direction. With these substitutions, the equations of projectile motion are as follows:

Horizontal Motion	Vertical Motion
$v_x = v_0 \cos\theta$	$v_y = v_0 \sin\theta - g\ t$
$x = x_0 + (v_0 \cos\theta)t$	$y = y_0 + \frac{1}{2}(v_0 \sin\theta + v_y)\ t$
	$v_y^2 = v_0^2 \sin^2\theta - 2g(y - y_0)$
	$y = y_0 + (v_0 \sin\theta)\ t - \frac{1}{2}gt^2$

This set of equations describes projectile motion for both zero and nonzero launch angles.

Example 4–3 Over the Edge Wagging his tail, a dog knocks an object horizontally off the edge of a table that is 0.750 m high. If the object hits the floor 0.995 m away from the edge of the table, with what speed did the object leave the table?

Picture the Problem The sketch shows the object leaving the edge of the table and the path it takes. The positive coordinate directions are also indicated.

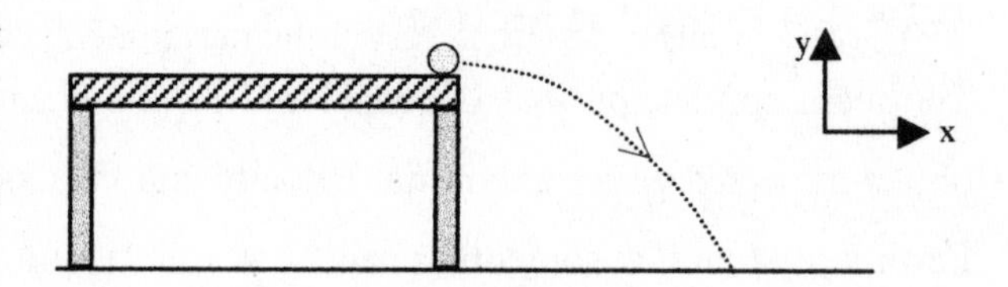

Strategy Let's choose $(x_0, y_0) = (0, 0)$. Also notice that $v_{0y} = 0$, and $\cos\theta = 1$. The only equation involving v_0 is the one giving x as a function of time.

Solution

1. Solve the horizontal equation for v_0: $v_0 = \frac{x}{t}$

2. Use the vertical equation with all known quantities to get an expression for t: $y = -\frac{1}{2}gt^2 \Rightarrow t = \sqrt{\frac{-2y}{g}}$

3. Put these results together to get an expression for v_0 in terms of known quantities and solve: $v_0 = x\sqrt{\frac{-g}{2y}} = (0.995\text{ m})\sqrt{\frac{-9.81\text{ m/s}^2}{2(-0.750\text{ m})}} = 2.54\text{ m/s}$

Insight This was an example of a projectile with horizontal launch. Here, no values were calculated at an intermediate step. Instead, the formula was carried through to the end. This approach is better for handling round-off error, but problems can sometimes get messy this way.

Example 4–4 Toss Me That Wrench A person stands 12.5 m from the base of a building that is 40.0 m tall. The person wants to toss a wrench to his coworker who is working on the roof. If he releases the tool at a height of 1.00 m above the ground, what velocity must he give the tool if it is to just make it onto the roof?

Picture the Problem Our sketch shows the wrench being tossed so that it just barely makes it onto the roof.

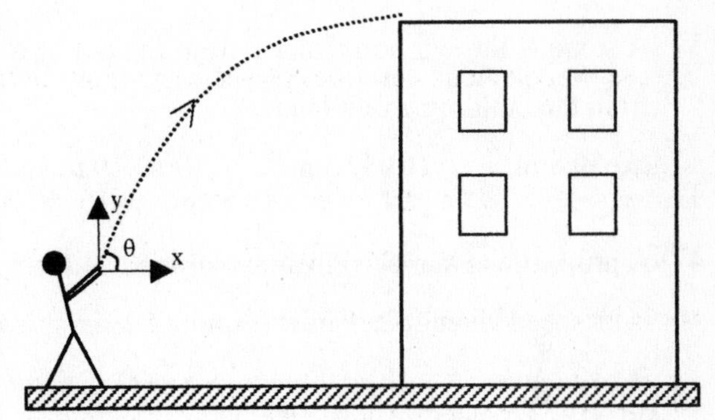

Strategy Since the wrench just barely makes it onto the roof, the top of the roof is the maximum height of the wrench, where $v_y = 0$. Treating the motions separately, we need to find the components of v_0. From them we can then determine the magnitude and direction.

Solution

Since we choose the origin at the point of launch, we take $x_0 = 0$, $y_0 = 0$, and $y = 39.0$ m.

1. Choose an equation containing v_{0y} with all other quantities known and solve:

$$v_y^2 = v_{0y}^2 - 2g(y - y_0) \Rightarrow v_{0y} = \sqrt{2g(y - y_0)}$$

$$\therefore\ v_{0y} = \sqrt{2(9.81\ \text{m/s}^2)(39.0\ \text{m})} = 27.66\ \text{m/s}$$

2. The only equation involving v_{0x} requires the time. Choose a vertical equation involving time with all other quantities known:

$$v_y = v_{0y} - gt \quad \Rightarrow \quad t = \frac{v_{0y}}{g} = \frac{27.66\ \text{m/s}}{9.81\ \text{m/s}^2} = 2.820\ \text{s}$$

3. Solve the horizontal equation for v_{0x}:

$$v_{0x} = \frac{x - x_0}{t} = \frac{12.5\ \text{m}}{2.820\ \text{s}} = 4.433\ \text{m/s}$$

4. Calculate the magnitude of the initial velocity:

$$v_0 = \sqrt{v_{0x}^2 + v_{0y}^2} = \sqrt{(4.433\ \text{m/s})^2 + (27.66\ \text{m/s})^2}$$
$$= 28.0\ \text{m/s}$$

5. Calculate the direction of the initial velocity:

$$\theta = \tan^{-1}\left(\frac{v_{oy}}{v_{0x}}\right) = \tan^{-1}\left(\frac{27.66\ \text{m/s}}{4.433\ \text{m/s}}\right) = 80.9^\circ$$

Insight In both this example and in the previous one, the origin was taken to be the initial position. Although this is not necessary, it is sometimes useful to adopt a convention such as this and not have to worry about what choices to make for every problem.

Practice Quiz

3. A stone is thrown horizontally from the top of a 50.0-m building with a speed of 12.3 m/s. How far from the building will it land?

   (a) 30.8 m **(b)** 39.3 m **(c)** 3.19 m **(d)** 615 m

4. A projectile is launched from the ground with a speed of 23.7 m/s at 33.0° above the horizontal. What is its speed when it is at its maximum height above the ground?

   (a) 19.9 m/s **(b)** 0.00 m/s **(c)** 12.9 m/s **(d)** 23.7 m/s

5. A projectile is launched from the ground with a speed of 53.4 m/s at 68.0° above the horizontal. Take its initial position as the origin, what are its (x, y) coordinates after 8.24 seconds?

   (a) (440 m, 440 m) **(b)** (440 m, 107 m) **(c)** (408 m, 333 m) **(d)** (165 m, 74.9 m)

4–5 Projectile Motion: Key Characteristics

The equations for projectile motion can be used to derive several important properties of the motion. The key characteristics and symmetries are as follows:

* The path, or trajectory, that a projectile follows is a parabola.

* If the initial and final elevations are the same, the **range** (R) of a projectile, which is the horizontal distance it travels before landing, is given by

$$R = \frac{v_0^2}{g}\sin 2\theta$$

* If the initial and final elevations are the same, the launch angle that produces maximum range is $\theta = 45°$.

* If the initial and final elevations are the same, the amount of time the projectile spends in the air, sometimes called the *time-of-flight*, is given by

$$t = \frac{2v_0}{g}\sin\theta$$

* If the initial and final elevations are the same, the time it takes a projectile launched at some upward angle to reach its maximum height equals the time it takes to fall back down from its maximum height.

* The maximum height that a projectile will reach above its initial height is given by

$$y_{max} = \frac{(v_0 \sin\theta)^2}{2g}$$

* The speed that a projectile has at a given height on its way up is equal to the speed it will have at that same height on its way back down.

* At a given height the angle of the velocity of a projectile above the horizontal on its way up equals the angle of the velocity below the horizontal on its way down.

Notice that all these characteristics are determined by the initial velocity given to the projectile.

Exercise 4–5 At the Driving Range A golf ball sitting on level ground is struck and given an initial velocity of 41.2 m/s at an angle of 58.0°. **(a)** How high does the ball go into the air? **(b)** How far does it travel? **(c)** How long is the ball in the air?

Solution Try to sketch a picture for this problem; the ball moves in a parabolic path starting and ending on the ground. The following information is supplied in the problem

Given: $v_0 = 41.2$ m/s, $\theta = 58.0°$; **Find**: **(a)** y_{max}, **(b)** R, **(c)** t

We are given the initial velocity and we know that it completely specifies the motion. Making use of the known results we can directly solve for each of these quantities.

(a) $$y_{max} = \frac{(v_0 \sin\theta)^2}{2g} = \frac{\left[(41.2\ \text{m/s})\sin(58.0°)\right]^2}{2(9.81\ \text{m/s}^2)} = 62.2\ \text{m}$$

(b) $$R = \frac{2v_0^2}{g}\sin 2\theta = \frac{(41.2\ \text{m/s})^2}{9.81\ \text{m/s}^2}\sin(116°) = 156\ \text{m}$$

(c) $$t = \frac{2v_0}{g}\sin\theta = \frac{2(41.2\ \text{m/s})}{9.81\ \text{m/s}^2}\sin(58.0°) = 7.12\ \text{s}$$

The questions asked in this problem are some of the basic things you might want to know about a projectile. This exercise illustrates the utility of working out equations for certain quantities.

Practice Quiz

6. In general, what is the shape of the path of a projectile?

   (a) a hyperbola (b) a parabola (c) a straight line (d) a circle

7. A projectile is launched from the ground with an initial velocity of 44.0 m/s at a launch angle of 26.0°. How long will this projectile be in the air?

   (a) 8.97 s (b) 8.06 s (c) 1.97 s (d) 3.93 s

8. A projectile reaches a height h when launched at an angle θ with speed v_0. If this projectile is launched from the same level at the same angle with speed $2v_0$, how high will it go?

   (a) h (b) 2h (c) 4h (d) h/2

9. A projectile remains airborne for a time t when launched at an angle θ with speed v_0. If this projectile is launched from the same level at the same angle with speed $2v_0$, how long will it be airborne?

   (a) t (b) 2t (c) 4t (d) t/2

10. A projectile travels a distance R when launched at an angle θ with speed v_0. If this projectile is launched from the same level at the same angle with speed $2v_0$, how far will it go?

   (a) R (b) 2R (c) 4R (d) R/2

Reference Tools and Resources

I. Key Terms and Phrases

projectile motion the motion of an object that is projected with an initial velocity and then moves under the influence of gravity only

launch angle the angle of the initial velocity of a projectile measured relative to the horizontal

range the horizontal distance traveled by a projectile before it lands

II. Important Equations

Name/Topic	Equation	Explanation
constant velocity	$x = x_0 + v_{0x}t$ $y = y_0 + v_{0y}t$	Each direction obeys the constant velocity equation independently.
constant acceleration: horizontal component	$v_x = v_{0x} + a_x t$ $x = x_0 + \frac{1}{2}(v_{0x} + v_x)t$ $v_x^2 = v_{0x}^2 + 2a_x(x - x_0)$ $x = x_0 + v_{0x}t + \frac{1}{2}a_x t^2$	The constant acceleration equations for horizontal components.
constant acceleration: vertical component	$v_y = v_{0y} + a_y t$ $y = y_0 + \frac{1}{2}(v_{0y} + v_y)t$ $v_y^2 = v_{0y}^2 + 2a_y(y - y_0)$ $y = y_0 + v_{0y}t + \frac{1}{2}a_y t^2$	The constant acceleration equations for vertical components.
range of a projectile	$R = \frac{v_0^2}{g}\sin 2\theta$	The horizontal distance traveled if the initial height equals the final height.
time-of-flight of a projectile	$t = \frac{2v_0}{g}\sin\theta$	The total time in the air if the initial height equals the final height
maximum height of a projectile	$y_{max} = \frac{(v_0 \sin\theta)^2}{2g}$	The maximum height above the initial height of launch.

III. Tips

As discussed in Chapter 2, you may find it useful to add a fifth equation to the list of equations describing motion with constant acceleration. This equation does not contain the initial velocity. In the context of two-dimensional motion the additional equations are

$$x = x_0 + v_x t - \frac{1}{2}a_x t^2$$

$$y = y_0 + v_y t - \frac{1}{2}a_y t^2$$

or, in the context of projectile motion,

$$y = y_0 + v_y t + \tfrac{1}{2} g t^2$$

In addition to the equations derived in your text and listed below for projectile motion, it may also be useful to know the equation for the path of the projectile, that is, y as a function of x. This equation can be determined by solving for t from the horizontal equation $x - x_0 = (v_0 \cos\theta)t$, and substituting this result into the vertical equation $y - y_0 = (v_0 \sin\theta)t - \tfrac{1}{2} g t^2$. After some manipulations the result is

$$y - y_0 = (\tan\theta)\ (x - x_0) - \frac{g}{2 v_0^2 \cos^2\theta}(x - x_0)^2$$

Try to rework Example 4–4 using this equation.

Practice Problems

1. A baseball is hit off a 1-meter-high tee at an angle of 24 degrees and at a speed of 26.6 m/s. How far, to the nearest tenth of a meter, does it travel horizontally?
2. A golf ball is chipped at an angle of 45 degrees and with a speed of 9.7 m/s. How far does it travel, to the nearest tenth of a meter?
3. A monkey swings on a vine and at the bottom of the swing, while moving horizontally, lets go of the vine, aiming for another vine 26.5 meters away horizontally and whose end is 5.6 meters below the monkey. How fast must the monkey be moving, to the nearest tenth of a m/s, to catch the end of the vine?
4. In problem 3 what would the monkey's speed have to be if he, moving at 45° when he was released, and the horizontal distance was the same but the other vine's end was the stated distance above him?
5. Two cars lost in a blinding snowstorm are traveling across a large field, with each driver he is are on the road, as shown in the figure. The cars collide. If the distance x is 115 m and the car on the left is travelling at 11.2 mi/h, how fast, to the nearest hundredth of a mi/h, was the car on the right car traveling?

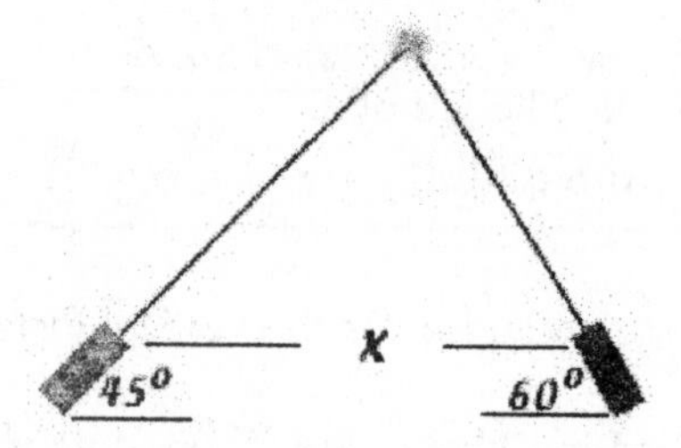

6. In problem 5, how much time passes before the cars collide, to the nearest tenth of a second?
7. A baseball is thrown horizontally off a cliff with a speed of 27 m/s. What is the horizontal distance, to the nearest tenth of a meter, from the face of the cliff after 2.7 seconds?
8. In problem 7, how far has it fallen in that time?

9. A ball is thrown with a speed of 30 m/s at an angle of 42 degrees. How high does it go above the cliff, to the nearest tenth of a meter?

10. In problem 9, if the ball hits the ground 7.1 seconds later, how high, to the nearest tenth of a meter, is the cliff?

Puzzle

THE LONG SHOT

You should know by now, from reading the text and from working the projectile motion problems, that the maximum range for a projectile is achieved when the projectile is fired at 45°. This is true if the launch starts and ends at the same altitude. What happens if you fire at a target at a lower elevation? Is the optimal angle still 45°? Is it more? Is it less? Answer these questions in words, not equations, briefly explaining how you obtained your answers.

Selected Solutions

3. (a) If we take the car's horizontal motion to be the positive x direction, the x component of its acceleration is

$$a_x = \left(2.0\ \frac{\text{m}}{\text{s}^2}\right)\cos 5.5° = 1.99\ \frac{\text{m}}{\text{s}^2}$$

Since it starts from rest, $v_0 = 0$, the equation for the horizontal distance traveled is

$$x = \tfrac{1}{2}a_x t^2 = \tfrac{1}{2}\left(1.99\ \frac{\text{m}}{\text{s}^2}\right)(10\ \text{s})^2 = \boxed{100\ \text{m}}$$

(b) The car's vertical acceleration is

$$a_y = \left(2.0\ \frac{\text{m}}{\text{s}^2}\right)\sin 5.5° = 0.192\ \frac{\text{m}}{\text{s}^2}$$

The expression for the vertical distance traveled is

$$y = \tfrac{1}{2}a_y t^2 = \tfrac{1}{2}\left(0.192\ \frac{\text{m}}{\text{s}^2}\right)(10\ \text{s})^2 = \boxed{9.6\ \text{m}}$$

11. (a) Since there is no horizontal acceleration, the time it takes the ball to reach the catcher is

$$t = \frac{x}{v_x} = \frac{18\ \text{m}}{32\ \text{m/s}} = 0.5625\ \text{s}$$

The ball is thrown horizontally, so $v_{0y} = 0$. The expression for the distance fallen is, therefore,

$$y - y_0 = -\tfrac{1}{2}gt^2 = -\tfrac{1}{2}\left(9.81\ \frac{\text{m}}{\text{s}^2}\right)(0.5625\ \text{s})^2 = -1.6\ \text{m}$$

so $\boxed{\text{it falls a distance of 1.6 m before reaching the catcher}}$.

(b) If v_x increases, then t will decrease, since $t \propto 1/v_x$. If the time the ball travels is less, $\boxed{\text{the drop distance decreases}}$ since $\Delta y \propto t^2$.

(c) The Moon's gravity is less, therefore the drop distance would $\boxed{\text{decrease}}$.

25. Since we know the horizontal distance that the cork travels and the time, we can determine its horizontal speed:

$$v_x = \frac{x}{t} = \frac{1.50\ \text{m}}{1.25\ \text{s}} = 1.20\ \frac{\text{m}}{\text{s}} = v_{0x}$$

This x-component of the initial velocity, together with its direction, can be used to find the initial speed

$$v_0 = \frac{v_{0x}}{\cos\theta} = \frac{1.20\ \text{m/s}}{\cos 40.0°} = \boxed{1.57\ \text{m/s}}$$

37. If we take her initial vertical position to be $y_0 = 0$ and note that $v_{0y} = v_0 \sin\theta$, her final position is given by

$$y = (v_0 \sin\theta)t - \tfrac{1}{2}gt^2 = \left(2.25\ \frac{\text{m}}{\text{s}}\right)\sin(35.0°)(1.60\ \text{s}) - \tfrac{1}{2}\left(9.81\ \frac{\text{m}}{\text{s}^2}\right)(1.60\ \text{s})^2 = -10.5\ \text{m}$$

Therefore, the girl was $\boxed{10.5\ \text{m}}$ above the water when she let go.

47. **(a)** At the lava's maximum height, $v_y = 0$. We can use the equations of accelerated motion assume that the lava is ejected vertically,

$$v_y^2 = v_{0y}^2 - 2a\Delta y = 0 \quad \Rightarrow \quad v_{0y}^2 = 2a\Delta y$$

$$\therefore \quad v_{0y} = \sqrt{2a\Delta y} = \sqrt{2\left(1.80\ \text{m/s}^2\right)\left(2.00 \times 10^5\ \text{m}\right)} = \boxed{849\ \text{m/s}}$$

(b) Since we know that $y_{max} \propto 1/g$, the maximum height reached by the lava would be $\boxed{\text{less}}$ on Earth due to Earth's greater gravity.

Answers to Practice Quiz

1. (c) **2.** (a) **3.** (b) **4.** (a) **5.** (d) **6.** (b) **7.** (d) **8.** (c) **9.** (b) **10.** (c)

Answers to Practice Problems

1.	55.8 m	**2.**	9.6 m
3.	24.8 m/s	**4.**	18.2 m/s
5.	9.14 mi/h	**6.**	20.6 s
7.	72.9 m	**8.**	35.8 m
9.	20.5 m	**10.**	104.9 m

CHAPTER 5
NEWTON'S LAWS OF MOTION

Chapter Objectives

After studying this chapter, you should:

1. be able to state, and understand the meaning of, Newton's three laws of motion.
2. be able to apply Newton's laws to simple situations in one and two dimensions.
3. be able to draw free-body diagrams.
4. know the difference between weight and mass.
5. be able to apply Newton's laws on inclined surfaces.

Warm-Ups

1. The word *push* is a reasonably good synonym for the word force.
   Is *hold* a synonym for force?
   How about *support*?
   How many others can you think of?
2. The engine on a fighter airplane can exert a force of 105,840 N (24,000 lb). The take-off mass of the plane is 16,875 kg. (It weighs 37,500 lb.). If you mounted this aircraft's engine on your car, what acceleration would you get? (Use metric units. The data in pounds are given for comparison. Use a reasonable estimate for the mass of your car. One kilogram of mass weighs 2.2 lb.)
3. A 72-kg skydiver is descending with a parachute. His speed is still increasing at 1.2 m/s^2. What are the magnitude and direction of the force of the parachute harness on the skydiver? What are the magnitude and direction of the net force on the skydiver?
4. Suppose you run into a wall at 4.5 m/s (about 10 mi/h). Let's say the wall brings you to a complete stop in 0.5 s. Find your deceleration, and estimate the force (in newtons) that the wall exerted on you while you were stopping. Compare that force with your weight.

Chapter Review

In the previous chapters we studied kinematics, which you may recall is the description of motion. In this chapter we begin our study of **dynamics**. To begin this study we start with the concept of **force**. Force relates not only to a description of motion but to the causes of various types of motion. The concept of force is governed by the three laws known as Newton's laws of motion; these three laws and how to apply them are the principal focus of this chapter.

5–1 Force and Mass

A force is a push or a pull. When a nonzero net force is applied to an object the object's response to this force can be detected in its subsequent motion. Precisely how an object moves in response to a force depends on the object's **mass**. Mass can be thought of as a measure of the matter content of an object. More importantly for motion, though, mass is a measure of an object's *natural* tendency to move with constant velocity, referred to as its **inertia**. That is, **mass is a measure of inertia**. As discussed in Chapter 1, the SI unit of mass is the kilogram (kg). We haven't used this unit much so far; but from now on we'll use it a lot.

5–2 – 5–5 Newton's Laws of Motion

Newton's first law of motion is also known as the *law of inertia*. This law essentially defines the concept of inertia in the context of motion.

> **Newton's First Law**
>
> *An object moving with constant velocity continues to do so unless acted upon by a nonzero net force.*

This law says that the natural state of motion is that of constant velocity. Since the law does not distinguish the constant velocity of zero (at rest) from any other constant velocity, it says that all constant velocities are equivalent.

Often, we loosely refer to anything that can detect an object as an observer. Any observer can be treated as the origin of a coordinate system in which it makes measurements. Moving observers carry their coordinate systems with them. We refer to an observer's coordinate system as its *frame of reference*. An **inertial frame of reference** (or just *inertial frame* for short) is one in which the law of inertia holds. Inertial frames move with constant velocity relative to other inertial frames. Newton's three laws of motion are true with respect to these constant-velocity reference frames, and it is in this sense that all constant velocities are equivalent.

The law of inertia says that an object's velocity will change (i.e., it will accelerate) if a nonzero force is applied to it; therefore, **force causes acceleration**. Newton's second law picks up on this theme and tells us, quantitatively, how the acceleration relates to the net force.

Newton's Second Law

The acceleration of an object equals the ratio of the net force on the object to its mass.

As an equation this law is most commonly written

$$\mathbf{a} = \frac{\sum \mathbf{F}}{m} \quad \text{or} \quad \sum \mathbf{F} = m\mathbf{a}$$

In these expressions $\Sigma\mathbf{F}$ represents the net force as the vector sum of all forces acting on the object. From this form of Newton's second law we can also see, from the fact that the acceleration is inversely proportional to the mass, why mass is interpreted as a measure of inertia. Saying that inertia is the tendency to maintain a constant velocity is the same as saying that it is the resistance to acceleration. Newton's second law tells us that the greater an object's mass, the smaller its acceleration will be for a given force, and vice versa. Thus, mass is what determines how strongly an object "wants" to keep its velocity constant.

The SI unit of force is the newton (N). The mathematical statement of Newton's second law makes it clear that the unit of force is the product of the units of mass and acceleration: 1 N = 1 kg·m/s^2. In the application of Newton's second law it will frequently be convenient to resolve the vector equation into scalar components

$$\sum F_x = ma_x \qquad \sum F_y = ma_y \qquad \sum F_z = ma_z$$

In all situations involving force, each of these equations must be satisfied independently. Another very important aspect of force analysis is the use of a **free-body diagram**. In such a diagram, we isolate the pertinent body, making it a "free body," then we draw all the force vectors acting on this body. In cases of nonrotational motion we will usually (but not always) idealize the body as a point particle without loss of accuracy. As detailed in your text, a good general strategy for doing force analysis is as follows:

1. Sketch the physical picture with forces drawn.
2. Draw a free-body diagram of the object(s) of interest.
3. Choose a convenient coordinate system.
4. Resolve the forces on your free-body diagram into components.
5. Apply Newton's second law to each coordinate direction.

Example 5–1 Skydiving Skydiving equipment, including the parachute, typically has a mass of about 9.1 kg. At one point during a jump, an 80.0-kg skydiver is accelerating downward at 1.50 m/s^2. Determine the net force on the skydiver in vector notation.

Picture the Problem The leftmost sketch shows the skydiver in the air and the two main forces involved. The rightmost sketch is a free-body diagram of an idealized skydiver. The choice of coordinate system is indicated between them.

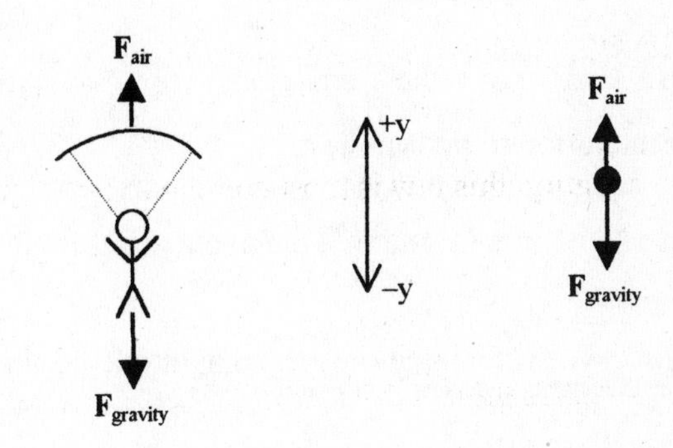

Strategy We need only the net force on the skydiver, so we should draw a free-body diagram and go straight to Newton's second law for the vertical direction, with up as +y.

Solution

1. Determine the total mass, M: $M = 80.0 \text{ kg} + 9.1 \text{ kg} = 89.1 \text{ kg}$

2. By the choice of coordinate system the acceleration is in the negative y direction: $\mathbf{a} = -(1.50 \text{ m/s}^2)\hat{\mathbf{y}}$

3. Use Newton's second law to get the net force: $\mathbf{F}_{net} = M\mathbf{a} = -(89.1 \text{ kg})(1.50 \text{ m/s}^2)\hat{\mathbf{y}} = -(134 \text{ N})\hat{\mathbf{y}}$

Insight The acceleration and net force are written as negative only by choice. We could easily have chosen downward to be the positive direction, and no negative sign would have been needed.

Example 5–2 Vehicle Performance With a full tank of gas the Pontiac Grand AM SE has a mass of about 1500 kg; its 0 → 60 mi/h time is listed as 6.7 seconds. What net forward force must have been exerted on the vehicle during this performance test?

Picture the Problem The sketch shows the car and the net forward force on it.

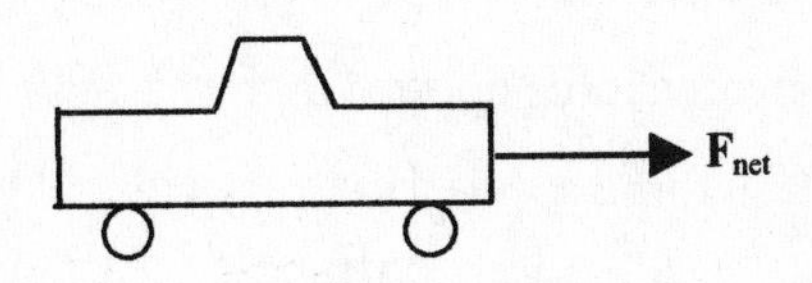

Strategy Since we don't know for certain that the acceleration is uniform, we can determine only the average net force on the vehicle from the average acceleration.

Solution

1. Convert the speed to SI units: $v = 60\text{ mi/h} \times \dfrac{1\text{ m/s}}{2.24\text{ mi/h}} = 26.8\text{ m/s}$

2. Use one of the equations for constant acceleration to determine a: $a_{av} = \dfrac{v - v_0}{\Delta t} = \dfrac{26.8\text{m/s}}{6.7\text{s}} = 4.0\text{m/s}^2$

3. Use Newton's laws to get the average net force: $F_{av} = ma_{av} = (1500\text{kg})(4.0\text{m/s}^2) = 6000\text{N}$

Insight We calculated only the magnitude because we already stipulated that the direction is "forward."

Newton's third law of motion is commonly known as the law of action and reaction. The words *action* and *reaction* refer to forces and so *force* is the word we'll use:

Newton's Third Law

For every force that an agent applies to an object, there is a reaction force equal in magnitude and opposite in direction applied by the object to the original agent.

An important point to remember here is that a force and its reaction *always act on different objects*; therefore, these forces *never* cancel each other. Basically, this law says that a single object cannot act on others without being acted on; two objects always interact, applying equal and opposite forces to each other.

Practice Quiz

1. If the same force is applied to two different objects, the object with greater inertia will have
   (a) greater acceleration **(b)** smaller acceleration **(c)** the same acceleration as the other object
   (d) zero acceleration **(e)** none of the above

2. If two different objects have the same acceleration, the object with greater inertia has
   (a) greater force applied to it
   (b) less force applied to it
   (c) the same force applied to it as the object has applied to it
   (d) zero force applied to it
   (e) none of the above

3. An object is observed to have an acceleration of 8.3 m/s^2 when a net force of 12.2 N is applied to it. What is its mass?

   (a) 100 kg **(b)** 0.68 kg **(c)** 1.5 kg **(d)** 21 kg **(e)** 3.9 kg

4. Under the action of a constant net force, a 3.1-kg object moves a distance of 15 m in 8.99 s starting from rest. What is the magnitude of this net force?

   (a) 0.37 N **(b)** 1.2 N **(c)** 420 N **(d)** 5.2 N **(e)** 0.58 N

5–6 – 5–7 Weight, Normal Force, and Inclined Surfaces

Weight

We all live on the surface of Earth. As a result, our everyday lives are partly governed by the fact that everything has **weight**. This weight is a direct result of Earth's gravitational pull; in fact, on Earth, a body's weight is the downward gravitational force exerted on the body by Earth. You may recall that near Earth's surface all bodies fall with the gravitational acceleration, $g = 9.81\ m/s^2$. In accordance with Newton's second law, therefore, the gravitational force on a body, its weight, equals the product of its mass and this gravitational acceleration:

$$\mathbf{W} = m\mathbf{g}$$

The direction of the vectors **W** and **g** is toward the center of Earth.

The sensation of having weight is clear to us because of the force of contact between our feet and the ground beneath us. Earth's gravity presses us into the floor, and the floor reacts by pressing back against us. This reaction is what feels like our weight. However, if the object on which we are standing is accelerating, this reaction force can trick us into feeling either heavier or lighter depending on the direction of the acceleration. This reaction force is called our **apparent weight** because it is the weight we perceive ourselves to have even when it differs from the actual force of gravity on us.

Normal Force

Another everyday force occurs when two surfaces come into direct physical contact. The force of contact between the surfaces can be resolved into components that are parallel and perpendicular to the surfaces. The perpendicular component of this force is called the **normal force** (*normal* means "perpendicular"). The normal force that the floor exerts on your feet when you are standing is what we just called the apparent weight. As you'll see in the next chapter, the normal force between two surfaces is an important factor in determining the friction between those surfaces.

Example 5–3 Going Up in an Elevator Low-acceleration elevators, such as might be found in a hospital, typically accelerate at about 3.00 ft/s^2 when first starting upward. If a person knows her weight to be 125 lb, what would a scale read if she is standing on it when the elevator starts upward?

Picture the Problem The lefthand sketch shows a person standing on a scale in an upward-accelerating elevator. The righthand sketch is the free-body diagram.

Strategy The problem asks for the reading on the scale, which is determined by the force that the person exerts on the scale. By Newton's third law we know that the force on the scale equals the normal force **N** of the scale on the person (her apparent weight, **N** = $\mathbf{W}_a$). So, instead, we draw a free-body diagram of the person because we know the person's true weight. Let's choose up to be +y.

Solution

1. Apply Newton's second law to the free-body diagram:

$$\sum F_y = N_y + W_y = N - W = ma$$
$$\Rightarrow \; N = W + ma$$

2. Solving for N gives us the apparent weight:

$$N = 125\text{ lb} + \left(\frac{125\text{ lb}}{32\text{ ft/s}^2}\right)(3.00\text{ ft/s}^2) = 137\text{ lb}$$

Insight Notice that in the sum of forces in step 1, $W_y = -W = -125$ lb.

Exercise 5–4 Rearranging While rearranging the living room a student pushes a 27.5-kg sofa across the room at constant speed by applying a force that makes an angle of 35.0° below the horizontal. If the floor opposes the sliding sofa (by friction) with a backward force of 160 N, then **(a)** what is the normal force of the floor on the sofa, and **(b)** what magnitude force does the student apply to the sofa?

Solution Try to sketch a physical picture for this problem; the free-body diagram for the sofa is shown.

Given: m = 27.5 kg, θ = –35.0°, f = 160 N; **Find**: **(a)** N, **(b)** F

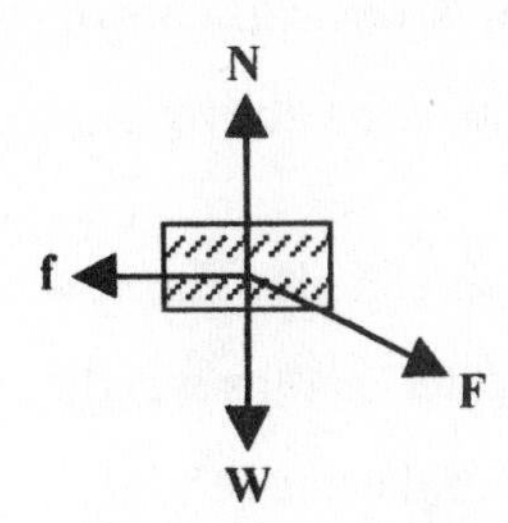

Taking the +x direction to the right and the +y direction up, we can apply Newton's second law to both the x and y directions. Since part (a) asks about the normal force, let's start with the y direction. The sofa does not accelerate vertically up or down, so the net vertical force must be zero:

$$\sum F_y^{all} = N_y + W_y + F_y = 0 \Rightarrow N - W - F_y = 0$$
$$\therefore N = W - F_y$$

We cannot solve this equation immediately for N because we don't yet know F_y. Notice, however, that since we know the direction of **F**, we can determine F_y provided we can figure out F_x. Thus, we next apply Newton's second law to the x direction. The sofa is being pushed at constant speed; therefore, the horizontal acceleration is zero, which means that the net horizontal force must also be zero,

$$\sum F_x^{all} = F_x + f_x = 0 \Rightarrow F_x - f = 0$$
$$\therefore F_x = f = 160 \text{ N}$$

Now that we know F_x we can determine F_y through the relation $\tan\theta = F_y / F_x$:

$$F_y = F_x \tan\theta = (160 \text{ N})\tan(-35.0^\circ) = -112 \text{ N}$$

The minus sign just means that F_y is along the –y direction. Having F_y, we are now ready to calculate the normal force on the sofa.

$$N = W - F_y = mg - F_y = (27.5 \text{ kg})(9.81 \text{ m/s}^2) + 112 \text{ N} = 382 \text{ N}$$

which is the final result for part (a).

For part (b) we seek the magnitude of **F**. Since we have already calculated the two components of **F**, we can determine its magnitude from the Pythagorean theorem.

$$F = \sqrt{F_x^2 + F_y^2} = \sqrt{(160 \text{ N})^2 + (-112 \text{ N})^2} = 195 \text{ N}$$

An important point here is that Newton's second law must be satisfied in every direction and that often a result from an equation for one direction is useful in solving an equation for a different direction.

Practice Quiz

5. An object is at rest on a horizontal table. The normal force exerted by the table on the object

(a) points vertically up (b) points vertically down (c) is parallel to the table
(d) is zero (e) none of the above

6. An object sits undisturbed on a horizontal table. The normal force exerted by the table on the object
(a) equals its weight (b) is greater than its weight (c) is less than its weight
(d) is zero (e) none of the above

7. An object has a mass of 3.25 kg. Its weight is
(a) 3.25 kg (b) 3.25 N (c) 3.02 N (d) 7.15 N (e) none of the above

8. A man of mass 55.0 kg sits in a chair of mass 62.0 kg. The normal force of the chair on the man is
(a) 540 N (b) 608 N (c) 1150 N (d) 117 N (e) 7.00 N

9. A woman stands in an elevator that is moving upward at constant speed. Compared to her actual weight on Earth's surface, her apparent weight is
(a) greater (b) less (c) the same (d) zero (e) none of the above

10. By pushing on the floor, a person accelerates a 120-lb box across a room at 1.3 m/s^2. The floor opposes the motion of the box with a frictional force of 110 N. What horizontal force does the person apply to the box?
(a) 230 N (b) 71 N (c) 180 N (d) 110 N (e) 423 N

Inclined Surfaces

Life isn't conveniently arranged to take place only on smooth, flat, horizontal surfaces. Sometimes we just have to go uphill or downhill. In these situations, it is best to choose a coordinate system with axes that are parallel and perpendicular to the surface. Generally, we take the x-axis to be parallel to the surface and the y-axis to be perpendicular to the surface as shown below.

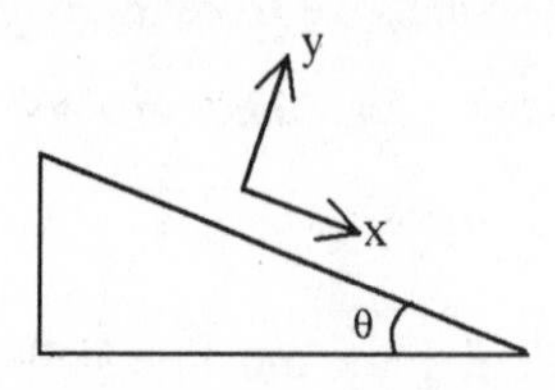

In this coordinate system the weight, which always acts vertically downward, can be resolved into x and y components. If θ is the angle that the surface makes with the horizontal, then

$$W_x = W \sin\theta = mg \sin\theta \qquad W_y = -W\cos\theta = -mg\cos\theta$$

Because the weight has components in both directions, the rest of the analysis on inclined surfaces follows precisely as it does on horizontal surfaces.

Example 5–5 Finding a Better Way A heavy suitcase that weighs 450 N and has wheels needs to be placed onto the back of a truck 0.65 m above the ground. Instead of lifting it straight up, you decide to get a thick piece of wood to use as a ramp and roll it up onto the truck. If the extended ramp makes an angle of 60° with the horizontal, **(a)** determine the minimum force needed to roll the suitcase up the ramp, and **(b)** compare this value with the minimum force needed to lift it up onto the truck.

Picture the Problem The upper sketch shows the truck with the wooden ramp and the suitcase. The lower sketch is a free-body diagram of the suitcase in which F is the applied force to pull the suitcase up the ramp.

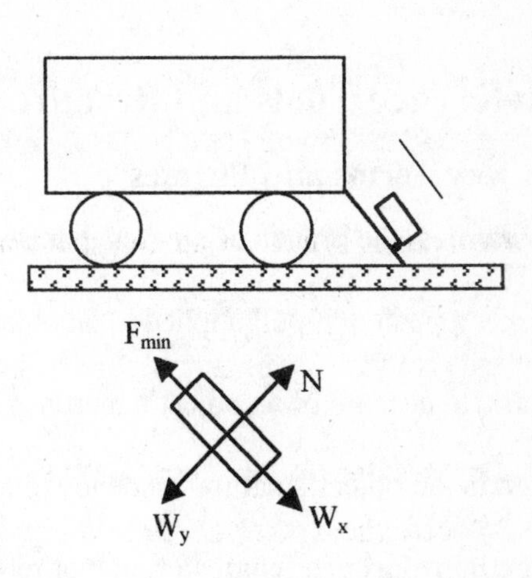

Strategy For part (a) we first recognize that the *minimum* force to roll the suitcase up must be parallel to the ramp (so all the force is used) and just enough to balance the suitcase's tendency to roll down the ramp (due to gravity). Therefore, the suitcase will roll up at constant speed. Likewise, for part (b) the minimum force just balances the weight of the suitcase.

Solution

Part (a)

1. Apply Newton's second law to the x direction: $\Sigma F_x = F_{min,x} + W_x = 0 \Rightarrow -F_{min} + W\sin\theta = 0$

2. Solve this equation for F_{min}: $F_{min} = W\sin\theta = (450\text{ N})\sin(60^\circ) = 390\text{ N}$

Part (b)

3. Compare the result of (a) with the weight of the suitcase: $W - F_{min} = 60\text{ N}$

4. Calculate the percent difference: $\frac{60\text{ N}}{450\text{ N}} \times 100\% = 13\%$

Insights In the free-body diagram for this problem the components of the weight were drawn directly on the diagram, rather than the vertically downward weight vector. This procedure is sometimes more convenient. Either way the vectors are resolved into components, as part of the recommended strategy for force analysis, whether before or after the free-body diagram is drawn.

Practice Quiz

11. An object is held at rest on a surface inclined at an angle θ with the horizontal. If the force holding it in place is parallel to the surface, the normal force exerted by the surface on the object

 (a) equals its weight **(b)** is greater than its weight **(c)** is less than its weight

 (d) is zero **(e)** none of the above

Reference Tools and Resources

I. Key Terms and Phrases

dynamics the branch of physics that studies force and the causes of various types of motion

force a push or a pull applied to an object

mass a measure of an object's inertia

inertia an object's natural tendency to move with constant velocity

inertial reference frame a frame of reference in which the law of inertia holds

free-body diagram a diagram of an isolated object showing all the force vectors acting on the object

weight the downward force due to gravity

apparent weight the perceived weight of an object as its force of contact with the ground or a scale

normal force the component of the contact force on a surface that is perpendicular to the surface

II. Important Equations

Name/Topic	Equation	Explanation
Newton's second law	$F_{net} = \sum F = ma$	The vector sum of all forces acting on an object equals its mass times its acceleration.
inclined surfaces	$W_x = mg \sin\theta$ $W_y = -mg \cos\theta$	The components of an object's weight on a surface inclined at an angle θ to the horizontal.

III. Know Your Units

Quantity	Dimension	SI Unit
force	$[M]\cdot[L]/[T^2]$	N

IV. Tips

You may have noticed that in the free-body diagrams in the prceding examples the forces were always drawn with their tails on the idealized body and their heads pointing away. This is the usual tradition and it is better to be consistent; so always draw the forces on your free-body diagrams with tails on the body and heads pointing away from the body.

Practice Problems

1. A plane traveling at 18.9 m/s is brought to rest in a distance of 182 meters. To the nearest hundredth of a m/s^2, what is the magnitude of its acceleration?
2. If the mass of the plane in problem 1 is 1.75×10^5 kg, what is the magnitude of the retarding force, to the nearest newton?
3. Two skaters are in the exact center of a circular frozen pond. Skater 1 pushes skater 2 off with a force of 100 N for 1.5 seconds. If skater 1 has a mass of 41 kg, and skater 2 has a mass of 77 kg, what is the relative velocity ($\mathbf{v}_1-\mathbf{v}_2$) after the push, to the nearest hundredth of a m/s?
4. After reaching the other shore, how fast, to the nearest tenth of a m/s, must skater 1 run around the lake to meet skater 2 at the opposite shore?
5. The left box has a mass of 27.3 kg, the right box has a mass of 6.2 kg, and the force is 150 N. To the nearest tenth of a m/s^2, what is the acceleration of the combination?

6. To the nearest newton, in problem 5, what force does the right box exert on the left box?
7. A child pulls a wagon with a force of 32 N by a handle making an angle of 27 degrees with the horizontal. If the wagon has a mass of 5 kg, to the nearest hundredth of a m/s^2, what is the acceleration of the wagon?
8. To the nearest newton, what would be the minimum force applied at that angle that would lift the wagon off the ground?
9. A student stands on a scale in an elevator that is accelerating at 0.8 m/s^2. If the student has a mass of 86 kg, to the nearest newton, what is the scale reading?

10. A 31-kg child sits on a 5-kg sled and slides down a 96-meter 30° slope, to the nearest m/s, what is his or her speed at the bottom?

Puzzle

STRETCHING

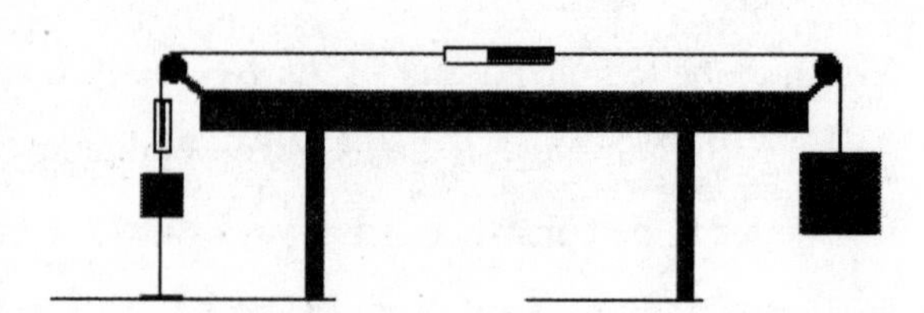

A 10-kg mass (right) and a 5-kg mass (left) are suspended from pulleys as shown in the diagram. The lower mass is tied to the floor by another string. Two small metric scales (calibrated in newtons) are spliced into the rope. One of them is between the two pulley discs, the other just above the smaller mass. Assume that the masses of the scales are small enough to be neglected.

What do the scales read?

The string under the lighter mass is now cut. What do the scales read after the masses start to move.

Selected Solutions

9. **(a)** According to Newton's 2nd law, $\mathbf{F}_{av} = m\mathbf{a}_{av}$, where $\mathbf{a}_{av} = \dfrac{\Delta \mathbf{v}}{\Delta t}$. Therefore,

$$\begin{aligned}\mathbf{F}_{av} &= m\frac{\Delta \mathbf{v}}{\Delta t} = m\frac{\mathbf{v}_f - \mathbf{v}_i}{\Delta t} = (950\text{ kg})\frac{(9.50\text{ m/s} - 16.0\text{ m/s})}{1.20\text{ s} - 0}\hat{\mathbf{x}} \\ &= (950\text{ kg})(-5.42\text{ m/s}^2)\,\hat{\mathbf{x}} = -5.15\times 10^3\text{N}\,\hat{\mathbf{x}}\end{aligned}$$

Therefore, the average force is $\boxed{\text{5.15 kN opposite the direction of motion}}$.

(b) Since we know the average acceleration from part (a), we can treat the motion as if the acceleration is constant, having the value a_{av}. From the equations for motion with constant acceleration we can write

$$v_f^2 = v_0^2 + 2a_{av}(\Delta x) \Rightarrow \Delta x = \frac{v_f^2 - v_0^2}{2a_{av}} = \frac{(9.50\text{ m/s})^2 - (16.0\text{ m/s})^2}{2(-5.42\text{ m/s}^2)} = \boxed{15.3\text{ m}}$$

15. (a) By Newton's 3rd law (of action and reaction), the parent and the child apply equal and opposite forces to each other. Therefore, the force experienced by the child is $\boxed{\text{the same}}$ as the force experienced by the parent.

(b) The acceleration of the child is $\boxed{\text{more}}$ than the acceleration of the parent. The child, who has less mass than the parent, must have a larger acceleration to keep the forces equal.

(c) Since $F_{child} = F_{parent}$, then $m_{child}a_{child} = m_{parent}a_{parent}$. This fact means that

$$a_{parent} = \frac{m_{child}}{m_{parent}}a_{child} = \left(\frac{18\text{ kg}}{54\text{ kg}}\right)\left(2.6\text{ m/s}^2\right) = \boxed{0.87\text{ m/s}^2}$$

25. (a) The net force on the skier results in the skier's motion. This motion is parallel to the slope downhill, thus $\boxed{\text{the direction of the net force is also parallel to the slope downhill}}$. To get the magnitude we multiply the skier's mass times the parallel component of acceleration:

$$F_{net} = ma_{\parallel} = mg\sin\theta = (65\text{ kg})\left(9.81\text{ m/s}^2\right)\sin(22^\circ) = \boxed{240\text{ N}}$$

(b) As the slope becomes steeper, the net force $\boxed{\text{increases}}$. As the incline increases, sinθ approaches 1. In this case, the skier is falling. As the incline decreases, sinθ approaches zero, and the skier is standing still. In this case, the force of gravity is counteracted by the force of the ground on the skier.

35. (a) The direction of acceleration is $\boxed{\text{upward}}$. An upward acceleration results in an apparent weight greater than the actual weight because the force that the scale applies must exceed the weight to accelerate you upward.

(b) The magnitude of the acceleration can be determined by applying Newton's 2nd law to the y direction

$$\sum F_y = W_a - W = ma \quad \therefore \quad a = \frac{W_a - W}{m} = \frac{W_a - W}{W/g}$$

$$a = \left(\frac{730\text{ N} - 610\text{ N}}{(610\text{ N})/\left(9.81\text{ m/s}^2\right)}\right) = \boxed{1.9\text{ m/s}^2}$$

39. (a) The free-body diagram for the child is shown on the right.

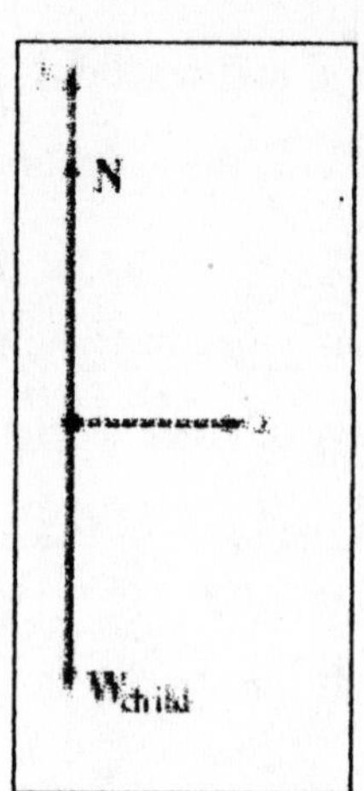

The normal force is found by applying Newton's 2nd law to the child in the vertical direction, noting that a = 0.

$$\sum F_y = N - mg = 0 \quad \therefore \quad N = mg$$

$$N = (9.0\text{ kg})\left(9.81\ \frac{m}{s^2}\right) = \boxed{88\text{ N}}$$

(b) The free-body diagram for the chair is shown on the right.

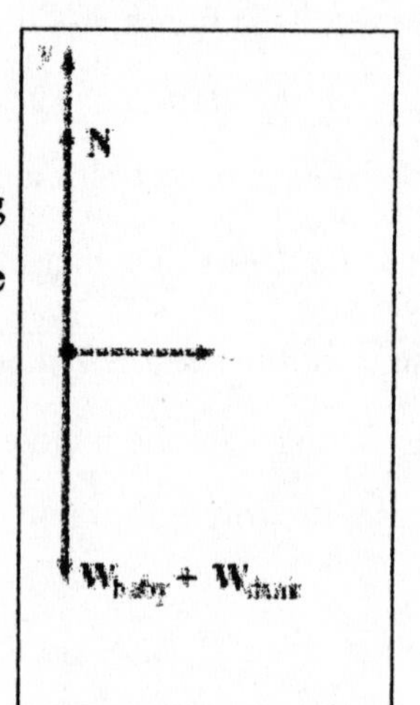

The normal force is found by applying Newton's 2nd law to the chair in the vertical direction, noting that a = 0.

$$\sum F_y = N - (m_{baby} + m_{chair})g = 0$$

$$\therefore \quad N = (m_{baby} + m_{chair})g$$

$$N = (9.0\text{ kg} + 2.3\text{ kg})\left(9.81\ \frac{m}{s^2}\right) = \boxed{110\text{ N}}$$

Answers to Practice Quiz

1. (b) **2.** (a) **3.** (c) **4.** (b) **5.** (a) **6.** (a) **7.** (e) **8.** (a) **9.** (c) **10.** (c) **11.** (c)

Answers to Practice Problems

1. 0.98 m/s^2 2. 171500 N

3. 5.61 m/s 4. 13.1 m/s

5. 4.5 m/s^2 6. 28 N

7. 5.7 m/s^2 8. 108 N

9. 912 N 10. 31 m/s

CHAPTER 6

APPLICATION OF NEWTON'S LAWS

Chapter Objectives

After studying this chapter, you should:

1. be able to perform force analysis in situations involving both static and kinetic friction.

2. be able to perform force analysis in situations involving string tensions and spring forces.

3. have a thorough understanding of translational equilibrium.

4. understand the roles of force and acceleration in circular motion.

Warm-Ups

1. People could not walk without friction. Which type of friction are we referring to here, kinetic friction or static friction?

2. The coefficient of friction between a "safe" walking surface and your shoes is supposed to be about 0.6. Estimate the maximum acceleration you could attain under these conditions.

3. List all the forces acting on a car negotiating a turn at constant speed on a banked road. What should be the magnitude of the vector sum of all the forces acting on this car?

4. Estimate the coefficient of friction between skis and snow that would allow a skier to move down a 30 degree slope with constant speed if other factors were neglected. Is it reasonable to neglect other factors, such as air drag on the skier?

Chapter Review

This chapter is a continuation of Chapter 5. Here we discuss further details in the application of Newton's laws by adding several specific forces, namely friction, tension, and the Hooke's law force. We also introduce the concepts of centripetal force and the associated centripetal acceleration that relate to circular motion.

6–1 Frictional Force

When two surfaces are in direct contact, and one surface either slides or attempts to slide across the other, a force, called **friction**, that opposes the motion (or attempted motion) is generated between the surfaces. When the surfaces are sliding across each other we call the force **kinetic friction**. When there is only attempted motion that is halted by friction we call it **static friction**. For example, if you apply a small horizontal force to a heavy object in an attempt to slide it, the object may not move at all because of the friction between it and the floor. In this case there is definitely friction but no motion, i.e., static friction.

The origin of friction is based, in complicated ways, on the microscopic structure of the surfaces involved. Often in physics we handle these types of situations by identifying those factors on which the dependence is fairly simple and representing the rest of the complicated physics by one measured quantity. For kinetic friction we note that the frictional force, f_k, is directly proportional to the normal force, N, between the surfaces. The measured proportionality factor, μ_k, is called the **coefficient of kinetic friction**. Thus, we have

$$f_k = \mu_k N$$

Note that this equation relates only the magnitudes of $\mathbf{f}_k$ and **N** because they are not in the same direction; they are perpendicular to each other.

For static friction the force, f_s, takes on a range of values depending on the strength of the force it opposes. Static friction will cancel out any force trying to slide two surfaces across each other up to some maximum value beyond which static friction is overcome. This maximum force of static friction is also directly proportional to the normal force. The measured proportionality factor, μ_s, is called the **coefficient of static friction**. Thus, we have

$$f_{s,max} = \mu_s N$$

which, for the more general case yields

$$f_s \leq \mu_s N$$

Typically, for a given pair of surfaces we have $\mu_k < \mu_s$.

Exercise 6–1 Getting It Going A 105-lb crate sits on a floor. If the minimum horizontal force required to start it sliding is 33 lb, what is the coefficient of static friction between the crate and the floor?

Solution: The free-body diagram is shown below. The following information is given in the problem:

Given: W = 105 lb, F_{min} = 33 lb; **Find**: μ_s

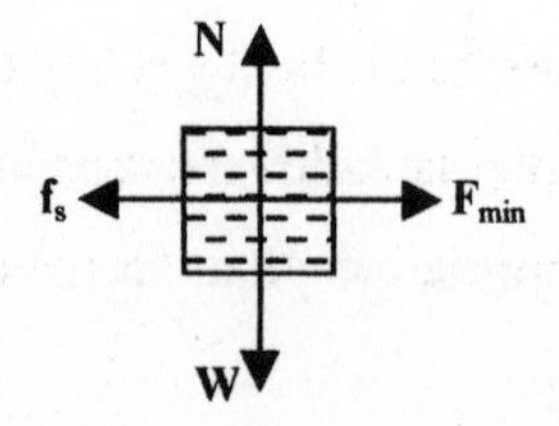

The fact that we seek the minimum force needed to get the crate moving means that we want the force that is just barely large enough to overcome static friction. The force right at this critical condition equals the maximum force of static friction

$$F_{min} = f_{s,max} = \mu_s N$$

To determine N we apply Newton's second law in the y-direction (recognizing that the vertical acceleration is zero): $N - W = 0$. This tells us that $N = W$. Hence, we have

$$F_{min} = \mu_s W \quad \therefore \quad \mu_s = \frac{F_{min}}{W} = \frac{33\text{ lb}}{105\text{ lb}} = 0.31$$

Practice Quiz

1. A person pushes an object across a room, on a horizontal floor, by applying a horizontal force. If, instead, he pushed it with a force that made an angle θ below the horizontal, the object would have experienced

   (a) greater friction **(b)** less friction **(c)** the same frictional force

   (d) zero friction **(e)** none of the above

2. A person pulls an object across a room, on a horizontal floor, by applying a horizontal force. If, instead, she pulled it with a force that made an angle θ above the horizontal, the object would have experienced

   (a) greater friction **(b)** less friction **(c)** the same frictional force

   (d) zero friction **(e)** none of the above

3. The coefficient of kinetic friction represents

   (a) the force that one surface applies to another when they slide across each other

   (b) the force that one surface applies to another when they do not slide across each other

   (c) the force that one surface applies to another that is perpendicular to the interface between them

   (d) the ratio of the normal force to the force of friction between two sliding surfaces

   (e) none of the above

6–2 Strings and Springs

A common way in which forces are applied is by the action of a string (or any rope, cord, or cable wire). The force transmitted through the string is called the **tension** in the string. Unless otherwise stated, we shall, for now, consider strings of negligible mass for which the tension can be considered constant throughout. Therefore, the force applied on one end of a string gets transmitted through to objects attached to the other end.

Another important force in physics is governed by a principle known as **Hooke's law**. The prototype object that applies a Hooke's law force is an ideal spring. When a spring is deformed from its equilibrium length, that is, when it is stretched or compressed, it responds to this deformation by applying a force that tries to restore itself to equilibrium. For this reason a Hooke's law force is also referred to as a **restoring force**. The precise strength of this restoring force is found to be directly proportional to the amount of deformation (provided the spring is not stretched beyond a certain limit). The proportionality factor between the force and the amount of deformation is called the **force constant** k. Taking the direction along which the string is deformed as the x direction, we write Hooke's law as

$$F_x = -kx$$

where the minus sign indicates that the force is exerted by the spring in the direction opposite the deformation x. If x is in the positive direction, then F_x points in the negative direction, and vice versa.

Example 6–2 Rearranging (Again) Suppose the student in Exercise 5–4 doesn't like the way he rearranged his room and decides to try again. This time he attaches a cord to the sofa and pulls the 27.5-kg sofa across the room, accelerating it at 0.150 m/s^2. If the cord makes an angle of 55.0° above the horizontal, and the coefficient of kinetic friction between the sofa and the floor is 0.240, what is the tension in the cord?

Picture the Problem The top sketch shows the sofa with the cord being pulled by the student. Below it is the free-body diagram of the sofa.

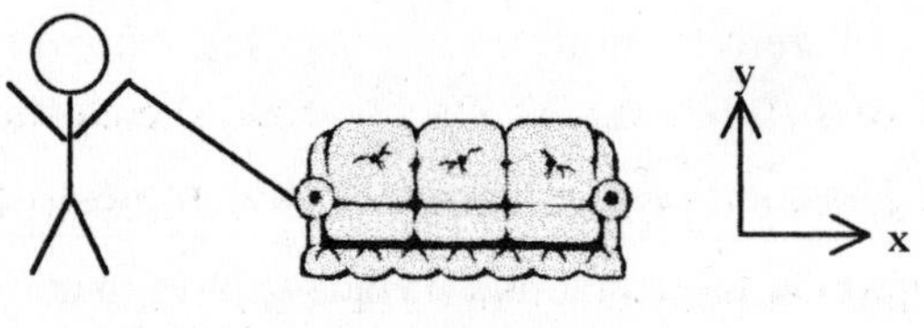

y
x

N
T
f_k
W

Strategy In this example we follow the same steps for force analysis as in the previous chapter. First we resolve the forces into their x and y components and then apply Newton's second law to each direction.

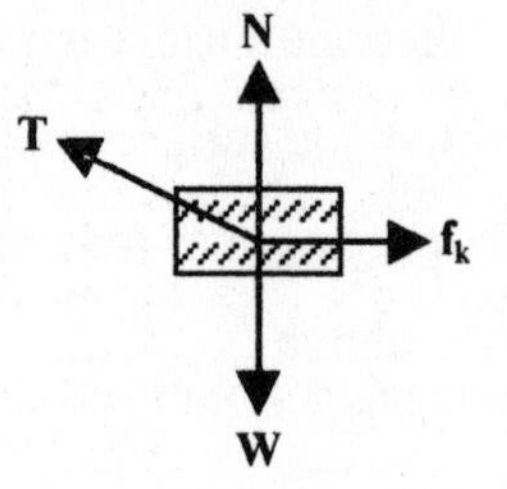

Solution

1. Resolve the components of the tension: $T_x = -T\cos\theta, \quad T_y = T\sin\theta$

2. Apply Newton's law to the x direction: $\sum F_x = f_{k,x} + T_x = ma \;\Rightarrow\; \mu_k N - T\cos\theta = ma$

3. Apply Newton's law to the y direction (there is no y component of acceleration):

$$\sum F_y = N_y + W_y + T_y = 0 \Rightarrow N - mg + T\sin\theta = 0$$

4. Solve for N using the equation for the y direction:

$$N = mg - T\sin\theta$$

5. Place N into the equation for the x direction:

$$\mu_k(mg - T\sin\theta) - T\cos\theta = ma$$

6. Solve for T and evaluate (note that the acceleration is in the negative direction):

$$\mu_k mg - T(\mu_k \sin\theta + \cos\theta) = ma \quad \therefore$$

$$T = \frac{m(\mu_k g - a)}{\mu_k \sin\theta + \cos\theta}$$

$$= \frac{27.5\text{ kg}\left[0.240(9.81\text{ m/s}^2) + 0.150\text{ m/s}^2\right]}{0.240\sin(55.0^\circ) + \cos(55.0^\circ)} = 89.4\text{ N}$$

Insights Notice here that we had to be careful with signs (as always). Both T_x and a were explicitly given minus signs because of the choice of coordinate system. If the problem asked for the force applied by the student instead, we would still work the problem exactly the same way using the approximation that the student's force is transmitted unchanged throughout the length of the cord.

Example 6–3 Industrial Strength Certain industrial springs, used to absorb shock and minimize damage, have force constants of about 225 N/cm. If such a spring breaks when stretched more than 5.5 cm, what is the maximum mass that can be hung from this spring without breaking it?

Picture the Problem The left-hand sketch shows a mass hanging from a spring. The right-hand sketch is the free-body diagram of the mass.

Strategy The maximum mass that can be hung from the spring without breaking it is a mass whose weight stretches it 5.5 cm. When such a weight is hung from the spring it sits motionless with an acceleration of zero.

Solution

1. Apply Newton's second law to the forces in the free-body diagram:

$$\mathbf{F}_s + \mathbf{W} = 0 \quad \Rightarrow \quad kx - mg = 0$$

2. Rearrange and solve for mass:

$$mg = kx \Rightarrow m = \frac{kx}{g}$$

$$\therefore m = \frac{(225\ \text{N/cm})(5.5\ \text{cm})}{9.81\ \text{m/s}^2} = 130\ \text{kg}$$

Insight This mass of 130 kg weighs approximately 285 lb.

Practice Quiz

4. Two people pull on opposite ends of a rope. If each person pulls with 30 N, what is the tension in the rope?

   (a) 60 N **(b)** 30 N **(c)** 15 N **(d)** 7.5 N **(e)** 0 N

5. If a spring stretches 3.5 cm from equilibrium when a 2.7-kg mass is hung from its end, what force is required to compress the spring 1.8 cm from its equilibrium length?

   (a) 26 N **(b)** 760 N **(c)** 17 N **(d)** 14 N **(e)** 130 N

6–3 Translational Equilibrium

A particularly important special case arises when the net force on an object is zero. Objects for which this is true are said to be in **translational equilibrium**. Since the net force equals zero, the acceleration is also zero. Why is this case important? Look around you; objects in translational equilibrium are everywhere. Notice that being in equilibrium (for short) does not mean motionless. Exercise 5–4 and Example 5–5 were cases of translational equilibrium in which the object moved with constant velocity, whereas Example 6–3 is one in which the object of interest was motionless.

Example 6–4 Held in Place A 5.32-kg box is held stationary on a ramp inclined at 40.0 degrees to the horizontal by a cord that is attached to a vertical wall. The length of the cord is parallel to the ramp, and it provides just enough force to hold the box in place. If the coefficient of static friction between the box and the ramp is 0.101, what is the tension in the cord?

Picture the Problem The left-hand diagram shows the box on the ramp being held in place by the cord. The right-hand sketch is the free-body diagram of the box.

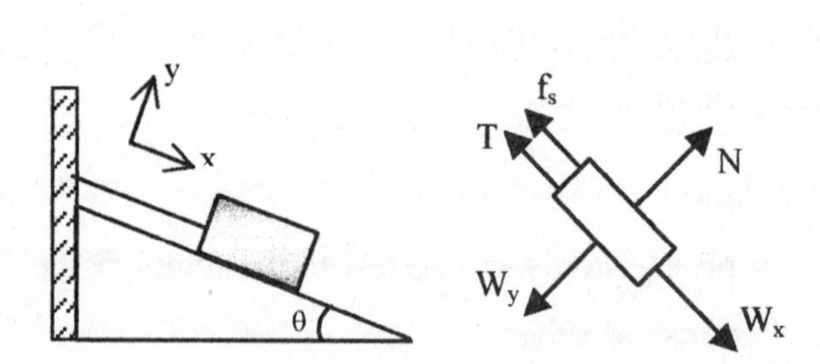

Strategy The case of translational equilibrium

has only one condition to apply, $\Sigma\mathbf{F} = 0$, so we apply this condition for each direction if necessary. Since the tension is just enough to balance the forces, static friction must have its maximum value.

Solution

1. Apply the equilibrium condition to the x direction: $W_x + T_x + f_{s,max} = 0 \Rightarrow W_x - T - \mu_s N = 0$

2. Solve this equation for T: $T = mg\sin\theta - \mu_s N$

3. Apply the equilibrium condition to the y-direction: $N_y + W_y = 0 \Rightarrow N - mg\cos\theta = 0$

   $\therefore N = mg\cos\theta$

4. Substitute N into the equation for T: $T = mg\sin\theta - \mu_s mg\cos\theta = mg(\sin\theta - \mu_s\cos\theta)$

5. Solve for the tension: $T = (5.32\text{ kg})(9.81\text{ m/s}^2)[\sin(40^\circ) - 0.101\cos(40^\circ)]$

   $= 29.5\text{ N}$

Insights Notice that for translational equilibrium in any direction the condition is always that the sum of the forces equal zero.

Practice Quiz

6. If an object is in translational equilibrium
   (a) it must be at rest
   (b) there are no forces acting on it
   (c) it is not rotating
   (d) it moves with constant acceleration
   (e) none of the above

7. Can an object be in translational equilibrium if only one force acts on it?
   (a) Yes, if it is a gravitational force
   (b) No, because even one force can have more than one component
   (c) Yes, trying to lift an object that is too heavy for you to lift is an example
   (d) No, an object with only one force on it must accelerate
   (e) none of the above

6–5 Circular Motion

In the previous chapter we noted that life doesn't take place only on smooth, flat, horizontal surfaces, which is why we needed to study inclined surfaces. Likewise, life doesn't always take place in straight-line paths. Therefore, we need to study how to handle motion along curves. Specifically, this section focuses on circular motion, or at least motion along a circular section. If the object that moves along this circular path does so at constant speed, then the acceleration must be perpendicular to the velocity and always points toward the center of the path. For this reason the acceleration is called **centripetal acceleration**; its magnitude is given by

$$a_{cp} = \frac{v^2}{r}$$

where v is its speed, and r is the radius of the circular path.

According to Newton's second law, where there is an acceleration there must be a force that causes it. In the case of circular motion the force must also point toward the center of the circular path and is therefore called a **centripetal force**. Also in accordance with Newton's law, the magnitude of this centripetal force equals the product of the mass and the centripetal acceleration:

$$f_{cp} = m\frac{v^2}{r}$$

It is important to recognize that centripetal force is not a new kind of force. The above expression is merely a condition that must be met by whatever force holds the object in this *uniform circular motion*, as it is often called. In your studies, and in everyday life, you will come across circular motion caused by many different sources such as a normal force, tension, static friction, gravity, and in later chapters, electric and magnetic forces as well.

Example 6–5 Making a Turn Suppose that the combined mass of you and your bike is 75 kg. You are riding down the street and have to make a turn in a circular section of road whose radius is approximately 6.5 m. The speedometer on your bike reads 20 km/h. If the coefficient of static friction between your wheels and the road is 0.79, are you going too fast to make the turn safely?

Picture the Problem The sketch shows a side view for the free-body diagram of you and your bike. The center of the circular path is to the right.

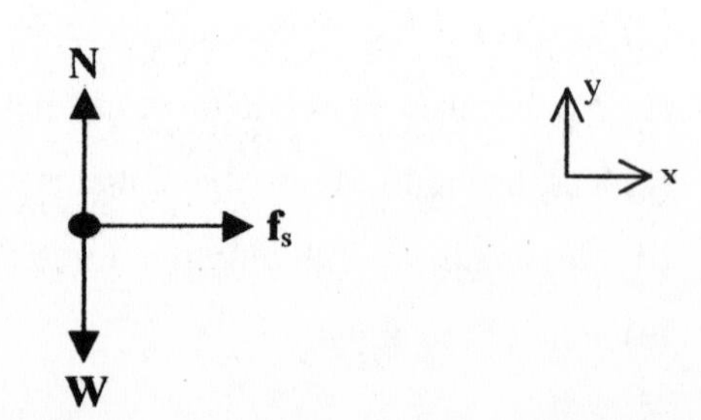

Strategy The first thing to recognize is that static

friction must supply the centripetal force to sustain the circular motion, so the problem is really asking for a comparison between the required centripetal force and $f_{s,max}$.

Solution

1. Convert the speed to m/s: $v = 20\ \text{km/h}\left(\frac{1000\ \text{m}}{\text{km}}\right)\left(\frac{1\ \text{h}}{3600\ \text{s}}\right) = 5.56\ \text{m/s}$

2. Evaluate the centripetal force: $f_{cp} = m\frac{v^2}{r} = (75\ \text{kg})\frac{(5.56\ \text{m/s})^2}{6.5\ \text{m}} = 360\ \text{N}$

3. Determine the maximum force of static friction: $f_{s,max} = \mu_s N = \mu_s mg = (0.79)(75\ \text{kg})(9.81\ \text{m/s}^2)$
$= 580\ \text{N}$

Insights The results show that the needed centripetal force is well within the range of static friction, so you're safe. Notice that in step 3, we set N = mg. We did this, without having to write out a separate Newton's second law equation for the y direction, by inspecting the free-body diagram and recognizing that **N** and **W** must balance. As you do more and more problems you should begin to gain a similar feel for the free-body diagrams.

Practice Quiz

8. For an object undergoing circular motion at constant speed
   (a) the velocity is constant
   (b) the acceleration is constant
   (c) the direction of the velocity is constant
   (d) the direction of the acceleration is constant
   (e) none of the above

9. Which of the following correctly identifies the relationship between the directions of the velocity and acceleration for objects in circular motion at constant speed?
   (a) They are in the same direction.
   (b) They are in opposite directions.
   (c) They are perpendicular to each other.
   (d) Their directions may differ by any angle depending on the curve of the arc.
   (e) none of the above

10. With what constant speed must an object traverse a circular path of radius 1.34 m in order to have a centripetal acceleration equal to the acceleration of gravity?

(a) 3.13 m/s **(b)** 13.1 m/s **(c)** 1.34 m/s **(d)** 9.81 m/s **(e)** 3.63 m/s

Reference Tools and Resources

I. Key Terms and Phrases

kinetic friction the contact force between two sliding surfaces that opposes their motion

static friction the contact force between two nonsliding surfaces that opposes their attempt to slide

tension the force transmitted through a string or taut wire

Hooke's law the force law for an ideal spring

force constant the proportionality factor between the force and the deformation in Hooke's law

translational equilibrium the situation in which the net force on an object is zero

centripetal acceleration the center-pointing acceleration of objects in circular motion

centripetal force the center-pointing force on objects in circular motion

II. Important Equations

Name/Topic	Equation	Explanation
kinetic friction	$f_k = \mu_k N$	The force of kinetic friction is directly proportional to the normal force.
static friction	$0 \le f_s \le \mu_s N$ $f_{s,max} = \mu_s N$	The force of static friction takes on a range of values up to a maximum.
Hooke's law	$F_x = -kx$	The force law for an ideal spring.
centripetal acceleration	$a_{cp} = \frac{v^2}{r}$	The magnitude of the acceleration for circular motion at a constant speed v.
centripetal force	$f_{cp} = m\frac{v^2}{r}$	The magnitude of force on mass m moving in circular motion at constant speed v.

III. Know Your Units

Quantity	Dimension	SI Unit
coefficient of friction	—	dimensionless

Practice Problems

1. A skier traveling at 22.1 m/s encounters a 27.4 degree slope. If you could ignore friction, to the nearest meter, how far up the hill does he go?
2. If the coefficient of kinetic friction in problem 1 was actually 0.16, and the slope 30 degrees, to the nearest meter, how far up the hill does he go?
3. You have a mass of 82 kg and are on a 30 degree slope hanging on to a cord with a breaking strength of 194 newtons. What must be the coefficient of static friction, to two decimal places, between you and the surface for you to be saved from the fire?

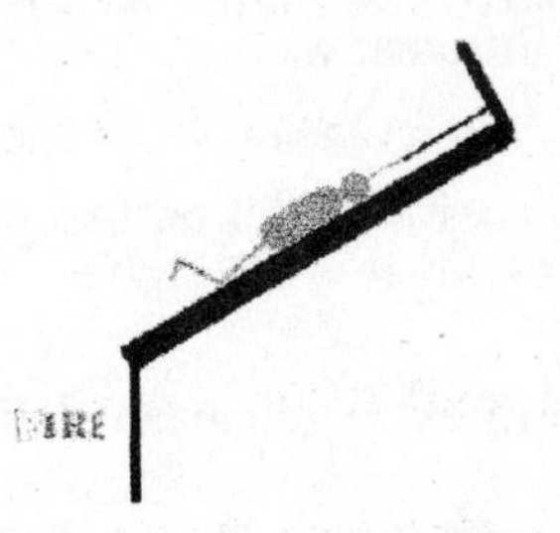

4. In problem 3, if the coefficient of static friction is zero, to the nearest tenth of a degree, what would the incline angle have to be in order for the cord not to break?
5. The box in the figure has a mass of 45 kg. To the nearest newton, what is the tension in the rope?
6. If the spring in the figure has a spring constant of 518 N/m, to the nearest tenth of a cm how far is it stretched?

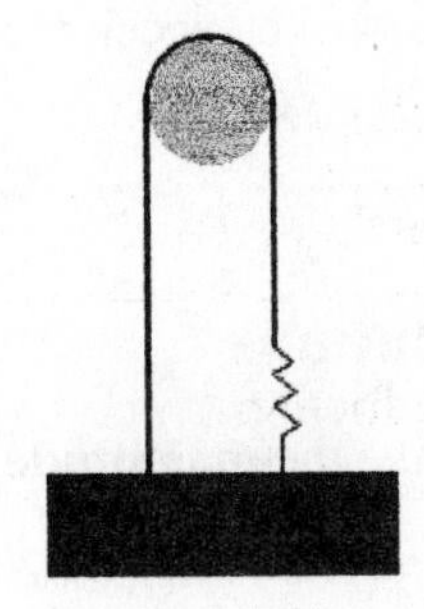

7. The 50-kg man in the roller-coaster car is sitting on a bathroom scale. If he is traveling at 35.6 m/s at the point shown, and the radius of the vertical coaster track is 50 meters, to the nearest newton, what does the scale read?
8. What would be the answer to problem 7 if the roller-coaster car was at the bottom of the track?
9. What would be the answer to problem 7 if the mass of the rider was 72 kg?

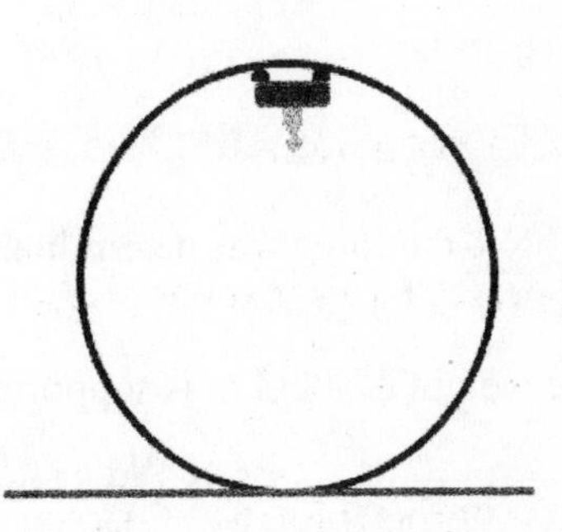

10. Three ropes are pulling on a ring. If $F_1 = 83$ N and $F_2 = 60$ N, to the nearest tenth of a newton, what must F_3 equal if the ring is not to move?

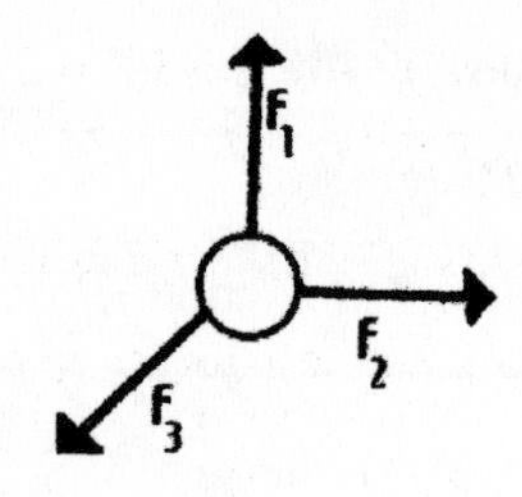

Puzzle

DOUBLE THE FUN?

Experimenting with his toy rockets, Danny fires his model rockets three different ways.

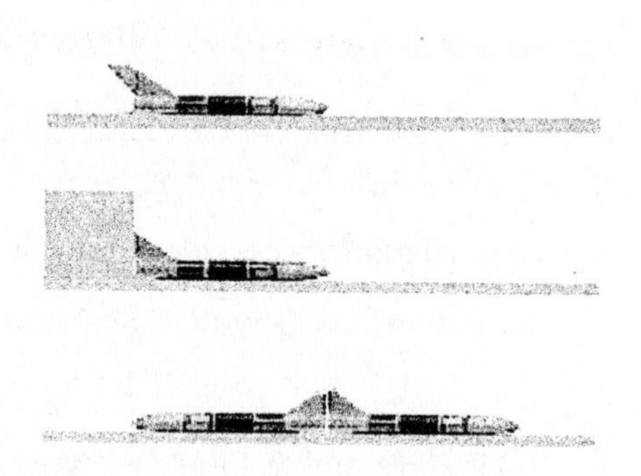

1. First, he places a model on the ground, free of obstructions, and fires.
2. Next, he braces an identical model against a stiff wall and fires.
3. Third, he positions two identical models on the ground with their tails butting, and fires them simultaneously.

   How do the rocket accelerations in the three experiments compare?

Selected Solutions

17. **(a)** Determine the **magnitude** (always positive) of the force required to stretch the spring 2.00 cm.

$$F = kx = (150\ \text{N/m})(0.0200\ \text{m}) = 3.00\ \text{N}$$

Using Newton's 2nd law we can calculate the magnitude of the force of static friction. Note that $f_{s,x} = f_s$

$$F_x - f_{s,x} = ma = 0 \therefore F - f_s = 0 \Rightarrow f_s = F = \boxed{3.00\ \text{N}}$$

(b) $\boxed{\text{No}}$, the force was determined using only the spring constant and the extension of the spring.

29. The weight of mass m is supported by the tension in the rope. Therefore,

$$T_1 = T_2 = mg = (2.50\ \text{kg})(9.81\ \text{m/s}^2) = 24.525\ \text{N}$$

The force F that the leg exerts on the pulley is balanced by the net force along the central line between T_1 and T_2

$$F = T_1\cos 30^\circ + T_2\cos 30^\circ = 2(24.525\ \text{N})\cos 30^\circ = \boxed{42.5\ \text{N}}$$

39. (a) The *x*-axis is in the direction of the force. Let $m_1 = 1.5$ kg, $m_2 = 0.93$ kg.

Newton's 2nd law for m_1 gives the equation

$$\sum F_{1x} = F - T = m_1 a$$

Newton's 2nd law for m_2 gives the equation

$$\sum F_{2x} = T = m_2 a$$

Adding these two equations eliminates T to give

$$F = (m_1 + m_2)a$$

We can now solve this for the acceleration:

$$a = \frac{F}{m_1 + m_2} = \frac{6.4\ \text{N}}{1.5\ \text{kg} + 0.93\ \text{kg}} = \boxed{2.6\ \text{m/s}^2}$$

(b) Since the tension is the only force on m_2, we can write

$$T = m_2 a = \frac{m_2 F}{m_1 + m_2} = \frac{(0.93\ \text{kg})(6.4\ \text{N})}{1.5\ \text{kg} + 0.93\ \text{kg}} = \boxed{2.4\ \text{N}}$$

(c) By examining the preceding equation in part (b), we see that increasing m_1 would increase the denominator and therefore $\boxed{\text{decrease}}$ the tension.

47. (a) The rider moves in uniform circular motion, so the net force on the rider provides a centripetal force. Applying Newton's 2nd law at the top of the Ferris wheel (where all the forces are vertical) we get:

$$\sum F_y = N - mg = ma_y = -m\frac{v^2}{r}$$

$$\therefore\ N = mg - m\frac{v^2}{r} = m\left(g - \frac{v^2}{r}\right)$$

You can see that the normal force exerted on a rider is less than that rider's weight by an amount mv^2/r, which results in an $\boxed{\text{apparent weight less than the rider's actual weight}}$.

Applying Newton's 2nd law at the bottom of the Ferris wheel gives us

$$\sum F_y = N = mg = ma_y = m\frac{v^2}{r}$$

$$\therefore\ N = mg + m\frac{v^2}{r} = m\left(g + \frac{v^2}{r}\right)$$

Here you can see that the normal force exerted on a rider is greater than that rider's weight by an amount mv^2/r, which results in an apparent weight greater than the rider's actual weight.

(b) Let's first determine the speed at which we move around the circular path. This speed, being constant, can be calculated as distance divided by time:

$$v = \frac{C}{t} = \frac{2\pi r}{t} = \frac{2\pi(7.2\text{ m})}{28\text{ s}} = 1.616\text{ m/s}$$

As discussed in part (a), the apparent weight at the top equals the normal force:

$$W_{top} = m\left(g - \frac{v^2}{r}\right) = (55\text{ kg})\left[9.81\text{ m/s}^2 - \frac{(1.616\text{ m/s})^2}{7.2\text{ m}}\right] = \boxed{520\text{ N}}$$

The apparent weight at the bottom equals the normal force at the bottom given in part (b)

$$W_{bot} = m\left(g + \frac{v^2}{r}\right) = (55\text{ kg})\left[9.81\text{ m/s}^2 + \frac{(1.616\text{ m/s})^2}{7.2\text{ m}}\right] = \boxed{560\text{ N}}$$

63. (a) Let T be the tension in the slanting stretch of rope. Then $T\sin(45^\circ)$ is the tension in the rope supporting mass B, and $T\cos(45^\circ)$ is the tension in the rope pulling on mass A. We can use the fact that $\sin(45^\circ) = \cos(45^\circ)$. At the knot where the three ropes intersect we apply Newton's 2nd law.

For the y direction:

$$\sum F_y = T\sin 45^\circ - m_B g = 0 \quad \Rightarrow \quad T\sin 45^\circ = m_B g$$

For the x-direction:

$$\sum F_x = T\cos 45^\circ - f_s = 0 \quad \Rightarrow \quad T\cos 45^\circ = T\sin 45^\circ = f_s$$

$$\therefore \quad f_s = m_B g = (2.25\text{ kg})(9.81\text{ m/s}^2) = \boxed{22.1\text{ N}}$$

If mass A is truly in equilibrium, we must have that $f_s \le f_{s,max}$:

$$f_{s,max} = \mu_s m_A g = 0.320(8.50\text{ kg})(9.81\text{ m/s}^2) = 26.7\text{ N}$$

so the condition is satisfied.

(b) If mass A is doubled, $f_{s,max}$ will double, but this will not affect f_s.

Answers to Practice Quiz

1. (a) **2.** (b) **3.** (e) **4.** (b) **5.** (d) **6.** (e) **7.** (d) **8.** (e) **9.** (c) **10.** (e)

Answers to Practice Problems

1.	54 m	**2.**	39 m
3.	0.3	**4.**	14.0°
5.	221 N	**6.**	42.7 cm
7.	777 N	**8.**	1758 N
9.	1119 N	**10.**	102.4

CHAPTER 7

WORK AND KINETIC ENERGY

Chapter Objectives

After studying this chapter, you should:

1. understand the value of the concepts of work and kinetic energy.

2. be able to calculate the work done by constant forces and approximate the work of variable forces.

3. be able to determine the kinetic energy of a moving object.

4. know how to calculate the average power delivered when work is done.

Warm-Ups

1. A weight lifter picks up a barbell and (a) lifts it chest high, (b) holds it for 5 minutes, then (c) puts it down. Rank the amounts of work W the weight lifter performs during these three operations. Label the quantities W_1, W_2, and W_3. Justify you ranking order.
2. Estimate the amount of work the engine performed on a 1200-kg car as it accelerated at 1.2 m/s^2 over a 150-meter distance.
3. A real-world roller-coaster released at point A and coasting without external power would traverse a track somewhat like the figure below. Friction is not negligible in the real world.
   Does the roller-coaster have the same energy at points B and C? Is the total energy conserved during the coaster ride? Can you account for all the energy at any point on the track?

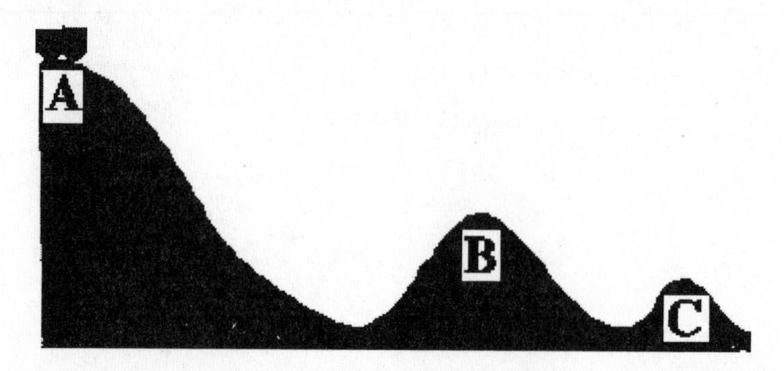

4. A gallon of gasoline contains about 1.3×10^8 joules of energy. A 2000-kg car traveling at 20 m/s skids to a stop. Estimate how much gasoline it will take to bring the car back to its original speed. To

complicate matters further, consider that only about 15% of the energy extracted from gasoline actually propels the car. The rest gets exhausted as heat and unburned fuel.

Chapter Review

In Chapters 5 and 6 you learned how to analyze mechanical situations by direct use of Newton's laws of motion. In this chapter you begin to learn about other concepts that are consistent with Newton's laws that can often make such analysis easier to perform. This is especially true when the number of applied forces is large, when they vary with distance or time, or when they are not accurately known.

7–1 Work Done by a Constant Force

When a force acts on an object that undergoes a displacement we say that the force does a certain amount of **work** on the object. This work can be positive, negative, or zero depending on how the direction of the force relates to the direction of the displacement.

If we consider the case in which a force **F** and the displacement **d** are in the same direction, then we can consider the force as acting both through and *with* the displacement to its fullest extent. In this special case the work done by the force, W, will be positive and equal to the product of the magnitude of the force and the magnitude of the displacement: $W = Fd$. This result represents the maximum amount of positive work that the force **F** can do on the object. The opposite situation occurs when the force and the displacement are in opposite directions (as is often the case with kinetic friction). Here, the force acts through and *against* the displacement to its fullest extent. In this latter case, the work done by the force will be negative $W = - Fd$. A third special case occurs when the direction of **F** is perpendicular to **d** (as is the case with the normal force on an object moving across a surface). When **F** and **d** are perpendicular, **F** neither works with nor against the displacement, and the work done by **F** on the object is zero.

In the most general cases of work done by a constant force, the force and displacement vectors are at arbitrary angles with each other. Consider the following figure:

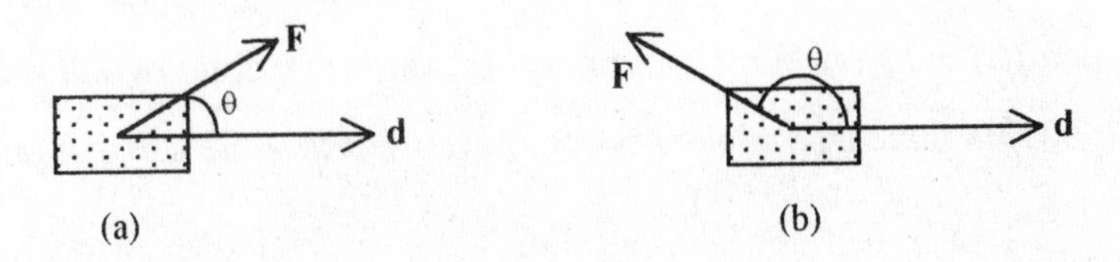

In part (a) we can see that **F** only partly acts with **d**, and in part (b) it only partly acts against **d**. To determine the work done by **F** in such cases, we need to use only that part of **F** that acts with or against the displacement; that is, we need the component of **F** along the direction of **d** — F_d. Based on the above diagrams and what you know about vector components, try to convince yourself that $F_d = F \cos \theta$. In part

(a) of the figure F_d will be positive, and in part (b) it will be negative. The work done by **F** in this more general case, therefore, is given by

$$W = F_d d = (F\cos\theta)d$$

Notice that this general result for the work done by a constant force includes all three of the special cases discussed in the previous paragraph. As you can see in the preceding equation, the units of work must result from the product of the units of force and displacement. This combination, N·m, when applied to work, is called a **joule** (J).

It is useful to recognize that the work done can also be viewed as the component of the displacement in the direction of the force multiplied by the magnitude of the force. Algebraically, the difference amounts only to a regrouping of the factors involved:

$$W = F(d\cos\theta)$$

Try to convince yourself that the component of **d** along **F** really is $d_F = d\cos\theta$. The important concept here is that work depends only on the extent to which the force and the displacement act together (i.e., along the same direction).

When several forces act on an object while it is being displaced each force does work on the object according to the above discussion. The total work done on the object is the sum of all these contributions. An alternative way to approach this calculation is to first determine the total force acting on the object, $\mathbf{F}_{\text{total}}$, and then calculate the work done by this force. Therefore, we have

$$W_{\text{total}} = W_1 + W_2 + W_3 + \cdots = (F_{\text{total}}\cos\theta)d$$

where the total force, also called the net force, is given by the vector sum of all the forces

$$\mathbf{F}_{\text{total}} = \mathbf{F}_1 + \mathbf{F}_2 + \cdots$$

Example 7–1 Shopping for Groceries A person pushes a 751-N shopping cart full of groceries down the aisle at the local store. The person applies a force of 112 N at an angle of 40.0 degrees below the horizontal. The cart is pushed the full length of a 15.5-m aisle at constant speed. Determine **(a)** the work done by the shopper, **(b)** the work done by gravity, **(c)** the work done by the normal force of the floor on the cart, and **(d)** the work done by various frictional forces during the cart's motion down the aisle.

Picture the Problem The diagram shows the shopping cart; the force exerted by the shopper **F**; the weight **mg**; the normal force **N**; and the displacement **d**.

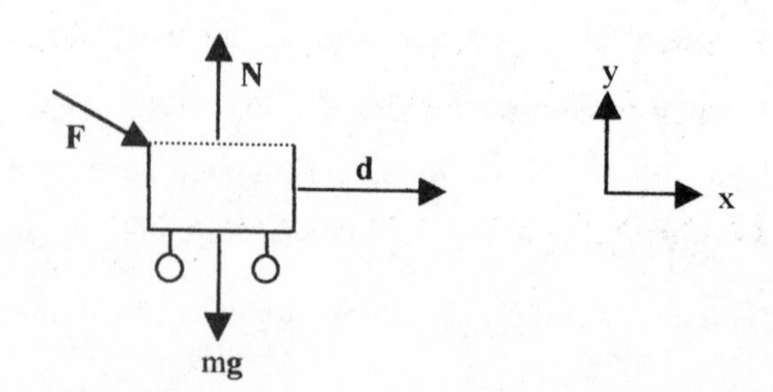

Strategy To calculate the work done by the individual forces listed we need to identify the magnitude of each force and its direction relative to the displacement. We can determine the work done by frictional forces from the total work on the cart.

Solution

Part (a)

1. From given information, the angle between **F** and **d** must be: $\theta_F = 40.0°$

2. The work done by **F** then is: $W_F = Fd\cos\theta_F = (112\ \text{N})(15.5\ \text{m})\cos(40.0°) = 1330\ \text{J}$

Part (b)

3. From the given situation the angle between the cart's weight and **d** is: $\theta_g = -90.0°$

4. The work done by gravity is: $W_g = mgd\cos\theta_g = (751\ \text{N})(15.5\ \text{m})\cos(90.0°) = 0\ \text{J}$

Part (c)

5. From the given information, the angle between **N** and **d** is: $\theta_N = 90.0°$

6. The work done by **N** then is: $W_N = Nd\cos\theta_N = Nd\cos(90.0°) = 0\ \text{J}$

Part (d)

7. Since the cart moves with constant speed the total force on the cart is: $F_{total} = ma = m(0\ \text{m/s}^2) = 0$

8. The total work done on the cart then is: $W_{total} = F_{total}d\cos(\theta_{total}) = 0$

9. Since $W_{total} = 0$, the sum of the work done by each force must be zero: $W_{total} = W_F + W_g + W_N + W_{fric} = 0$

10. Since $W_g = W_N = 0$, this implies that: $W_{fric} = -W_F = -1330\text{J}$

Insight Recognizing that both the normal force and the weight of the cart are perpendicular to the displacement, we could have immediately concluded that the work done by these forces is zero. However, the preceding approach shows that even if you don't recognize this, you can arrive at the correct result by performing the general calculation. Notice that the phrase "constant speed" in the statement of the problem was crucial to our ability to solve the problem. Be careful to pay attention to such phrases. Finally, take note that we were able to find the work done by friction forces without knowing anything about those forces. This point has caused many students to get stuck on similar problems. Keep in mind that there are many ways to get at information once you put all the physics together.

Exercise 7–2 Sliding Down A 3.5-kg object slides 1.7 m down a ramp that is inclined at 35 degrees to the horizontal. If the coefficient of kinetic friction between the object and the ramp is 0.12, **(a)** how much work is done by gravity, and **(b)** how much work is done by kinetic friction.

Solution: The following information is given in the problem:

Given: $m = 3.5$ kg, $d = 1.7$ m, $\alpha = 35°$, $\mu_k = 0.12$; **Find**: **(a)** W_g, **(b)** W_f

The diagram shows the object on the ramp and the forces acting on it. Notice that the weight has been divided into its components parallel and perpendicular to the incline.

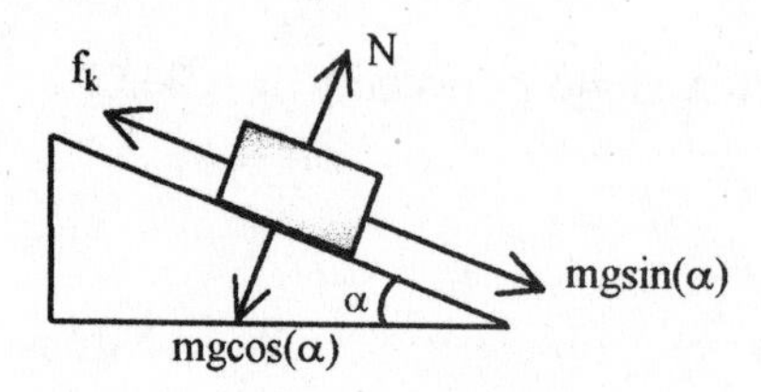

For part (a) we see that mgsin(α) is the component of the gravitational force in the direction of the displacement. Thus, we can immediately calculate the work done by gravity as

$$W_g = (mg\sin\alpha)d = (3.5\text{ kg})(9.81\text{ m/s}^2)(1.7\text{ m})\sin(35^\circ) = 33\text{ J}$$

For part (b) we need to determine the value of the force of kinetic friction. Since the object does not accelerate in the direction perpendicular to the incline, the two forces along that direction must cancel. This implies that

$$N = mg\cos\alpha$$

The force of friction then is given by

$$f_k = \mu_k N = \mu_k mg\cos\alpha$$

Since f_k always opposes the direction of motion, the angle between $\mathbf{f_k}$ and $\mathbf{d}$ is 180°. This gives

$$W_f = f_k d\cos(180^\circ) = -\mu_k mgd\cos\alpha = -(0.12)(3.5\,\text{kg})(9.81\,\text{m/s}^2)(1.7\,\text{m})\cos(35^\circ) = 5.7\,\text{J}$$

There were two angles in this calculation: the angle of inclination of the ramp and the angle between the force and displacement. A common mistake is to get them confused and use the 35 degrees as the angle between the force and the displacement. Be careful about this situation. Also notice that unlike Example 7–1, this time we determined the force of friction before calculating the work it did. Can you tell why it was more convenient to treat these two cases differently?

Practice Quiz

1. A person pushes an object across a room, on a horizontal floor, by applying a horizontal force. If, instead, he pushed it through the same horizontal distance with the same magnitude of force at an angle θ below the horizontal, he would do

   (a) greater amount of positive work **(b)** lesser amount of positive work **(c)** zero work

   (d) negative work **(e)** none of the above

2. A person pulls a 3.0-kg object across a room, on a horizontal floor, by applying a force of 13 N that makes an angle of 77° above the horizontal. If she does a total of 26 J of work, how far does she pull the object?

   (a) 2.0 m **(b)** 230 m **(c)** 3.9 m **(d)** 10 m **(e)** 8.9 m

3. A person pushes an object across a room, on a horizontal floor, by applying a horizontal force. If, instead, he pushed the object through the same horizontal distance with the same magnitude of force at an angle θ above the horizontal, friction would do

   (a) a greater amount of positive work **(b)** a greater amount of negative work

   (c) a lesser amount of positive work **(d)** a lesser amount of negative work **(e)** zero work

7–2 Kinetic Energy and the Work-Energy Theorem

One of the reasons that the concept of work has proved to be very important is that when a nonzero amount of work is done on an object that work must manifest itself as some sort of observable effect on the object. A clearly observable effect of work is the change in speed that results. However, we cannot say that work is directly converted into speed because work and speed are different types of quantities (work is measured in joules, and speed in m/s); however, work must convert into something closely related to speed. We call that "something" the **kinetic energy,** K, of the object. Kinetic energy is given by the equation

$$K = \tfrac{1}{2}mv^2$$

The relationship between the total work done on an object and its kinetic energy is given by the following expression known as the **work-energy theorem:**

$$W_{total} = K_f - K_i = \Delta K$$

The above equation shows that kinetic energy must have the same unit as work, namely, the joule. You can think of kinetic energy as a measure of the amount of work that goes into the motion of an object. Thus, if you see an object with 10 J of kinetic energy, you know that it took 10 J of work to get it up to the speed that it has, and it will likewise take 10 J of work to bring it to rest.

Exercise 7–3 Skidding to a Stop While driving at 12.0 m/s in a car of mass 1470 kg you notice a squirrel running into the street in front of you. You hit the brakes suddenly and come to a stop after skidding for 3.20 m. How much work is done by friction in bringing you to a stop?

Solution: The following information is given in the problem:

Given: $m = 1470$ kg, $d = 3.20$ m, $v_i = 12.0$ m/s, $v_f = 0$; **Find**: W_f

Because we know the mass of the car and its initial and final speeds, we can find the initial and final kinetic energies and therefore the change in kinetic energy:

$$\Delta K = K_f - K_i = \tfrac{1}{2}mv_f^2 - \tfrac{1}{2}mv_i^2 = \tfrac{1}{2}m\left(v_f^2 - v_i^2\right) = \frac{1470 \text{ kg}}{2}\left[0 - (12.0 \text{ m/s})^2\right] = -1.06\times10^5 \text{ J}$$

Since the car stops because of friction only, we know that the total work done is the work done by friction that we seek. Therefore,

$$W_f = \Delta K = -1.06\times10^5 \text{ J}$$

From the motion alone we determined the work done by friction without knowing the force at all.

Practice Quiz

4. When a net amount of positive work is done on an object, you expect the object to

   (a) speed up **(b)** slow down **(c)** stop **(d)** maintain a constant speed **(e)** none of these

5. If a 93 g bullet has a kinetic energy of 4200 J, what is the muzzle velocity of the gun?

   (a) 390 m/s **(b)** 45 m/s **(c)** 300 m/s **(d)** 210 m/s **(e)** 77 m/s

7–3 Work Done by a Variable Force

When the force that is applied to an object varies with distance, we can use a graphical technique to either approximate or exactly determine the amount of work done by this force. The key to this technique is that any force that varies with distance can be treated approximately constant over small enough displacements. For these small displacements we can approximate the amount of work done using the expression for the work done by a constant force. For variable forces we shall consider only cases for which the force and displacement are along the same direction. As shown in the following graph of force versus displacement on the left, when the force can be treated as constant, the work done by the force equals the area of the rectangle whose width is the displacement Δx and whose height is the value of the force F.

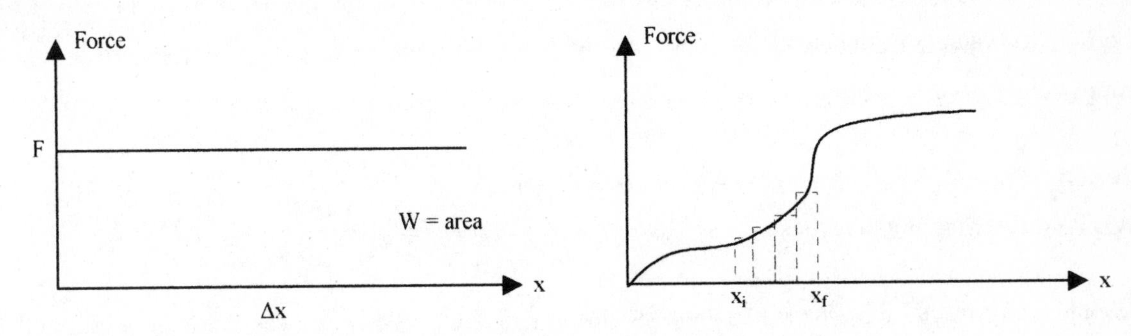

In the graph on the right, we see how small rectangles, representing small displacements, can be used to approximate the area, and therefore the work done, for a given overall displacement from x_i to x_f. Notice that the height of the left-hand sides of the rectangles tend to overestimate the force, whereas the right-hand sides tend to underestimate. This compensation helps make this approximation more accurate. The smaller the rectangles, the better the approximation.

A variable force of special interest is the Hooke's law force, $F = -kx$, that exactly describes an ideal spring and approximately describes the behavior of any solid object when it is deformed slightly. Since the force varies linearly with x, the area under the curve is that of a simple triangle: area = 1/2 base × height. In this case we can use the described graphical technique to exactly (not approximately) determine the work done in stretching or compressing the spring. We find that the minimum amount of work *required* to stretch or compress a spring an amount x from its equilibrium length is

$$W = \tfrac{1}{2}kx^2$$

The work done by the spring while it is being deformed is the opposite of this value $W_{spring} = -\tfrac{1}{2}kx^2$.

Example 7–4 Spring Cushion A spring with a force constant of 2000 N/m is used to cushion objects of moderate weight. If a 75-lb object is placed on one of these springs, how much work does the spring do before it balances the weight of the object?

Picture the Problem Part (a) of the figure shows the object being balanced by the spring. Part (b) is a free-body diagram of the object. The coordinate system is at the far right.

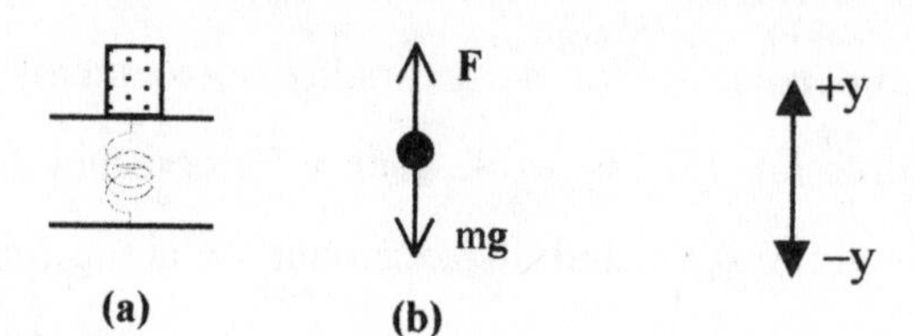

Strategy Based on the information given here, we need to know how much the spring compresses in order to calculate the work it does while compressing.

Solution

1. Convert the weight to newtons: $\text{weight} = 75\ \text{lb}\left(\frac{4.45\ \text{N}}{1\ \text{lb}}\right) = 333.75\ \text{N}$

2. Apply Newton's 2nd law to the vertical direction: $\sum F = kx - mg = 0$

3. Solve this equation for x: $kx = mg \;\Rightarrow\; x = \frac{mg}{k}$

4. Find the work done by the spring:

$$W = -\frac{1}{2}kx^2 = -\frac{k}{2}\left(\frac{mg}{k}\right)^2$$

$$= -\frac{2000\ \text{N/m}}{2}\left(\frac{333.75\ \text{N/m}}{2000\ \text{N/m}}\right)^2 = -28\ \text{J}$$

Insight Understand that the work done by the spring is negative because as it compresses downward the spring itself applies an upward force on the object.

Practice Quiz

6. A spring is held in a compressed position, then released. As the spring uncoils to its equilibrium length it does

   (a) negative work. **(b)** positive work. **(c)** zero work **(d)** none of the above

7. It requires a minimum of 0.72 J of work to stretch a certain spring 1.5 cm, what is its force constant?

   (a) 1.1 N/m **(b)** 0.011 N/m **(c)** 96 N/m **(d)** 6400 N/m **(e)** none of the above

8. If it requires an amount of work W to stretch a certain spring by an amount x, how much work would it take to stretch it by an amount 2x?

   (a) 2W **(b)** $\sqrt{2}$W **(c)** $\frac{1}{2}$W **(d)** $\frac{1}{4}$W **(e)** 4W

7–4 Power

From a practical standpoint, the mere fact that a certain amount of work is done is not always good enough. The question of how long it takes to do this work often determines the practical values of certain devices. For example, everyone wants a car that can go from 0 to 60 mi/h, but no one wants it if it takes 5 min. The quantity we use to measure how rapidly work is done is called **power**.

Power is the rate at which work is done. For the applications of interest to us, we will mainly use the average power, which is the amount of work done, W, divided by the amount of time, t, t required to do it:

$$P = \frac{W}{t}$$

The unit of power must be the unit of work divided by the unit of time, or joules per second (J/s). In SI the unit J/s is given the name **watt** (W). You are probably familiar with this unit, as it is common to give the power rating of light bulbs in watts. For cases when work is being done by a constant force on an object moving at constant speed, the power delivered by that force is given by

$$P = Fv$$

Verify that the product Fv has the SI unit watt. Also be aware that the above equation can be useful even if the force or the speed is not constant. In these cases the average power can be calculated by substituting the average force or the average speed: $P = F_{av}v = Fv_{av}$. If both F and v vary, use $P = F_{av}v_{av}$.

Example 7–5 An Industrial Pulley The motor of a chain-linked industrial pulley designed to lift heavy weights is rated to deliver 3000 watts of power on average. If the mechanism is only 80% efficient, how long will it take to raise a 55-kg crate a height of 12 m at a constant speed?

Picture the Problem The diagram shows part of the pulley system lifting the crate with a constant upward velocity **v**.

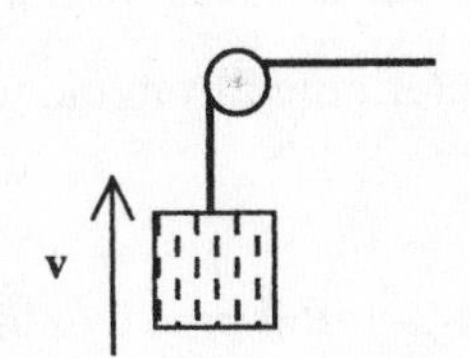

Strategy The amount of time it takes is related to

how much work is done and the rate at which it is done (the power). Therefore, we will determine the work done and the actual power output.

Solution

1. Obtain the expression for the time t: $P_{out} = \frac{W_{pulley}}{t} \Rightarrow t = \frac{W_{pulley}}{P_{out}}$

2. Since the pulley applies a force in the same direction as the displacement: $W_{pulley} = F_{pulley}d$

3. Since its velocity is constant the force of the pulley must exactly balance the weight: $F_{pulley} = mg \therefore W_{pulley} = mgd$

4. Using the fact that the motor is 80% efficient, find the time t: $t = \frac{mgd}{P_{out}} = \frac{mgd}{0.8P} = \frac{(55\text{ kg})(9.81\text{ m/s}^2)(12\text{ m})}{0.8(3000\text{ W})} = 2.7\text{ s}$

Insight We could also have solved this problem by finding the velocity and calculating the time for the crate to move the 12 m distance. Try it.

Exercise 7–6 Test Driving The manufacturer of a 1500-kg automobile advertises that the engine delivers 175 hp. If all of this power is transferred to the motion of the car with 100% efficiency, what 0-to-60 mi/h time would you expect the car to achieve?

Solution: The following information is given in the problem:

Given: $m = 1500$ kg, $P = 175$ hp, $v_i = 0.0$ m/s, $v_f = 60$ mi/h; **Find**: t

We have a mix of units here, so let's convert everything to SI.

$$P = 175\text{ hp}\left(\frac{746\text{ W}}{1\text{ hp}}\right) = 1.306\times10^5\text{ W} \quad\text{and}\quad v_f = 60\text{ mi/h}\left(\frac{0.447\text{ m/s}}{1\text{ mi/h}}\right) = 26.82\text{ m/s}$$

The time for the car to reach 60 mi/h depends on the amount of work done: t = W/P. The work done by the car can be determined from the work-energy theorem

$$W = K_f - K_i = K_f = \tfrac{1}{2}mv_f^2$$

So, the expected 0- to-60 time would be

$$t = \frac{mv_f^2}{2P} = \frac{(1500\,\text{kg})(26.82\,\text{m/s})^2}{2(1.306\times10^5\,\text{W})} = 4.1\,\text{s}$$

Look up some typical 0- to-60 times. What does this say about the efficiency at which the car converts the engine's power into the motion of the car?

Practice Quiz

9. If motor A delivers more power than motor B then

   (a) motor A can do more work than motor B

   (b) motor A takes more time to do the same amount of work as motor B

   (c) motor A takes less time to do the same amount of work as motor B

   (d) both motors take the same amount of time to do the same amount of work

   (e) motor A is more efficient than motor B

10. A lift mechanism delivers a power of 85 W in lifting a 250-N crate at constant speed. What is the speed?

    (a) 3.4 m/s **(b)** 0.29 m/s **(c)** 6.0 m/s **(d)** 9.8 m/s **(e)** 0.35 m/s

Reference Tools and Resources

I. Key Terms and Phrases

work work is done when a force acts through a displacement

kinetic energy the quantity that measures the amount of work that goes into the motion of an object

work-energy theorem the expression $W_{total} = \Delta K$ that relates work and kinetic energy

power the rate at which work is done

II. Important Equations

Name/Topic	Equation	Explanation
work	$W = Fd\cos\theta$	The work done by a constant force F on an object that undergoes a displacement of magnitude d.
kinetic energy	$K = \frac{1}{2}mv^2$	The kinetic energy of a mass m moving at speed v.

work-energy theorem	$W_{total} = \Delta K$	The relationship between work and kinetic energy change.
work by a variable force	$W = \frac{1}{2}kx^2$	The work required to deform a spring an amount x from equilibrium.
power	$P = \frac{W}{t}$	The average power delivered when work is done over a time period t.

III. Know Your Units

Quantity	Dimension	SI Unit
work	$[M]\cdot[L^2]/[T^2]$	J
kinetic energy	$[M]\cdot[L^2]/[T^2]$	J
power	$[M]\cdot[L^2]/[T^3]$	W

Practice Problems

1. A man pulls a 10-kg box across a smooth floor with a force of 72 newtons at an angle of 10 degrees for a distance of 100 meters. How much work, to the nearest joule, does he do?
2. If the floor in problem 1 is angled upward at 11.4 degrees and the man pulls the box up the floor at constant speed, what work does he do to the nearest joule?
3. In problem 2, to the nearest joule, what is the work done by gravity?
4. In problem 1 if the box starts from rest, what is its final kinetic energy to the nearest joule?
5. If the mass of the box in problem 1 is 13 kilograms, what is its final speed to the nearest tenth of a meter/second?
6. You are traveling in your 2000-kg car at 10 m/s and wish to accelerate to 18.8 m/s in 4 seconds, how much work, to the nearest joule, is required?
7. In problem 6 what is the average power, to the nearest watt?
8. What is the equivalent (nearest) horsepower in problem 7?
9. Assuming you have a bow that behaves like a spring with a spring constant of 169 N/m and you pull it to a draw of 52 cm, to the nearest joule, how much work do you perform?
10. In problem 9, to the nearest tenth of a m/s, what is the speed of the 120-gram arrow when it is released?

Puzzle

WHO IS RIGHT?

Bill is riding in a railroad car. He throw a ball towards the back wall of the car. The train is moving at a constant 20 m/s to the right. The ball is flying away from Bill at 8 m/s. According to Bill, his 0.145-kg baseball has 4.64 joules of kinetic energy. His brother is standing on the ground disagreeing. According to him the baseball has 10.44 joules of kinetic energy. Who is right? Answer this question in words, not equations, briefly explaining how you obtained your answer.

Selected Solutions

9. Since the crate moves horizontally, the angle that the rope makes with the horizontal is the same as the angle between the force and the displacement. Therefore, the work done by the tension is

$$W = (T\cos\theta)d = (125\text{ N})(\cos 40.0°)(5.0\text{ m}) = \boxed{480\text{ J}}$$

19. **(a)** We can use the work-energy theorem to determine the total work done on the player: $W_{total} = \Delta K$.

$$W_{total} = \tfrac{1}{2}m(v_f^2 - v_i^2) = \tfrac{1}{2}(60.0\text{ kg})\left[0 - (4.2\text{ m/s})^2\right] = \boxed{-530\text{ J}}$$

(b) All of the work was done by kinetic friction, so we can write

$$W_{total} = -f_k d = -\mu_k N d = -\mu_k mgd$$

$$\therefore\ \mu_k = -\frac{W_{total}}{mgd} = \frac{529\text{ J}}{(60.0\text{ kg})(9.81\text{ m/s}^2)(3.00\text{ m})} = \boxed{0.30}$$

25. The initial kinetic energy of the block is stored as potential energy in the spring. Therefore,

$$\tfrac{1}{2}mv^2 = \tfrac{1}{2}kx^2 \Rightarrow k = \frac{mv^2}{x^2}$$

$$\therefore\ k = \frac{(1.8\text{ kg})(2.1\text{ m/s})^2}{(0.35\text{ m})^2} = \boxed{65\text{ N/m}}$$

39. The minimum power needed equals the power required to lift 10.0 lb a distance of 2.00 m in 1 second at constant speed.

$$P = \frac{W}{t} = \frac{Fh}{t} = \left[\frac{(10.0\text{ lb})\left(\dfrac{1\text{ N}}{0.2248\text{ lb}}\right)(2.00\text{ m})}{1\text{ s}}\right] = 88.97\text{ W}\left(\frac{1\text{ hp}}{746\text{ W}}\right) = \boxed{0.119\text{ hp}}$$

51. **(a)** The puck slows because of work done by friction. We can use the work-energy theorem to relate the reduction in speed to the coefficient of friction. The work done is

$W = -f_k d = -\mu_k N d = -\mu_k mgd$; therefore, $-\mu_k mgd = \frac{1}{2}m(v_f^2 - v_i^2)$

Solving this for μ_k gives us

$$\mu_k = \frac{-\frac{1}{2}m(v_f^2 - v_i^2)}{mgd} = \frac{v_i^2 - v_f^2}{2gd} = \frac{(45\text{ m/s})^2 - (44\text{ m/s})^2}{2(9.81\text{ m/s}^2)(25\text{ m})} = \boxed{0.18}$$

(b) Now that we know the coefficient of friction we can solve for the final speed using 44 m/s as the initial speed. This gives

$$v_f = \sqrt{v_i^2 - 2\mu_k gd} = \sqrt{(44\text{ m/s})^2 - 2(0.181)(9.81\text{ m/s}^2)(25\text{ m})} = \boxed{42.98\text{ m/s} < 43\text{ m/s}}$$

Answers to Practice Quiz

1. (b) **2.** (e) **3.** (d) **4.** (a) **5.** (c) **6.** (b) **7.** (d) **8.** (e) **9.** (c) **10.** (a)

Answers to Practice Problems

1. 7091 J 2. 1939 J

3. −1939 J 4. 7091 J

5. 33.0 m/s 6. 253440 J

7. 63360 W 8. 85 hp

9. 23 J 10. 19.6 m/s

CHAPTER 8

POTENTIAL ENERGY AND CONSERVATIVE FORCES

Chapter Objectives

After studying this chapter, you should:

1. understand the differences between conservative and nonconservative forces.

2. understand the concept of potential energy.

3. be able to apply the principle of the conservation of mechanical energy.

4. know how to handle energy considerations when work is done by nonconservative forces.

5. be able to read information off potential energy curves.

Warm-Ups

1. Do frictional forces always cause a loss of mechanical energy?
2. A good professional baseball pitcher throws a ball straight up in the air. Estimate how high the ball will go. (A good throw can reach 90 mi/h.)
3. Which potential energy U(x)-versus-x graph corresponds to the force F(x)-versus-x graph shown?

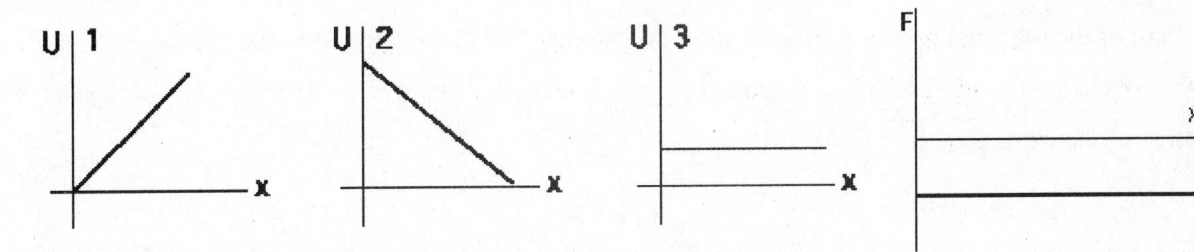

4. Estimate the energy burst you would need to clear a 1-meter hurdle.

Chapter Review

In this chapter we continue the study of work and energy and encounter another form of energy called potential energy. Most importantly, in this chapter we introduce the concept of the conservation of energy, which is one of the most important concepts in science.

8–1 Conservative and Nonconservative Forces

Forces are generally divided into two categories depending on properties of the work that a force does. A force can either be a **conservative force** or a **nonconservative force**. A force is called *conservative* if the work it does on any object is independent of the path that object takes during its displacement. Equivalently, a force is conservative if the work it does around any closed path is zero. Any force that does not meet this condition is a *nonconservative* force.

The primary reason for the two classifications is that when work is done by a conservative force, that work is in some sense stored within the system and can be easily recovered, usually as kinetic energy. Gravity is a conservative force. When an object is lifted against gravity a certain amount of negative work is done by gravity on the object. When that object is released, or lowered back down, gravity does an equal amount of positive work. Kinetic friction is a nonconservative force. If you slide a block across a table, friction does a certain amount of negative work on it. When you slide the block back to its original place, friction does even more negative work on it — the work is not recovered in this case.

8–2 Potential Energy and the Work Done by Conservative Forces

In Chapter 7 you were introduced to the idea that when work is done it goes into changing the kinetic energy of the object or system on which the work is done. In that case it was the *total* work done by all forces acting. A similar concept can be applied to specific, or individual, forces as well. When work is done by a conservative force, this work goes into changing the **potential energy**, U, of the system. You can think of the potential energy as representing the "storage" of this work within the system.

The mathematical definition of potential energy is that the change in potential energy equals the negative of the work done by conservative forces

$$\Delta U = -W_c$$

Notice that it is only the change in potential energy that has physical meaning. Particular values of potential energy are defined relative to a chosen reference. This reference can be chosen arbitrarily, although in some cases certain choices are more convenient and usually adopted by convention. Based on the preceding definition, it is clear that the SI unit of potential energy is the same as for work and kinetic energy — the joule.

Gravity

Because of the connection with conservative forces, potential energies are always associated with specific forces. Since gravity is a conservative force, we can define gravitational potential energy. Near Earth's surface, where we can treat the force of gravity as constant, the gravitational potential energy is

$$\Delta U = mg(\Delta y)$$

where Δy is the change in vertical position. The height at which U = 0 can always be chosen to be the initial height. Therefore, we often write the gravitational potential energy as just

$$U = mgy$$

The above equation is written with the understanding that it is only used this way when the reference height has been specified.

Springs

Besides gravity, a second conservative force that we have encountered is that of Hooke's law; therefore, we can define a potential energy to be associated with this force as well. We call this the **spring potential energy**, also often called the *elastic potential energy*. We saw in Chapter 7 that the work done by a spring while being deformed by an amount x from equilibrium is $W = -\frac{1}{2}kx^2$, so the ΔU associated with this force will be of opposite sign:

$$\Delta U = \tfrac{1}{2}kx^2$$

In the case of Hooke's law, it is almost always most convenient to take the equilibrium configuration (x = 0) as the reference level for potential energy (U = 0). With this understanding the spring potential energy is written as

$$U = \tfrac{1}{2}kx^2$$

Example 8–1 Rolling Down A 0.25-kg ball rolls 2.1 m down a ramp that is inclined at 32 degrees to the horizontal. Determine the change in gravitational potential energy.

Picture the Problem The diagram shows the ball rolling down the ramp.

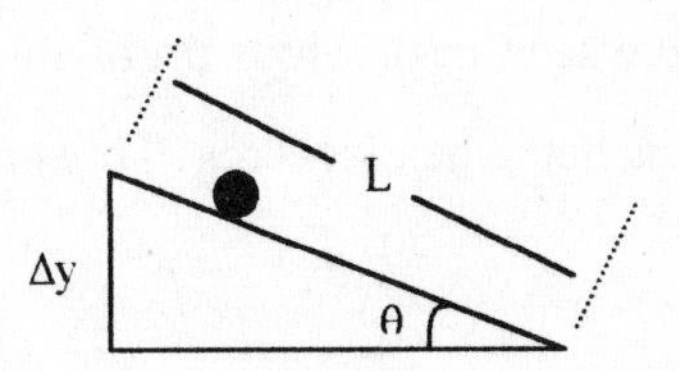

Strategy To determine the change in potential energy we need to find the change in height and use it in the expression for gravitational potential energy.

Solution

1. From trigonometry we see that: $|\Delta y| = L\sin\theta$
2. Since the ball rolls downhill: $\Delta y = -L\sin\theta$

3. The change in potential energy is:

$$\Delta U = mg\Delta y = -mgL\sin\theta$$
$$= -(0.25\text{ kg})(9.81\,\text{m/s}^2)(2.1\text{ m})\sin(32^\circ) = -2.7\text{ J}$$

Insight Notice that we had to be careful to notice that Δy should be negative. The mathematics alone dose not give the minus sign.

Exercise 8–2 Hanging from a Spring A spring of force constant 21 N/cm hangs vertically from a ceiling. A 1.2-kg mass is attached to its end and lowered until it hangs motionless from the end of the spring. If we take the potential energy of the Earth-spring-mass system to be zero initially, what is it in the final configuration?

Solution The following information is given in the problem:

Given: $m = 1.2$ kg, $k = 21$ N/cm, $U_i = 0$; **Find**: U_f

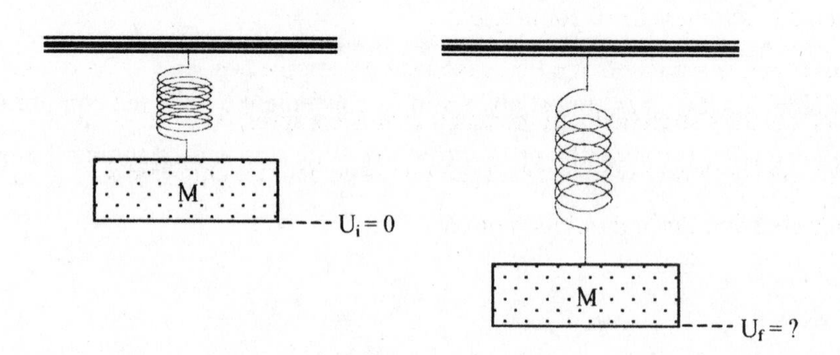

Here we have changes in both the gravitational potential energy and the potential energy of the spring. The change in each depends on how much the spring stretches under the weight of the mass; i.e., $x = |\Delta y|$. As we have seen in previous problems, equilibrium between the gravitational force and the Hooke's law force is reached when $kx = mg$. This implies that

$$x = |\Delta y| = \frac{mg}{k}$$

The change in potential energy is then

$$\Delta U = \Delta U_{spring} + \Delta U_{grav} = \tfrac{1}{2}kx^2 + mg\Delta y$$

Inserting the expressions for x and Δy (noting that Δy is negative), we get

$$\Delta U = \frac{(mg)^2}{2k} - \frac{(mg)^2}{k} = -\frac{(mg)^2}{2k}$$

Since $\Delta U = U_f - U_i = U_f$ we can conclude that

$$U_f = -\frac{(mg)^2}{2k} = -\frac{\left[(1.2\text{ kg})(9.81\text{ m/s}^2)\right]^2}{2(21\text{ N/cm})(100\text{ cm/m})} = -0.033\text{ J}$$

The important thing here was to remember to account for both forms of potential energy.

Practice Quiz

1. When an object is thrown up into the air, the gravitational potential energy
   (a) increases on the way up and decreases on the way down
   (b) decreases on the way up and increases on the way down
   (c) changes the same way going up as it does going down
   (d) remains constant throughout its motion
   (e) none of the above

2. The potential energy of a spring that is *stretched* an amount x from equilibrium versus one that is *compressed* an amount x from equilibrium
   (a) increases as it is stretched and decreases as it is compressed
   (b) decreases as it is stretched and increases as it is compressed
   (c) changes the same way whether the spring is stretched or compressed
   (d) remains constant throughout its motion
   (e) none of the above

3. A 2.0-kg object is lifted 2.0 m off the floor, carried horizontally across the room, and finally placed down on a table that is 1.5 m high. What is the overall change in gravitational potential energy for the Earth-object system?
   (a) 6.0 J **(b)** 20 J **(c)** 29 J **(d)** 9.8 J **(e)** 59 J

4. If a 0.75-kg block is rested on top of a vertical spring of force constant 45 N/m, how much potential energy is stored in the spring relative to its equilibrium configuration?
   (a) 0 J **(b)** 14 J **(c)** 60 J **(d)** 34 J **(e)** 0.60 J

8–3 Conservation of Mechanical Energy

Up to this point we have discussed two forms of energy kinetic and potential. Within any system we refer to the sum of these two forms of energy as the **mechanical energy**, E = K + U. The most important thing about this mechanical energy is that for systems in which only conservative forces do work the total mechanical energy is conserved, that is, remains constant. The energy within the system may change forms between kinetic and potential, but the sum of the two does not change. Three equivalent mathematical statements of this are

$$E = \text{constant}, \qquad E_f = E_i, \qquad K_f + U_f = K_i + U_i$$

Example 8–3 Rolling Down II In Example 8–1 we had a 0.25-kg ball rolling 2.1 m down a ramp that is inclined at 32 degrees to the horizontal. If the ball starts at rest, calculate its speed at the bottom of the ramp.

Picture the Problem The diagram is the same as in Example 8–1.

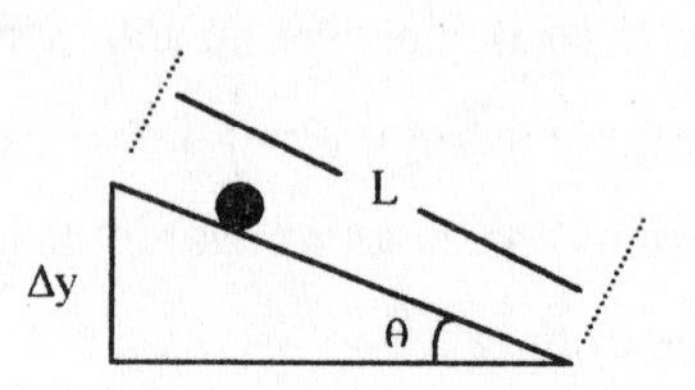

Strategy To solve this problem using the conservation of mechanical energy we shall write expressions for both the initial and final energies and set them equal. We set the reference for gravitational potential energy to be at the bottom of the ramp.

Solution

1. Write out the initial mechanical energy. Use the fact that $v_i = 0$ means $K_i = 0$: $E_i = K_i + U_i = 0 + U_i = mg|\Delta y| = mgL\sin(\theta)$

2. Write out the final mechanical energy: $E_f = K_f + U_f = \frac{1}{2}mv_f^2 + 0 = \frac{1}{2}mv_f^2$

3. Use energy conservation, $E_f = E_i$, to solve for v_f: $\frac{1}{2}mv_f^2 = mgL\sin(\theta) \Rightarrow v_f = \sqrt{2gL\sin(\theta)}$

4. Calculate the numerical result: $v_f = \sqrt{2(9.81\text{ m/s}^2)(2.1\text{ m})\sin(32°)} = 4.7\text{ m/s}$

Insight We already worked out what Δy should be in Example 8–1.

Example 8–4 Spring Loaded At a party, a 0.63-kg ball is going to be shot vertically upward using a spring-loaded mechanism. The spring has a force constant of 188 N/m and is initially compressed 45 cm from equilibrium. **(a)** How high will the ball go above the initial position, and **(b)** what will its speed be at half of this height? (Assume the ball leaves contact with the spring when the spring reaches its usual equilibrium position.)

Picture the Problem The sketch on the left shows the ball loaded onto the spring, and the sketch on the right shows it after launch.

Strategy To determine the height of the ball we can use energy conservation, since only conservative forces are doing work here. Set U_{grav} = 0 in the initial position of the ball.

Solution

Part (a)

1. Write the total energy for the initial situation where it is all in the spring: $E_i = \tfrac{1}{2}kx^2$

2. Write the total energy for the final situation where it is all in the position of the ball: $E_f = mgh$

3. Set these equal and solve for h: $mgh = \frac{1}{2}kx^2 \Rightarrow h = \frac{kx^2}{2mg}$

4. Calculate the numerical result for h: $h = \frac{(188\text{ N/m})(0.45\text{ m})^2}{2(0.63\text{ kg})(9.81\text{ m/s}^2)} = 3.1\text{ m}$

Part (b)

5. Write the total energy for system when the ball is at half the calculated height in part (a): $E_f = \tfrac{1}{2}mv^2 + \tfrac{1}{2}mgh$

6. Set this expression equal to E_i and solve for v: $\tfrac{1}{2}kx^2 = \tfrac{1}{2}mv^2 + \tfrac{1}{2}mgh \therefore v = \left(\frac{kx^2}{m} - gh\right)^{1/2}$.

$$v = \left(\frac{(188\text{ N/m})(0.45\text{ m})^2}{(0.63\text{ kg})} - (9.81\text{ m/s}^2)(3.08\text{ m})\right)^{1/2}$$

$$= 5.5\text{ m/s}$$

Insight This problem is a good example of one that would have been very difficult to solve just using Newton's second law. It demonstrates why the concepts of work and energy are useful.

Practice Quiz

5. If mechanical energy is conserved within a system,

   (a) all objects in the system move at constant speed

   (b) there is no change in the potential energy of the system

   (c) only gravitational forces act within the system

   (d) all the energy is in the form of potential energy

   (e) none of the above

6. Which of the following statements is more accurate?

   (a) Mechanical energy is always conserved.

   (b) Mechanical energy is conserved only when potential energy is transformed into kinetic energy.

   (c) Mechanical energy is conserved when no nonconservative forces are applied.

   (d) Mechanical energy is conserved when no work is done by nonconservative forces.

   (e) Mechanical energy is never conserved.

7. A horizontal spring of force constant k = 100 N/m is compressed by 5.6 cm and placed against a 0.25-kg ball (that rests on a frictionless, horizontal surface) then released. How fast will the ball be moving if it loses contact with the spring when the spring passes its equilibrium point?

   (a) 1.3 m/s **(b)** 1.1 m/s **(c)** 1.4 m/s **(d)** 140 m/s **(e)** 2.5 m/s

8–4 Work Done by Nonconservative Forces

In the previous section we discussed how mechanical energy is conserved when only conservative forces do work. It follows that when work is done by nonconservative forces mechanical energy is not conserved. Instead, the work done by nonconservative forces goes into changing the mechanical energy by an amount equal to the work done:

$$W_{nc} = \Delta E$$

If W_{nc} is positive then we say that energy is being added to the system. If W_{nc} is negative, then energy is being lost, or dissipated, from the system.

Exercise 8–5 The Luge In the 1998 Winter Olympic games in Nagano, Japan, Georg Hackl of Germany won the gold medal in the luge. The total course length was 1326 m with a vertical drop of 114 m. If Georg made a starting leap that gave him an initial speed of 1.00 m/s, and the coefficient of kinetic friction was 0.0222, what was his speed at the finish line?

Solution: The luge track is a winding and uneven surface, so a straight inclined plane would not be an accurate picture. However, on average, the results work out to be the same as you would get on a straight inclined plane. Thus, we work the problem as if it were a straight inclined path with the understanding that intermediate results can be considered to be only average values. Try to sketch a diagram for the luge.

Given: $L = 1326$ m, $\Delta y = -114$ m, $v_i = 1.00$ m/s, $\mu_k = 0.0222$; **Find**: v_f

In this problem we cannot apply the conservation of mechanical energy because work is being done by kinetic friction (a nonconservative force), and the amount of work done by friction goes into a change in the total mechanical energy. The work done by friction is

$$W_f = -f_k L = -\mu_k NL = -\mu_k mg\cos(\alpha)L$$

where the angle is $\alpha = \sin^{-1}(|\Delta y|/L) = 4.932°$. The minus sign is from cos(180°).

If we take the potential energy to be zero at the bottom of the track, the change in mechanical energy is

$$\Delta E = \Delta K + \Delta U = \tfrac{1}{2}mv_f^2 - \tfrac{1}{2}mv_i^2 + mg\Delta y$$

Putting it all together, we have

$$-\mu_k mgL\cos(\alpha) = \tfrac{1}{2}mv_f^2 - \tfrac{1}{2}mv_i^2 + mg\Delta y$$

Solving for v_f gives

$$v_f = \left[v_i^2 - 2g(\Delta y) - 2\mu_k gL\cos\alpha\right]^{1/2}$$

$$= \left[(1.00 \text{ m/s})^2 - 2(9.81 \text{ m/s}^2)(-114 \text{ m}) - 2(0.0222)(9.81 \text{ m/s}^2)(1326 \text{ m})\cos(4.932°)\right]^{1/2} = 40.8 \text{ m/s}$$

In this problem every quantity requires knowledge of Georg's mass, which was not given. Precisely because every term required it we were able to cancel it in the equation. Don't necessarily think you're stuck if you don't know something that appears to be required; it may cancel out in the long run.

Example 8–6 Assembly Line A conveyer belt in a factory carries a part down an assembly line to be checked by different workers. At the end of the line the part is boxed and has a total mass of 2.35 kg. The belt then slides it down a 20.0° ramp to the shippers waiting 3.00 meters below. If the coefficient of kinetic friction between the box and the ramp is 0.377, and the box reaches the shippers with a speed of 0.100 m/s, with what speed did the conveyer belt carry the part?

Picture the Problem The diagram shows the box moving along the conveyer belt toward the ramp.

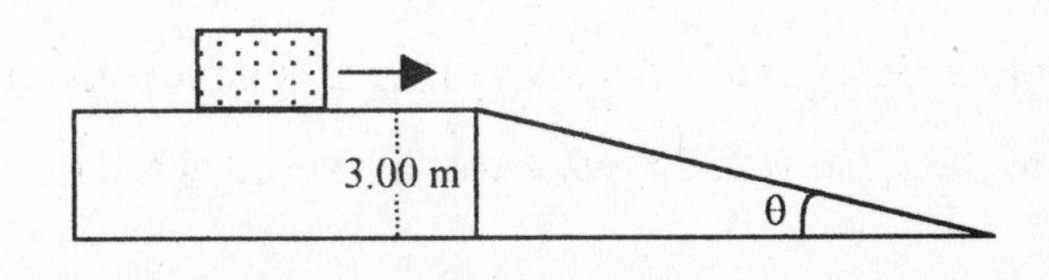

Strategy As in the above example, kinetic friction does work here and changes the mechanical energy. To solve this problem, we write out the work done by friction and set it equal to ΔE.

Solution

1. The work done by friction is: $W_f = -f_k L = -\mu_k mg\cos(\theta)L$

2. The length of the ramp is (h = height): $L = h/\sin(\theta)$

3. Write the work as: $W_f = -f_k L = -\mu_k mgh\dfrac{\cos(\theta)}{\sin(\theta)} = -\dfrac{\mu_k mgh}{\tan\theta}$

4. Taking U = 0 at the bottom of the ramp, calculate the change in mechanical energy: $\Delta E = \Delta K + \Delta U = \frac{1}{2}mv_f^2 - \frac{1}{2}mv_i^2 - mgh$

5. Set the frictional work equal to ΔE: $-\dfrac{\mu_k mgh}{\tan\theta} = \frac{1}{2}mv_f^2 - \frac{1}{2}mv_i^2 - mgh$

6. Solve for v_i:

$$v_i = \left[v_f^2 - 2gh + \frac{2\mu_k gh}{\tan\theta}\right]^{1/2} = \left[v_f^2 - 2gh\left(1 - \frac{\mu_k}{\tan\theta}\right)\right]^{1/2}$$

$$= \left[(0.100\ \text{m/s})^2 - 2(9.81\ \text{m/s}^2)(3.00\ \text{m})\left(1 - \frac{0.377}{\tan(20.0^\circ)}\right)\right]^{1/2}$$

$$= 1.46\ \text{m/s}$$

Insight Having worked through this example and the previous one, you should begin to become comfortable with applying the work done by nonconservative forces, especially friction, to work and energy problems.

Practice Quiz

8. Work done by a nonconservative force

(a) can change only the kinetic energy

(b) can change only the potential energy

(c) can change either the kinetic or potential energy, or both

(d) Work cannot be done by a nonconservative force.

(e) none of the above

9. In which of the following cases does work done by a nonconservative force cause a change in potential energy?

   (a) A spring launches a ball into the air.

   (b) A person throws a ball into the air.

   (c) A ball falls vertically downward with negligible air resistance.

   (d) A sliding block on a horizontal surface comes to rest.

   (e) none of the above

10. A block given an initial speed of 5.7 m/s comes to rest due to friction after sliding 2.4 m upward along a surface that is inclined at 30 degrees. What is the coefficient of kinetic friction?

    (a) 0.22 **(b)** 0.42 **(c)** 0.58 **(d)** 0.50 **(e)** 0.87

8–5 Potential Energy Curves and Equipotentials

A plot of the potential energy function of a system, known as a potential energy curve, can provide a lot of useful information about the behavior of the system. Consider the following potential energy curve for a case in which the mechanical energy is conserved.

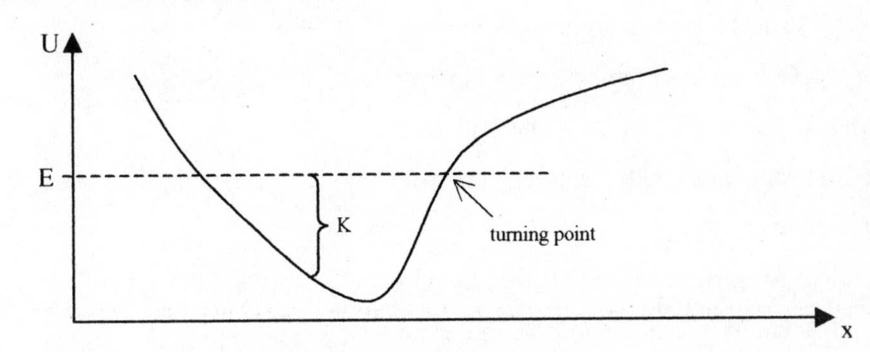

The dashed horizontal line represents the value of the mechanical energy of the system. The difference between the potential energy curve and the mechanical energy must equal the kinetic energy, K. Where the horizontal line intersects the curve is where the energy of the system is completely in the form of potential energy, so that K = 0; the motion of the system must stop here. These points are called **turning points** because objects within the system stop moving in one direction and begin moving in the opposite direction.

Within the *potential well* between the two turning points the motion will be oscillatory, going back and forth between the two turning points. Regions of the curve that are higher than E would correspond to negative kinetic energy, which is not possible. These are the *forbidden regions* of the motion. The above plot is of a one-dimensional potential energy function. In many situations U is a function of three-dimensional space. In these cases it can be useful to plot just the lines of constant potential energy — the **equipotentials**. This type of analysis is used widely in chemistry.

Example 8–7 Potential Energy Curves A ball, starting at point A, moves under the influence of the potential energy curve shown. Describe the final motion of the ball.

Picture the Problem The diagram shows the potential energy curve labeled as U, the mechanical energy E and five specific points labeled $a \rightarrow e$.

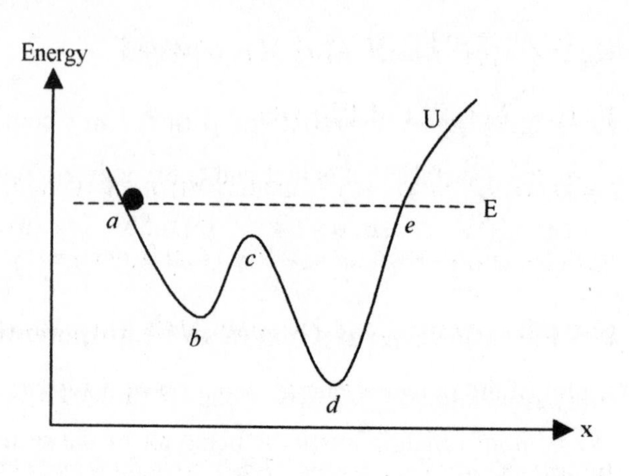

Strategy/Solution

At point a U = E, so the ball is at rest. It will gain kinetic energy as it moves toward b so its speed increases from a to b. As the ball moves from b to c it loses kinetic energy as it gains potential energy, so it slows down; however, it does not lose all its kinetic energy so it does not stop at c. In moving from c to d it speeds up again, this moving faster at d than it did at b. On its way to e the ball gets slower and slower eventually coming to rest, instantaneously, at e.

After coming to rest at e, the ball's motion reverses, and it has the same speeds at the same points along its path. The ball eventually comes back to rest at point a. Since there is no work done to change the mechanical energy, this cycle continues with the ball going back-and-forth between the turning points a and e.

Insight What would happen it there were a small amount of friction consuming energy from the ball's motion?

Practice Quiz

11. When an object reaches a turning point in its motion

(a) it must stop and remain at rest

(b) it must continue moving past this point

(c) all of its mechanical energy is in the form of kinetic energy

(d) none of its mechanical energy is in the form of kinetic energy

(e) none of the above

Reference Tools and Resources

I. Key Terms and Phrases

conservative force any force for which the work done is independent of path

nonconservative force any force that is not a conservative force

potential energy a representation of the extent to which work is stored in the configuration of a system

mechanical energy the sum of the kinetic and potential energies in a system

turning point the position on a potential energy curve at which an object will stop and reverse direction

equipotential a region in which every point has an equal value of potential energy

II. Important Equations

Name/Topic	Equation	Explanation
potential energy	$\Delta U = -W_c$	The definition of potential energy.
gravitational potential energy	$U = mgy$	The gravitational potential energy, assuming $U = 0$ at $y = 0$.
spring potential energy	$U = \frac{1}{2}kx^2$	The spring potential energy assuming $U = 0$ at $x = 0$.
mechanical energy	$E = K + U$	The definition of the mechanical energy.
work by nonconservative forces	$W_{nc} = \Delta E$	The work done by nonconservative forces changes the mechanical energy.

III. Know Your Units

Quantity	Dimension	SI Unit
potential energy	$[M]\cdot[L^2]/[T^2]$	J

Practice Problems

1. A block is falling at a speed of 18 m/s and is 5 meters above a spring. The spring constant is 4000 N/m. To the nearest tenth of a cm how far does the spring compress?
2. In problem 1 to the nearest hundredth of a meter to what height will the block rise after it hits and leaves the spring?
3. In the figure $h_1 = 2.3$ and $h_2 = 2.5$. If the floor is the zero of potential energy, to the nearest tenth of a joule what is the potential energy of a 1-kg test mass placed at point A?

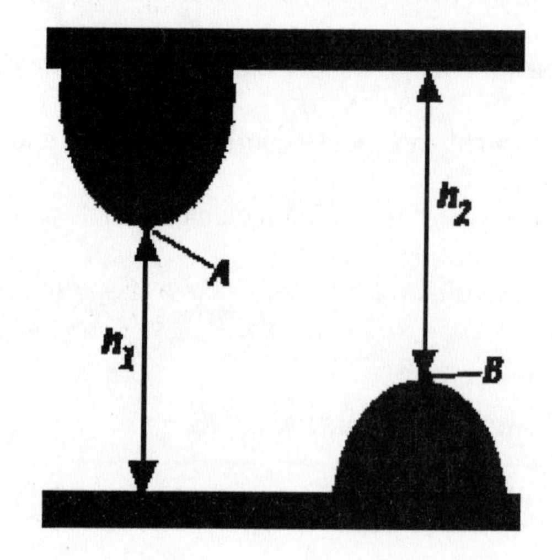

4. In problem 3, taking the zero of potential energy at the ceiling, what is the potential energy of the test mass at point B?
5. In problem 3, using the floor as the zero of potential energy, to the nearest tenth of a joule what is the difference in potential energy between points A and B?
6. In problem 3, using the ceiling as the zero of potential energy, to the nearest tenth of a joule what is the difference in potential energy between points A and B?
7. A 1-kg ball starting at h = 8.3 meters slides down a smooth surface and encounters a rough surface and is brought to rest at B a distance 29.8 meters away. To the nearest joule what is the work done by friction?

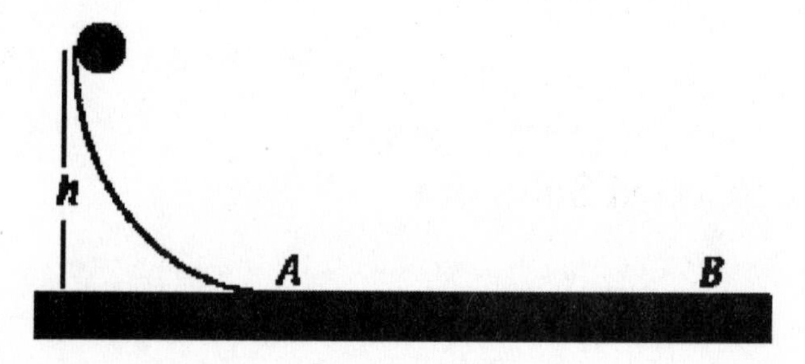

8. In problem 7, to two decimal places, what is the coefficient of friction?

9. The potential energy of a particle traveling along the x axis is given by PE = $x^2 + 4.5x$. If the total energy is 25 joules, to the nearest hundredth of a meter where is the positive turning point?

10. In problem 9 what is the negative turning point?

Puzzle

ENERGY DROP

Trapeze artist Lea stands at point A in the picture (marked by the square). Under her knees she attaches a taut 10-meter rope, which is tied to the ceiling at the point marked with the dot. She lets herself go. At the bottom of the quarter circle she picks up her twin sister, who is standing at point B. Together they continue on the 10-meter-radius circle until they come to a stop at the platform marked C. Between platforms B and C they rise vertically 2.5 meters. Can you account for all the energy transformations in this stunt?

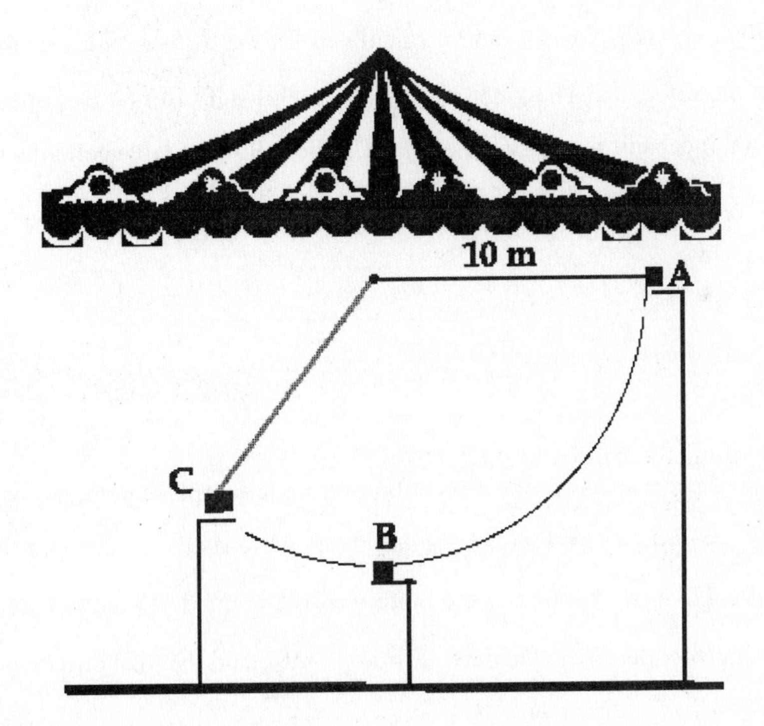

Selected Solutions

1. Path #1 has five segments, the work done by each is calculated as follows:

$$W_1 = -mg\left(\Delta y_1 + \Delta y_2 + \Delta y_3 + \Delta y_4 + \Delta y_5\right)$$

$$W_1 = -(2.6\text{ kg})\left(9.81\text{ m/s}^2\right)\left[4.0\text{ m} + 0 - 1.0\text{ m} + 0 - 1.0\text{ m}\right] = \boxed{-51\text{ J}}$$

Path #2 has three segments:

$$W_2 = -(2.6\text{ kg})\left(9.81\text{ m/s}^2\right)\left[0 + 2.0\text{ m} + 0\right] = \boxed{-51\text{ J}}$$

Path #3 has three segments:

$$W_3 = -(2.6\text{ kg})\left(9.81\text{ m/s}^2\right)\left[-1.0\text{ m} + 0 + 3.0\text{ m}\right] = \boxed{-51\text{ J}}$$

7. The potential energy of the spring is given by $U = \frac{1}{2}ky^2$, and we seek the compression, $y = \sqrt{2U/k}$.

We can determine k from the given information:

$$k = \frac{2U}{y^2} = \frac{2(0.0025\text{ J})}{(0.0050\text{ m})^2} = 200\text{ N/m}$$

Given k, we can now calculate y:

$$y = \sqrt{\frac{2U}{k}} = \sqrt{\frac{2(0.0084\text{ J})}{200\text{ N/m}}} = \boxed{0.92\text{ cm}}$$

15. The gravitational potential energy can be calculated for the different heights using $U = mgy$, taking the ground as the reference. The kinetic energy at each height can be determined using $\Delta K = -\Delta U$. Finally, the total mechanical energy is $E = K + U$. The following table contains the results:

y (m)	4.0	3.0	2.0	1.0	0
U (J)	8.2	6.2	4.1	2.1	0
K (J)	0	2.1	4.1	6.2	8.2
E (J)	8.2	8.2	8.2	8.2	8.2

29. Since the force of resistance is in the direction opposite the displacement, the work done by this force is given by $W_{nc} = -Fd$, where F is the magnitude of the resistive force. The potential energy of the rock is given by $U = mgh$ relative to the bottom of the pond. The kinetic energy of the rock can be calculated from the work-energy theorem, $K = W_{nc} - \Delta U$, and the total energy is $E = U + K$.

For d = 0:

$$W_{nc} = -4.10\text{ N }(0) = \boxed{0}$$

$$U = (1.75\text{ kg})\left(9.81\text{ m/s}^2\right)(1.00) = \boxed{17.2\text{ J}}$$

$$K = 0 - 0 = \boxed{0}$$

$$E = 17.2\text{ J} + 0 = \boxed{17.2\text{ J}}$$

For d = 0.500 m:

$$W_{nc} = -4.10 \text{ N } (0.500 \text{ m}) = \boxed{-2.05 \text{ J}}$$

$$U = (1.75 \text{ kg})(9.81 \text{ m/s}^2)(0.500) = \boxed{8.58 \text{ J}}$$

$$K = -2.05 \text{ J} - (8.58 \text{ J} - 17.17 \text{ J}) = \boxed{6.54 \text{ J}}$$

$$E = 8.58 \text{ J} + 6.54 \text{ J} = \boxed{15.1 \text{ J}}$$

For d = 1.00 m:

$$W_{nc} = -4.10 \text{ N } (1.00 \text{ m}) = \boxed{-4.10 \text{ J}}$$

$$U = (1.75 \text{ kg})(9.81 \text{ m/s}^2)(0) = \boxed{0}$$

$$K = -4.10 \text{ J} - (0 - 17.2 \text{ J}) = \boxed{13.1 \text{ J}}$$

$$E = 0 + 13.1 \text{ J} = \boxed{13.1 \text{ J}}$$

49. Coming off the slide, the person becomes a projectile launched horizontally. The landing site of such a projectile is determined by $x = v_x t$, with $t = \sqrt{2y/g}$. Therefore,

$$x = v\sqrt{\frac{2y}{g}} \quad \Rightarrow \quad v^2 = \frac{x^2 g}{2y}$$

We can relate this velocity to the height of the slide using energy conservation:

$$mgh = \frac{1}{2}mv^2 \quad \Rightarrow \quad h = \frac{v^2}{2g}$$

Substituting in the result for v^2 gives us

$$h = \frac{1}{2g}\left(\frac{x^2 g}{2y}\right) = \frac{x^2}{4y} = \frac{(2.50 \text{ m})^2}{4(1.50 \text{ m})} = \boxed{1.04 \text{ m}}$$

Answers to Practice Quiz

1. (a) **2**. (c) **3**. (c) **4**. (e) **5**. (e) **6**. (d) **7**. (b) **8**. (c) **9**. (b) **10**. (a) **11**. (d)

Answers to Practice Problems

1. 45.9 cm 2. 21.47 m

3. 22.6 J 4. −24.5 J

5. 12.8 J 6. 12.8 J

7. −81 J 8. 0.28

9. 3.23 m 10. −7.73 m

CHAPTER 9

LINEAR MOMENTUM AND COLLISIONS

Chapter Objectives

After studying this chapter, you should:

1. know the definition of *linear momentum* and how it relates to force.
2. know the meaning of *impulse* and how it relates to linear momentum.
3. be able to use linear momentum to analyze elastic and inelastic collisions.
4. be able to determine the location of the center of mass of a system.
5. be able to apply Newton's laws to a system of particles.
6. understand the basic principles behind rocket propulsion.

Warm-Ups

1. The left ball in the following diagram is chasing after the right ball. The speed of the left ball is 3 m/s, and the speed of the right ball is 2 m/s. After the collision the left ball rebounds at 2 m/s, and the right ball speeds away at 3 m/s. What was the change in velocity experienced by the left ball during the collision? What was the change in velocity experienced by the right ball during the collision? If the mass of the left ball is the same as the mass of the right ball, was momentum conserved during this collision?

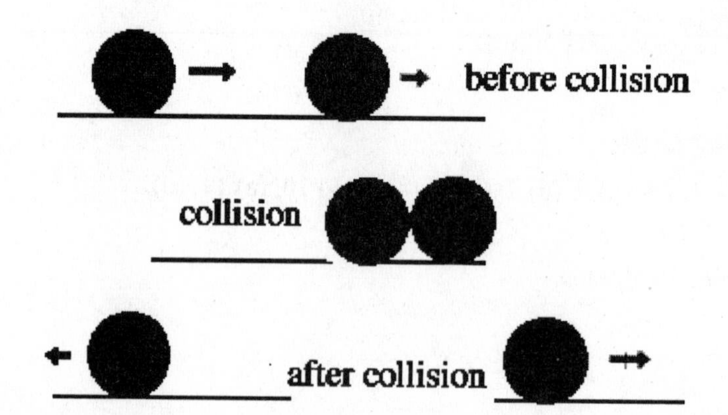

2. Estimate the force on a 145-gram baseball if it hits the bat at 50 miles per hour, stays in contact with the bat for 0.1 second during which time it reverses direction and leaves the bat at 75 miles per hour. Express the result in newtons and in pounds.
3. Suppose the shuttle were launched from a launch pad on the Moon. What changes would we observe? Suppose it were launched from a launch pad in the vacuum of space? Would it work? What changes would we observe?
4. Ball A is propelled forward and collides with ball B which is initially at rest. After the collision, the balls' trajectories, A′ and B′, are as shown in the diagram. How do the masses of the two balls compare? Briefly explain your answer.

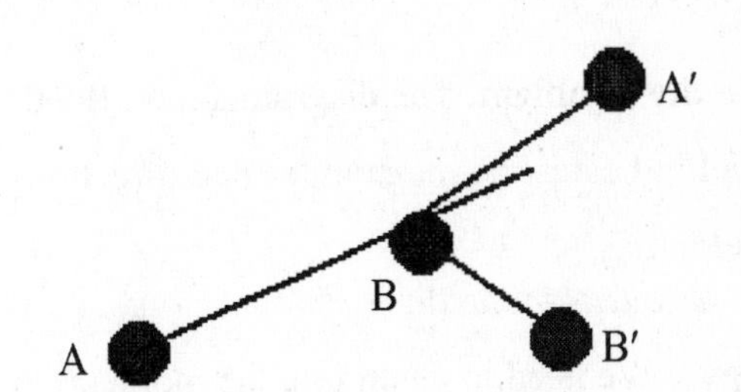

Chapter Review

In Chapters 7 and 8 we saw that defining concepts beyond just force, velocity, acceleration, and displacement made analyzing some situations much easier, especially the conservation of energy. In this chapter we do more of the same by introducing the concept of linear momentum. In a way similar to work and energy, the idea of linear momentum often provides a more direct path to understanding physical interactions than working with forces only. Furthermore, as with energy, there is a conservation principle for linear momentum that is of tremendous value in physics.

9–1 Linear Momentum

The **linear momentum** of an object is defined as the product of its mass and its velocity

$$\mathbf{p} = m\mathbf{v}$$

where **p** is the symbol for linear momentum. Often, we refer to just **p** as the momentum and leave off the word linear. Notice that momentum is a vector quantity whose direction is that of the velocity **v** and whose units are just the mass unit times velocity units, kg·m/s. In one sense you can think of linear momentum as a measure of the effect the motion of an object has when it interacts with other objects (you'll see this more clearly in the discussion of collisions). In another sense, linear momentum can be viewed as an alternative measure of an object's inertia, i.e., its tendency to maintain a constant state of motion (you'll see this more clearly in the discussion of momentum conservation).

For a system of particles we often speak of the *total momentum* of the system. This total momentum is the vector sum of the momenta of every particle in the system:

$$\mathbf{p}_{total} = \mathbf{p}_1 + \mathbf{p}_2 + \mathbf{p}_3 + \cdots$$

We shall see later that the total momentum of a system tells us a lot about how a system behaves as a whole.

Example 9–1 The Total Momentum Three balls of mass 2.0 kg, 3.0 kg, and 4.0 kg move with the velocities shown below. Determine the magnitude and direction of the total momentum of the balls.

Picture the Problem The diagram shows the three balls and indicates the magnitudes and directions of their velocities.

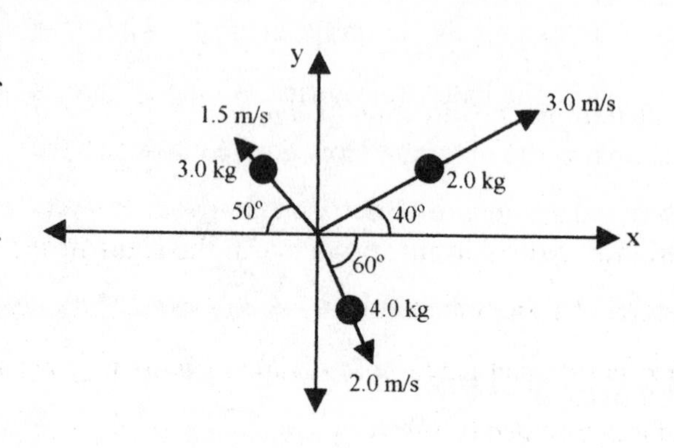

Strategy We need to determine the momentum of each particle, then form a vector sum to get the total momentum. Let's take m_1 = 2.0 kg, v_1 = 3.0 m/s, m_2 = 3.0 kg, v_2 = 1.5 m/s, m_3 = 4.0 kg, and v_3 = 2.0 m/s.

Solution

1. Determine the components of $\mathbf{p}_1$:

$$p_{1x} = m_1 v_1 \cos(40^\circ)$$
$$= (2.0\,\text{kg})(3.0\,\text{m/s})\cos(40^\circ) = 4.60\,\text{kg}\cdot\text{m/s}$$

$$p_{1y} = m_1 v_1 \sin(40^\circ)$$
$$= (2.0\,\text{kg})(3.0\,\text{m/s})\sin(40^\circ) = 3.86\,\text{kg}\cdot\text{m/s}$$

2. Determine the components of $\mathbf{p}_2$:

$$p_{2x} = -m_2 v_2 \cos(50^\circ)$$
$$= -(3.0\,\text{kg})(1.5\,\text{m/s})\cos(50^\circ) = -2.89\,\text{kg}\cdot\text{m/s}$$

$$p_{2y} = m_2 v_2 \sin(50^\circ)$$
$$= (3.0\,\text{kg})(1.5\,\text{m/s})\sin(50^\circ) = 3.45\,\text{kg}\cdot\text{m/s}$$

3. Determine the components of $\mathbf{p}_3$:

$$p_{3x} = m_3 v_3 \cos(-60^\circ)$$
$$= (4.0\,\text{kg})(2.0\,\text{m/s})\cos(-60^\circ) = 4.00\,\text{kg}\cdot\text{m/s}$$

$$p_{3y} = m_3 v_3 \sin(-60^\circ)$$
$$= (4.0\,\text{kg})(2.0\,\text{m/s})\sin(-60^\circ) = -6.93\,\text{kg}\cdot\text{m/s}$$

4. Determine the components of $\mathbf{p}_{total}$:

$$p_{total,x} = p_{1x} + p_{2x} + p_{3x}$$
$$= (4.60 - 2.89 + 4.00)\ \text{kg}\cdot\text{m/s} = 5.71\ \text{kg}\cdot\text{m/s}$$

$$p_{total,y} = p_{1y} + p_{2y} + p_{3y}$$
$$= (3.86 + 3.45 - 6.93)\ \text{kg}\cdot\text{m/s} = 0.380\ \text{kg}\cdot\text{m/s}$$

5. Determine the magnitude of $\mathbf{p}_{total}$:

$$p_{total} = \left[p_{total,x}^2 + p_{total,y}^2\right]^{1/2}$$
$$\left[(5.71\ \text{kg}\cdot\text{m/s})^2 + (0.380\ \text{kg}\cdot\text{m/s})^2\right]^{1/2} = 5.7\ \text{kg}\cdot\text{m/s}$$

6. Determine the direction of $\mathbf{p}_{total}$:

$$\theta_{total} = \tan^{-1}\left(\frac{p_{total,y}}{p_{total,x}}\right) = \tan^{-1}\left(\frac{0.380\ \frac{\text{kg}\cdot\text{m}}{\text{s}}}{5.71\ \frac{\text{kg}\cdot\text{m}}{\text{s}}}\right) = 3.8^\circ$$

Insights Note that for $\mathbf{p}_2$ we put in the signs of the components from our knowledge of their directions instead of using the angle with the +x axis. This way of doing it is common, so be sure you understand it. Also notice that since both components of $\mathbf{p}_{total}$ are positive, the 3.8° is the angle from the +x axis, and nothing else needs to be said.

Practice Quiz

1. If two objects have the same velocity but one has twice the mass of the other, the object with larger mass

   (a) has half the momentum of the other

   (b) has twice the momentum of the other

   (c) has the same momentum as the other

   (d) must move slower than the other

   (e) none of the above

2. If two objects have the same mass and speed but move in opposite directions, then

   (a) they have momenta of equal magnitude and opposite directions

   (b) they have equal momenta

   (c) each has zero momentum

   (d) one must move faster than the other

   (e) none of the above

3. A 3.2-kg object has a velocity of 1.8 m/s in the x direction. What is its linear momentum?

   (a) $5.0\ \text{kg}\cdot\text{m/s}\ \hat{\mathbf{x}}$ **(b)** $1.8\ \text{kg}\cdot\text{m/s}\ \hat{\mathbf{x}}$ **(c)** $0.56\ \text{kg}\cdot\text{m/s}\ \hat{\mathbf{x}}$ **(d)** $5.8\ \text{kg}\cdot\text{m/s}\ \hat{\mathbf{x}}$ **(e)** $1.4\ \text{kg}\cdot\text{m/s}\ \hat{\mathbf{x}}$

9–2 Momentum and Newton's Second Law

The form of Newton's second law that we have been using until now, $\mathbf{F}_{\text{net}} = m\mathbf{a}$, applies only to circumstances in which the mass remains constant; however, in the most general cases, as with rockets, the mass may change during the motion. The most general form of Newton's second law is expressed in terms of momentum:

$$\mathbf{F}_{\text{net}} = \frac{\Delta \mathbf{p}}{\Delta t}$$

Thus, forces equals the rate at which momentum changes whether the change is in the mass, the velocity, or both. For cases when mass is constant, this equation reduces to $\mathbf{F}_{\text{net}} = m\mathbf{a}$.

Example 9–2 Newton's Second Law A 0.25-kg object moves due east at 2.1 m/s. What force is needed to cause it to move due north at 3.6 m/s in 1.52 s?

Solution Even though this is a case of constant mass, let's work it in terms of momentum to illustrate that approach.

Given: $m = 0.25$ kg, $\mathbf{v}_i = 2.1\text{ m/s}\,\hat{\mathbf{x}}$, $\mathbf{v}_f = 3.6\text{ m/s}\,\hat{\mathbf{y}}$, $t = 1.52$ s; **Find**: **F**

The initial and final momenta of the object are

$$\mathbf{p}_i = m\mathbf{v}_i = 0.25\text{ kg}(2.1\text{ m/s})\,\hat{\mathbf{x}} = 0.525\text{ kg}\cdot\text{m/s}\,\hat{\mathbf{x}}$$

$$\mathbf{p}_f = m\mathbf{v}_f = 0.25\text{ kg}(3.6\text{ m/s})\,\hat{\mathbf{y}} = 0.900\text{ kg}\cdot\text{m/s}\,\hat{\mathbf{y}}$$

The change in momentum, therefore, is

$$\Delta\mathbf{p} = \mathbf{p}_f - \mathbf{p}_i = \left(-0.525\,\hat{\mathbf{x}} + 0.900\,\hat{\mathbf{y}}\right)\text{kg}\cdot\text{m/s}$$

The force, then, is given by

$$\mathbf{F} = \frac{\Delta\mathbf{p}}{t} = \frac{\left(-0.525\,\hat{\mathbf{x}} + 0.900\,\hat{\mathbf{y}}\right)\text{kg}\cdot\text{m/s}}{1.52\text{ s}} = \left(-0.35\,\hat{\mathbf{x}} + 0.59\,\hat{\mathbf{y}}\right)\text{N}$$

Insight If you are curious, work this out as $\mathbf{F} = m\mathbf{a}$, and check that you get the same result.

Practice Quiz

4. When does the equation $\mathbf{F} = m\mathbf{a}$ not accurately describe the dynamics of an object?

   (a) when gravity is not present

   (b) when more than one force acts on the object

   (c) when kinetic friction does work on the object

   (d) when the mass of the object is constant

   (e) none of the above

5. If the linear momentum of a 1.6-kg object is decreasing at a rate of 5.0 $\frac{\text{kg·m/s}}{\text{s}}$, what is the magnitude of the force on the object?

(a) 8.0 N **(b)** 3.4 N **(c)** 5.0 N **(d)** 6.6 N **(e)** 3.1 N

9–3 Impulse

As mentioned previously, the concept of momentum is important when two objects interact. In this chapter, the interaction we focus on is called a **collision**. A collision occurs when the forces of interaction between two objects are large for a finite period of time. The average force applied to an object times the amount of time this force is applied is called the **impulse, I**

$$\mathbf{I} = \mathbf{F}_{av}\,\Delta t$$

The SI unit of impulse is the N·s, which has no special name. The same impulse can be delivered by a weak force acting for a long period of time or a strong force acting for a short period of time. The most common usage of impulse is for the latter case. The concept of impulse is closely related to momentum by what is often called the *impulse-momentum theorem*

$$\mathbf{I} = \Delta\mathbf{p}$$

The above expression is really just a restatement of Newton's second law in a form that is convenient to describe interactions, like collisions, for which it is difficult to know precise values of the forces involved.

Example 9–3 Ricochet The velocity of a rock of mass 0.24 kg moving with a speed of 3.33 m/s makes a 60° angle with the normal to a brick wall. The rock is in contact with the wall for only 0.032 s. If the velocity of the rock makes an angle of 40° with the normal to the wall after it strikes and has a magnitude of 2.68 m/s, **(a)** what impulse does the wall apply to the rock, and **(b)** what average force causes this impulse?

Picture the Problem The diagram shows the initial and final momenta of the rock as it bounces off the wall.

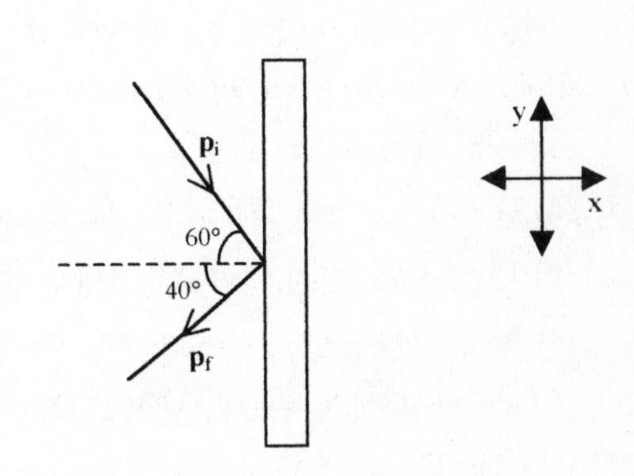

Strategy To solve for the impulse, we can find the change in momentum that results from the bounce. Once we the know the impulse, we'll use it to get the average force.

Solution

Part (a)

1. Determine the components of the initial and final momenta of the rock:

$$p_{ix} = mv_i \cos(60^\circ) = 0.24\,\text{kg}\,(3.33\,\text{m/s})\cos(60^\circ) = 0.3996\,\text{kg}\cdot\text{m/s}$$

$$p_{iy} = -mv_i \sin(60^\circ) = -0.24\,\text{kg}\,(3.33\,\text{m/s})\sin(60^\circ) = -0.6921\,\text{kg}\cdot\text{m/s}$$

$$p_{fx} = -mv_f \cos(40^\circ) = -0.24\,\text{kg}(2.68\,\text{m/s})\cos(40^\circ) = -0.4927\,\text{kg}\cdot\text{m/s}$$

$$p_{fy} = -mv_f \sin(40^\circ) = -0.24\,\text{kg}\,(2.68\,\text{m/s})\sin(40^\circ) = -0.4134\,\text{kg}\cdot\text{m/s}$$

2. Determine the impulse as the change in momentum:

$$\mathbf{I} = \mathbf{p}_f - \mathbf{p}_i = (p_{fx} - p_{ix})\hat{\mathbf{x}} + (p_{fy} - p_{iy})\hat{\mathbf{y}} = [(-0.4927 - 0.3996)\hat{\mathbf{x}} + (-0.4134 + 0.6921)\hat{\mathbf{y}}]\,\text{kg}\cdot\text{m/s} = (-0.892\,\hat{\mathbf{x}} + 0.279\,\hat{\mathbf{y}})\,\text{N}\cdot\text{s}$$

Part (b)

3. Use the definition of impulse to obtain an expression for the average force:

$$\mathbf{I} = \mathbf{F}_{av} t \Rightarrow \mathbf{F}_{av} = \mathbf{I}/t$$

4. Calculate the numerical value of the force:

$$\mathbf{F}_{av} = \frac{(-0.892\,\hat{\mathbf{x}} + 0.279\,\hat{\mathbf{y}})\,\text{N}\cdot\text{s}}{0.032\,\text{s}} = (-28\,\hat{\mathbf{x}} + 8.7\,\hat{\mathbf{y}})\,\text{N}$$

Insight Be sure that the signs and relative magnitudes of the components of **I** and $\mathbf{F}_{av}$ make sense to you.

Practice Quiz

6. During one trial, a force F is applied to a ball for an amount of time t. During a second trial a force of 3F is applied for a time of t/2. Which of the following is true concerning the magnitude of the change in momentum of the ball?

   (a) There's a greater change for the second trial.

   (b) There's a smaller change for the second trial.

   (c) We get the same nonzero momentum change in each trial.

   (d) The momentum doesn't change in either trial.

   (e) none of the above

7. An object of mass 2.8 kg moves at 1.1 m/s in the +x direction. If an impulse of $1.3\ \text{N}\cdot\text{s}\,\hat{\mathbf{x}}$ is applied, what is its final momentum?

(a) 3.1 kg·m/s **(b)** 4.0 kg·m/s **(c)** 5.2 kg·m/s **(d)** 4.4 kg·m/s **(e)** 4.2 kg·m/s

9–4 Conservation of Linear Momentum

In the previous section we discussed the fact that an impulse applied to an object causes a change in the object's momentum. Consequently, if there is no impulse, then there is no change in momentum. The net impulse on an object will be zero when the net force on that object is zero. This result leads to the following statement:

When the net force on an object is zero, its linear momentum is conserved.

The situation becomes a little more interesting for a system of objects such as the billions of air molecules in an enclosed container such as your bedroom. In this latter case you have billions of particles interacting with one another by collisions. By the law of action and reaction all these particles apply equal and opposite forces to each other that cancel out when the system is considered as a whole. That is, the sum of all forces internal to the system is zero: $\Sigma \mathbf{F}_{\text{int}} = 0$.

We are left, therefore, with only the external forces applied to a system to make up the overall net force: $\mathbf{F}_{\text{net}} = \Sigma \mathbf{F}_{\text{ext}}$. By applying Newton's second law, we can see that the net impulse on a system equals the change in its total momentum: $\mathbf{p}_{\text{total}}$ (or $\mathbf{p}_{\text{net}}$)

$$\Delta \mathbf{p}_{\text{total}} = \left(\sum \mathbf{F}_{\text{ext}}\right)\Delta t$$

By reasoning similar to that in the single-particle case, we can see from this expression that

if the net external force on a system is zero, then the total linear momentum of the system is conserved.

When the total momentum of a system is conserved we can often apply this result by setting the final total momentum equal to the initial total momentum:

$$\mathbf{p}_{1\text{f}} + \mathbf{p}_{2\text{f}} + \mathbf{p}_{3\text{f}} + \cdots = \mathbf{p}_{1\text{i}} + \mathbf{p}_{2\text{i}} + \mathbf{p}_{3\text{i}} + \cdots$$

Notice that because momentum is a vector quantity it, unlike energy (a scalar), must be conserved in both magnitude and direction.

Practice Quiz

8. Which of the following statements is most accurate?

(a) The linear momentum of a system is always conserved.

(b) The linear momentum of a system is conserved only if no external forces act on the system.

(c) The linear momentum of a system is conserved if the net external force on the system is zero.

(d) The linear momentum of a system is not conserved.

(e) none of the above

9–5 Inelastic Collisions

When studying collisions, we often divide them into two categories according to whether or not the total kinetic energy of the colliding bodies is conserved. If the total kinetic energy is not conserved, the collision is called an **inelastic collision**. In analyzing these types of collisions we usually just apply the conservation of the total linear momentum to the system: $\mathbf{p}_f = \mathbf{p}_i$. A special case of an inelastic collision occurs when the colliding objects stick together and emerge from the collision effectively as one object. This latter case is called a *completely* inelastic collision because in this case the system loses the maximum amount of kinetic energy it can lose while still conserving momentum.

Example 9–4 One-dimensional collision The driver of a 1485-kg car is driving along a street at 25 mi/h. Another driver, coming from directly behind (and not paying attention), is driving a 1520-kg car at 38 mi/hr and hits the slower car. If both drivers slam the brakes at the moment of impact and their fenders catch on each other, with what combined speed do they begin to skid?

Picture the Problem The upper sketch shows the heavier and faster car (m_1) catching up to the lighter car (m_2). The lower sketch shows them stuck together after the collision.

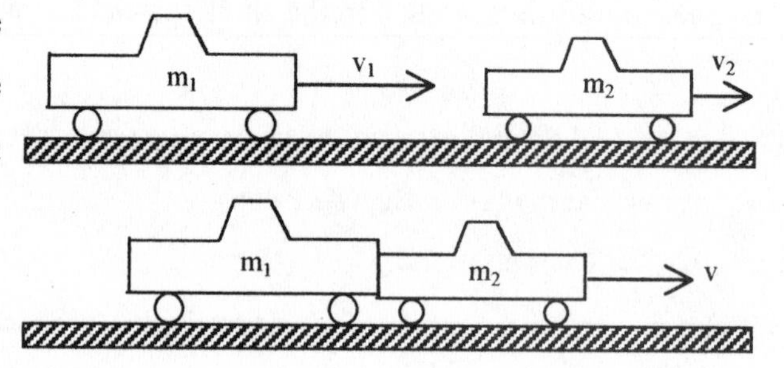

Strategy This is a completely inelastic collision. We apply the conservation of momentum by setting the total momenta before and after the collision equal. Take the direction of motion to be the x direction.

Solution

1. The total momentum before the collision is: $p_{total} = m_1 v_1 + m_2 v_2$

2. The total momentum after the collision is: $p_{total} = (m_1 + m_2)v$

3. Set them equal: $m_1 v_1 + m_2 v_2 = (m_1 + m_2)v \therefore v = \dfrac{m_1 v_1 + m_2 v_2}{m_1 + m_2}$

4. Calculate the numerical result:

$$v = \frac{1520\ \text{kg}(38\ \frac{\text{mi}}{\text{h}}) + 1485\ \text{kg}(25\ \frac{\text{mi}}{\text{h}})}{1520\ \text{kg} + 1485\ \text{kg}}\left(\frac{0.447\ \text{m/s}}{1\ \text{mi/h}}\right)$$

$$= 14\ \text{m/s} \approx 32\ \text{mi/h}$$

Insight Since this was just a one-dimensional problem, we dispensed with full vector notation.

Practice Quiz

9. A cart is rolling along a horizontal floor when a box is dropped vertically into it. After catching the box the cart will move
   (a) more slowly
   (b) more quickly
   (c) at the same speed as before
   (d) in a different direction
   (e) The cart will come to a stop as a result of catching the box.

10. Two objects of equal mass of 1.7 kg move directly toward each other with equal speed of 2.2 m/s. If they stick together after the collision their speed will be
    (a) 0.65 m/s **(b)** 2.2 m/s **(c)** 1.1 m/s **(d)** 3.7 m/s **(e)** 0 m/s

9–6 Elastic Collisions

A second category of collisions considers those for which the total kinetic energy is conserved. These collisions are called **elastic collisions**. Keep in mind that, as with momentum, it is the *total* kinetic energy of the system, not that of any individual particle, that we are considering

$$K_{1f} + K_{2f} + K_{3f} + \cdots = K_{1i} + K_{2i} + K_{3i} + \cdots$$

In everyday life few collisions are precisely elastic. However, many everyday collisions are approximately elastic because such a small fraction of the kinetic energy is lost. Microscopically, perfectly elastic collisions are commonplace. Along with the conservation of kinetic energy, we still must apply the conservation of momentum to these collisions as well.

In one dimension (a head-on collision), we can combine the equations for the conservation of momentum and the conservation of kinetic energy and solve them for the final velocities of the two particles. For particles m_1 and m_2 with m_2 initially at rest and m_1 moving with speed v_0 directly toward m_2, the result is

$$v_{1f} = \left(\frac{m_1 - m_2}{m_1 + m_2}\right) v_0, \qquad v_{2f} = \left(\frac{2m_1}{m_1 + m_2}\right) v_0$$

Example 9–5 Head-on elastic collision A 1.3-kg ball initially moving with a speed of 3.1 m/s strikes a stationary 2.2-kg ball head-on. What are the final velocities of the two balls?

Solution This is precisely the situation just discussed. Let's identify the masses and velocities.

Given: $m_1 = 1.3$ kg, $v_0 = 3.1$ m/s, $m_2 = 2.2$ kg; **Find:** v_{1f}, v_{2f}

Making direct use of the equations listed above we have

$$v_{1f} = \frac{1.3\,\text{kg} - 2.2\,\text{kg}}{1.3\,\text{kg} + 2.2\,\text{kg}}(3.1\,\text{m/s}) = -0.80\,\text{m/s}$$

$$v_{2f} = \frac{2(1.3\text{kg})}{1.3\text{kg} + 2.2\,\text{kg}}(3.1\text{m/s}) = 2.3\,\text{m/s}$$

Insight Notice that after the collision the balls move in opposite directions.

Example 9–6 Elastic Collision in Two Dimensions Two objects, one (m_2) of mass 1.50 kg moving due south at 15.3 m/s and the other (m_1) of mass 1.19 kg moving northeast at 11.7 m/s, collide elastically. After they collide, m_2 moves with a velocity of 12.3 m/s at 10.9° north of east. What is the final velocity of m_1?

Picture the Problem The diagram shows the two masses moving toward each other before colliding as well as the speed and direction of m_2 afterward.

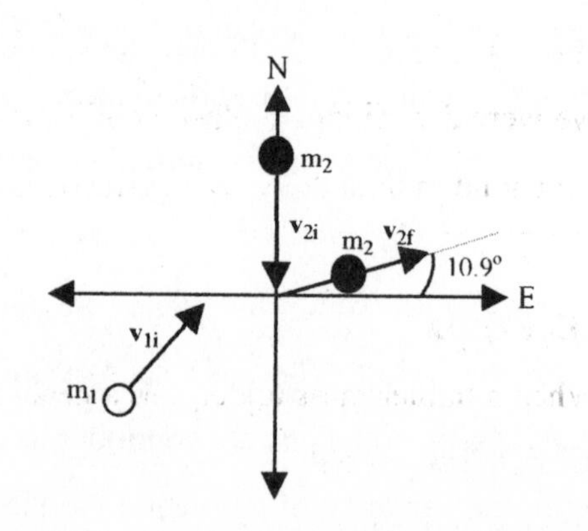

Strategy

To perform the analysis we use the conservation of kinetic energy together with the conservation of momentum in both the x- and y-directions.

Solution

1. Apply the conservation of kinetic energy:

$$K_{total}^{before} = K_{total}^{after}$$
$$\therefore \tfrac{1}{2}m_1v_{1i}^2 + \tfrac{1}{2}m_2v_{2i}^2 = \tfrac{1}{2}m_1v_{1f}^2 + \tfrac{1}{2}m_2v_{2f}^2$$
$$\therefore m_1v_{1i}^2 + m_2v_{2i}^2 = m_1v_{1f}^2 + m_2v_{2f}^2$$

2. Insert numerical values and solve for v_{1f}:

$$514.0\,J = (1.19\,kg)v_{1f}^2 + 226.9\,J$$
$$\therefore \quad v_{1f} = \sqrt{\frac{287.1\,J}{1.19\,kg}} = 15.5\,m/s$$

3. Apply the conservation of momentum along the x direction:

$$p_{total,x}^{before} = p_{total,x}^{after}$$
$$\therefore \quad m_1v_{1i}\cos(45^\circ) = p_{1f,x} + m_2v_{2f}\cos(10.9^\circ)$$
$$\therefore \quad p_{1f,x} = m_1v_{1i}\cos(45^\circ) - m_2v_{2f}\cos(10.9^\circ)$$

4. Evaluate $p_{1f,x}$:

$$p_{1f,x} = 9.845\,kg \cdot m/s - 18.12\,kg \cdot m/s = -8.272\,kg \cdot m/s$$

5. Apply the conservation of momentum along the y direction:

$$p_{total,y}^{before} = p_{total,y}^{after}$$
$$m_1v_{1i}\sin(45^\circ) - m_2v_{2i} = p_{1f,y} + m_2v_{2f}\sin(10.9^\circ)$$
$$p_{1f,y} = (9.845 - 22.95 - 3.489)\,kg \cdot m/s = -16.59\,kg \cdot m/s$$

6. Determine the reference angle:

$$\theta_{1f,ref} = \tan^{-1}\left(\frac{p_{1f,y}}{p_{1f,x}}\right) = \tan^{-1}\left(\frac{-16.59\,kg \cdot m/s}{-8.272\,kg \cdot m/s}\right) = 63.5^\circ$$

7. Since both components are negative $\theta_{1f,ref}$ is measured from the negative x-axis in the 4th quadrant. Therefore, the final velocity of m_1 is:

15.5 m/s at 63.5° south of west

Insight This problem had many steps but we started with the conservation of kinetic energy to ensure that we were describing an elastic collision. Also, it was helpful to use the equations for the conservation of momentum in both the x and y directions to make sure that we had the correct orientation of $\mathbf{v}_{1f}$.

Practice Quiz

11. When a lighter mass undergoes a head-on, elastic collision with a heavier mass that is initially at rest, the lighter mass will

(a) stop and come to rest

(b) continue forward at a slower speed

(c) recoil and move in the direction opposite its original motion

(d) cannot answer definitively; it depends on the masses and the initial speed of the lighter mass

(e) none of the above

9–7 Center of Mass

At many times in our study we have treated large objects such as balls, cars, rockets, etc., as if they were point particles even though they clearly are not. The reason we have been able to do that is because in many situations systems of particles behave (as a whole) just like point particles. The concept that helps us see this fact most clearly is the **center of mass**.

The center of mass of a system is the average location of mass in that system. This average is a weighted average in that each particle contributes more or less to the average according to its mass. Another way to think of the center of mass of a solid object is as the point on the object at which it can be balanced near Earth's surface where we treat Earth's gravity as constant.

The location of the center of mass can be determined by the following expressions:

$$X_{cm} = \frac{m_1 x_1 + m_2 x_2 + \cdots}{m_1 + m_2 + \cdots} = \frac{\sum m_i x_i}{M}$$

$$Y_{cm} = \frac{m_1 y_1 + m_2 y_2 + \cdots}{m_1 + m_2 + \cdots} = \frac{\sum m_i y_i}{M}$$

In the above equations X_{cm} and Y_{cm} are the x and y coordinates of the center of mass, respectively, x_i and y_i are the x and y coordinates of the individual masses, and M is the total mass of the system. Be aware that the center of mass is just an average location; there does not need to be any mass at that location. For a uniform circular ring, the center of mass is right at the center of the ring, where no mass is located. In fact we can take it as given that for uniform, symmetric, and continuous distributions of matter the center of mass is located at the geometric center of the system.

If we examine the motion of the center of mass of a system, we can see the sense in which the system as a whole behaves like a point particle. The velocity of the center-of-mass point is given by $\mathbf{V}_{cm} = (\Sigma m_i \mathbf{v}_i)/M$. Slight manipulation of this equation shows that

$$\mathbf{P}_{total} = M\mathbf{V}_{cm}$$

This is just what we would have for the momentum of a single particle of mass M located at the center of mass. Similarly, the acceleration of the center of mass is given by $\mathbf{A}_{cm} = (\Sigma m_i \mathbf{a}_i)/M$. Slight manipulation of this equation and cancellation of the internal forces (by action and reaction) shows that

$$\mathbf{F}_{net,ext} = M\mathbf{A}_{cm}$$

This is just what we would have for the force on a single particle of mass M located at the center of mass. The two results for the total momentum and force on a system show that when considering the system as a whole, as we have been in many cases such as a moving car, we can treat the system as if it were a single particle with all its mass located at its center of mass.

Example 9–7 Center of Mass Three objects are located at the following positions in a two-dimensional (x, y) coordinate system: m_1 at (1.3 m, 5.4 m), m_2 at (−2.2 m, 9.4 m), and m_3 at (4.1 m, −0.77 m). If the masses are 10 kg, 15 kg, and 20 kg, respectively, what is the position of the center of mass of this system?

Solution Sketch a diagram of the three masses on a coordinate system.

Given: $m_1 = 10$ kg, $m_2, = 15$ kg, $m_3 = 20$ kg, $x_1 = 1.3$ m, $y_1 = 5.4$ m, $x_2 = -2.2$ m, $y_2 = 9.4$ m, $x_3 = 4.1$ m, $y_3 = -0.77$ m;

Find: X_{cm}, Y_{cm}

We make direct use of the expression for calculating the center of mass:

$$X_{cm} = \frac{\sum m_i x_i}{M} = \frac{m_1 x_1 + m_2 x_2 + m_3 x_3}{m_1 + m_2 + m_3} = \frac{10\,\text{kg}(1.3\,\text{m}) + 15\,\text{kg}(-2.2\,\text{m}) + 20\,\text{kg}(4.1\,\text{m})}{45\,\text{kg}} = 1.4\,\text{m}$$

$$Y_{cm} = \frac{\sum m_i y_i}{M} = \frac{m_1 y_1 + m_2 y_2 + m_3 y_3}{m_1 + m_2 + m_3} = \frac{10\,\text{kg}(5.4\,\text{m}) + 15\,\text{kg}(9.4\,\text{m}) + 20\,\text{kg}(-0.77\,\text{m})}{45\,\text{kg}} = 4.0\,\text{m}$$

Insight Notice that no mass is actually located at this position.

Practice Quiz

12. For two objects of slightly different mass, the center of mass will be
    (a) just slightly away from the heavier mass, on the opposite side from the lighter one
    (b) between the two, just slightly closer to the lighter mass
    (c) very close to the lighter mass, on the opposite side from the heavier one
    (d) between the two, just slightly farther away from the lighter object
    (e) between the two, very close to the heavier mass

9–8 Systems with Changing Mass: Rocket Propulsion

As mentioned previously, Newton's second law in terms of momentum is a more general form for this law because it is not limited to systems with constant mass. A rocket is a perfect example of a system with changing mass because at a typical launch most of the rocket's mass is fuel, and the amount of this fuel decreases rapidly as it is exhausted. For a rocket whose exhaust is expelled at a speed v, relative to the rocket, the magnitude of the forward force exerted on the rocket is

$$F = \left|\frac{\Delta m}{\Delta t}\right| v$$

this force defines the **thrust** of the rocket. Here we can see that attempting to understand rocket propulsion using **F** = m**a** would be problematic because this equation assumes that there is just one value of m for the entire acceleration.

Example 9–8 An Accelerating Rocket At liftoff an advanced rocket has a total mass (payload + fuel) of 1.8×10^5 kg. If the rocket burns fuel at a rate of 3000 kg/s with an exhaust velocity of 4500 m/s, what is its acceleration 12.0 seconds after liftoff?

Picture the Problem The sketch shows the rocket accelerating upward after liftoff.

Strategy
We need to determine the thrust that propels the rocket upward. The net force on the rocket from the thrust and gravity will then allow us to determine the acceleration at that instant.

Solution

1. Determine the thrust of the rocket:

$$F_{thrust} = \left|\frac{\Delta m}{\Delta t}\right| v = (3000\,\text{kg/s})(4500\,\text{m/s}) = 1.35 \times 10^7\ \text{N}$$

2. Determine the mass of the rocket after 12 s:

$$m = m_i - \left|\frac{\Delta m}{\Delta t}\right| t = 1.8 \times 10^5\ \text{kg} - (3000\ \text{kg/s})(12.0\ \text{s})$$
$$= 1.44 \times 10^5\ \text{kg}$$

3. Find the net force at t = 12.0 sec:

$$F_{net} = F_{thrust} - mg$$
$$= 1.35 \times 10^7\ \text{N} - (1.44 \times 10^5\ \text{kg})(9.81\ \text{m/s}^2)$$
$$= 1.21 \times 10^7\ \text{N}$$

4. Finally calculate the acceleration of the rocket:

$$a = \frac{F_{net}}{m} = \frac{1.21 \times 10^7\ \text{N}}{1.44 \times 10^5\ \text{kg}} = 84\ \text{m/s}^2$$

Insight Notice the roles of the two forms of Newton's second law here. To obtain the thrust on the rocket we need to account for changing mass. In a case like this, the F = ma version is applicable only at a particular instant in time when the mass has a particular value.

Practice Quiz

13. The force that thrusts a rocket forward comes from

 (a) pushing against the air

 (b) the pull of gravity

 (c) the reaction force from the exhaust

 (d) pushing against the ground at liftoff

 (e) none of the above

Reference Tools and Resources

I. Key Terms and Phrases

linear momentum the product of the mass and the velocity of an object

impulse the product of force and the amount of time the force acts

collision an interaction in which forces are exerted for a finite period of time

conservation of linear momentum the principle that the total linear momentum of a systems remains constant unless a nonzero external net force is applied

inelastic collision a collision in which kinetic energy is not conserved

elastic collision a collision in which kinetic energy is conserved

center of mass the average location of mass within a system

thrust the forward force exerted by the expelled mass in rocket exhaust

II. Important Equations

Name/Topic	Equation	Explanation
linear momentum	$p = mv$	The definition of linear momentum.

Newton's second law	$\mathbf{F}_{net} = \dfrac{\Delta \mathbf{p}}{\Delta t}$	The most general form of Newton's second law.
impulse	$\mathbf{I} = \mathbf{F}_{av} \Delta t = \Delta \mathbf{p}$	The definition of impulse and its relationship to linear momentum.
center of mass	$X_{cm} = \dfrac{m_1 x_1 + m_2 x_2 + \cdots}{m_1 + m_2 + \cdots} = \dfrac{\sum m_i x_i}{M}$ $Y_{cm} = \dfrac{m_1 y_1 + m_2 y_2 + \cdots}{m_1 + m_2 + \cdots} = \dfrac{\sum m_i y_i}{M}$	The x and y coordinates of the center of mass of a system.
thrust	$F_{thrust} = \left\lvert \dfrac{\Delta m}{\Delta t} \right\rvert v$	The force exerted by rocket exhaust.

III. Know Your Units

Quantity	Dimension	SI Unit
linear momentum	$[M]\cdot[L]/[T]$	kg·m/s
impulse	$[M]\cdot[L]/[T]$	N·s (= kg·m/s)
thrust	$[M]\cdot[L]/[T^2]$	N

Practice Problems

1. For a collision in which a ball collides with a fixed wall, the time of impact is 39.96 milliseconds and the average force is 503.51 newtons. To the nearest hundredth of a joule·sec what is the impulse?
2. If the mass of the ball in problem 1 is 3.41, and its velocity before collision is 3.93 m/s, to the nearest hundredth of a m/s, what is its velocity after collision?
3. One lump of clay traveling at 27 m/s overtakes a second lump traveling at 8.8 m/s. After collision they are stuck together. To the nearest tenth of a m/s, what is their common velocity?

2 kg ⟶ *1 kg* ⟶

4. If the second lump in problem 3 was moving to the left before the collision, what would be the answer?
5. One ball traveling at 15 m/s strikes a second ball at rest in an elastic collision. If the mass of the first ball is 27 kg, and the mass of the second is 1.1 kg, to the nearest tenth of a m/s what is the velocity of the first ball after the collision?

6. In problem 5 what is the velocity of the second ball after collision?

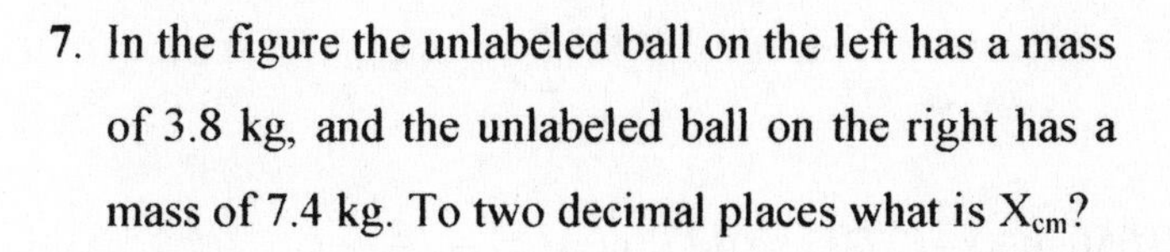

7. In the figure the unlabeled ball on the left has a mass of 3.8 kg, and the unlabeled ball on the right has a mass of 7.4 kg. To two decimal places what is X_{cm}?

8. In problem 7 what is Y_{cm}?

9. The 2-kg ball in the figure has a speed of 1.59 m/s, and the 1-kg ball is at rest prior to an elastic glancing collisions. After the collision the 2-kg ball has a speed of 1.24 m/s. To the nearest tenth of a degree, measured counterclockwise from east, at what angle does it scatter if the 1-kg ball is scattered at 280°?

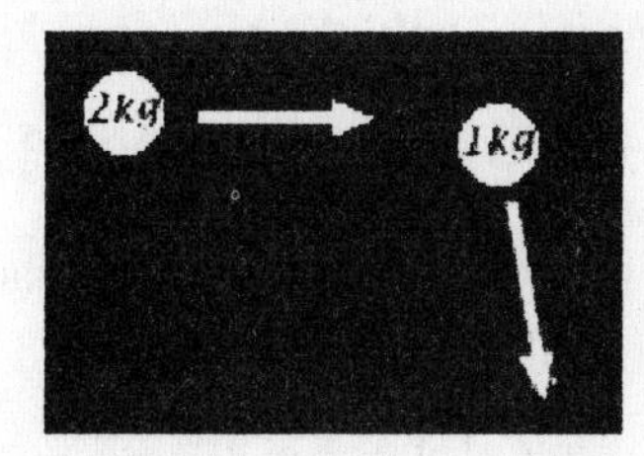

10. A rocket has an exhaust gas velocity of 2289 m/s and a burn rate of 8802 kg/s. If it has a mass of 1,000,000 kg, what is its initial acceleration, to two decimal places?

Puzzle

KNOCK-OFF

You are standing on a log and a friend is trying to knock you off. He throws a ball at you. You can either catch it or let it bounce off you. Which is more likely to topple you, catching the ball or letting it bounce off? Briefly explain what physics you used to reach your conclusion.

Selected Solutions

15. The sum of the skaters' momenta must be zero. The setup is similar to Example 9–3. Let $m_1 = 45$ kg, $v_1 = -0.62$ m/s and $v_2 = 0.89$ m/s.

$$p_{1x} + p_{2x} = m_1 v_{1x} + m_2 v_{2x} = 0$$

$$\therefore \quad m_2 = \frac{-m_1 v_{1x}}{v_{2x}} = \frac{-(45\text{ kg})(-0.62\ \frac{\text{m}}{\text{s}})}{0.89\ \frac{\text{m}}{\text{s}}} = \boxed{31\text{ kg}}$$

25. **(a)** Use momentum conservation. Let the subscripts b and B denote the bullet and the block, respectively.

$$m_b v_{bi} + m_B v_{Bi} = m_b v_{bf} + m_B v_{Bf}$$

$$m_b v_{bi} + 0 = m_b v_{bf} + m_B v_{Bf}$$

$$\therefore \quad v_{bf} = \frac{m_b v_{bi} - m_B v_{Bf}}{m_b} = \frac{(0.0040 \text{ kg})(650 \text{ m/s}) - (0.095 \text{ kg})(23 \text{ m/s})}{0.004 \text{ kg}} = \boxed{100 \text{ m/s}}$$

(b) The final kinetic energy is less than the initial kinetic energy because energy is lost to the heating and deformation of the bullet and block.

(c) Initially, only the bullet has kinetic energy. Therefore,

$$K_i = \tfrac{1}{2} m_b v_{bi}^2 = \tfrac{1}{2}(0.0040 \text{ kg})(650 \text{ m/s})^2 = 850 \text{ J}$$

Finally, both the bullet and the block have kinetic energy. Therefore,

$$K_f = \tfrac{1}{2} m_b v_{bf}{}^2 + \tfrac{1}{2} m_B v_{BF}{}^2 = \tfrac{1}{2}(0.0040 \text{ kg})(103.75 \tfrac{\text{m}}{\text{s}})^2 + \tfrac{1}{2}(0.095 \text{ kg})(23 \tfrac{\text{m}}{\text{s}})^2 = 47 \text{ J}$$

We see that $\boxed{K_f < K_i,}$ which verifies the answer to part (b).

29. Let's make the following definitions:

m_1 = the mass of the truck

m_2 = the mass of the car

v_0 = the initial speed of the truck

Use conservation of momentum to find an equation for the final speed of the truck:

$$m_1 v_0 = m_1 v_{1f} + m_2 v_{2f} \quad \Rightarrow \quad m_1 v_{1f} = m_1 v_0 - m_2 v_{2f}$$

$$\therefore \quad v_{1f} = v_0 - \frac{m_2}{m_1} v_{2f}$$

There is one equation and two unknowns. Use conservation of kinetic energy to find a second equation relating v_{1f} and v_{2f}.

$$\tfrac{1}{2} m_1 v_0^2 = \tfrac{1}{2} m_1 v_{1f}^2 + \tfrac{1}{2} m_2 v_{2f}^2 \quad \Rightarrow \quad m_1 v_{1f}^2 = m_1 v_0^2 - m_2 v_{2f}^2$$

Substitute for v_{1f} and solve for v_{2f}.

$$m_1\left(v_0 - \frac{m_2}{m_1}v_{2f}\right)^2 = m_1 v_0^2 - m_2 v_{2f}^2$$

$$\Rightarrow \quad m_1 v_0^2 - 2m_2 v_0 v_{2f} + \frac{m_2^2}{m_1}v_{2f}^2 = m_1 v_0^2 - m_2 v_{2f}^2$$

$$\Rightarrow \quad -2v_0 v_{2f} + \frac{m_2}{m_1}v_{2f}^2 = -v_{2f}^2$$

$$\Rightarrow \quad \left(\frac{m_1 + m_2}{m_1}\right)v_{2f}^2 = 2v_0 v_{2f}$$

$$\Rightarrow \quad v_{2f} = \left(\frac{2m_1}{m_1 + m_2}\right)v_0$$

Substitute for v_{2f} in the equation for v_{1f}:

$$v_{1f} = v_0 - \frac{m_2}{m_1}\left(\frac{2m_1}{m_1 + m_2}\right)v_0 = \left(1 - \frac{2m_2}{m_1 + m_2}\right)v_0 = \left(\frac{m_1 - m_2}{m_1 + m_2}\right)v_0$$

Substitute the given information, $m_1 = 1620$ kg, $m_2 = 722$ kg, and $v_0 = 14.5$ m/s. The final speeds of the truck and car are $v_{truck} = \boxed{5.56\ \text{m/s}}$ and $v_{car} = \boxed{20.1\ \text{m/s}}$.

39. Due to symmetry, $X_{cm} = 0$. Therefore, we need only to calculate Y_{cm}.

$$Y_{cm} = \frac{\Sigma my}{M} = \frac{my_1 + my_2 + m_s y_s}{2m + m_s} = \frac{2my_1 + m_s y_s}{2m + m_s}$$

$$= \frac{2(16\,u)(0.143\ \text{nm})\sin 30° + (32\,u)(0)}{2(16\,u) + 32\,u} = 3.6 \times 10^{-11}\ \text{m}$$

Therefore, $\boxed{(X_{cm}, Y_{cm}) = (0,\ 3.6 \times 10^{-11}\ \text{m})}$

55. Use conservation of momentum to determine the horizontal speed of the bullet and block.

m = the mass of the bullet

M = the mass of the block

v = the initial speed of the bullet

v_f = the final speed of the bullet + block

$$mv + M(0) = (m + M)v_f \quad \Rightarrow \quad v_f = \left(\frac{m}{m + M}\right)v$$

Recall that for a mass that is initially stationary.

$$\Delta y = h = \tfrac{1}{2}gt^2 \quad \Rightarrow \quad t = \sqrt{2h/g}$$

Substitute for v_f and t to find the horizontal distance:

$$x = v_f t = \left(\frac{m}{m+M}\right) v \sqrt{\frac{2h}{g}}$$

$$= \left(\frac{0.0100\ \text{kg}}{0.0100\ \text{kg} + 1.30\ \text{kg}}\right)\left(725\ \frac{\text{m}}{\text{s}}\right)\sqrt{\frac{2(0.750\ \text{m})}{9.81\ \text{m/s}^2}} = \boxed{2.16\ \text{m}}$$

Answers to Practice Quiz

1. (b) **2**. (a) **3**. (d) **4**. (e) **5**. (c) **6**. (a) **7**. (d) **8**. (c) **9**. (a) **10**. (e) **11**. (c) **12**. (d) **13**. (c)

Answers to Practice Problems

1. 20.12 J   **2**. −1.97 m/s

3. 20.9 m/s   **4**. 15.1 m/s

5. 13.8 m/s   **6**. 28.8 m/s

7. 3.52   **8**. 2.74

9. 34.0°   **10**. 10.34 m/s^2

CHAPTER 10

ROTATIONAL KINEMATICS AND ENERGY

Chapter Objectives

After studying this chapter, you should:

1. know the meanings of angular position, velocity, and acceleration.
2. be able to describe rotational motion.
3. understand the connection between rotational quantities and corresponding linear quantities.
4. understand and be able to describe rolling motion.
5. understand the concept of the moment of inertia and its role in rotational motion.
6. be able to use rotational kinetic energy and apply the conservation of energy to rotating and rolling objects.

Warm-Ups

1. When an audio or videotape rewinds, why does the tape wind up faster at the end than at the beginning?
2. Estimate the magnitude of the tangential velocity of an object in your hometown due to the rotation of Earth.
3. A small object is sliding inside a circular frictionless track. At point A another frictionless track switches it to a smaller circle (see the picture). At all times the tracks can exert only forces that are perpendicular to the motion of the little slug. What happens to the linear velocity of the slug when it switches tracks? What about its angular velocity?

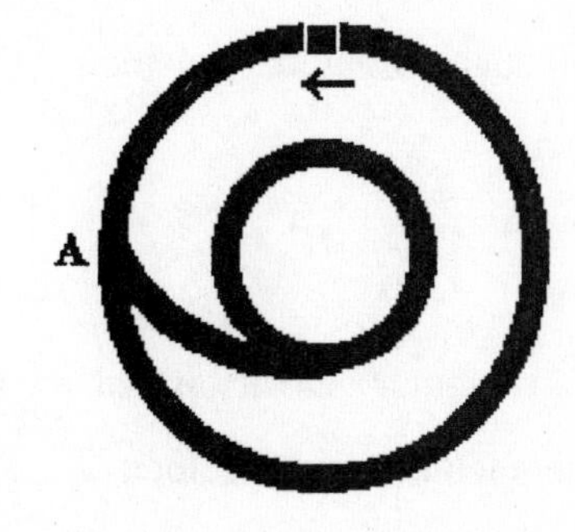

4. A skater is spinning with his arms outstretched. He has a 2-lb weight in each hand. In an attempt to change his angular velocity he lets go of both weights. Does he succeed in changing his angular velocity? If yes, how does his angular velocity change?

Chapter Review

In this chapter and the next we study rotational kinematics and dynamics. Rotational motion is every bit as important as the linear (or translational) motion that we've been studying thus far. As you read these chapters try to notice how the study of rotational motion parallels that of translational motion. Many of the concepts as well as the mathematical treatments are the same, we need only to alter the physical interpretation of these concepts from that of a translating body to a rotating one.

10–1 Angular Position, Velocity, and Acceleration

Just as with linear motion we describe rotational motion with three basics quantities analogous to displacement, velocity, and acceleration. We shall call these quantities **angular position**, **angular velocity**, and **angular acceleration**.

When we describe the rotation of an object we think of the object as rotating about an axis. If you consider a line drawn from the axis of rotation to the object, the angular position of the object, denoted by θ, is the angle between that line and an arbitrarily chosen reference line. The reference line plays the same role as the arbitrarily chosen origin of a coordinate system. In SI the unit of angular measure, for rotational motion is the *radian*. Angular measure is actually dimensionless, but radian measure works better in describing motion than do degrees. Angular position plays the same role for rotational motion that position does for translational motion. In keeping with the usual convention, if θ is measured counterclockwise, it is taken to be a positive angle, if it is measured clockwise it is taken to be negative.

To describe how rapidly an object rotates about an axis we use the quantity angular velocity, ω. This quantity plays the same role for rotational motion that velocity does for linear motion. Angular velocity is the rate of change of angular position. Often it is useful to use the average rate of change, or average angular velocity:

$$\omega_{av} = \frac{\Delta\theta}{\Delta t}$$

The SI unit of angular velocity is rad/s = s^{-1} (since rad is dimensionless). It is also sometimes useful to use the instantaneous angular velocity:

$$\omega = \lim_{\Delta t \to 0} \frac{\Delta\theta}{\Delta t}$$

To maintain consistency with the sign convention for angular displacement $\Delta\theta$, angular velocity is negative if the object rotates clockwise, and positive if it rotates counterclockwise.

To describe how the angular velocity of an object changes we use the quantity angular acceleration, α. This quantity plays the same role for rotational motion that acceleration, a, does for linear motion.

Angular acceleration is the rate of change of angular velocity. Often it is useful to use the average rate of change, or average angular acceleration:

$$\alpha_{av} = \frac{\Delta\omega}{\Delta t}$$

The SI unit of angular acceleration is rad/s^2 = s^{-2} (since rad is dimensionless). It is also sometimes useful to use the instantaneous angular acceleration

$$\alpha = \lim_{\Delta t \to 0} \frac{\Delta\omega}{\Delta t}$$

The sign of α depends on the sign of $\Delta\omega$. If $\Delta\omega$ is negative then α is negative; if $\Delta\omega$ is positive, then α is positive.

Example 10–1 Turn on the Fan On a hot day you turn on a small fan to help cool you off. The tip on one of the blades of the fan might go through 1.25 revolutions at an average angular velocity of 7.85 rad/s. **(a)** What is the angular displacement of the tip of the blade, and **(b)** how much time does it take to go through that displacement?

Picture the Problem The sketch shows the blades of a fan with one tip indicated by the black dot. In this diagram the fan is rotating clockwise.

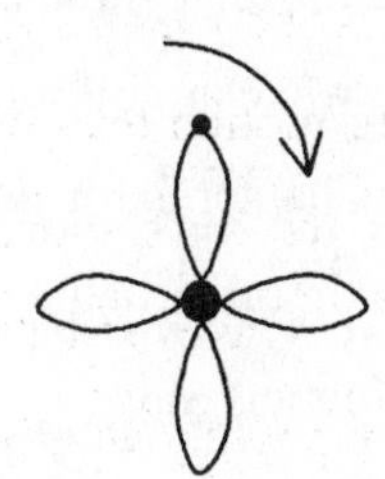

Strategy For part (a) we must translate between the number of revolutions and the number of radians to get the SI angular displacement. For part (b) we'll make use of the definition of average angular velocity.

Solution

Part (a)

1. Convert from revolutions to radians: $\Delta\theta = -1.25\,\text{rev}\left(\frac{2\pi\,\text{rad}}{\text{rev}}\right) = -7.85\,\text{rad}$

Part (b)

2. Use the definition of ω_{av} to solve for Δt: $\omega_{av} = \frac{\Delta\theta}{\Delta t} \Rightarrow \Delta t = \frac{\Delta\theta}{\omega_{av}}$

3. Calculate the numerical result: $$\Delta t = \frac{-7.854\text{ rad}}{-7.85\text{ rad/s}} = 1.00\text{ s}$$

Insight The minus signs appear by convention because the blade is rotating clockwise.

Practice Quiz

1. Which quantity is used to describe how rapidly an objects rotates?

   (a) angular position **(b)** angular velocity **(c)** angular acceleration **(d)** none of the above

2. If an object is seen to rotate faster and faster, which quantity best describes this aspect of its motion?

   (a) angular position **(b)** angular velocity **(c)** angular acceleration **(d)** none of the above

10–2 Rotational Kinematics

One thing you may have noticed about the three quantities described in the previous section is that the mathematical relationships among them are exactly the same as the relationships among the corresponding quantities for translational motion. This means that when looking for a way to describe rotational motion we can follow the prescription already laid out for translational motion. All the equations used in Chapter 2 for describing motion with constant velocity and constant acceleration also apply to rotational motion. The only mathematical difference is that the names of the variables are changed in the following way:

$$x \to \theta,\ v \to \omega,\ a \to \alpha,\ t \to t$$

Of course, along with changing the variables, you should also change the physical picture in your mind of the type of motion you are describing. Nevertheless, the mathematical descriptions are identical.

Thus, to describe rotational motion we make the above replacements and reuse the equations in Chapter 2 for one-dimensional kinematics (assuming $t_0 = 0$, as is customary).

Motion with Constant Angular Velocity

$$\Delta\theta = \omega\,\Delta t, \quad \omega_f = \omega_i$$

Motion with Constant Angular Acceleration

$$\omega = \omega_0 + \alpha t$$

$$\theta = \theta_0 + \tfrac{1}{2}(\omega_0 + \omega)t$$

$$\theta = \theta_0 + \omega_0 t + \tfrac{1}{2}\alpha t^2$$

$$\omega^2 = \omega_0^2 + 2\alpha(\theta - \theta_0)$$

Example 10–2 Turn on the Fan II Suppose that the blades of the fan discussed in Example 10–1 rotate with a constant angular acceleration for the first 1.00 seconds. **(a)** What is the instantaneous angular velocity of the blade at t = 1.00 seconds after being turned on, and **(b)** what is the angular acceleration?

Picture the Problem The sketch shows the blades of a fan with one tip indicated by the black dot. This time the fan is rotating counterclockwise.

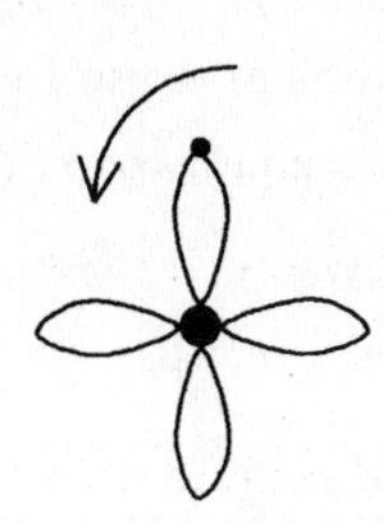

Strategy In both parts we make use of the expressions for uniformly accelerated motion to obtain a solution. Let's make use of our freedom to define $\theta_0 = 0$ and $t_0 = 0$. Also notice that since the fan is just being turned on $\omega_0 = 0$.

Solution

Part (a)

1. Choose an appropriate expression for finding ω based on the given information: $\theta = \theta_0 + \tfrac{1}{2}(\omega_0 + \omega)t \Rightarrow \theta = \tfrac{1}{2}\omega t$

2. Solve for angular velocity at the appropriate time: $\omega = \dfrac{2\theta}{t} = \dfrac{2(7.854\text{ rad})}{1.00\text{ s}} = 15.7\text{ rad/s}$

Part (b)

3. Choose an appropriate expression for finding α based on the given information: $\omega = \omega_0 + \alpha t \Rightarrow \alpha = \dfrac{\omega}{t}$

4. Use the known data to get the numerical result: $\alpha = \dfrac{15.71\text{ rad/s}}{1.00\text{ s}} = 15.7\text{ rad/s}^2$

Insight In this example all the values are positive: for θ and ω because the blade is rotating counterclockwise, and for α because the angular speed increases counterclockwise. Try to see how a similar translational problem could have been solved way back in Chapter 2. If you think about it, you actually have already done this stuff!

Exercise 10–3 Turn off the Fan Now, suppose that the blades of our fan have a top rotation rate of 25 revolutions per second. Once we've cooled down enough we turn off the power, and it takes 6.65 seconds for the blades to come to rest with a constant angular acceleration. **(a)** What is the angular acceleration of the blades, and **(b)** through what angular displacement do the blades turn while coming to rest?

Solution: The problem doesn't gives us a direction of rotation so we are free to choose. Let's choose counterclockwise as a default.

Given: $\omega_0 = 25$ rev/s, $\omega = 0$, $t = 6.65$ s;   **Find**: (a) α, (b) θ

First, we'll convert the initial angular velocity to SI units:

$$\omega_0 = 25\ \text{rev/s}\left(\frac{2\pi\ \text{rad}}{\text{rev}}\right) = 157\ \text{rad/s}$$

Now, for part (a), we can determine α by using

$$\omega = \omega_0 + \alpha t \quad \Rightarrow \quad \alpha = -\frac{\omega_0}{t} = -\frac{157\ \text{rad/s}}{6.65\,\text{s}} = -24\ \text{rad/s}^2$$

Calculating the answer to part (b) is also a straightforward use of the equations. The most convenient one is

$$\theta = \theta_0 + \tfrac{1}{2}(\omega_0 + \omega)\ t \Rightarrow \theta = \tfrac{1}{2}\omega_0\ t = \tfrac{1}{2}(157\ \text{rad/s})(6.65\ \text{s}) = 520\ \text{rad}$$

Notice that because of the original rotation rate the answers are given to only two significant digits.

Practice Quiz

3. In comparing linear and rotational motion, angular position is most similar to which linear variable?

   (a) x **(b)** v **(c)** a **(d)** t **(e)** none of the above

4. How long would it take an object rotating at a constant speed of 17.0 rad/s to rotate through 235 radians?

   (a) 0.0723 s **(b)** 3.4 N **(c)** 5.0 N **(d)** 13.8 s **(e)** 3.1 N

5. If an object rotates at 2.50 rpm, what is the equivalent angular speed in rad/s?

   (a) 2.50 **(b)** 15.7 **(c)** 0.262 **(d)** 0.0417 **(e)** 25.0

10–3 Connections Between Linear and Rotational Quantities

When an object is moving in a circular path (pure rotation) we can describe its motion using the rotational quantities discussed above. However, we can also describe the object's motion using the linear quantities. It is very useful to know how to relate these two ways to describe the motion.

With pure rotation, the relationship between the angular distance θ and the linear distance s along the arc is well known to be

$$s = r\theta$$

when θ is measured in radians, and r is the radius of the circular path (r would be the distance of the object from the axis of rotation if the motion were not circular). Notice that the above equation makes it clear that angular measure is dimensionless because both s and r have dimension [L].

The velocity of the object in pure rotation will be tangent to the circular path of its motion. The magnitude of this velocity is called the *tangential speed* of the object, v_t. The relationship between v_t and ω can be determined directly from the preceding expression for the arc length to be

$$v_t = r\omega$$

The relationship between the tangential acceleration and the angular acceleration can similarly be determined from the expression for tangential speed:

$$a_t = r\alpha$$

Since the object is rotating, it will also have an inward-pointing component to its acceleration, the centripetal acceleration discussed in Chapter 6. Recall that a_{cp} depends on the tangential speed $a_{cp} = v_t^2/r$. Using the above expression for v_t we can get the following relationship between a_{cp} and the rotational quantities:

$$a_{cp} = r\omega^2$$

Example 10–4 Turn off the Fan II Suppose that the distance from the central axis of the blade unit of the fan to the tip of each blade is 7.00 cm. Given the information for turning off the fan in Example 10–3, calculate **(a)** the linear distance the tip of the blade traversed while coming to rest, **(b)** the maximum tangential speed of the tip of the blade, **(c)** the tangential component of the acceleration of the tip of the blade, and **(d)** the centripetal acceleration of the tip of the blade.

Picture the Problem The sketch shows the fan (and the tip of one blade) rotating counterclockwise.

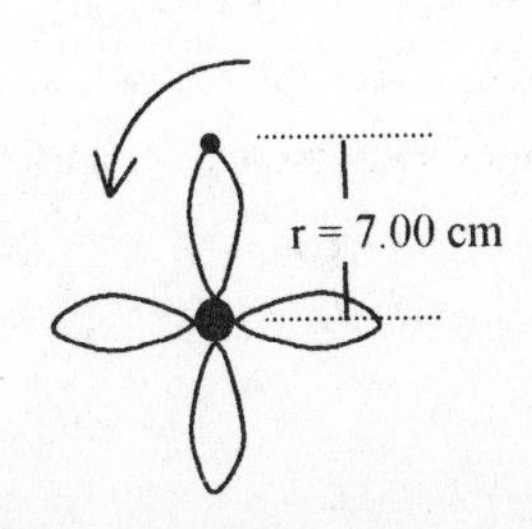

Strategy To solve for the desired information, we use the results of Example 10–3 and the known relations between the linear and rotational quantities.

Solution

Part (a)

1. We know θ to three significant figures from part (b) in Exercise 10–3, so:

$$s = r\theta = (0.0700\,\text{m})(522\,\text{rad}) = 37\,\text{m}$$

Part (b)

2. We know ω from Exercise 10–3:

$$v_t = r\omega = (0.0700\,\text{m})(157\,\text{rad/s}) = 11\,\text{m/s}$$

Part (c)

3. We make direct use of the known result, to three significant figures, from Exercise 10–3:

$$a_t = r\alpha = (0.0700\,\text{m})(23.6\,\text{rad/s}^2) = 1.7\,\text{m/s}^2$$

Part (d)

4. Again, we make direction use of the known result:

$$a_{cp} = r\omega^2 = (0.0700\,\text{m})(157\,\text{rad/s})^2 = 1700\,\text{m/s}^2$$

Insight Make sure you understand why the units of the final answers don't include the radian.

Practice Quiz

6. If a wheel of radius 1.3 m rotates at 2.2 rad/s, which is the most accurate way to write the linear speed of a point on the rim of this wheel?

   (a) 2.9 rad·m/s **(b)** 1.7 rad·m/s **(c)** 1.7 m/s **(d)** 2.9 m/s **(e)** 0.59 m/s

7. If a wheel of radius 1.3 m rotates at a constant speed of 2.2 rad/s, a point on the rim of the wheel accelerates tangentially at

   (a) 6.3 m/s² **(b)** 2.9 m/s² **(c)** 1.7 m/s² **(d)** 3.7 m/s² **(e)** none of the above

10–4 Rolling Motion

Rolling motion represents an important application that combines both rotational and translational motion. This motion also illustrates the importance of understanding the connection between linear and rotational quantities discussed in the previous section.

For a wheel that is free to roll, without slipping, each point on the wheel rotates about the wheel's axle while also being carried forward from the overall translation of the wheel. As a result of this combination of rotational and translational speeds, the point of contact between the wheel and the ground is instantaneously at rest (otherwise there would be slipping), the center of mass of the wheel moves forward at $v_{cm} = r\omega$ (r – radius, ω – angular velocity), and the point at the top of the wheel moves at twice this speed: $v_{top} = 2r\omega$. Armed with just these few facts, much can be understood about rolling motion.

Exercise 10–5 A Bicycle Wheel The tires of an 18-speed mountain bike have a typical radius of about 34 cm. If you are riding hard and your tires are rotating at an angular velocity of 33 rad/s, **(a)** how fast are you traveling in m/s, and **(b)** how far will you travel after 15 min at this pace?

Solution: The following information is given in the problem.

Given: $r = 34$ cm, $\omega = 33$ rad/s, $t = 15$ min; **Find**: **(a)** v_{trans}, **(b)** x

The key to solving for the speed at which you are moving is to notice that your speed is that of the center of mass of the wheel. Therefore,

$$v_{trans} = v_{cm} = r\omega = (0.34\,\text{m})(33\,\text{rad/s}) = 11\,\text{m/s}$$

Since you travel at this constant speed for 15 min we can determine the distance using

$$x = v_{cm}t = (11.22\,\text{m/s})(15\,\text{min})(60\,\text{s/min}) = 1.0\times10^4\ \text{m}$$

Practice Quiz

8. If a bicycle moves at 6.5 m/s and its wheels have a diameter of 70 cm, how fast are the wheels rotating?

   (a) 9.3 rad/s **(b)** 19 rad/s **(c)** 4.5 rad/s **(d)** 2.3 rad/s **(e)** 0.093 rad/s

10–5 – 10–6 Rotational Kinetic Energy, Moment of Inertia, and Energy Conservation

When studying translational motion we introduced the idea of kinetic energy which is an energy associated with "mass in motion." Objects that are rotating but not translating also have mass in motion.

We should, therefore, be able to write its kinetic energy in terms of the rotational quantities. In doing this, we find that the rotational kinetic energy of a system is

$$K_{rot} = \tfrac{1}{2}\left(\sum m_i r_i^2\right)\omega^2$$

Comparison between this expression and the translational form, $K = \frac{1}{2}mv^2$, suggests that the quantity Σmr^2 plays a role similar to mass. We call this quantity the **moment of inertia,** I of the system (also called *rotational inertia*):

$$I = \sum m_i r_i^2$$

The SI unit of the moment of inertia is $kg{\cdot}m^2$.

The above expression for the moment of inertia is convenient for calculating I when the number of particles in the system is small. However, for systems with so many particles that we should treat the system as a continuous distribution of matter a more involved calculation is required. Therefore, some of the more common results for the moments of inertia of continuous systems are listed for you in Table 10–1 of your textbook. Notice that I depends on the axis about which the object is rotating, so that the same object can have many different I values associated with it. You should become very familiar with Table 10-1.

The conservation of total mechanical energy applies to objects that are rolling without slipping. This is true because even though friction is present (to prevent slipping) it is *static* friction and therefore does no work. The key to applying energy conservation to rolling objects is to remember that the kinetic energy is taken up by both the rotational and translational motions,

$$K = K_{trans} + K_{rot} = \tfrac{1}{2}mv_{cm}^2 + \tfrac{1}{2}I_{cm}\omega^2$$

Notice here that the translational kinetic energy comes from treating the system as a single particle located at, and moving with, the center of mass of the object. Also recognize that the moment of inertia in this expression is calculated about an axis passing through the center of mass (perpendicular to the direction of motion).

Example 10–6 The Energy of the Fan An employee for the manufacturer of our favorite fan needs to estimate the kinetic energy of the fan when the blades are rotating at top speed. She takes, as a working model, **(a)** each blade to be a thin rod of length 7.00 cm and mass 0.113 kg, and as an alternative model, **(b)** each blade as a point mass with all its mass located at its center of mass. Assuming that the central axle is of negligible mass, calculate the rotational kinetic energy in each case. Which do you think is more accurate? **(c)** If she requires this fan to reach top speed 2.50 seconds after being turned on, how powerful of a motor should she use?

Picture the Problem The sketch shows the blades of the fan rotating counterclockwise.

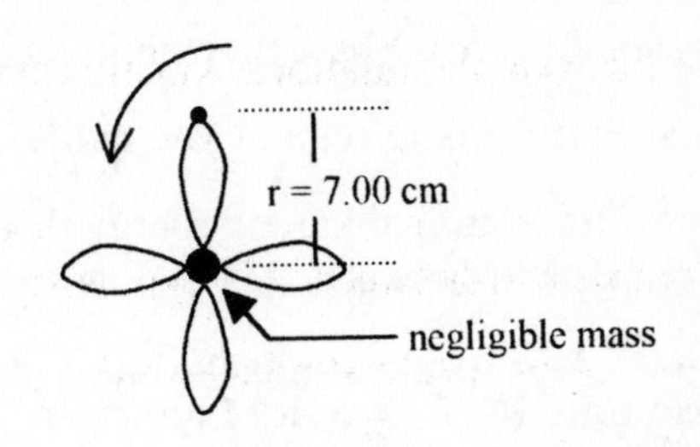

Strategy Since the first model is of a continuous thin rod, we will make use of Table 10–1 in the text in calculating the moment of inertia. We already know the top angular velocity from previous examples.

Solution

Part (a)

1. Each blade is a thin rod rotating about an axis through one end; from Table 10–1 the moment of inertia of one blade is: $I_1 = \frac{1}{3}ML^2 = \frac{1}{3}(0.113\,\text{kg})(0.0700\,\text{m})^2 = 1.846\times10^{-4}\ \text{kg}\cdot\text{m}^2$

2. The total moment of inertia of all 4 blades is: $I = 4I_1 = 7.383\times10^{-4}\ \text{kg}\cdot\text{m}^2$

3. The rotational kinetic energy then is: $K = \frac{1}{2}I\omega^2 = \frac{1}{2}\left(7.383\times10^{-4}\ \text{kg}\cdot\text{m}^2\right)\left(157\ \frac{\text{rad}}{\text{s}}\right)^2 = 9.1\ \text{J}$

Part (b)

4. Each blade is a point mass located at its center 3.50 cm from the axis. The moment of inertia is: $I = 4I_1 = 4mr^2 = 4(0.113\,\text{kg})(0.035\,\text{m})^2 = 5.537\times10^{-4}\ \text{kg}\cdot\text{m}^2$

5. The rotational kinetic energy then is: $K = \frac{1}{2}I\omega^2 = \frac{1}{2}\left(5.537\times10^{-4}\ \text{kg}\cdot\text{m}^2\right)\left(157\ \frac{\text{rad}}{\text{s}}\right)^2 = 6.8\ \text{J}$

6. The thin-rod model is a closer approximation to the actual blade and is more accurate.

Part (c)

7. The average power of the motor is given by: $P = \dfrac{W}{t}$

8. By the work–energy theorem: $W = \Delta K = K$

9. The power needed, using part (a), is: $P = \dfrac{K}{t} = \dfrac{9.099\ \text{J}}{2.50\ \text{s}} = 3.6\ \text{W}$

Insight As can be seen from the differing results of parts (a) and (b) it is not always valid to treat a system as a point particle with its mass located at the center of mass. You have to be careful when doing this. Part (c) is to remind you that the work-energy theorem is equally valid for rotational kinetic energy as for translational kinetic energy that we used to introduce the theorem.

Example 10–7 Gravity Launcher A solid ball of mass 1.40 kg and radius 0.391 m is released from rest and caused (by gravity) to roll from the top of a track 10.0 m high down to a height of 0.500 m above the ground, where it is launched vertically upward. How high in the air will the ball go assuming it continues to spin after leaving the track (neglect any friction)?

Picture the Problem The sketch shows the odd-shaped track with the ball at rest at the top.

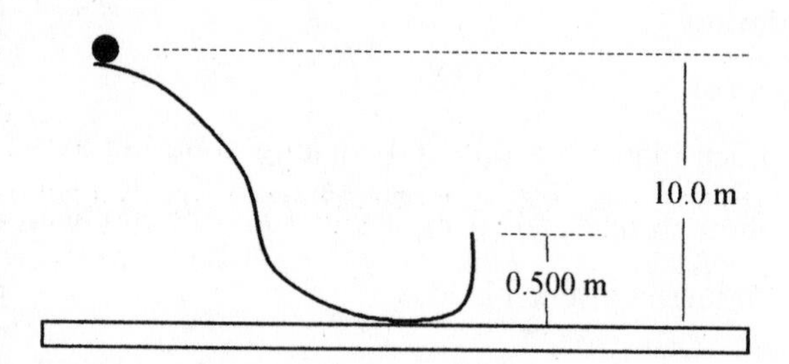

Strategy Because the ball rolls without slipping we can apply mechanical energy conservation. Let's take the bottom of the track (the ground) as the reference level for gravitational potential energy. The greater height will be called H and the lesser, h.

Solution

1. The mechanical energy before the ball is released is in the form of only gravitational potential energy: $E = mgH$ Eq. (1)

2. The final energy, at height y, is a combination of potential and rotational kinetic energy: $E = mgy + \frac{1}{2}I\omega^2$ Eq. (2)

3. We need to determine ω when the ball leaves the track. The energy when it leaves is: $E = mgh + \frac{1}{2}mv_{cm}^2 + \frac{1}{2}I\omega^2$ Eq. (3)

4. From Table 10–1 I is given by: $I = \frac{2}{5}mr^2$

5. The angular speed relates to the linear speed as: $v_{cm} = r\omega$

6. Substituting the results for I and ω into Eq. (3): $E = mgh + \frac{1}{2}mr^2\omega^2 + \frac{1}{5}mr^2\omega^2 = mgh + \frac{7}{10}mr^2\omega^2$

7. Set Eq. (3) equal to Eq. (1) and solve for ω: $mgH = mgh + \frac{7}{10}mr^2\omega^2 \Rightarrow \omega = \left[\frac{g(H-h)}{\frac{7}{10}r^2}\right]^{1/2}$

$$= \left[\frac{(9.81\,\text{m/s}^2)(9.50\,\text{m})}{\frac{7}{10}(0.391\,\text{m})^2}\right]^{1/2} = 29.51\,\text{rad/s}$$

8. Set Eq. (2) equal to Eq. (1) and solve for the final height, y: $mgH = mgy + \frac{1}{5}mr^2\omega^2 \;\therefore$

$$y = H - \frac{r^2\omega^2}{5g} = 10.0\,\text{m} - \frac{(0.391\,\text{m})^2(29.51\,\text{rad/s})^2}{5(9.81\,\text{m/s}^2)}$$

$$= 7.29\,\text{m}$$

Insight Check that you can complete the mathematical steps for items 7 and 8 and understand the choice for I.

Practice Quiz

9. Two solid wheels have the same mass, but one has twice the radius of the other. If they both rotate about axes through their centers, the one with the larger radius will have

   (a) half the rotational inertia of the other one

   (b) the same rotational inertia as the other one

   (c) twice the rotational inertia of the other one

   (d) four times the rotational inertia of the other one

   (e) eight times the rotational inertia of the other one

10. A thin rod of mass 1.0 kg and length 1.0 m rotates about an axis through its center at 1.0 revolutions per second. What is its kinetic energy?

    (a) 1.6 J **(b)** 1.0 J **(c)** 0.083 J **(d)** 39 J **(e)** 20 J

11. A hollow sphere of mass 0.22 kg and radius 0.16 m rolls along the ground with a constant linear speed of 2.0 m/s. What is its kinetic energy?

    (a) 0.0075 J **(b)** 0.29 J **(c)** 0.070 J **(d)** 0.73 J **(e)** 0.44 J

Reference Tools and Resources

I. Key Terms and Phrases

angular position the angle measured from a chosen reference line

angular velocity the rate of change of angular position

angular acceleration the rate of change of angular velocity

rolling without slipping a form of rolling motion in which the point of contact between the wheel and the surface is instantaneously at rest

moment of inertia a quantity that represents the inertial property of a rotating object or system

II. Important Equations

Name/Topic	Equation	Explanation
angular velocity	$\omega_{av} = \frac{\Delta\theta}{\Delta t}$	The average angular velocity of a rotating object.
angular acceleration	$\alpha_{av} = \frac{\Delta\omega}{\Delta t}$	The average angular acceleration of a rotating object.
constant angular acceleration	$\omega = \omega_0 + \alpha t$ $\theta = \theta_0 + \frac{1}{2}(\omega_0 + \omega)t$ $\theta = \theta_0 + \omega_0 t + \frac{1}{2}\alpha t^2$ $\omega^2 = \omega_0^2 + 2\alpha(\theta - \theta_0)$	The system of equations that describe motion with constant angular acceleration.
linear quantities	$s = r\theta$ $v_t = r\omega$ $a_t = r\alpha$ $a_{cp} = r\omega^2$	The 4 equations that relate the linear and angular quantities in rotational motion.
rotational kinetic energy	$K = \frac{1}{2}I\omega^2$	The rotational kinetic energy of a system.
moment of inertia	$I = \sum m_i r_i^2$	The expression for the moment of inertia of a system of particles.

III. Know Your Units

Quantity	Dimension	SI Unit
angular position	—	rad
angular velocity	$[T^{-1}]$	rad/s
angular acceleration	$[T^{-2}]$	rad/s^2
moment of inertia	$[M]\cdot[L^2]$	kg·m^2

Practice Problems

1. The wheel of fortune is divided into 12, 30° (0.5236 radians) sections. When the wheel is spun it will have a negative angular acceleration of 0.535 radians per s^2. If the jackpot is in position 9, what is the minimum angular velocity, to the nearest hundredth of a radians/s, you need to give it in order to win the jackpot?

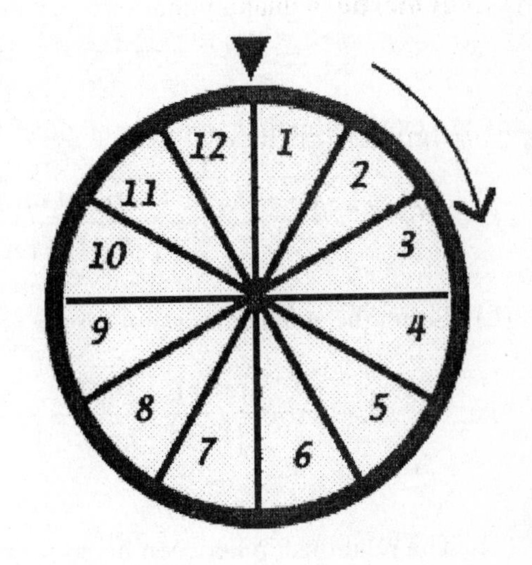

2. Now you are eligible to try for the grand prize. It is located at position 5. The catch is, you must go by 5 one full revolution before stopping at it. What is the needed minimum angular velocity?
3. A solid disk rolls up a hill. If it is revolving at 269 rpm when it starts up the hill, what is its angular velocity, to two decimal places in radians/s?
4. If the disk in problem 3 has a negative angular acceleration of 0.41 rad/s^2, how long will it take to stop, to the nearest hundredth of a second?
5. If the radius of the disk in problem 4 is 0.42 meters, what is the magnitude of its linear acceleration, to three decimal places.
6. Using conservation of energy, to the nearest hundredth of a meter, what is the final height of the disk?
7. How far up the hill does the disk travel before stopping, to the nearest hundredth of a meter?
8. To the nearest thousandth of a degree, what is the angle of the incline in problem 3?
9. A 1-kg mass hangs by a string from a disk with radius 15 cm that has a rotational inertia of 5×10^{-5} kg·m^2. After it falls a distance of 1.4 meters how fast is it going, to the nearest hundredth of a m/s?
10. In problem 9 what is the angular velocity of the disk, to the nearest tenth of a radians per second?

Puzzle

LOOKALIKES

You are give two identical-looking metal cylinders and a long rope. The cylinders have the same size and shape, and they weigh the same. You are told that one of them is hollow; the other is solid. How would you determine which is which using only the rope and the two cylinders?

Selected Solutions

15. Since the fan slows uniformly to a stop, we can take $\omega_f = 0$. It will be useful for both parts (a) and (b) to first determine the angular acceleration α:

$$\omega_f = \omega_0 + \alpha t \Rightarrow \alpha = \frac{\omega_f - \omega_0}{t} \quad \therefore \quad \alpha = \frac{0 - 0.50 \text{ rev/s}}{12 \text{ s}} = -0.0417 \text{ rev/s}^2$$

(a) An expression that gives the angular displacement (or number of revolutions) is

$$\theta - \theta_0 = \frac{\omega^2 - \omega_0^{\,2}}{2\alpha} = \frac{0 - (0.50 \text{ rev/s})^2}{2(-0.0417 \text{ rev/s}^2)} = \boxed{3.0 \text{ rev}}.$$

(b) Using the same expression for the angular displacement, except with $\omega_f = 0.25$ rev/s, we get

$$\theta - \theta_0 = \frac{(0.25 \text{ rev/s})^2 - (0.50 \text{ rev/s})^2}{2(-0.0417 \text{ rev/s}^2)} = \boxed{2.2 \text{ rev}}$$

25. (a) The relationship between linear (or tangential) and angular speed is $\omega = v_t/r$. This gives

$$\omega = \frac{v_t}{r} = \frac{8.50 \text{ m/s}}{7.20 \text{ m}} = \boxed{1.18 \text{ rad/s}}$$

(b) The centripetal acceleration is

$$a_{cp} = \frac{v_t^2}{r} = \frac{(8.50 \text{ m/s})^2}{7.20 \text{ m}} = \boxed{10.0 \text{ m/s}^2}$$

41. (a) The angular acceleration is given by $\alpha = a_t/r$. To find a_t we use

$$a_t = \frac{\Delta v}{\Delta t} = \frac{8.90 \text{ m/s} - 0}{12.2 \text{ s}} = 0.7295 \text{ m/s}^2$$

Therefore, we have

$$\alpha = \frac{a_t}{r} = \frac{0.7295 \text{ m/s}^2}{(36 \text{ cm})\left(\frac{1 \text{ m}}{100 \text{ cm}}\right)} = \boxed{2.0 \text{ rad/s}^2}$$

(b) Since $\alpha \propto \frac{1}{r}$, a smaller r would result in an angular acceleration $\boxed{\text{greater than}}$ that in part (a).

57. (a) We can use the conservation of mechanical energy of the system: $U_i + K_i = U_f + K_f$. This includes the rotational kinetic energy of the pulley. Initially, there is no kinetic energy, $K_i = 0$, and we can take the reference of potential energy such that the final potential energy is zero, $U_f = 0$. Therefore,

$$m_b gh + 0 = 0 + \tfrac{1}{2}m_b v^2 + \tfrac{1}{2}I_p\omega^2 \Rightarrow m_b gh = \tfrac{1}{2}m_b v^2 + \tfrac{1}{2}\left(\tfrac{1}{2}m_p r^2\right)\left(\frac{v}{r}\right)^2$$

$$\therefore \quad m_b gh = \tfrac{1}{2}m_b v^2 + \tfrac{1}{4}m_p v^2 \Rightarrow m_b gh = v^2\left(\frac{m_b}{2} + \frac{m_p}{4}\right)$$

$$\therefore \quad v = \sqrt{\frac{m_b gh}{\frac{m_b}{2} + \frac{m_p}{4}}} = \sqrt{\frac{(1.3\text{ kg})(9.81\text{ m/s}^2)(0.50\text{ m})}{\frac{1.3\text{ kg}}{2} + \frac{0.31\text{ kg}}{4}}} = \boxed{3.0\text{ m/s}}$$

(b) The speed will decrease because I_p is increased, so more of the kinetic energy is used up in I_p rather than in v.

59. (a) Use the conservation of energy

$$U_i + K_i = U_f + K_f \Rightarrow mgh + 0 = 0 + K_f$$

$$\therefore \quad K_f = mgh = (2.0\text{ kg})\left(9.81\frac{\text{m}}{\text{s}^2}\right)(0.75\text{ m}) = \boxed{15\text{ J}}$$

(b) To determine the rotational kinetic energy, we need to know the cylinder's speed, which we can now determine from the above result for K_f:

$$K_f = \tfrac{1}{2}mv^2 + \tfrac{1}{2}I\omega^2 \Rightarrow K_f = \tfrac{1}{2}mv^2 + \tfrac{1}{2}\left(\tfrac{1}{2}mr^2\right)\left(\frac{v}{r}\right)^2 = \tfrac{3}{4}mv^2$$

$$\therefore \quad v = 2\sqrt{\frac{K_f}{3m}} = 2\sqrt{\frac{14.7\text{ J}}{3(2.0\text{ kg})}} = 3.13\text{ m/s}$$

Now we can use

$$K_r = \tfrac{1}{2}I\omega^2 = \tfrac{1}{2}\left(\tfrac{1}{2}mr^2\right)\left(\frac{v}{r}\right)^2 = \tfrac{1}{4}mv^2 = \tfrac{1}{4}(2.0\text{ kg})\left(3.13\frac{\text{m}}{\text{s}}\right)^2 = \boxed{4.9\text{ J}}$$

(c) The translational kinetic energy is

$$K_t = \tfrac{1}{2}mv^2 = \tfrac{1}{2}(2.0\text{ kg})\left(3.13\frac{\text{m}}{\text{s}}\right)^2 = \boxed{9.8\text{ J}}$$

Answers to Practice Quiz

1. (b) **2**. (c) **3**. (a) **4**. (d) **5**. (c) **6**. (d) **7**. (e) **8**. (b) **9**. (d) **10**. (a) **11**. (d)

Answers to Practice Problems

1.	2.12 rad/s	**2**.	2.99 rad/s
3.	28.17 rad/s	**4**.	68.71 s
5.	0.172 m/s^2	**6**.	10.70 m
7.	406.80 m	**8**.	1.51°
9.	5.24 m/s	**10**.	34.9 rad/s

CHAPTER 11

ROTATIONAL DYNAMICS AND STATIC EQUILIBRIUM

Chapter Objectives

After studying this chapter, you should:

1. know how to calculate the torque due to a given force about a given axis.
2. understand the role of torque for rotational motion and its relationship to angular acceleration.
3. be able to analyze situations of static equilibrium.
4. know how to calculate angular momentum, understand its relationship to torque, and be able to apply angular momentum conservation.
5. be able to use the rotational forms of work and power.
6. be able to determine the spatial direction of the rotational vector quantities.

Warm-Ups

1. Two solid disks are linked with a belt. The diameter of the larger disk is twice as large as the diameter of the smaller disk. If the smaller disk is rotating at 500 revolutions per minute (rpm), what is the rotational speed of the larger disk? Express your answer in rpm and in radians per second.

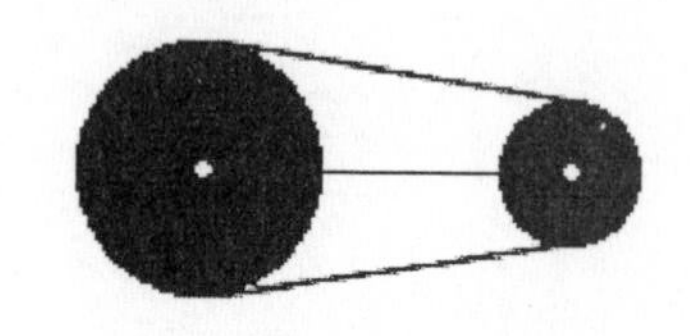

2. Estimate the angular acceleration of a small pebble stuck to a bicycle tire as the bicycle accelerates from rest to 10 mi/h (4.47 m/s) in 2 seconds.
3. A book can be rotated about many different axes. The moment of inertia of the book will depend on the axis chosen. Rank the choices A to C in the sketch in order of increasing moments of inertia.

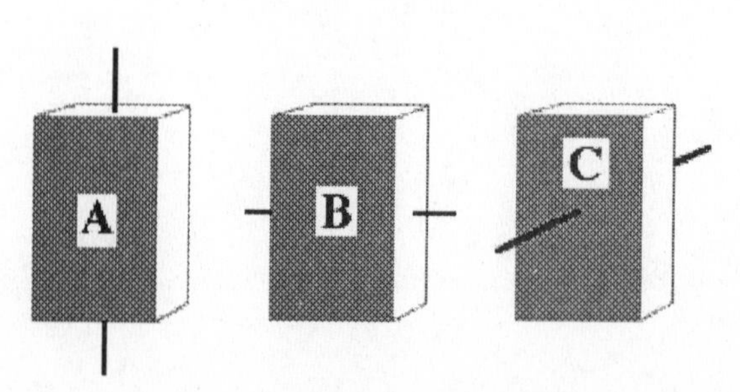

4. A hoop and a solid disk are released from rest at the top of an incline and allowed to roll down the incline without slipping. Which of the following is correct?
   a. The hoop has a larger moment of inertia than the disk
   b. Gravity is the only force that exerts a torque on the hoop and the disk.
   c. Both gravity and friction exert a torque on the hoop and the disk.
   d. Gravity exerts a greater torque on the hoop than on the disk.
   e. Gravity exerts a greater torque on the disk than on the hoop.
   f. Gravity torque on the hoop has the same magnitude as gravity torque on the disk.
   g. If the net torque on the hoop has the same magnitude as the net torque on the disk, the angular acceleration of the hoop is the same as the angular acceleration of the disk.
   h. If the net torque on the hoop has the same magnitude as the net torque on the disk, the angular acceleration of the hoop is smaller then the angular acceleration of the disk.
   i. If the angular acceleration is smaller, the tangential acceleration is smaller.
   j. If the tangential acceleration is smaller the center-of-mass acceleration is smaller.
   k. The object with smaller tangential acceleration will take longer time to reach the bottom of the incline.

Chapter Review

In this chapter we continue the study of rotation by looking at the dynamics of rotational motion. Here is where we introduce rotational versions of concepts like force, linear momentum, and Newton's second law.

11–1 – 11–2 Torque and Angular Acceleration

If you want to cause a nonrotating object to rotate, you must apply a force. However, not just any force will result in a rotation. Crucial to determining the rotation that results from the force are two factors: (a) how the applied force is directed relative to the axis of rotation and (b) the distance of the point at which the force is being applied from the axis. These two factors combine with the magnitude of the force to form a quantity called **torque**. It is through the application of a torque that an object will begin to rotate. The magnitude of the torque τ that results from the application of a force **F** is given by

$$\tau = rF \sin \theta$$

where r is the magnitude of a radial vector **r** from the axis of rotation to the point of action of the force, and θ is the smallest angle between **r** and **F**. The SI unit of torque is the N·m. *Note that in calculations dealing with torque the N·m is **not** called a joule.*

For convenience, the expression for torque is often looked at in two ways. The quantity $F\sin(\theta)$ equals the component of **F** tangential to a circle of radius r centered on the axis of rotation, F_t, so we can write $\tau = rF_t$. Grouped the other way, the quantity $r\sin(\theta)$ equals the component of **r** perpendicular to the line of force, $r_\perp$, called the **moment arm** (or *lever arm*) of the force, so we can also write $\tau = r_\perp F$.

All the above equations are for the magnitude of the torque only. Torque is a vector quantity, and we'll briefly discuss how to find its spatial direction later. In our applications torque will be a one-dimensional vector, so we can account for its direction with just an algebraic sign. The convention is as follows:

> *A torque is positive if its tendency is to rotate an object counterclockwise and negative if its tendency is to rotate an object clockwise.*

Example 11–1 Tether Ball A child is playing tether ball; she has a 2.0-m-long cord with one end connected to a vertical pole. Attached to the other end of the cord is a light ball of mass 0.43 kg. As the ball whirls around in a horizontal circle, the string makes an angle of 35° with the pole. The child then hits the ball with a force of 3.6 N that lies in the horizontal plane of the ball's motion and makes an angle of $\phi = 30°$ with the tangent to the ball's path (away from the pole). What magnitude of torque does the child's force apply to the ball about an axis through the pole?

Picture the Problem Figure **(a)** shows the ball connected to the pole by a cord of length L. Figure **(b)** is a top view showing the ball's circular path and the force **F** applied by the child.

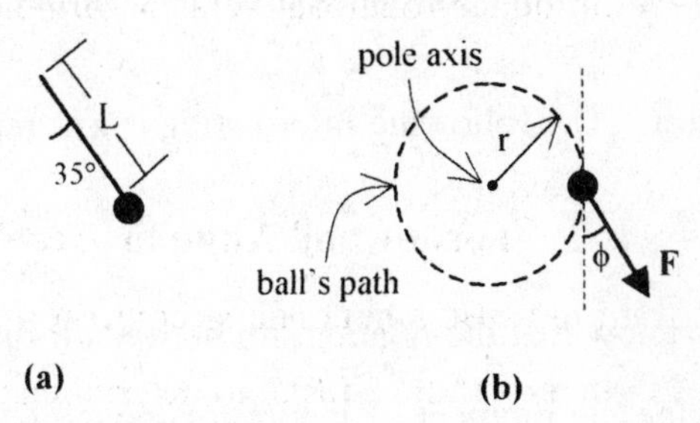

Strategy To complete this calculation we must correctly determine the quantities that are relevant to calculating torque: r, F, and θ.

Solution

1. In figure **(a)** r equals the horizontal distance between the ball and the pole's axis: $\sin(35^\circ) = \frac{r}{L} \quad \therefore \quad r = L\sin(35^\circ)$

2. Since the vector **r** is perpendicular to the tangent line, the angle θ can be found from ϕ: $\theta = 90° - \phi = 90° - 30° = 60°$

3. Directly calculate the magnitude of the torque:

$$\tau = rF\sin\theta = LF\sin(35^\circ)\sin\theta$$
$$= (2.0\text{ m})(3.6\text{ N})\sin(35^\circ)\sin(60^\circ) = 3.6\text{ N}\cdot\text{m}$$

Insight We needed to calculate only the magnitude of the torque in this problem. If we wanted to associate a sign or direction to this torque, according to the sketch it would be negative because the torque contributes to a clockwise rotation of the ball.

Torque is the quantity that plays the role of force for rotational motion. Just as force causes translational acceleration, torque causes angular acceleration. The relationship between τ and α is very similar to that between force and linear acceleration:

$$\tau = I\alpha$$

Comparing this equation with F = ma, we can see how strong the similarity is by recalling that the moment of inertia, I, is the rotational quantity that acts like mass does for translational motion. The above equation is referred to as Newton's second law for rotation.

Exercise 11–2 Engine Specifications The engine specifications on a Pontiac Grand Am SE state that it supplies 225 ft·lb of torque at 4200 rpm. If this same torque is applied to a solid cylinder of mass 3.3 kg and radius 32 cm that rotates about its central axis, what angular acceleration will result?

Solution: The following information is given in the problem.

Given: $\tau = 225$ ft·lb, $M = 3.3$ kg, $r = 32$ cm, $\omega = 4200$ rpm; **Find**: α

We know that the relationship between torque and angular acceleration is $\tau = I\alpha$. Therefore, if we can determine the moment of inertia for the cylinder we can use it to determine α. From Table 10–1 in the text we can see that the moment of inertia for a solid cylinder about an axis through its center is

$$I = \tfrac{1}{2}Mr^2.$$

Substituting this into Newton's second law for rotation and solving for α yields

$$\alpha = \frac{\tau}{I} = \frac{2\tau}{Mr^2}$$

Inserting numerical values and converting, we obtain the final result:

$$\alpha = \frac{2\tau}{Mr^2} = \frac{2(225\,\text{ft}\cdot\text{lb})\left(\dfrac{1\,\text{N}\cdot\text{m}}{0.738\,\text{ft}\cdot\text{lb}}\right)}{(3.3\,\text{kg})(0.32\,\text{m})^2} = 1800\,\text{rad/s}^2$$

Notice that to answer the question of this problem we did not need the angular velocity. For the engineers working on this car the angular velocity is important because it is at this angular speed that the result applies. You should always try to understand the relevance of a calculated result; otherwise, you will greatly diminish its usefulness to you.

Practice Quiz

1. In trial 1 a force **F** is applied tangentially to the rim of a wheel, causing it to rotate about its center. In trial 2 an equal force is applied tangentially at a point halfway out from the center to the rim, also causing a rotation about the center. How do the torques applied in each trial relate to each other?

   (a) Trial 2 has twice the torque of trial 1.

   (b) Trial 1 has twice the torque of trial 2.

   (c) Trial 1 has half the torque of trial 2.

   (d) The torques are equal.

2. If an object has a constant net torque applied to it, this implies that the object must also have
   (a) constant moment of inertia
   (b) constant angular velocity
   (c) constant angular acceleration
   (d) constant angular displacement
   (e) none of the above

11–3 – 11–4 Static Equilibrium and Balance

In many applications, especially in the design of buildings and other structures, objects need to be balanced so that they do not fall down or tip over on people. This means that the objects must be in **static equilibrium**. We have seen examples for which an equilibrium of forces means that an object will not translate; however, to be in complete static equilibrium the object must not rotate either. For the object to be nonrotating, there must be an equilibrium of the torques on the object. The conditions for static equilibrium, then, are that both the net force and the net torque on an object must be zero:

$$\sum F_i = 0, \qquad \sum \tau_i = 0$$

When doing an analysis for equilibrium it is useful to remember that the center-of-mass concept is often very useful here. The weight of an object can exert a torque about any axis of rotation that does not pass through the center of mass. Therefore, with regard to torque, we can treat rigid objects like point masses whose mass is located at the center of mass, as is illustrated by the following example.

Example 11–3 Hanging the Lights A horizontally suspended 2.7-m track for fluorescent lights is held in place by two vertical beams that are connected to the ceiling. The track and lights jointly weigh 15 N, and each beam is attached 60 cm from opposite ends of the track. What upward force does each beam apply to the track?

Picture the Problem The top diagram shows the light (at the bottom) attached to a track that hangs from the ceiling by two vertical beams.

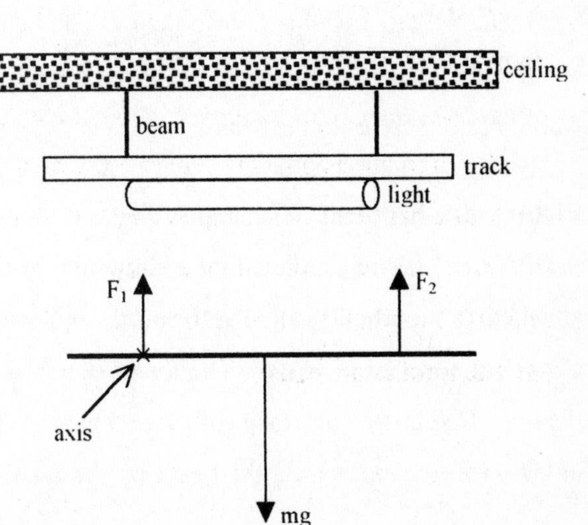

Strategy To solve this problem we will apply the equilibrium conditions. The first step is to draw a free-body diagram of the track (shown in the lower sketch) with the left beam applying a force labeled F_1 and the right beam applying a force F_2. The axis for calculating torques is located at the left beam.

We will calculate torque as force times lever arm.

Solution

1. Write out the torque due to F_1: $\tau_1 = \ell_1 F_1 = (0)F_1 = 0$

2. Write out the torque due to F_2: $\tau_2 = \ell_2 F_2 = (2.7\,\text{m} - 1.2\,\text{m})F_2 = (1.5\,\text{m})F_2$

3. Write out the torque due to gravity: $\tau_G = \ell_G mg = (\frac{2.7\,\text{m}}{2} - 0.60\,\text{m})\,mg = (0.75\,\text{m})\,mg$

4. Apply the equilibrium condition for torque: $\sum \tau_i = \tau_1 + \tau_2 - \tau_G = 0$

$$(1.5\,\text{m})\,F_2 - (0.75\,\text{m})\,mg = 0 \quad \therefore \quad F_2 = \frac{0.75\,\text{m}}{1.5\,\text{m}}mg$$

$$F_2 = \tfrac{1}{2}(15\,\text{N}) = 7.5\,\text{N}$$

5. Apply the equilibrium condition for force: $\sum F_i = 0 \quad \therefore \quad F_1 + F_2 - mg = 0$

$$F_1 = mg - F_2 = 15\,\text{N} - 7.5\,\text{N} = 7.5\,\text{N}$$

Insight Each beam contributes equally to supporting the track. Once you become more used to this type of analysis you'll be able to solve a problem like this without any calculations. The symmetry of the problem shows that F_1 and F_2 have equal lever arms about an axis through the center of the track. Since they apply torques of equal magnitude about this axis, they must also apply forces of equal magnitude. Since the lever arms share the 15 N equally, each must be 7.5 N.

Example 11–4 Decorating A decorative ornament is hung from the end of an 82.3-cm-long horizontal beam attached to a wall as shown below. The beam weighs 12.1 N, and the ornament weighs 7.62 N. **(a)** What is the tension in the cord? **(b)** What are the horizontal and vertical components of the force that the wall directly applies to the beam?

Picture the Problem The upper diagram shows the ornament being supported by a horizontal beam attached to a vertical wall. The beam is supported by a cord attached to its end. The lower sketch is a *partial* free-body diagram of the beam (the unknown force exerted on the beam by the wall is

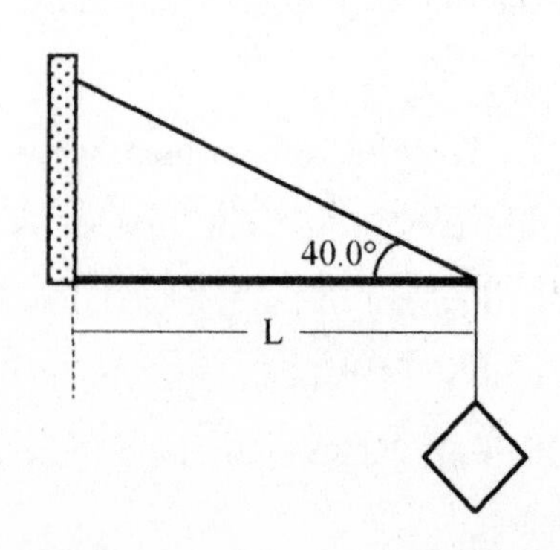

not shown). The lever arm for the tension T is also shown on the diagram as ℓ_T.

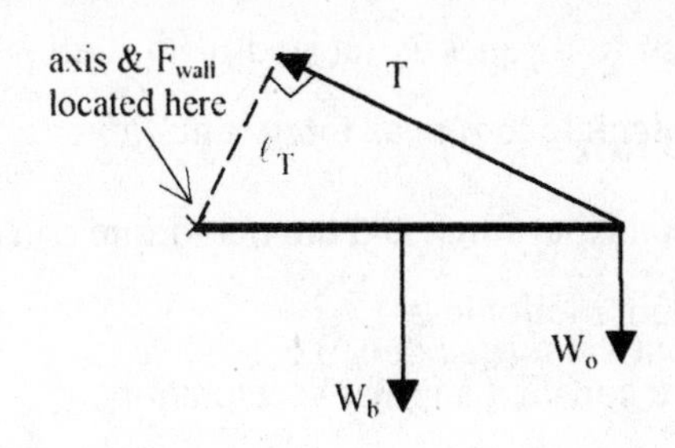

Strategy F_{wall} is unknown at this stage, so it is not explicitly shown in the lower diagram. We chose the axis to be at that location to simplify the equation for the net torque. Let's first apply the torque condition and see where it leads us. (When needed we shall choose up as the +y direction, and right as the +x direction)

Solution

Part (a)

1. Write the equilibrium condition for the torques: $\sum \tau_i = T\ell_T - W_b \frac{L}{2} - W_o L = 0$

2. From the geometry of the problem write the lever arm for the tension in terms of L: $\sin(40.0^\circ) = \frac{\ell_T}{L} \quad \therefore \quad \ell_T = L\sin(40.0^\circ)$

3. Divide through by L: $T\sin(40.0^\circ) - \frac{1}{2}W_b - W_o = 0$

4. Solve for T: $T = \frac{W_o + W_b/2}{\sin(40.0^\circ)} = \frac{7.62\text{ N} + 12.1\text{ N}/2}{\sin(40.0^\circ)} = 21.3\text{ N}$

Part (b)

5. Apply the equilibrium condition for the x components of the forces:

$$\sum F_x = F_{wall,x} - T_x = 0$$
$$\therefore F_{wall,x} = T_x = T\cos(40.0^\circ)$$
$$F_{wall,x} = (21.27\text{ N})\cos(40.0^\circ) = 16.3\text{ N}$$

6. Apply the equilibrium condition for the y components of the forces:

$$\sum F_y = F_{wall,y} + T_y - W_b - W_o = 0$$
$$F_{wall,y} = W_b + W_o - T_y = W_b + W_o - T\sin(40.0^\circ)$$
$$F_{wall,y} = 12.1\text{ N} + 7.62\text{ N} - (21.27\text{ N})\sin(40.0^\circ)$$
$$= 6.05\text{ N}$$

Insight The signs of the results in part (b) tell us that $F_{wall,x}$ is to the right, and $F_{wall,y}$ is upward. We could have discovered these facts without calculations, however, by looking carefully at the equilibrium situation. This will be discussed in more detail in the Tips section. Also, the length of the beam does not enter into the calculations.

Practice Quiz

3. If both the net force and the net torque on an object is zero, the object will
   (a) remain motionless
   (b) have constant angular acceleration
   (c) have constant linear acceleration
   (d) have constant angular velocity
   (e) have no forces applied to it

4. A 1.5-m-long, uniform beam weighing 5.5 N is supported at one end by a fulcrum as shown. What force F is needed at the other end to maintain static equilibrium?

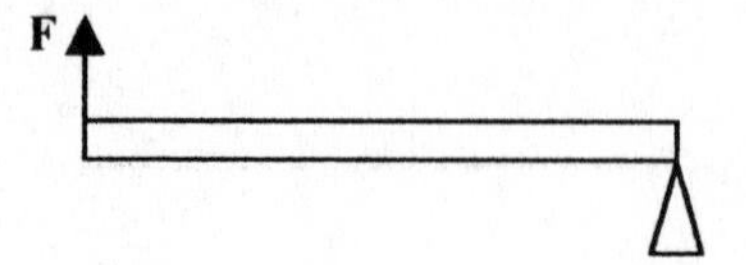

   (a) 2.8 N (b) 5.5 N (c) 3.0 N
   (d) 0.75 N (e) 8.2 N

11–5 Dynamic Applications of Torque

The previous section dealt with static applications of torque. In this section we consider how to handle cases for which the net torque is not zero, resulting in angular acceleration.

Exercise 11–5 A Design Flaw Suppose the person who designed the ornament display in Example 11–4 neglected to do the calculation and used a cord that could not withstand the required tension, and it broke. What was the initial angular acceleration of the beam when the cord broke? (Assume the end attached to the beam remained attached.)

Solution: The following information is known.

Given: $L = 82.3$ cm, $W_b = 12.1$ N, $W_o = 7.62$ N; **Find**: α

Since the net torque is not equal to zero in this case, we use Newton's second law, $\tau_{net} = I\alpha$. The moment of inertia is given by both the beam and the ornament: $I = I_b + I_o$. For the beam we use the results in Table 10–1

$$I_b = \frac{1}{3} M_b L^2$$

and for the ornament we have just $I_o = M_o L^2$. Therefore,

$$I = \left(M_o + \frac{1}{3}M_b\right)L^2$$

The net torque can be calculated from the results of Example 11–4 without the tension,

$$\tau_{net} = -W_b\frac{L}{2} - W_o L = -(W_o + W_b/2)L\,.$$

Using these results in Newton's second law to solve for α we get

$$\alpha = \frac{\tau_{net}}{I} = -\frac{W_o + W_b/2}{(M_o + M_b/3)L}$$

Making use of the fact that M = W/g we can solve for the angular acceleration,

$$\alpha = -\frac{7.62\,N + (12.1\,N)/2}{[7.62\,N + (12.1\,N)/3](0.823\,m)/9.81\,m/s^2} = -14.0\,rad/s^2$$

Remember, this result is only the initial angular acceleration of the beam when the cord breaks.

Practice Quiz

5. A uniform hoop of radius 0.75 m and mass 2.25 kg rotates with an angular acceleration of 44 rad/s^2 about an axis through its center. What is the net torque on this hoop?

   (a) 0 N·m **(b)** 33 N·m **(c)** 1.3 N·m **(d)** 74 N·m **(e)** 56 N·m

11–6 – 11–7 Angular Momentum and its Conservation

In studying translational motion we found that the concept of linear momentum was very useful and important to understanding the behavior of objects, especially systems of particles. For similar reasons we also have a rotational form of this concept called **angular momentum**, L. Just as for linear momentum, there is a conservation principle for angular momentum.

The angular momentum of a rotating object can be calculated as

$$L = I\omega$$

where I is calculated about the axis of rotation. In making the comparison with linear momentum, p = mv, we see that angular momentum results by replacing the linear quantities with angular ones: p → L, m → I, and v → ω. For a particle of mass m, the angular momentum can be written in terms of the linear momentum in analogy with how torque is written in terms of force:

$$L = rp\sin\theta$$

where, as with torque, r is the magnitude of a radial vector **r** from the axis of rotation to the particle, and θ is the smallest angle between **r** and **p**. The quantity p sin(θ) equals the component of the linear momentum tangential to a circle of radius r centered on the axis of rotation, p_t. The quantity r sin(θ) equals the component of **r** perpendicular to the direction of **p**, $r_\perp$. Therefore, we also have

$$L = rp_t = r_\perp p$$

From the above expressions we can determine that the SI unit of angular momentum is kg·m²/s = J·s.

Recall that with translational motion the concept of linear momentum led us to a better form for Newton's second law because F = ma works only when mass is constant. The same is true for rotational motion. The expression τ = Iα works only when I remains constant. It is much more common with rotational motion to see situations with varying I than it is to see situations with varying mass in linear motion. The concept of angular momentum solves this problem because the best rotational form of Newton's second law, valid for situations in which I is either constant or changing, is that torque equals the rate of change of angular momentum:

$$\tau = \frac{\Delta L}{\Delta t}$$

Example 11–6 Thrown from a Ledge A person throws a 0.336-kg object vertically downward from a 7.83-m-high ledge with an initial speed of 2.25 m/s. What average torque is exerted on this object by gravity during its fall to the ground relative to a point *O* on the ground that is a distance x = 5.92 m from where it strikes?

Picture the Problem The diagram shows the object at its initial height h with a downward initial velocity v_i. The point *O* relative to which the torque will be calculated is also shown.

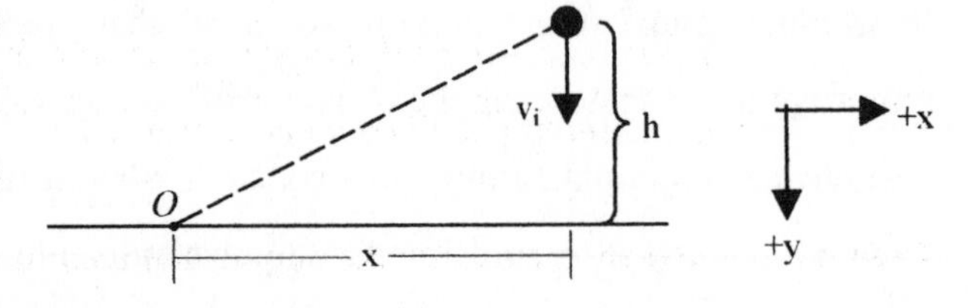

Strategy The average torque can be determined from the average rate of change of angular momentum, ΔL/Δt, so we can determine τ_{av} by calculating the angular momentum at the top and the bottom as well as the time it takes the object to reach the ground.

Solution

1. Obtain the magnitude of the initial angular momentum:

$$L_i = (r_{i\perp})p_i = xmv_i = (5.92\text{ m})(0.336\text{ kg})(2.25\text{ m/s})$$
$$= 4.476\text{ J}\cdot\text{s}$$

2. Obtain the magnitude of the final velocity using the free-fall equations:

$$v_f = \sqrt{v_i^2 + 2gh} = \sqrt{(2.25\ \tfrac{m}{s})^2 + 2(9.81\ \tfrac{m}{s^2})(7.83\ m)} = 12.60\ m/s$$

3. The final angular momentum is:

$$L = (r_{f\perp})p_f = xmv_f = (5.92\ m)(0.336\ kg)(12.60\ \tfrac{m}{s}) = 25.06\ J \cdot s$$

4. Setting $\Delta t = t$, calculate the time it takes to reach the ground from the equations for free-fall:

$$v_f = v_i + gt \quad \therefore \quad t = (v_f - v_i)/g$$

$$t = \frac{12.60\ m/s - 2.25\ m/s}{9.81\ m/s^2} = 1.055\ s$$

5. Calculate the average torque:

$$\tau_{av} = \frac{L_f - L_i}{t} = \frac{25.06\ J \cdot s - 4.476\ J \cdot s}{1.055\ s} = 19.5\ N \cdot m$$

Insight This problem asked only for the average torque, so we used only the total change in angular momentum over a period of time and not the instantaneous rate of change.

Newton's second law for rotation, as just discussed, implies that torque causes a change in the angular momentum of a system. Consequently, if there is no net torque on a system, there will be no change in the angular momentum; this is the principle of the conservation of angular momentum. A more complete statement of this principle is the following:

> *If the net external torque on a system is zero, then the total angular momentum of the system is conserved.*

The total angular momentum is the sum of the angular momenta of every object in the system.

Example 11–7 The Merry-go-round A merry-go-round with a moment of inertia of 750 kg·m^2 and a radius R of 2.55 m is rotating with an angular velocity of 9.42 rad/s clockwise (as viewed from above). A 334-N child runs at 2.76 m/s tangent to the rim of the merry-go-round and jumps onto it in the direction opposite its sense of rotation. With what angular velocity does the merry-go-round rotate after the child jumps onto it?

Picture the Problem The diagram shows a top view of the rotating merry-go-round and the running child's path before the child jumps on.

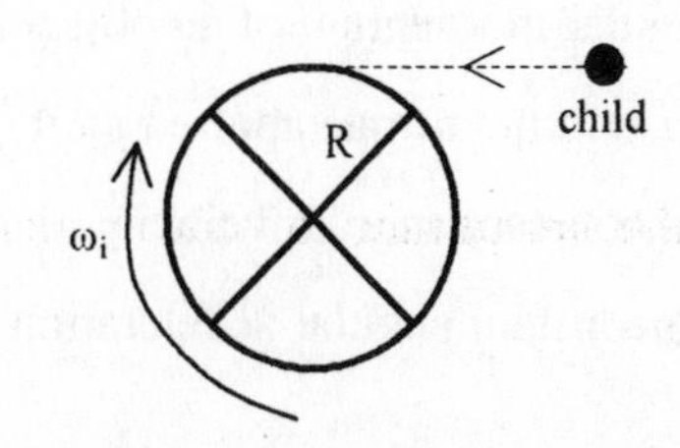

Strategy We can solve this problem by using the conservation of angular momentum. We will equate the total angular momentum before the child jumps on with the total angular momentum after the child jumps on.

Solution

1. The initial angular momentum at the instant the child is about to jump on is:

$$L_{total} = L_c - L_{mgr} = Rp_c - I_{mgr}\omega_i$$

2. The final angular momentum after the child jumps on is:

$$L_{total} = I_{total}\omega_f = \left(I_{mgr} + m_c R^2\right)\omega_f$$

3. The mass of the child is:

$$m_c = W_c / g = 334\ \text{N}/(9.81\ \text{m/s}^2) = 34.05\ \text{kg}$$

4. Set the initial and final angular momenta equal to each other and solve for ω_f:

$$Rp_c - I_{mgr}\omega_i = \left(I_{mgr} + m_c R^2\right)\omega_f$$

$$\omega_f = \frac{Rm_c v_c - I_{mgr}\omega_i}{I_{mgr} + m_c R^2}$$

5. Obtain the numerical result:

$$\omega_f = \frac{(2.55\ \text{m})(34.05\ \text{kg})(2.76\ \tfrac{\text{m}}{\text{s}}) - (750\ \text{kg}\cdot\text{m}^2)(9.42\ \tfrac{\text{rad}}{\text{s}})}{750\ \text{kg}\cdot\text{m}^2 + 34.05\ \text{kg}(2.55\ \text{m})^2}$$

$$= -7.03\ \text{rad/s}$$

Insight The negative sign on ω_f results because the rotation is clockwise. The initial angular velocity, ω_i, is also clockwise, and its negative sign was accounted for by subtracting L_{mgr} from L_c.

Practice Quiz

6. A uniform hoop of radius 0.75 m and mass 2.25 kg rotates with an angular velocity of 44 rad/s about an axis through its center. What is the magnitude of its angular momentum?

   (a) 0 J·s **(b)** 74 J·s **(c)** 1.3 J·s **(d)** 33 J·s **(e)** 56 J·s

7. If the angular momentum of an object is conserved

   (a) it has zero net torque applied on it

   (b) it has constant angular velocity

   (c) it has constant angular acceleration

(d) it has constant moment of inertia

(e) The angular momentum of every object is conserved

8. If the angular momentum of a system of several objects is conserved,

(a) there is a large net external torque on the system

(b) no torque acts on any object in the system

(c) the net external torque on the system is zero

(d) the system rotates as a rigid body

(e) none of the above

11–8 Rotational Work (and Power)

So far we have written rotational forms for several important quantities that we studied in translational motion, namely, displacement, velocity, acceleration, inertia, kinetic energy, force, and momentum. Two quantities that we have not addressed are work and power.

When a force gives rise to a torque that rotates an object, the force acts through a displacement and therefore does work. An analysis of this work shows that it can be written in the following rotational form

$$W = \tau\,\Delta\theta\,.$$

Once again, you should compare this equation with the one-dimensional linear version, $W = F\Delta x$. Keep in mind that in the above expression, the angular displacement is measured in radians. The average power consumed when a torque does work to rotate an object can also be written in rotational form. Working from the definition of average power, $P = W/\Delta t$, we obtain

$$P = \tau\omega$$

This expression compares directly with $P = Fv$ for the case of translational motion.

Exercise 11–8 Tool Sharpening To sharpen a tool, the toolmaker uses a grinding wheel of radius 0.695 m with a coarse rim. The tool to be sharpened is pressed against the rim of the wheel with a force of 25.9 N (directed toward the center) as the wheel undergoes exactly 12 rotations. If the coefficient of kinetic friction between the tool and the rim is 0.644, how much work is done by friction in sharpening the tool?

Solution: The following information is given in the problem:

Given: $N = 25.9$ N, $r = 0.695$ m, # of rev = 12, $\mu_k = 0.644$; **Find**: W

In this situation torque is applied by the force of kinetic friction. This force acts tangentially along the rim of the grinding wheel. The torque about the center of the wheel is then given by

$$\tau = rf_k = r\mu_k N$$

The number of rotations of the wheel corresponds directly to the angular displacement:

$$\Delta\theta = 12\ \text{rev}\left(\frac{2\pi\ \text{rad}}{\text{rev}}\right) = 75.40\ \text{rad}$$

Therefore, the work done by friction is

$$W = \tau\Delta\theta = r\mu_k N\Delta\theta = (0.695\ \text{m})(0.644)(25.9\ \text{N})(75.40\ \text{rad}) = 874\ \text{J}$$

Practice Quiz

9. If a constant torque of 5.2 N·m rotates an object through 136°, how much work is done by the torque?

(a) 5.2 J **(b)** 26 J **(c)** 710 J **(d)** 3.9 J **(e)** 12 J

11–9 The Vector Nature of Rotational Motion

Many of the rotational quantities that we have discussed are vector quantities and are therefore associated with particular directions in space. The quantities that must be considered are angular velocity, ω; angular acceleration α; torque τ; and angular momentum, **L**. The directions of rotating objects are determined by right-hand rules. We need to state only two right-hand rules, and we can use them to get the directions of all the quantities.

Angular Velocity

The direction of the angular velocity vector is perpendicular to the plane in which the rotation takes place. To determine the direction in which ω points we use the following right-hand rule:

Curl the fingers of your right hand along the direction of rotation; your extended thumb now indicates the direction of the angular velocity.

Once we know how to determine the direction of ω, we can also determine the direction of angular momentum because the vector equation $\mathbf{L} = I\omega$ tells us that **L** must point along ω.

Torque

The direction of the torque vector is perpendicular to the plane formed by the vectors **r** and **F**. To determine the side of this plane to which τ points we use the following right-hand rule:

Curl the fingers of your right hand along the direction (in the **r-F** *plane) in which the torque tends to rotate the object; your extended thumb now indicates the direction of the torque.*

Once we know how to determine the direction of τ we can also determine the direction of the angular acceleration resulting from this torque because the vector equation $\tau = I\alpha$ tells us that τ and α must point in the same direction.

Example 11–9 Getting the Directions **(a)** A solid circular disk rotates about a vertical axis as shown. What is the direction of the angular momentum of the disk about its central axis? **(b)** A beam sits on a fulcrum located at its center. If a force **F** is then applied to the end of the beam as shown, what is the direction of the torque due to this force about the center of the beam? **(c)** A particle of mass m moves with momentum **p** relative to a point *O* as shown. Determine the direction of the angular momentum of the particle relative to *O*.

Picture the Problem The diagram shows the three situations described in the statement of the problem.

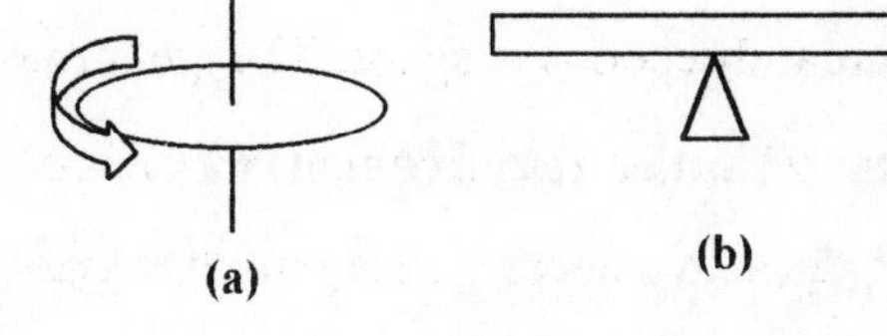

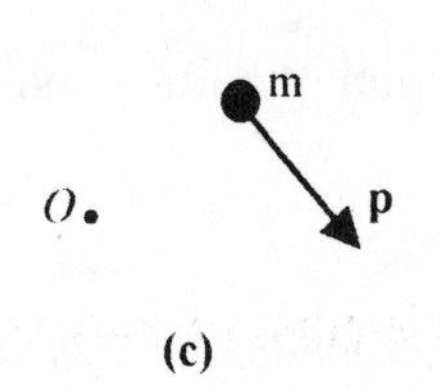

Strategy In each case we use the relevant right-hand rule.

Solution

Part (a)

1. Curl the fingers of your right hand along the direction of rotation. Your thumb should indicate that **L** is along the axis pointing upward.

Part (b)

2. The vector **r** points horizontally to the right, so the **r-F** plane is the plane of the page. The force tries to rotate the beam counterclockwise. Curl your fingers this way. Your thumb should indicate that **L** points directly out of the page (perpendicularly) toward you.

Part (c)

3. The mass m can be viewed as instantaneously rotating around point O with an angular velocity ω in the plane of the page. This rotation is clockwise in the figure shown. Curling your fingers clockwise should indicate that **L** points directly into the page (perpendicularly) away from you.

Insight Try reversing the directions of rotation for part (a), **F** for part (b), and **p** for part (c), and verify that you will get **L** pointing in the opposite directions.

Practice Quiz

10. The axis of a rotating sphere is aligned along the north-south direction. If the sphere rotates clockwise as viewed from the southern side, what is the direction of its angular momentum?
(a) north (b) south (c) east (d) west (e) upward toward the sky

Reference Tools and Resources

I. Key Terms and Phrases

torque the combination of a force and the distance at which it is applied from an axis that causes angular acceleration

moment arm the perpendicular distance from the axis of rotation to the line of force used to calculate torque

static equilibrium the state of motion in which an object neither translates nor rotates

angular momentum a vector quantity equal to $\Sigma r\ p \sin(\theta)$ that represents a rotational analog of linear momentum

angular momentum conservation the principle that the total angular momentum of a systems remains constant unless a nonzero external net torque is applied

right-hand rule the rules for using your right hand to determine the direction of rotational vector quantities

II. Tips

Sometimes, when you are performing an analysis of a static equilibrium situation, it is not immediately clear in which direction some of the forces are pointing at the time you draw a free-body diagram. In Example 11–4, we didn't determine the direction of the forces exerted by the wall on the beam until the mathematical solution was completed. You can often get around this problem by considering different locations for the axis about which torques are being considered. Remember, if there is equilibrium of torques about one axis, then there is equilibrium about every axis.

Let's consider the partial free-body diagram from Example 11–4, except this time we place the axis on the right end of the beam instead of on the left end. Now, consider the torques about this new axis. Neither T nor W_o will exert a torque about this axis. The force W_b will try to rotate the beam counterclockwise about this axis. Since we know the beam is in equilibrium, the y component of F_{wall} must be such as to balance the torque of W_b. Therefore, we can conclude that $F_{wall,y}$ must point vertically upward. With a little practice you can do this kind of determination in your head fairly quickly. Afterward, you'll have a complete free-body diagram and can solve the numerical problem with the axis placed anywhere you want it.

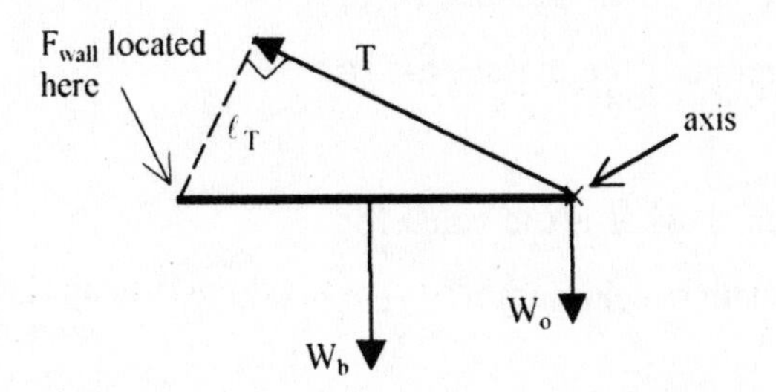

Another tactic is to put the unknown forces in positive directions and solve the equations as you normally would. If your result comes out positive, you guessed correctly; if the result comes out negative, then you should reverse the direction.

III. Important Equations

Name/Topic	Equation	Explanation
torque	$\tau = rF\sin\theta$	The magnitude of the torque due to a force **F**.
rotational dynamics	$\tau = I\alpha$ $\tau = \dfrac{\Delta L}{\Delta t}$	The rotational forms of Newton's second law. The lower equation is more general.

angular momentum	$L = I\omega$ $L = rp\sin\theta$	The magnitude for the angular momentum. The lower equation applies to point particles.
work	$W = \tau\,\Delta\theta$	Work written in terms of rotational quantities.
power	$P = \tau\omega$	Power written in terms of rotational quantities.

IV. Know Your Units

Quantity	Dimension	SI Unit
torque	$[M]\cdot[L^2]/[T^2]$	N·m
angular momentum	$[M]\cdot[L^2]/[T]$	$kg\cdot m^2/s$ (= J·s)

Practice Problems

1. In the mobile at the right $m_1 = 0.23$ kg, and $m_2 = 0.59$ kg. What must the unknown distance be to the nearest tenth of a cm be if the masses are to be balanced?
2. In the mobile of problem 1 what is the value for m_3 to the nearest hundredth of a kilogram?

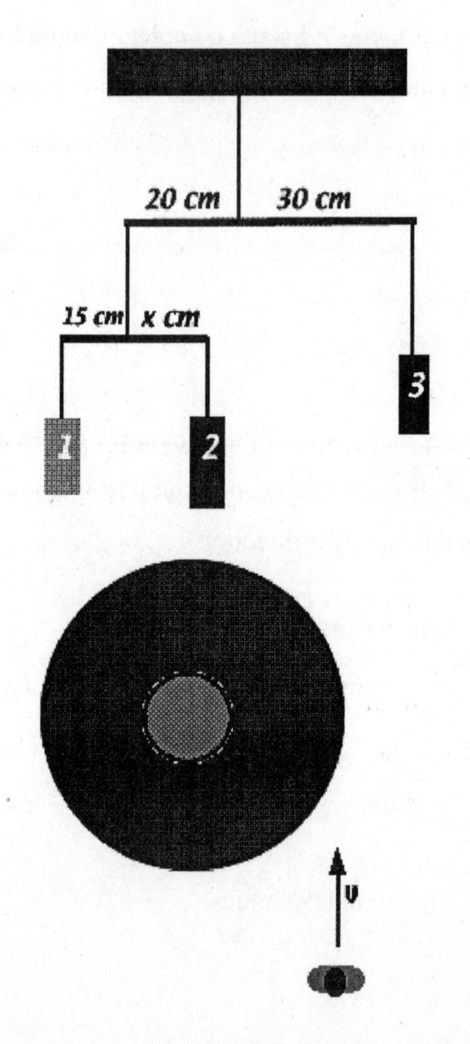

3. A runner of mass m = 49 kg and running at 3.3 m/s runs as shown and jumps on the rim of a playground merry-go-round that has a moment of inertia of 400 $kg\cdot m^2$ and a radius of 2 m. Assuming the merry-go-round is initially at rest, calculate its final angular velocity to three decimal places

4. A disk of radius 2.46 cm and mass 1 kg is pulled by a string wrapped around its circumference with a constant force of 0.5 newtons. What is its angular acceleration, to three decimal places?
5. In problem 4 what is the angular velocity of the disk, to three decimal places, after it has been turned through 0.71 of a revolution?
6. A large uniform 'butcher block' of mass 100 kg rests on two supports and has a weight hanging from its end. The block has a mass of 100 kg and a length of 2 meters. If L = 1.66 meters and the hanging weight is 149 newtons what is the force on the left support, to the nearest newton?

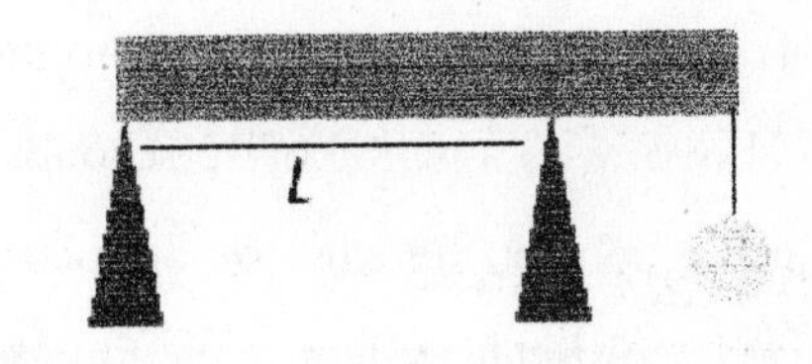

7. In problem 6 what is the force on the right support?
8. In problem 6 if the hanging weight is 157 newtons what is the minimum value for L for the configuration to remain stable, to the nearest hundredth of a meter?
9. In problem 6 what is the minimum value for the hanging weight, to the nearest tenth of a kilogram if L = 1.21 meters and the configuration is to remain stable?
10. A ladder can fall for two reasons: if it is set too steep and the climber gets his or her mass to the left (in the diagram) of the ladder's base, the ladder likely will fall over backward. If the ladder is set at too shallow an angle, the required force of friction between the ladder and the ground may be too great, and the base of the ladder will slip. Assume that there is no friction between the ladder and the wall and that the ladder is effectively weightless. The coefficient of friction between the base of the ladder and the ground is 0.27. The person using the ladder will be 3/4 of the way up the ladder. If the person climbing the ladder has a weight of 980 N and the ladder is 3.69 m long, how far from the wall can the base of the ladder be placed, to the nearest hundredth of a meter, and not slip?

Puzzle

DIZZYBALL

You practice throwing baskets from a spot 15 feet from the pole until you can make a basket every time. The ball leaves your hands the same way every time. The initial velocity of the ball is exactly correct. You visit an amusement park where there is a merry-go-round with a 15 feet diameter platform. A standard-size basketball post is fastened to the rim of the platform. The platform is rotating clockwise (as seen from above) with a period of 15 seconds.

Step onto a spot at the edge of the platform, diametrically opposite the basketball post. Imagine a ground-based coordinate system, with you at the origin, the positive x-axis drawn to your right, the y-axis drawn away from you toward the post, and the z-axis drawn straight up from the spot you are standing. When the platform is standing still you can always make the basket with the v_y and v_z components of the initial velocity you practiced on the ground.

How will you have to adjust the initial velocity of the ball to make a basket on the moving platform? Do you need to add an x component to the velocity? Do you have to adjust the y component and the z components in any way?

Selected Solutions

11. (a) For a disk, $I = \frac{1}{2}mr^2$. We can use the work-energy theorem to determine the average torque:

$$W = \Delta K \;\Rightarrow\; -\tau_{av}\Delta\theta = 0 - \tfrac{1}{2}I\omega_0^2 \;\therefore\; -\tau_{av}\Delta\theta = -\tfrac{1}{4}mr^2\omega_0^2 \;\Rightarrow\; \tau_{av} = \frac{mr^2\omega_0^2}{4\Delta\theta}$$

Therefore,

$$\tau_{av} = \frac{(6.4\text{ kg})(0.71\text{ m})^2(1.22\text{ rad/s})^2}{4(0.75\text{ rev})(2\pi\text{ rad/rev})} = \boxed{0.25\text{ N}\cdot\text{m}}$$

(b) Doubling the mass and halving the radius reduces I by a factor of 2 therefore reducing its kinetic energy. The same torque brings the wheel to rest in a $\boxed{\text{decreased}}$ angle of rotation.

27. (a) The rod is in equilibrium, so the net torque must be zero. The two contributions to the torque come from the tension in the cord and the weight of the rod.

$$\sum\tau = r_{\perp-\text{wire}}T - r_{\perp-\text{weight}}mg = 0 \;\Rightarrow\; T = \frac{r_{\perp-\text{weight}}}{r_{\perp-\text{wire}}}mg$$

For the given setup the lever arms are $r_{\perp-\text{weight}} = \frac{1}{2}L\cos 25^\circ$ and $r_{\perp-\text{wire}} = 0.51\text{ m}$, where L is the length of the rod. Therefore,

$$T = \frac{r_{\perp-\text{weight}}}{r_{\perp-\text{wire}}} mg = \frac{\frac{1}{2}(1.2\text{ m})\cos 25°}{0.51\text{ m}}(3.1\text{ kg})\left(9.81\ \frac{\text{m}}{\text{s}^2}\right) = \boxed{32\text{ N}}$$

(b) The horizontal component of the hinge force is equal to, and opposes, the horizontal wire tension: $\boxed{32\text{ N}}$. The vertical component of the hinge force is equal to, and opposes, the rod weight: (3.1 kg)(9.81 m/s^2) = $\boxed{30\text{ N}}$.

41. Since the force is tangential, the torque is $\tau = rF = I\alpha$, where $I = \frac{1}{2}mr^2$, and α can be written as $\alpha = \Delta\omega/\Delta t$. Combining these equations we get

$$rF = \frac{mr^2\Delta\omega}{2\Delta t} \Rightarrow m = \frac{2F\Delta t}{r\Delta\omega} = \frac{2(40.0\text{ N})(3.50\text{ s})}{(2.40\text{ m})(0.0870\text{ rev/s})(2\pi\text{ rad/rev})} = \boxed{213\text{ kg}}$$

59. The initial angular velocity is

$$\omega_i = \left(0.641\ \frac{\text{rev}}{\text{s}}\right)\left(2\pi\ \frac{\text{rad}}{\text{rev}}\right) = 4.0275\ \frac{\text{rad}}{\text{s}}$$

By the conservation of angular momentum,

$$L_{\text{disk}} + L_{\text{person}} = L_{\text{final}} \Rightarrow \left(\tfrac{1}{2}MR^2\right)\omega_i + mvR = \left(\tfrac{1}{2}MR^2 + mR^2\right)\omega_f$$

$$\therefore\ \omega_f = \frac{MR\omega_i + 2mv}{MR + 2mR} = \frac{(155\text{ kg})(2.63\text{ m})(4.0275\text{ rad/s}) + 2(59.4\text{ kg})(3.41\text{ m/s})}{(155\text{ kg})(2.63\text{ m}) + 2(59.4\text{ kg})(2.63\text{ m})} = \boxed{2.84\text{ rad/s}}$$

69. One complete turn = 2π rad; therefore,

$$W = \tau\Delta\theta = (3.3\text{ N}\cdot\text{m})(2\pi\text{ rad}) = \boxed{21\text{ J}}$$

Answers to Practice Quiz

1. (b) **2**. (c) **3**. (d) **4**. (a) **5**. (e) **6**. (e) **7**. (a) **8**. (c) **9**. (e) **10**. (a)

Practice Problem Answers:

1. 5.8 cm
2. 0.55 kg
3. 0.543 rad/s
4. 40.65 rad/s^2
5. 19.044 rad/s
6. 360 N
7. 770 N
8. 1.14 m
9. 260.8 kg
10. 1.25 m

CHAPTER 12

GRAVITY

Chapter Objectives

After studying this chapter, you should:

1. understand and be able to use Newton's universal law of gravitation.

2. be able to apply the principle of superposition to gravitational forces and potential energies.

3. know Kepler's laws of orbital motion.

4. know how to apply Kepler's third law to circular orbits.

5. be able to calculate gravitational potential energy and apply it in the conservation of energy.

6. be able to calculate the escape speed from planets and other objects.

Warm-Ups

1. A planet's orbit around the Sun is an ellipse. Consider points A and B on the ellipse. How does the centripetal force exerted on the planet at point A compare with the centripetal force exerted on the planet at point B? How about the potential energies at A and B? Kinetic energies? How about angular momenta?

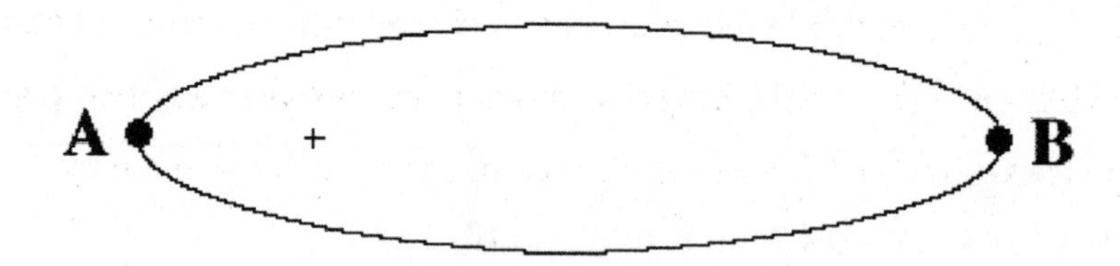

2. Estimate the force that the Moon exerts on you when it is directly overhead.
(You need some data from the text to answer this question.)

3. Two different planets are orbiting the same sun along two different orbits. The orbit containing points C and D is circular, the orbit containing points A and B is elliptical. Compare the speeds of the planet in the elliptical orbit at points A, B, and E. Compare the speeds of the planet in the circular orbit at points C, D, and E. The planet in the elliptical orbit is to be shifted to the circular orbit as it passes point E. Does it have to speed up or slow down?

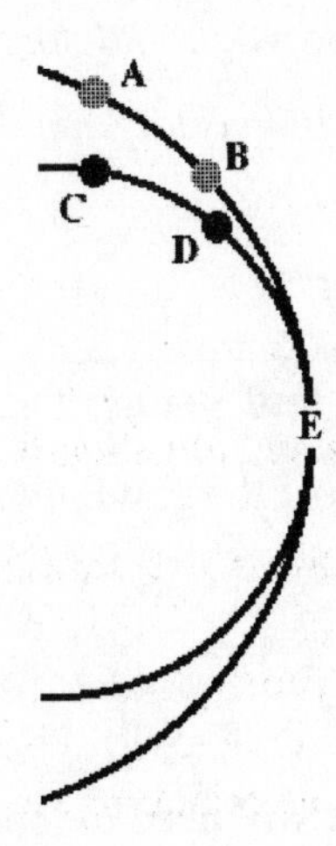

4. A space communication company is planning to take a spy satellite to a spot 35,800 km above Earth's surface and release into a geosynchronous orbit. (In a geosynchronous orbit a satellite orbits at the same rate as the points on the surface of Earth below it so as to appear to hover over the same spot.) Is this possible? If yes, how fast must the satellite be moving when it is released?

Chapter Review

In this chapter we study **gravity** — a fundamental force of nature. Of the four fundamental forces, gravity is the most familiar in everyday life. The other fundamental forces will be discussed later in the text.

12–1 – 12–2 Newton's Law of Universal Gravitation and the Attraction of Spherical Bodies

Gravitation is the phenomenon that between every two objects there is a force of attraction. **Newton's law of universal gravitation** describes the behavior of this force. Between any two point masses m_1 and m_2, the magnitude of the gravitational force on each mass due to the other is given by

$$F = G\frac{m_1 m_2}{r^2}$$

where r is the distance between the two masses, and G is a constant called the *universal gravitational constant*. The value of this constant is

$$G = 6.67 \times 10^{-11}\ \text{N}\cdot\text{m}^2/\text{kg}^2$$

The force on each mass points directly at the other mass because each mass attracts the other toward it.

For situations involving more than two masses we apply the **principle of superposition**:

The net gravitational force on any given mass due to two or more other masses is the vector sum of the gravitational forces due to each of the other masses individually.

Exercise 12–1 Earth and Venus On average Earth and Venus are separated by about 4.14×10^{10} m. What is the magnitude of the gravitational force between them at this separation?

Solution

In this problem we are given only the distance between Earth and Venus. To calculate the gravitational force between them we will need to look up their masses. The masses are as follows:

$$M_E = 5.97 \times 10^{24}\ \text{kg};\ M_V = 4.87 \times 10^{24}\ \text{kg}$$

Having all the data we need, the force can now be calculated.

$$F = \frac{GM_EM_V}{r^2} = \frac{(6.67\times10^{-11}\ \text{N}\cdot\text{m}^2/\text{kg}^2)(5.97\times10^{24}\ \text{kg})(4.87\times10^{24}\ \text{kg})}{(4.14\times10^{10}\ \text{m})^2} = 1.13\times10^{18}\ \text{N}$$

Remember, this is the magnitude of the force exerted on both Earth and Venus.

The above law of universal gravitation is stated for point masses. The detailed calculations for extended bodies can become complicated, but Newton figured out that the final result for spherical bodies becomes simple again. Basically, Newton showed that two completely separated, uniform spherical bodies attract each other as if they were point masses with all their mass located at their respective centers. This fact explains why the point-mass formula given above works so well for large objects like Earth and the Moon.

Treating the earth as a point mass M_E located at the center of the earth provides further insight into the acceleration due to gravity that we measure near Earth's surface. Objects near the surface are a distance from the center roughly equal to the radius of the earth, R_E. Using Newton's law of gravity we can conclude that the acceleration due to gravity near the surface is given by

$$g = \frac{GM_E}{R_E^2} \approx 9.8\ \text{m/s}^2$$

With this result you can also see that the higher you go above the surface, the farther you are from Earth's center, and therefore the weaker the effect of gravity, resulting in a smaller acceleration.

Example 12–2 Martian Gravity What is the acceleration due to gravity on the surface of Mars?

Picture the Problem The sketch is a representation of the planet Mars, with its radius extending from the center to the surface.

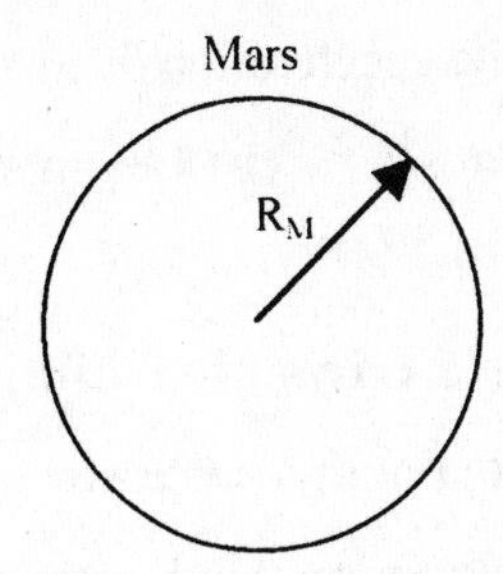

Strategy We need to adapt the expression for Earth's surface acceleration of gravity to Mars.

Solution

1. Look up the mass of Mars: $M_M = 6.4 \times 10^{23}$ kg

2. Look up the radius of Mars: $R_M = 3.4 \times 10^6$ m

3. Use the mass and radius of Mars in place of those for the Earth in the expression for g:

$$g_M = \frac{GM_M}{R_M^2} = \frac{(6.67\times10^{-11}\ \text{N}\cdot\text{m}^2/\text{kg}^2)(6.4\times10^{23}\ \text{kg})}{(3.4\times10^6\ \text{m})^2}$$
$$= 3.7\ \text{m/s}^2$$

Insight A similar substitution can be made for any of the spherical astronomical bodies.

Practice Quiz

1. Two objects are separated by a distance r. If the mass of each object is doubled, how does the gravitational force between them change?
   (a) It increases to twice as much.
   (b) It decreases to half as much.
   (c) It increases to four times as much.
   (d) It decreases to one-fourth as much.
   (e) None of the above

2. Two objects are separated by a distance r. If the distance between them is doubled, how does the gravitational force between them change?
   (a) It increases to twice as much.
   (b) It decreases to half as much.
   (c) It increases to four times as much.
   (d) It decreases to one-fourth as much.
   (e) None of the above

3. What is the acceleration due to gravity at a height of 1000 km above Earth's surface?

(a) 9.78 m/s^2 **(b)** 4.89 m/s^2 **(c)** 3.98 m/s^2 **(d)** 7.31 m/s^2 **(e)** None of the above

12–3 Kepler's Laws of Orbital Motion

One of the great early successes of Newton's work on gravity was his ability to accurately explain the laws of orbital motion that Kepler discovered from his observations of Mars and the other known planets. These laws of orbital motion are summarized as three statements. Kepler's first law is:

Planets follow elliptical orbits, with the Sun at one focus of the ellipse.

Kepler's second law is:

As a planet moves in its orbit, a line from the Sun to the planet sweeps out equal amounts of area in equal amounts of time.

And finally, the third law relates the amount of time it takes a planet to orbit the Sun, called its **orbital period**, T, to the planet's average distance from the Sun, r.

The orbital period of a planet is directly proportional to its average distance from the Sun raised to the 3/2 power. That is, $T \propto r^{3/2}$.

For the special case of circular orbits the constant of proportionality is relatively easy to determine. In this case the result of Kepler's third law can written in equation form as

$$T = \left(\frac{2\pi}{\sqrt{GM_S}}\right) r^{3/2}$$

Even though the above three laws are stated in terms of planetary orbits about the Sun, they also apply to the orbit of any body around a much more massive body such as the artificial satellites that orbit Earth.

Example 12–3 An Artificial Satellite If it is required to have a satellite in a circular orbit with an orbital period of 2 days, at what altitude h above the ground should it be placed in orbit?

Picture the Problem The sketch shows Earth and an artificial satellite a height h above Earth's surface.

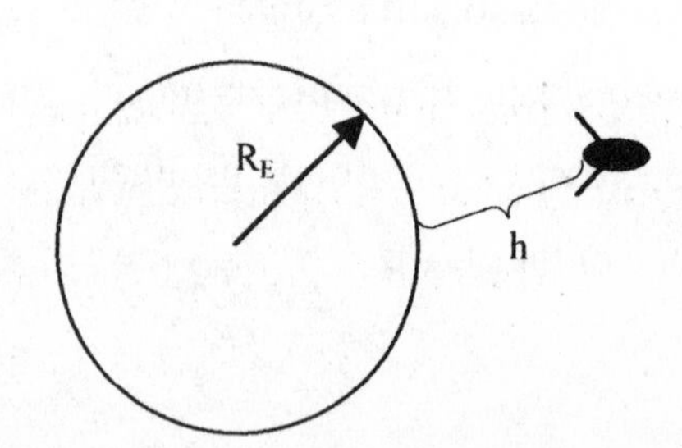

Strategy As mentioned above, Kepler's third law applies to satellites orbiting Earth as well, so we can use this law here by replacing M_S with M_E.

Solution

1. Use the equation for Kepler's third law to solve for r:

$$T\left(\frac{2\pi}{\sqrt{GM_E}}\right)^{-1} = r^{3/2} \quad\Rightarrow\quad r = \left(\frac{2\pi/T}{\sqrt{GM_E}}\right)^{-2/3}$$

2. Write the altitude h in terms of r:

$$h = r - R_E$$

3. Determine the period in seconds:

$$T = 2\,d\left(\frac{24\,h}{d}\right)\left(\frac{3600\,s}{h}\right) = 172{,}800\,s$$

4. Use the preceding results to determine the altitude:

$$h = \left(\frac{2\pi/T}{\sqrt{GM_E}}\right)^{-2/3} - R_E$$

$$= \left(\frac{2\pi/\left(1.728\times10^{5}\ s\right)}{\sqrt{\left(6.67\times10^{-11}\ N\cdot m^2/kg^2\right)\left(5.97\times10^{24}\ kg\right)}}\right)^{-2/3} - 6.38\times10^{6}\ m$$

$$= 6.07\times10^{7}\ m$$

Insight Since satellites are launched from Earth's surface, calculating the altitude is more practical than calculating the full distance from Earth's center.

Practice Quiz

4. Which of the following statements is true according to Kepler's laws?

   (a) Planets orbit the Sun at constant speed.

   (b) Planets move slower when they are closer to the Sun.

   (c) Planets move slower when they are farther away from the Sun.

   (d) A planet is always the same distance from the Sun.

   (e) None of the above

5. A satellite orbits Earth at a distance r from Earth's center. If you double the distance, by approximately what factor will the period change?

   (a) 2.8 **(b)** 0.35 **(c)** 2.0 **(d)** 0.5 **(e)** 4.0

12–4 – 12–5 Gravitational Potential Energy and Energy Conservation

In our previous look at gravitational potential energy we treated only the approximate case near Earth's surface where the gravitational force on objects is very nearly constant. In this chapter we see that, in general, the force of gravity behaves differently from the way we considered before and therefore the gravitational potential energy also behaves differently. For a system of two masses, m_1 and m_2, separated by a distance r, the gravitational potential energy is given by

$$U = -\frac{Gm_1m_2}{r}$$

where it is assumed that $U = 0$ at $r = \infty$ is the chosen reference.

The principle of superposition applies to the gravitational potential energy as a direct consequence of its application to the gravitational force; however, since potential energy is a scalar quantity, we can use simple algebra rather than dealing with vectors:

> *The total gravitational potential energy of a system of objects is the algebraic sum of the gravitational potential energies of each pair of objects.*

Example 12–4 A System of Two Masses In a binary star system two stars have elliptical orbits about the center of mass of the system. Consider a binary star system consisting of stars of masses 2.3×10^{30} kg and 3.1×10^{30} kg. If at one point in their orbit they are 4.4×10^{16} m apart, then they move closer together to become 1.8×10^{15} m apart, what is the change in the gravitational potential energy of the system?

Picture the Problem The diagram shows two stars separated by a center-to-center distance r.

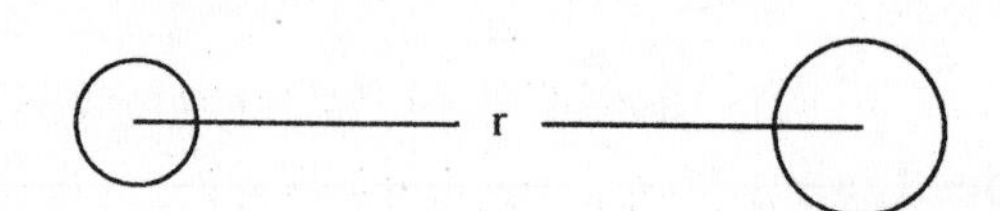

Strategy We need to determine the potential energy for each separation and find the difference.

Solution

1. The potential energy for the initial separation is:

$$U_i = -\frac{GM_1M_2}{r_i}$$

$$= -\frac{\left(6.67\times10^{-11}\ \frac{\text{N}\cdot\text{m}^2}{\text{kg}^2}\right)\left(2.3\times10^{30}\ \text{kg}\right)\left(3.1\times10^{30}\ \text{kg}\right)}{4.4\times10^{16}\ \text{m}}$$

$$= -1.08\times10^{34}\ \text{J}$$

2. The potential energy for the final separation is:

$$U_f = -\frac{GM_1M_2}{r_f}$$

$$= -\frac{\left(6.67\times10^{-11}\ \frac{N\cdot m^2}{kg^2}\right)\left(2.3\times10^{30}\ kg\right)\left(3.1\times10^{30}\ kg\right)}{1.8\times10^{15}\ m}$$

$$= -2.64\times10^{35}\ J$$

3. The change in potential energy is:

$$\Delta U = U_f - U_i = -2.64\times10^{35}\ J + 1.08\times10^{34}\ J$$

$$= -2.5\times10^{35}\ J$$

Insight As the stars get closer together the *magnitude* of the potential energy gets larger, but because it is negative there is a decrease in the potential energy ($\Delta U < 0$) rather than an increase.

That we can define a gravitational potential energy from Newton's universal law of gravitation means that we can also apply the conservation of mechanical energy to objects that interact only gravitationally. If, as is the case with most orbits that we will consider, we can treat one large object M as stationary while a smaller object m moves, then we can write the mechanical energy, E, as

$$E = K + U = \tfrac{1}{2}mv^2 - \frac{GmM}{r}$$

Notice that the kinetic energy is positive, as always, whereas the potential energy is negative. This allows for the case in which the two will cancel out numerically, producing a mechanical energy of zero. This case corresponds to the situation in which the motion of the smaller mass, m, has just barely enough kinetic energy to get infinitely far away from the larger mass, M, where it will come to rest. In such a case the speed of m is said to equal the **escape speed**, v_e, from M. The escape speed can be determined by setting $E = 0$ and solving for v; the result of this calculation is

$$v_e = \sqrt{\frac{2GM}{r}}$$

If m is moving slower than v_e, it will eventually come to rest momentarily, then fall back toward M, speeding up as it gets closer. If m is moving at a speed greater than v_e, it will continue moving away from M forever. In this latter case m will reach infinitely far away from M and still have speed left over.

Example 12–5 A Space Probe Boost Space probes are often set in orbit around Earth before receiving a *boost* into space. If we have a probe in a circular orbit 3550 km above Earth's surface, how much of a kick in speed (a "Δv") does it need to achieve escape speed at that position? (Assume that the boost is in the direction of orbital motion.)

Picture the Problem The diagram on the right shows a probe in orbit about Earth a height h above the surface.

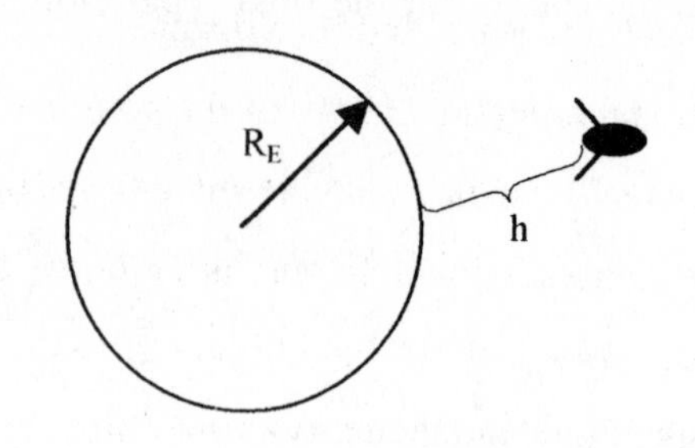

Strategy We need to determine the speed that the probe has in Earth orbit and the escape speed at the position of the probe.

Solution

1. In a circular orbit gravity provides the centripetal force:

$$\frac{GM_E m}{r^2} = m\frac{v^2}{r}$$

2. Solve for the orbital speed of the probe:

$$v = \sqrt{\frac{GM_E}{r}}$$

3. The boost is the difference between the escape speed and the orbital speed:

$$\Delta v = v_e - v = \sqrt{\frac{2GM_E}{r}} - \sqrt{\frac{GM_E}{r}} = \sqrt{\frac{GM_E}{r}}\left(\sqrt{2}-1\right)$$

4. Evaluate the numerical result:

$$\Delta v = \sqrt{\frac{GM_E}{r}}\left(\sqrt{2}-1\right)$$

$$= \sqrt{\frac{\left(6.67\times10^{-11}\ \text{N}\cdot\text{m}^2/\text{kg}^2\right)\left(5.97\times10^{24}\ \text{kg}\right)}{6.38\times10^{6}\ \text{m} + 3.55\times10^{6}\ \text{m}}}\left(\sqrt{2}-1\right)$$

$$= 2.62\times10^{3}\ \text{m/s}$$

Insight You can tell by examining this problem that, for circular orbits, the Δv needed to reach escape speed takes the general form determined here in all cases.

Practice Quiz

6. Two objects are separated by a distance r. If the mass of one object is reduced by half, how does the gravitational potential energy of the system change?
   (a) It increases, with U_f having half the magnitude of U_i.
   (b) It decreases, with U_f having half the magnitude of U_i.
   (c) It increases, with U_f having twice the magnitude of U_i.
   (d) It decreases, with U_f having twice the magnitude of U_i.
   (e) None of the above

7. Two objects are separated by a distance r. If the distance between them is tripled, how does the gravitational potential energy of the system change?
   (a) It increases, with U_f having three times the magnitude of U_i.
   (b) It decreases, with U_f having three times the magnitude of U_i.
   (c) It increases, with U_f having one-third the magnitude of U_i.
   (d) It decreases, with U_f having one-third the magnitude of U_i.
   (e) It decreases, with U_f having one-ninth the magnitude of U_i.

8. Assume that planet X has twice the mass and twice the radius of Earth. The escape speed from the surface of planet X, v_X, compared with that from the surface of Earth, v_E, is
   (a) $v_X = 2v_E$ **(b)** $v_X = \sqrt{2}v_E$ **(c)** $v_X = \frac{1}{2}v_E$ **(d)** $v_X = v_E/\sqrt{2}$ **(e)** $v_X = v_E$

Reference Tools and Resources

I. Key Terms and Phrases

gravity one of the fundamental forces of nature that represents the attraction between two objects with mass

Newton's law of universal gravitation between any two point masses there is an attractive force directly proportional to the product of the masses and inversely proportional to the square of the distance between them

principle of superposition (for gravity) the net result of the gravitational interaction within a system of many particles is the sum of the results for interactions between each pair of particles in the system

orbital period the amount of time it takes an object to execute one complete orbit

escape speed the speed at which a moving object can just barely escape the gravitational pull of another object by having the ability to get infinitely far away; at this speed the mechanical energy of the system is zero

II. Important Equations

Name/Topic	Equation	Explanation
gravitational force	$F = G\frac{m_1 m_2}{r^2}$	The magnitude of the gravitational force between two point masses.
gravitational acceleration	$g = \frac{GM_E}{R_E^2}$	The acceleration due to gravity near Earth's surface.
circular orbits	$T = \left(\frac{2\pi}{\sqrt{GM_S}}\right) r^{3/2}$	The equation form of Kepler's third law for the case of a circular orbit.
potential energy	$U = -\frac{Gm_1 m_2}{r}$	The gravitational potential energy of two point masses.
escape speed	$v_e = \sqrt{\frac{2GM}{r}}$	The escape speed of an object a distance r from an object of mass M.

III. Know Your Units

Quantity	Dimension	SI Unit
universal gravitational constant (G)	$\frac{[L^3]}{[M]\cdot[T^2]}$	$N\cdot m^2/kg^2$

Practice Problems

1. If D = 2399 meters and d = 1403 meters in the figure what is the force on the spaceship, to the nearest, meganewton? Consider all three to be point masses.

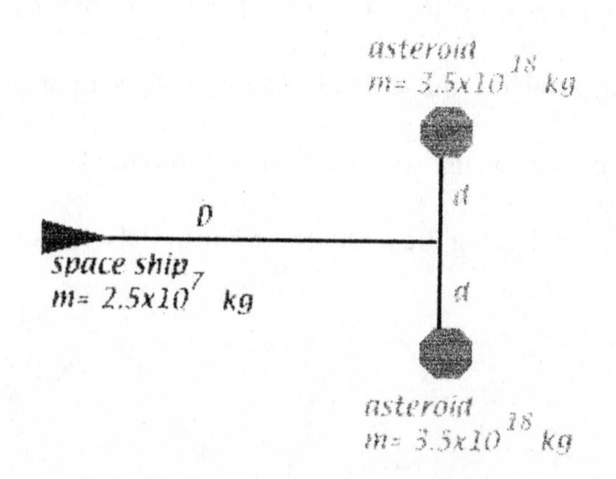

2. If D = 3819 meters and d = 345 meters to the nearest meganewton what is the force on the spaceship?

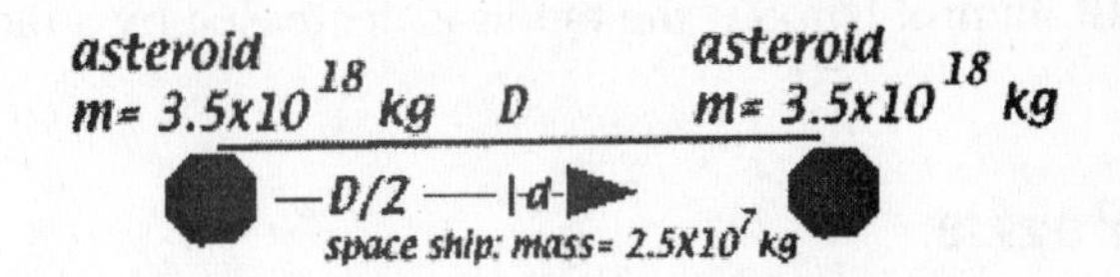

3. A planet/moon has a mass of 78.63 x 10^{23} kg and a radius of 10.69 x 10^{6} meters, what is g on the surface, to the nearest hundredth of a m/s^2?
4. In the figure the middle planet has a period of 1 year and an average distance from the Sun of 1.6 x 10^{11} meters. If the average distance from the Sun for the leftmost planet is 1.32 x 10^{11} meters what is its period, to the nearest hundredth of a year?
5. If the average distance of the rightmost planet in problem 4 is 2.24 x 10^{11} m what is its period?
6. If D = 3.13 x 10^{11} meters and d = 1.17 x 10^{11} meters and the rightmost planet sweeps an area of 1x10^{14} m^2 in a given short unit of time while traveling at 40000 m/s. to the nearest m/s what is its speed at d?

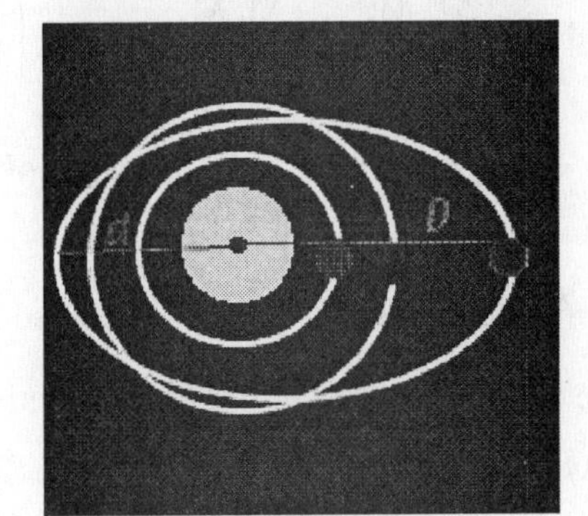

7. If m_1 = 2.62 kg, m_2 = 3.56 kg, and m_3 = 2.5 kg, to the nearest hundredth of a nJ, what is the potential energy of the configuration?

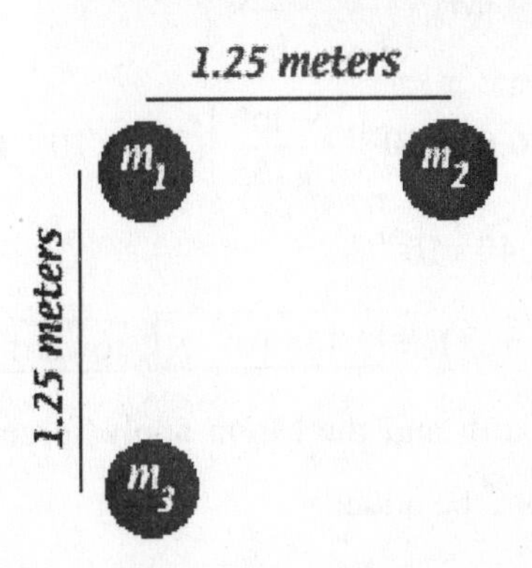

8. If a planet has a mass of 3.15 x 10^{24} kg and a radius of 5.92 x 10^{6} meters, to the nearest m/s what is its escape velocity?
9. In problem 8 if the mass is increased by a factor of 16, what is the new escape velocity?

10. In problem 8 if the radius is decreased by a factor 0.48, what is its new escape velocity?

Puzzle

THE ORBIT PARADOX

Consider a satellite in a circular orbit about Earth. If NASA scientists want to move the satellite into a higher (circular) orbit, they have to increase the satellite speed, and yet when the satellite is in the new (higher) orbit, its speed is actually slower than it was in the old (lower) orbit. Is this correct? If the answer is yes, can you explain why the satellite slows down? Answer this question in words, not equations, briefly explaining how you obtained your answer.

Selected Solutions

7. **(a)** Earth is pulled in the same direction by the Moon and the Sun, so we must add the two gravitational forces:

$$F = G\frac{M_E M_S}{r_{E\text{-}S}^2} + G\frac{M_E M_M}{r_{E\text{-}M}^2} = GM_E\left(\frac{M_S}{r_{E\text{-}S}^2} + \frac{M_M}{r_{E\text{-}M}^2}\right)$$

$$= \left(6.67\times10^{-11}\,\frac{\text{N}\cdot\text{m}^2}{\text{kg}^2}\right)(5.97\times10^{24}\text{ kg})\left(\frac{2.00\times10^{30}\text{ kg}}{(1.50\times10^{11}\text{ m})^2} + \frac{7.35\times10^{22}\text{ kg}}{(3.84\times10^{8}\text{ m})^2}\right) = 3.56\times10^{22}\text{ N}$$

Therefore, $\mathbf{F} = \boxed{3.56\times10^{22}\text{ N}}$, $\boxed{\text{toward the sun}}$.

(b) The Moon is pulled in opposite directions by Earth and the Sun. The Sun applies the larger force so we subtract Earth's force from the Sun's:

$$F = GM_M\left(\frac{M_S}{r_{S\text{-}M}^2} - \frac{M_E}{r_{E\text{-}M}^2}\right)$$

$$= \left(6.67\times10^{-11}\,\frac{\text{N}\cdot\text{m}^2}{\text{kg}^2}\right)(7.35\times10^{22}\text{ kg})\left(\frac{2.00\times10^{30}\text{ kg}}{(1.50\times10^{11}\text{ m} - 3.84\times10^{8}\text{ m})^2} - \frac{5.97\times10^{24}\text{ kg}}{(3.84\times10^{8}\text{ m})^2}\right)$$

$$= 2.40\times10^{20}\text{ N}$$

Therefore, $\mathbf{F} = \boxed{2.40\times10^{20}\text{ N}}$, $\boxed{\text{toward the Sun}}$.

(c) Both Earth and the Moon apply forces in the same direction on the Sun. The two forces must, therefore, be added.

$$F = GM_S\left(\frac{M_E}{r_{E\text{-}S}^2} + \frac{M_M}{r_{S\text{-}M}^2}\right)$$

$$= \left(6.67\times10^{-11}\ \frac{N\cdot m^2}{kg^2}\right)(2.00\times10^{30}\ kg)\left(\frac{5.97\times10^{24}\ kg}{(1.50\times10^{11}\ m)^2} + \frac{7.35\times10^{22}\ kg}{(1.50\times10^{11}\ m - 3.84\times10^{8}\ m)^2}\right)$$

$$= 3.58\times10^{22}\ N$$

Therefore, $\mathbf{F} = \boxed{3.58\times10^{22}\ N}$, $\boxed{\text{toward the Earth-Moon system}}$.

17. The expression for the gravitational acceleration at the surface of a spherical body of mass M and radius R is $g = GM/R^2$. Thus, for Titan,

$$g_T = \frac{GM_T}{R_T^2} = \frac{\left(6.67\times10^{-11}\ \frac{N\cdot m^2}{kg^2}\right)(1.35\times10^{23}\ kg)}{(2570\times10^3\ m)^2} = \boxed{1.36\ m/s^2}$$

27. (a) We can use the expression for the period of a satellite in circular orbit (around Earth) of radius r:

$$T = \left(\frac{2\pi}{\sqrt{GM_E}}\right)r^{3/2} = \left(\frac{2\pi}{\sqrt{\left(6.67\times10^{-11}\ \frac{N\cdot m^2}{kg^2}\right)(5.97\times10^{24}\ kg)}}\right)(2.0\times10^7\ m + 6.38\times10^6\ m)^{3/2} = \boxed{12\ h}$$

(b) Since the orbit is circular, and the satellite moves with a constant speed, the orbital speed is given by the circumference divided by the period. Therefore,

$$v = \frac{2\pi r}{T} = \frac{2\pi(2.638\times10^7\ m)}{4.266\times10^4\ s} = \boxed{3.9\ km/s}$$

37. The minimum kinetic energy needed to escape is that which gives $K_f = 0$. Therefore, by energy conservation (taking $U_f = 0$) we have

$$\Delta K + \Delta U = 0 \Rightarrow (K_f - K) + (U_f - U_i) = 0 \Rightarrow K = -U_i \quad \therefore \quad K = G\frac{Mm}{R}$$

(a) For the Moon:

$$K = G\frac{M_M m}{R_M} = \left(6.67\times10^{-11}\ \frac{N\cdot m^2}{kg^2}\right)\frac{(7.35\times10^{22}\ kg)(29{,}000\ kg)}{(1.74\times10^6\ m)} = \boxed{8.2\times10^{10}\ J}$$

(b) For Earth:

$$K = G\frac{M_E m}{R_E} = \left(6.67\times10^{-11}\ \frac{N\cdot m^2}{kg^2}\right)\frac{(5.97\times10^{24}\ kg)(29{,}000\ kg)}{(6.37\times10^6\ m)} = \boxed{1.8\times10^{12}\ J}$$

45. Taking h = 110 km, we can obtain the impact speed v_f by using energy conservation:

$$E_i = E_f \Rightarrow \tfrac{1}{2}mv_i^2 - \frac{GM_Mm}{R_M+h} = \tfrac{1}{2}mv_f^2 - \frac{GM_Mm}{R_M}$$

Multiplying through by $\frac{2}{m}$ gives: $v_i^2 - \frac{2GM_M}{R_M+h} = v_f^2 - \frac{2GM_M}{R_M}$

Solving this equation for v_f gives us the result

$$v_f = \sqrt{v_i^2 + 2GM_M\left(\frac{1}{R_M} - \frac{1}{R_M+h}\right)} = \sqrt{v_i^2 + \frac{2GM_Mh}{R_M(R_M+h)}}$$

$$= \sqrt{(1630\text{ m/s})^2 + \frac{2\left(6.67\times10^{-11}\,\frac{\text{N}\cdot\text{m}^2}{\text{kg}^2}\right)(7.35\times10^{22}\text{ kg})(110\times10^3\text{ m})}{(1.74\times10^6\text{ m})(1.85\times10^6\text{ m})}} = \boxed{1.73\text{ km/s}}$$

Answers to Practice Quiz

1. (c) **2**. (d) **3**. (d) **4**. (c) **5**. (a) **6**. (a) **7**. (c) **8**. (e)

Answers to Practice Problems

1. 1304 MN **2**. 1237 MN

3. 4.59 m/s^2 **4**. 0.75 yr

5. 1.66 yr **6**. 107009 m/s

7. −1.18 nJ **8**. 8425 m/s

9. 33700 m/s **10**. 12160 m/s

CHAPTER 13
OSCILLATIONS ABOUT EQUILIBRIUM

Chapter Objectives

After studying this chapter, you should:

1. know why periodic motion usually occurs.
2. know the characteristic quantities that describe periodic motion.
3. be able to analyze simple harmonic motion and understand its connection to uniform circular motion.
4. be able to apply the conservation of energy to systems undergoing simple harmonic motion.
5. understand the basic principles of both simple and physical pendulums.
6. gain good familiarity with the properties of damped and driven oscillations.

Warm-Ups

1. Consider a mass on a spring, oscillating under the influence of a nonconservative retarding force, such as air drag. How will the retarding force affect the period of the oscillations? In a sentence or two, justify your answer.
2. When you push a child on a swing, your action is most effective when your pushes are timed to coincide with the natural frequency of the motion. You are swinging a 30-kg child on a swing suspended from 5-meter cables. Estimate the optimum time interval between your pushes. Repeat the estimate for a 15-kg child.
3. Suppose you are told that the period of a simple pendulum depends only on the length of the pendulum L, and the acceleration due to gravity, g. Use dimensional analysis to show that the square of the period is proportional to the ratio L/g.
4. You want to construct a poor-man's amusement ride by mounting a seat on a large spring. Estimate the spring constant that will give you a ride with a period of 10 seconds.

Chapter Review

In this chapter we study what happens when objects in equilibrium (Chapter 11) are disturbed slightly away from this equilibrium. The most common result of such disturbances is that objects oscillate about (around) the equilibrium position (or configuration) from which it was disturbed. Here we study this oscillatory motion.

13–1 – 13–2 Periodic and Simple Harmonic Motion

An object's motion is referred to as **periodic motion** when it repeats itself over and over again. When the motion comes exactly back to its original state (of position and velocity) we say that the object has gone through a **complete cycle** or **complete oscillation**. The amount of time it takes for one complete cycle to occur is called the **period** of the motion, T.

The period is one way to characterize the speed of the periodic motion; another way is through the **frequency**, f. Frequency is a measure of how frequently the motion repeats; it is most commonly quoted as the number of oscillations (or *cycles* for short) per second. The frequency relates directly to the period by

$$f = \frac{1}{T}$$

Since period measures an amount of time, its SI unit is the second; therefore, the SI unit of frequency must be the inverse second, s^{-1}. When dealing with frequencies only, the special name hertz (Hz) is given to s^{-1}. One hertz refers to one oscillation cycle per second.

The above discussion applies to any type of periodic motion. One particular type of motion that is very important in physics is called **simple harmonic motion** (SHM). This latter type of periodic motion occurs as a result of a Hooke's law force. Hence, SHM is the motion experienced by an object on the end of a spring and, to a close approximation, by media that are disturbed slightly away from equilibrium.

The position of an object undergoing SHM changes sinusoidally with time. This fact means that it can be described using a sine or cosine function (we'll use cosine)

$$x = A\cos\left(\frac{2\pi}{T}t\right)$$

In the above equation, T is the period, and A is the **amplitude** of the motion. The amplitude represents the maximum distance the object gets from equilibrium as it moves back-and-forth. The above expression assumes that the object starts its motion at x = A. It is understood that the argument of the cosine function is an angle in radians; therefore, your calculator should be in radian mode when you are using the above expression.

Exercise 13–1 Simple Harmonic Motion An object undergoes simple harmonic motion with a frequency of 3.50 Hz and an amplitude of 0.850 m. If its equilibrium position is $x = 0$, where is it **(a)** when the motion starts, and **(b)** 1.055 seconds later? **(c)** Write the complete equation of the motion.

Solution We are given the following information.

Given: $f = 3.50$ Hz, $A = 0.850$ m; **Find**: (a) x_i (b) x_f (c) Write the equation.

(a) At the start of the motion is when we begin timing, so right at the beginning, $t = 0$. Since the problem does not state the initial conditions of the motion we use the expression for position versus time in SHM:

$$x = A\cos(0) = A = 0.850\ \text{m}.$$

(b) To get the position at $t = 1.055$ s, again we use the general expression for SHM together with the relationship between frequency and period:

$$x = A\cos\left(\frac{2\pi}{T}t\right) = A\cos(2\pi f t) = (0.850\ \text{m})\cos\left(2\pi[3.50\ \text{s}^{-1}][1.055\ \text{s}]\right) = -0.300\ \text{m}$$

(c) Here, we only need to insert all of the numerical values into the expression for x. First, let's find the period:

$$T = \frac{1}{f} = \frac{1}{3.50\ \text{s}^{-1}} = 0.286\ \text{s}$$

Now that we have the period, we can write out the full formula for the motion:

$$x = A\cos\left(\frac{2\pi}{T}t\right) = (0.850\ \text{m})\cos\left(\frac{2\pi}{0.286\ \text{s}}t\right)$$

In part (c) we also could have used the given value for the frequency instead of first finding the period.

Practice Quiz

1. *Most generally*, simple harmonic motion occurs
   (a) when a mass is attached to the end of a spring
   (b) when an object moves back-and-forth
   (c) when the motion of an object repeats itself
   (d) when an object is acted upon by a Hooke's law force
   (e) None of the above

2. If the period of an oscillating body is 2.3 seconds, what is its frequency?
   (a) 2.3 Hz **(b)** 0.43 Hz **(c)** 2.3 s **(d)** 0.43 s **(e)** 2.7 Hz

3. An object oscillates with SHM according to the equation $x = (0.15\ \text{m})\cos[(3.7\ \text{rad/s})t]$. What is the amplitude of this motion?

(a) 0.15 m **(b)** 3.7 m **(c)** 1.7 m **(d)** 0.94 m **(e)** 0.59 m

4. An object oscillates with SHM according to the equation $x = (0.15\ \text{m})\cos[(3.7\ \text{rad/s})t]$. What is the period of this motion?

(a) 0.15 s **(b)** 3.7 s **(c)** 1.7 s **(d)** 0.94 s **(e)** 0.59 s

13–3 – 13–4 Connections Between Uniform Circular Motion and Simple Harmonic Motion, and The Period of a Mass on a Spring

There is a basic connection between simple harmonic motion and uniform circular motion that proves to be very useful in understanding SHM. The connection between the two types of motion is the following:

For an object undergoing uniform circular motion, the projection of its motion onto any diameter of its path executes simple harmonic motion.

To see this connection more directly consider the figure below which shows the circular path of an object of mass m undergoing uniform circular motion. The radius of the path is r, and its center is at the origin of an x-y coordinate system.

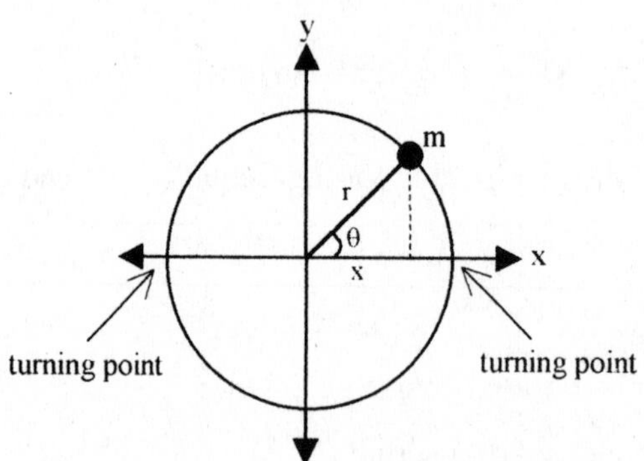

As the object moves around in a circular path its projection onto the x-axis oscillates back-and-forth along the central diameter of the path between the two turning points. From the geometry shown in the diagram we can see that the position of the projection is given by $x = r\cos\theta$. In uniform circular motion the angular position is given by $\theta = \omega t$. Also, the maximum displacement of the projection from the central position (the amplitude, A) equals the radius. Therefore, we see that the position can be written as $x = A\cos(\omega t)$. For constant angular velocity (uniform circular motion) the angular velocity is $\omega = 2\pi/T$, where T is the period of rotation; therefore, we can finally write

$$x = A\cos\left(\frac{2\pi}{T}t\right),$$

which is exactly the same expression that describes simple harmonic motion.

Because of the connection between uniform circular motion and SHM, any simple harmonic motion can be viewed as a projection of uniform circular motion. In such cases, the path is called the *reference circle* of the motion, and ω is called the **angular frequency**. Angular frequency relates to frequency f according to

$$\omega = 2\pi f$$

Following a similar approach as above, we can also use the connection between uniform circular motion and SHM to determine other relationships valid for SHM. The velocity oscillation in SHM can be determined to be

$$v = -A\omega\sin(\omega t)$$

This expression also establishes that the maximum speed in SHM, which occurs at equilibrium, is

$$v_{max} = A\omega$$

Similarly, we find that the acceleration oscillation in SHM is given by

$$a = -A\omega^2\cos(\omega t)$$

and the maximum acceleration, which occurs at the turning points, is given by

$$a_{max} = A\omega^2$$

Example 13–2 Kinematics of SHM An object, starting at its maximum displacement from an equilibrium situation, undergoes simple harmonic oscillations of amplitude 5.3 cm and frequency 1.7 Hz. After 0.82 sec have passed, determine its **(a)** position, **(b)** velocity, and **(c)** acceleration.

Picture the Problem The diagram shows the line along which the object oscillates with the object out at its initial position.

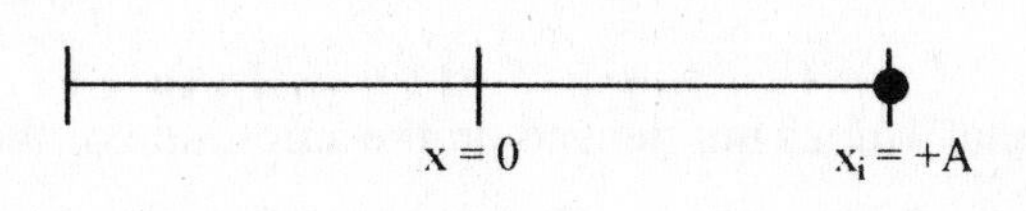

Strategy To solve this problem we make use of the preceding expressions for describing SHM.

Solution

Part (a)

1. Determine the angular frequency: $\omega = 2\pi f = 2\pi(1.7\text{ Hz}) = 10.68\text{ rad/s}$

2. Find the position of the object:

$$x = A\cos(\omega t) = (5.3\text{ cm})\cos[(10.68\text{ rad/s})(0.82\text{ s})]$$
$$= -4.2\text{ cm}$$

Part (b)

3. Use the known quantities to determine the velocity of the object:

$$v = -A\omega\sin(\omega t)$$
$$= -(5.3\text{cm})(10.68\text{ rad/s})\sin[(10.68\text{rad/s})(0.82\text{s})]$$
$$= -35\text{ cm/s}$$

Part (c)

4. Use the known quantities to determine the acceleration of the object:

$$a = -A\omega^2\cos(\omega t)$$
$$= -(5.3\text{cm})(10.68\text{ rad/s})^2\cos[(10.68\text{rad/s})(0.82\text{s})]$$
$$= 470\text{ cm/s}^2$$

Insight Take note of what the signs of these results mean. From part (a) we know that the object is 4.2 cm to the left of the equilibrium position. Part (b) tells us that it is moving in the negative direction (toward the turning point at $x = -A$). Part (c) tells us that it is slowing down, because the acceleration is positive (toward the central position).

Recall that Hooke's law is the force law that leads to simple harmonic motion

$$F = ma = -kx$$

Using the results for a and x in the preceding expression we obtain the relation

$$\omega = \sqrt{\frac{k}{m}}$$

This gives us a direct relationship between the physical system (k and m) and the motion (ω). Since $\omega = 2\pi f = 2\pi/T$, we can also write the above expression in terms of period:

$$T = 2\pi\sqrt{\frac{m}{k}}$$

Notice that T does not depend on the amplitude of the oscillation.

Example 13–3 Oscillations on a Spring Consider an industrial spring of force constant 225 N/cm that supports an object of mass 5.03 kg. If the system is disturbed and begins to oscillate, find the period, frequency, and angular frequency of the oscillation.

Picture the Problem The sketch shows the spring supporting the object.

Strategy We need only to use the relationships between the given quantities.

Solution

1. Convert the force constant: $k = 225\text{ N/cm}\left(\frac{100\text{ cm}}{\text{m}}\right) = 22500\text{ N/m}$

2. Calculate the period of the motion: $T = 2\pi\sqrt{\frac{m}{k}} = 2\pi\sqrt{\frac{5.03\text{ kg}}{22500\text{ N/m}}} = 0.0939\text{ s}$

3. Calculate the frequency of the motion: $f = \frac{1}{T} = \frac{1}{0.09394\text{ s}} = 10.6\text{ Hz}$

4. Calculate the angular frequency of the motion: $\omega = 2\pi f = 2\pi(10.64\text{ Hz}) = 66.9\text{ Hz}$

Insight Notice that in the calculations for f and ω, we used an extra figure in the previously obtained results.

For the case of a mass on a vertical spring, as in the above example, the force on the mass comes from both the spring and gravity. The motion, however, is still simple harmonic. The effect of gravity is essentially to shift the equilibrium position of the oscillation from the unstretched length of the spring alone to the new equilibrium length of the mass-spring system given by $y_0 = mg/k$. The mass then oscillates up-and-down about this new equilibrium position.

Practice Quiz

5. If the frequency, f, increases by a factor of 2, by what factor does the angular frequency increase?

   (a) 1/2 **(b)** 2π **(c)** 4 **(d)** 1/2π **(e)** 2

6. When the position of an object undergoing SHM is at its maximum displacement from equilibrium, the speed of the object is

   (a) at its maximum

   (b) zero

   (c) half its maximum value

   (d) twice its maximum value

   (e) none of the above

7. If a mass of 2.6 kg oscillates with a period of 1.73 seconds on the end of a vertical spring, what is the force constant of the spring?

   (a) 4.5 N/m **(b)** 34 N/m **(c)** 7.7 N/m **(d)** 0.029 N/m **(e)** 1.5 N/m

13–5 Energy Conservation in Oscillatory Motion

Because a force obeying Hooke's law is conservative, then the mechanical energy in SHM is conserved

$$E = K + U = \tfrac{1}{2}mv^2 + \tfrac{1}{2}kx^2 = \text{constant}$$

The energy oscillates between kinetic and potential energy. The energy is all in the form of potential energy when the mass is at its amplitude (x = A), where v = 0. Therefore, the mechanical energy is given by

$$E = \tfrac{1}{2}kA^2$$

Similarly, the energy is all in the form of kinetic energy when the mass is at the central position (x = 0), where $v = v_{max}$; therefore, we can also write the mechanical energy as $E = \tfrac{1}{2}mv_{max}^2$. Since $v_{max} = A\omega$, we can write

$$E = \tfrac{1}{2}mA^2\omega^2$$

Example 13–4 Using the Energy A spring of force constant 142 N/m is hung vertically. An object of mass 1.25 kg is attached to the end of this spring. The mass is then pulled down an additional 8.37 cm from the equilibrium position and released. Use energy conservation to determine the object's speed when it is halfway between its central position and its amplitude.

Picture the Problem The sketch shows the mass on the end of a vertical spring.

Strategy We must identify the most appropriate expressions for the mechanical energy of the system based on the given information to solve this problem.

Solution

1. Since the distance the mass is pulled down will be the amplitude A, and we know k one useful expression of the energy is: $E = \tfrac{1}{2}kA^2$

2. The energy expression involving velocity at an arbitrary position is: $E = \frac{1}{2}mv^2 + \frac{1}{2}kx^2$

3. Use $x = A/2$, and equate the two expressions: $\frac{1}{2}kA^2 = \frac{1}{2}mv^2 + \frac{1}{2}k\left(\frac{A}{2}\right)^2 = \frac{1}{2}mv^2 + \frac{1}{8}kA^2$

4. Solve this equation for velocity: $\frac{1}{2}mv^2 = \frac{1}{2}kA^2 - \frac{1}{8}kA^2 = \frac{3}{8}kA^2 \quad \therefore$

$$v = \sqrt{\frac{3k}{4m}}A = \sqrt{\frac{3(142 \text{ N/m})}{4(1.25 \text{ kg})}}(0.0837 \text{ m}) = 0.773 \text{ m/s}$$

Insight It was energy conservation that required that the energy at x = A is the same as when at x = A/2.

Practice Quiz

8. When the kinetic energy of an object undergoing SHM is maximum, its potential energy is

(a) maximum

(b) equal to the kinetic energy

(c) half its maximum

(d) twice the kinetic energy

(e) none of the above

9. An object of mass 3.3 kg oscillates with SHM according to the equation $x = (8.7 \text{ cm})\cos[(1.4 \text{ rad/s})t]$. What is the mechanical energy of this system?

(a) 29 J **(b)** 40 J **(c)** 3.6 J **(d)** 0.024 J **(e)** 0.0059 J

13–6 The Pendulum

A **simple pendulum** is a mass m (the *bob*) suspended by a cord or rod (of negligible mass) of length L. When a simple pendulum is displaced from equilibrium by small amounts its oscillation about equilibrium is very nearly simple harmonic motion. If we take the lowest point as the reference level for zero gravitational potential energy, then the potential energy of the simple pendulum is given by

$$U = mgL(1 - \cos\theta)$$

where θ is the angle the cord makes with the vertical. By analyzing the restoring force (due to gravity) for small-angle oscillations about the vertical line, we can see that the "force constant" for a simple pendulum works out to be $k = mg/L$. We can use this result for k to characterize the SHM of the simple pendulum in precisely the same way was used k previously. Therefore, substituting mg/L for k, we find the period of the simple pendulum (the primary useful quantity for a simple pendulum) to be

$$T = 2\pi\sqrt{\frac{L}{g}}$$

Notice that the period does not depend on the mass of the pendulum, just its length and the local acceleration of gravity.

For a simple pendulum we ignore the mass of the rod that holds the bob. When more precision is needed, this approximation may not be good enough. A pendulum for which the mass of the rod cannot be ignored is called a **physical pendulum**. For a physical pendulum we must consider it to be a rotational system and take into account its moment of inertia, I. When we do this analysis we find that the period of a physical pendulum is

$$T = 2\pi\sqrt{\frac{I}{mg\ell}}$$

where ℓ is the distance from the axis about which the pendulum swings to the center of mass of the system.

Exercise 13–5 On Top of Mount Everest Pendulums are sometimes used to measure the local acceleration of gravity. Mount Everest in Nepal is approximately 8850 m tall at its peak. If a simple pendulum of length 1.75 m set in small oscillations at the top of Mount Everest has a measured period of 2.64 s, what is the acceleration of gravity at the top of Mount Everest?

Solution: We are given the following information.

Given: h = 8850 m, L = 1.75 m, T = 2.64 s; **Find**: g

Since we seek to determine g, we first solve for g:

$$T = 2\pi\sqrt{\frac{L}{g}} \quad \Rightarrow \quad g = \frac{4\pi^2 L}{T^2}$$

Now we get the final numerical result:

$$g = \frac{4\pi^2 (1.75\ \text{m})}{(2.64\ \text{s})^2} = 9.91\ \text{m/s}^2$$

Practice Quiz

10. A meterstick is fixed to a horizontal rod at one end, and while the other end is allowed to swing freely. A small Styrofoam ball is glued to the free-swinging end of the stick. This system should be treated as

(a) a simple pendulum.

(b) a physical pendulum.

(c) none of the above

13–7 – 13–8 Damped and Driven Oscillations, Resonance

In realistic oscillating systems energy is lost during the motion. Because of this energy loss, the amplitude of the oscillation decreases. We call this type of motion **damped oscillation**. Frequently, damping forces (like air resistance) are proportional to the velocity of the object on which the force is applied, so that

$$\mathbf{F} = -b\mathbf{v}$$

where the coefficient b is called the **damping constant**.

The situation in which a system oscillates with decreasing amplitude occurs when the damping constant is small and is referred to as **underdamped** motion. A larger damping constant can lead to the case in which the system will no longer oscillate; this is called **critically damped** motion. Critically damped motion is on the borderline between a system that oscillates and one that just moves back to its equilibrium configuration. An even larger damping constant leads to **overdamped** motion. With overdamping the system relaxes to equilibrium more slowly than with critical damping and there is room to make the value of b smaller without causing oscillations.

The effects of damping forces can be overcome by putting energy into the system in order to maintain the oscillation. When this energy is input we say that the motion is a **driven oscillation**. With driven oscillations it is important to recognize that systems tend to have **natural frequencies** of oscillation, that is, frequencies at which they will oscillate if no driving force is present (consider the pendulum discussed above). When a system is being driven at a natural frequency of oscillation the result is oscillations of larger amplitude than occur when the system is driven at a different frequency. This phenomenon is called **resonance** (for this reason natural frequencies are sometimes called *resonant frequencies*). Resonance plays an important role in many applications and in the generation and detection of sound, which will be discussed in a later chapter.

Example 13–6 Swinging A child that weighs 45 lb enjoys swinging very high. If the length of the chain that holds the swing is 2.5 m, approximately how often should the child be pushed in order to get a good high swing?

Picture the Problem The ssketch shows the child on a swing.

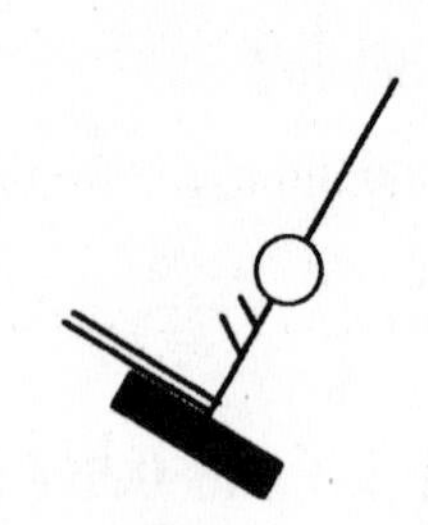

Strategy Since we seek only an approximate answer, we will make two approximations. We will treat the child and swing as a simple pendulum with the child as the bob, and we will use the SHM results and ignore any slight inaccuracies that result from the fact that the child swings through large angles.

Since we want the child's motion to have large amplitude, we should push the child at her resonant frequency.

Solution

1. Determine the natural frequency for the child: $f = \frac{1}{T} = \frac{1}{2\pi}\sqrt{\frac{g}{L}} = \frac{1}{2\pi}\sqrt{\frac{9.81\text{ m/s}^2}{2.5\text{ m}}} = 0.32\text{ Hz}$

2. Interpret the result: $f \approx \frac{1}{3\text{s}}$ $\therefore$ approximately 1 push every 3 sec.

Insight The main point of this problem was to recognize how the concepts of natural frequency and resonance fit together.

Reference Tools and Resources

I. Key Terms and Phrases

periodic motion motion that repeats itself

complete cycle this occurs when both the position and velocity of an oscillating object repeat themselves between two successive passes

period (for oscillations) the amount of time for one complete cycle

frequency the number of cycles per unit of time

simple harmonic motion the oscillatory motion that results from a force that obeys Hooke's law

amplitude the maximum displacement from equilibrium

angular frequency 2π times the frequency

simple pendulum a mass suspended by a cord or rod of negligible mass

physical pendulum a mass distribution that is suspended and free to oscillate

damped oscillation when an oscillating system looses energy

underdamped oscillation when a small damping constant causes an oscillating system to decrease in amplitude

critically damped oscillation when the damping constant is just large enough to prevent oscillations

overdamped oscillation when the damping constant is more than just large enough to prevent oscillations

driven oscillation when an external agent forces a system to oscillate

natural frequency a frequency at which a system will oscillate if no driving force is applied

resonance the phenomenon that large-amplitude oscillations occur when a system is driven at its natural frequency

II. Important Equations

Name/Topic	Equation	Explanation
frequency	$f = \frac{1}{T}$	The frequency equals the inverse of the period.
simple harmonic motion	$x = A\cos\left(\frac{2\pi}{T}t\right)$	Position as a function of time in SHM assuming the motion begins at the maximum displacement.
angular frequency	$\omega = 2\pi f$	The relationship between angular frequency and frequency.

period	$T = 2\pi\sqrt{\frac{m}{k}}$	The period of a mass m oscillating on a spring with force constant k.
mechanical energy	$E = \frac{1}{2}kA^2 = \frac{1}{2}mA^2\omega^2$	The mechanical energy in SHM
period	$T = 2\pi\sqrt{\frac{L}{g}}$	The period of a simple pendulum of length L.

III. Know Your Units

Quantity	Dimension	SI Unit
period	[T]	s
frequency (f)	$[T^{-1}]$	Hz
angular frequency (ω)	$[T^{-1}]$	rad/s
amplitude (A)	[L]	m

Practice Problems

1. The velocity of a simple harmonic oscillator is given by $v = -4.72 \sin(25.1\ t)$ (mks units). What is its angular frequency?
2. What is the amplitude of the motion in problem 1 in meters to two decimal places?
3. To the nearest hundredth of a meter, where is the mass in problem 1 at the time $t = 35.94$ seconds?
4. If the mass in problem 3 is 0.34 kg, what is the spring's potential energy, to the nearest tenth of a joule?
5. What is its kinetic energy to the nearest tenth of a joule?
6. A 0.29-kg mass is attached to a spring with spring constant 6.7 N/m, and let fall. To the nearest hundredth of a meter what is the point where it stops?
7. In problem 6 what is the amplitude of the resulting motion?
8. A 20-gram bullet traveling at 261 m/s is fired into a 0.482-kg wooden block anchored to a 100 N/m spring. How far is the spring compressed, to the nearest thousandth of a meter?
9. If the speed of the bullet in problem 8 is not known, but it is observed that the spring is compressed 55.4 cm, what was the speed of the bullet, to the nearest m/s?

10. A wooden rod of mass m = 0.51 kg and length L= 1.1 m is used as a physical pendulum. To the nearest tenth of a second what is the period of oscillation?

Puzzle

THE BUNGEE BUCKET

Little Danny stands next to a large amusement park bungee bucket. The empty bucket is hanging on two large, identical springs. It oscillates with an amplitude of 2 meters and a period of 2 seconds. It is shown at the equilibrium position in the sketch, 2.5 meters above the floor. Danny waits until the bucket comes to a momentary stop, ready to start its journey upward. He quickly takes a seat in the bucket. Danny's mass is equal to the mass of the empty bucket. What happens next? Does the bucket take Danny for a ride?

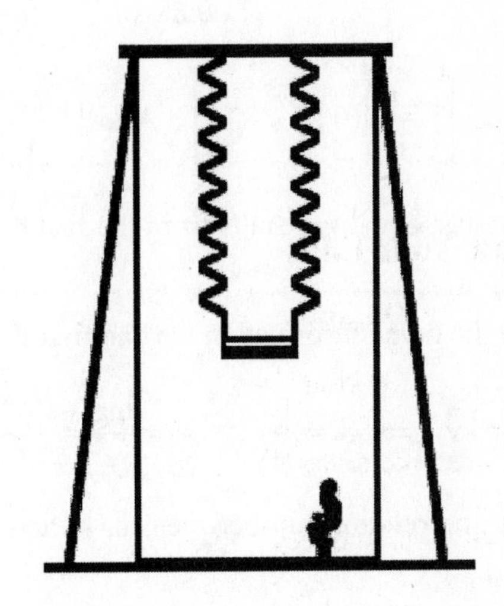

(You can treat the two springs as a single spring with the spring constant k equal to twice the k of each single spring. In this estimate, ignore the mass of the springs.)

Selected Solutions

7. **(a)** The frequency in hertz is the same as revolutions per second; therefore,

$$f = \left(2500\ \frac{\text{rev}}{\text{min}}\right)\left(\frac{60\ \text{s}}{\text{min}}\right) = 42\ \text{Hz}$$

The period in seconds is the inverse of the frequency,

$$T = \frac{1}{f} = \frac{1}{41.67\ \text{Hz}} = \boxed{0.024\ \text{s}}$$

(b) In revolutions per second, the frequency is given by f = 1/T. To convert to rpm we multiply by $\left(\frac{60\ \text{s}}{\text{min}}\right)$; therefore,

$$\#\ \text{of rpm} = \frac{1}{T}\left(\frac{60\ \text{s}}{\text{min}}\right) = \frac{60\ \text{s/min}}{0.034\ \text{s}} = \boxed{1800\ \text{rpm}}$$

13. (a) The general expression is $x = A\cos\left[\left(\frac{2\pi}{T}\right)t\right]$, so here $\boxed{T = 0.88\text{ s}}$.

(b) Being sure to calculate the cosine in radians, we get

$$x = (6.5\text{ cm})\cos\left[\left(\frac{2\pi}{0.88\text{ s}}\right)(0.25\text{ s})\right] = \boxed{-1.4\text{ cm}}$$

(c) $A\cos\left[\left(\frac{2\pi}{T}\right)(0.25\text{ s}+T)\right] = A\cos\left[\left(\frac{2\pi}{T}\right)(0.25\text{ s}) + 2\pi\right] = \boxed{A\cos\left[\left(\frac{2\pi}{T}\right)(0.25\text{ s})\right]}$

The last equality results from the fact that cosine is periodic with a period of 2π.

29. From the given information we can first determine the spring constant:

$$F = ky \Rightarrow k = \frac{F}{y} \quad \therefore \quad k = \frac{mg}{y} = \frac{(0.50\text{ kg})(9.81\text{ m/s}^2)}{(15\times 10^{-2}\text{ m})} = 32.7\ \frac{\text{N}}{\text{m}}$$

Using the relationship between m, k, and T, we can determine the mass that would produce a given period:

$$T = 2\pi\sqrt{\frac{m}{k}} \Rightarrow m = \left(\frac{T}{2\pi}\right)^2 k = \left(\frac{0.75\text{ s}}{2\pi}\right)^2\left(32.7\ \frac{\text{N}}{\text{m}}\right) = \boxed{0.47\text{ kg}}$$

37. The total energy of the system is given by $E = K + U$; therefore

$$K = E - U \Rightarrow \tfrac{1}{2}mv^2 = \tfrac{1}{2}kA^2 - \tfrac{1}{2}kx^2$$

$$\therefore \quad v = \sqrt{\frac{k(A^2 - x^2)}{m}} = \sqrt{\frac{(12.3\ \frac{\text{N}}{\text{m}})[(0.256\text{ m})^2 - (0.128\text{ m})^2]}{0.321\text{ kg}}} = \boxed{1.37\text{ m/s}}$$

55. The leg acts as a physical pendulum consisting of a uniform rod. For a uniform rod suspended at one end, $I = \tfrac{1}{3}mL^2$. The period of this physical pendulum then is

$$T = 2\pi\sqrt{\frac{L/2}{g}\left(\sqrt{\frac{\frac{1}{3}mL^2}{m(L/2)^2}}\right)} = 2\pi\sqrt{\frac{\frac{2}{3}L}{g}} = 2\pi\sqrt{\frac{\frac{2}{3}(0.50\text{ m})}{9.81\text{ m/s}^2}} = 1.2\text{ s}$$

If we assume the leg swings through about 1.0 radians per step, the distance traversed per step is

$$d = L\theta = 0.50\text{ m}$$

So that the estimated walking speed would be

$$v = \frac{d}{T/2} = \frac{0.50\text{ m}}{0.58\text{ s}} = \boxed{0.86\text{ m/s}}$$

The time used is T/2 because one forward step constitutes half a cycle for the swinging leg.

Answers to Practice Quiz

1. (d) **2**. (b) **3**. (a) **4**. (c) **5**. (e) **6**. (b) **7**. (b) **8**. (e) **9**. (d) **10**. (b)

Answers to Practice Problems

1. 25.1 rad/s **2**. 0.19 m

3. −0.17 m **4**. 3.1 J

5. 0.8 J **6**. 0.85 m

7. 0.43 m **8**. 0.737 m

9. 196 m/s **10**. 1.7 s

CHAPTER 14

WAVES AND SOUND

Chapter Objectives

After studying this chapter, you should:

1. know the two main types of waves.
2. know the main characteristics of waves and wave motion.
3. be able to determine the speed of a wave on a string.
4. understand the nature of sound waves.
5. understand the relationship between sound intensity and the human perception of sound.
6. understand the Doppler effect.
7. understand the superposition and interference of waves.
8. know how standing waves are generated and understand their modes of vibration on strings and in air columns.
9. know what beats are and what causes them.

Warm-Ups

1. A certain transverse traveling wave on a string can be represented by $y(x,t) = A\sin(kx + \omega t)$, where y is the displacement of the string, x is position of a point along the string, and t is the time. What are the units of the coefficients of the x term and the t term in the above expression? What do we call those coefficients?
2. Suppose you suspend a 3-meter nylon rope from a hook in the ceiling and tie a 10-kg ball to the end of the rope. The mass of the rope is 0.5 kg. Now, you pluck the rope. That sends a wave up and down the rope. Estimate the speed of propagation of that wave.
3. A certain transverse traveling wave on a string can be represented by $y(x,t) = A\sin(kx + \omega t)$ where y is the displacement of the string, x is position of a point along the string and t is the time. A wave can also be represented by a similar equation in which the two terms in the argument of the sine function

have opposite signs. Is this wave different from the one represented by the original expression? If the answer is yes, what is the difference?

4. How much faster does sound travel in steel than in water? What properties of the two materials are responsible for this difference?

Chapter Review

In this chapter we study **waves**. You can view a wave as resulting from the connection of a series of oscillators (oscillations were studied in the previous chapter) or as a propagating oscillation. Most generally, any disturbance that propagates can be called a wave. In this chapter we focus on **harmonic waves**, in which the oscillation that gives rise to the wave is a simple harmonic oscillation. The study of waves is important in almost every branch of physics and has many applications.

14–1 Types of Waves

There are two main types of waves. These types are distinguished by the relationship between the direction of the oscillation of the medium in which the wave is traveling and the direction of propagation of the wave. In a **transverse wave** the direction of oscillation is perpendicular (or transverse) to the direction of propagation. A wave on a string is a good example of a transverse wave. The other main type of wave is a *longitudinal wave*, in which the direction of oscillation is along the same line as the direction of propagation. A compression wave traveling along a spring (such as a slinky) is a good example of a longitudinal wave.

Since a harmonic wave results from simple harmonic motion within a medium, the main characteristics of waves are related to the cycle of this repeating motion. One of these characteristics relates to the minimum time it takes for a wave to repeat itself, the **period**, T. As with any simple harmonic motion the inverse of the period is called the **frequency**, f.

Waves also repeat themselves spatially; the minimum repeat length of a wave is called its **wavelength**, λ. If you consider the following diagram of a transverse wave, the wavelength equals the distance between successive crests, or troughs, of the wave (other corresponding points may also be used).

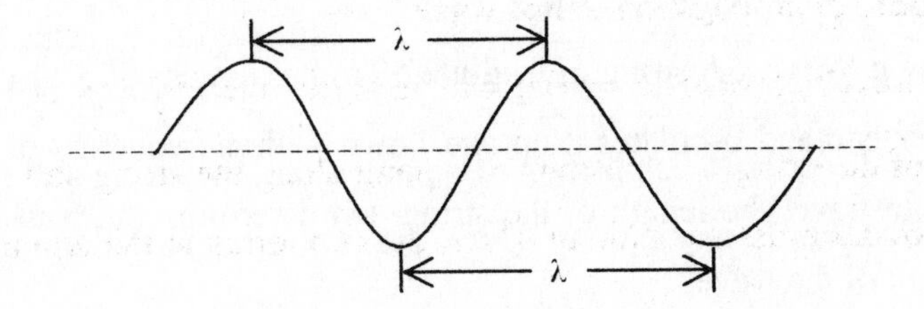

A third important characteristic of a wave is its speed of travel. This speed equals the distance the wave travels before it repeats (the wavelength) divided by the time it takes the wave to repeat (the period); therefore, the speed of a wave is given by

$$v = \frac{\lambda}{T} = \lambda f$$

Practice Quiz

1. The distance between two troughs of a wave of frequency 200 Hz is 1.7 m. What is the speed of this wave?

   (a) 5.0 mm/s **(b)** 340 m/s **(c)** 8.5 mm/s **(d)** 120 m/s **(e)** none of the above

2. If two waves traveling through the same medium have the same period, but wave 1 has half the wavelength of wave 2, how do the speeds of theses waves compare?

   (a) $v_1 = 2v_2$ **(b)** $v_1 = 4v_2$ **(c)** $v_2 = 4v_1$ **(d)** $v_2 = 2v_1$ **(e)** $v_2 = v_1$

14–2 Waves on a String

Waves traveling on a string (or any sort of linear cord), is a common way to generate sounds so it its important to understand this behavior. The speed at which a wave on a string travels is determined by two properties of the string: its tension and *linear density*, μ (mass per unit length). The more tightly pulled the string is, the more rapidly it will oscillate, and the faster the wave will travel. The heavier a segment of the string is, the more sluggishly (or slowly) it will move under a given tension, and the slower the wave will travel. Detailed analysis of this situation shows that the relationship between these three quantities is

$$v = \sqrt{\frac{F}{\mu}}$$

where F is the tension in the string (force transmitted through it), and μ is the mass per unit length, which is given by $\mu = m/L$, where m and L are the mass and length of the string, respectively.

Example 14–1 Sending a Wave A string of length 2.59 m has a mass of 5.11 g and is fixed at one end. A person takes the other end and oscillates it up and down with a frequency of 3.47 Hz. If it takes the resulting wave 0.862 s to travel the length of the string, **(a)** determine the tension in the string, and **(b)** determine the wavelength of the wave.

Picture the Problem The sketch shows the person oscillating one end of the string while the other end remains fixed.

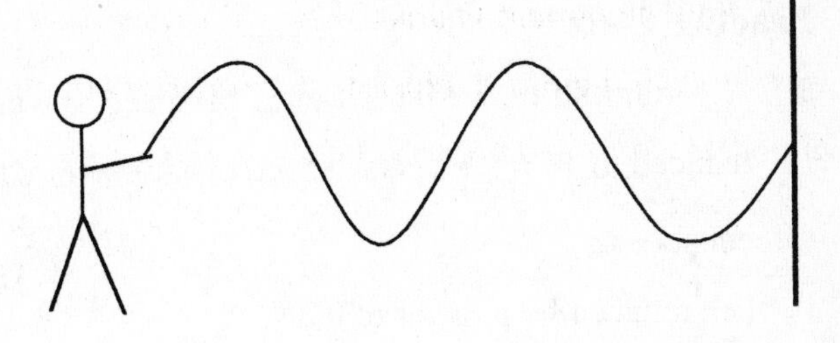

Strategy The given information allows us to determine both v and μ, so that we can use the relationship between v, μ, and F to solve part (a). Knowing v and frequency, we can then calculate the wavelength.

Solution

Part (a)

1. Determine mass per unit length:

$$\mu = \frac{m}{L} = \frac{0.00511\text{ kg}}{2.59\text{ m}} = 1.973\times10^{-3}\text{ kg/m}$$

2. Determine the speed of the wave:

$$v = \frac{L}{t} = \frac{2.59\text{ m}}{0.862\text{ s}} = 3.005\text{ m/s}$$

3. Solve for tension, F:

$$v = \sqrt{\frac{F}{\mu}} \Rightarrow F = \mu v^2$$

4. Obtain the numerical result:

$$F = \mu v^2 = \left(1.973\times10^{-3}\text{ kg/m}\right)\left(3.005\text{ m/s}\right)^2 = 0.0178\text{ N}$$

Part (b)

5. Use the relationship between v and λ to solve for λ:

$$v = \lambda f \Rightarrow \lambda = \frac{v}{f} = \frac{3.005\text{ m/s}}{3.47\text{ Hz}} = 0.866\text{ m}$$

Insight Notice that the wave speed connects the properties of waves (f, λ) with the properties of the medium (F, μ).

In the above example a wave was sent along a string with a fixed end. When the wave reaches that end it will be reflected back in the opposite direction along the string. Because the end is fixed it inverts each wave pulse upon reflection as a result of applying a force in a direction opposite that of the force applied by the string on the fixed connection. If the end of the string opposite the person was free to move, the reflected wave would not be inverted relative to the initial wave because the end oscillates along with the rest of the string. This issue of how a wave is reflected will become important in later discussions.

Practice Quiz

3. A certain string sustains waves of speed v when under a tension F. If the tension in the string is reduced to $F' = \frac{1}{3}F$, is the speed of the wave reduced or enhanced, and by what factor is it reduced or enhanced?

   (a) reduced by a factor of 0.58

   (b) reduced by a factor of 3

   (c) enhanced by a factor of 1.7

   (d) enhanced by a factor of 3

   (e) the speed stays the same

14-3* Harmonic Wave Functions

A traveling harmonic wave can be described by a reasonably simple functional form. For clarity we will consider a transverse wave for which the direction of oscillation is the y direction and the direction of propagation is the x direction. The position of a point on the wave (i.e., y) depends both on where you look in space (i.e., on x) and when you look in time (i.e., on t), so, the position y is a function of both x and t. A harmonic wave traveling in the +x direction can be described by the equation

$$y(x,t) = A\cos\left(\frac{2\pi}{\lambda}x - \frac{2\pi}{T}t\right)$$

where, A is the amplitude of the wave (its maximum displacement from equilibrium). Examination of this expression shows that for fixed t (a snapshot of the wave) the wave repeats whenever x increases by a distance λ. Similarly, for fixed x (at a given location) the wave repeats whenever t increases by a time T.

Exercise 14–2 A Harmonic Wave A transverse harmonic wave is described by the function

$$y = (3.2\text{ m})\cos\left[\left(0.25\text{ m}^{-1}\right)x - \left(1.7\text{ s}^{-1}\right)t\right].$$

What is the frequency of this wave?

Solution:

Comparing the given function with the general form shows us that $\frac{2\pi}{T} = 1.7\text{ s}^{-1}$. This means that

$$\frac{1}{T} = \frac{1.7\text{ s}^{-1}}{2\pi}$$

Since frequency is given by $f = 1/T$, we can conclude that

$$f = \frac{1.7\ \text{s}^{-1}}{2\pi} = 0.27\ \text{Hz}$$

Practice Quiz

4. What is the speed of a wave described by the expression $y = (0.21\ \text{m})\cos\left[(0.13\ \text{m}^{-1})x - (2.4\ \text{s}^{-1})t\right]$?

   (a) 48 m/s **(b)** 0.38 m/s **(c)** 18 m/s **(d)** 0.21 m/s **(e)** 2.4 m/s

14–4 – 14–5 Sound Waves and Sound Intensity

Sound is one of the most important everyday applications of waves. Sound waves are longitudal waves in the air (often, longitudinal waves in *any* medium are called sound waves). The precise speed of sound in air depends on the atmospheric conditions at the time and location of the wave; however, under normal conditions this speed approximately equals 343 m/s (or 770 mi/h). This is the value we will use unless otherwise specified.

The sensation of the pitch of a sound is determined by the frequency of the wave carrying the sound. A high-pitched sound has a relatively large (or high) frequency, and a low-pitched sound has a relatively small (or low) frequency. Just as the frequency of sound can take a wide range of values so can the wavelength. As mentioned previously, in air the speed of sound is roughly constant. Through the relation $v = \lambda f$, we can see that the wavelength takes on a range of values in association with the range of frequencies such that the product remains constant.

As discussed earlier, frequency is the physical property of a sound wave that determines the human perception of a sound's pitch. The human perception of the loudness of a sound is determined by the **intensity**, I, of the wave carrying the sound. The average intensity of a wave is the amount of energy that passes through a given area per unit time divided by the area, that is, energy per unit time per unit area. Since energy per unit time is power (in watts) delivered by the wave, on average the intensity is given by

$$I = \frac{P}{A}$$

where P is the power, and A is the area over which this power is spread. As can be seen from this relation, the SI unit of intensity is W/m^2. For a (point) source of sound whose power output P is dispersed equally in all directions (an isotropic source) the intensity of the sound decreases with the inverse square of the distance r from the source:

$$I = \frac{P}{4\pi r^2}$$

Example 14–3 Inverse Square Law The speakers of a small radio puts out 55 W of power. **(a)** What is the approximate intensity of its sound 3.5 m away? **(b)** How far from the radio would you have to be to receive one quarter of this intensity?

Picture the Problem The diagram shows a radio (modeled as a point source) giving off sound waves in all directions. The circles represent outward-moving crests of the wave and the dashed arrows indicate direction of propagation.

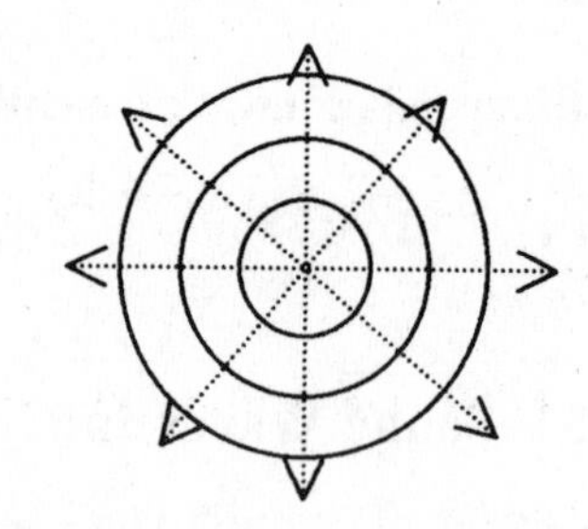

Strategy To solve this problem we make use of the expression for the variation of intensity varies with distance.

Solution

Part (a)

1. Use the expression for the intensity from an isotropic point source:

$$I = \frac{P}{4\pi r^2} = \frac{55\ \text{W}}{4\pi(3.5\ \text{m})^2} = 0.36\ \text{W/m}^2$$

Part (b)

2. Relate the two intensities:

$$\frac{I}{I_x} = \frac{P/4\pi r^2}{P/4\pi x^2} = \frac{x^2}{r^2}$$

3. Use the fact that $I_x = \frac{1}{4}I$ to solve for x:

$$4 = \frac{x^2}{r^2} \quad \Rightarrow \quad x = 2r = 2(3.5\ \text{m}) = 7.0\ \text{m}$$

Insight Try to think of two other approaches to solving part (b), one of which requires little or no calculations.

As mentioned above, human perception of the loudness of a sound is determined by the intensity of the sound. However, because the range of intensities that humans hear is very large we use a more convenient measure called the **intensity level,** β. The intensity level measures loudness by comparing the intensity of a sound to a standard reference intensity, $I_0 = 10^{-12}\ \text{W/m}^2$; this ratio is then rescaled by taking its logarithm. This quantity β is actually dimensionless, but we refer to the values of β as being measured in a "unit" called the decibel (dB). This situation is similar to quoting angular measures in radians even though angles are dimensionless. In mathematical terms, values of β can be obtained from the intensity I by the expression

$$\beta = 10\log\left(\frac{I}{I_0}\right)$$

When more than one source of sound contributes, the intensities of the individual sources add. The intensity level is determined from the resulting sum.

Exercise 14–4 The Desirable Range of Sound Given that the lower threshold intensity for human hearing is about 1.0×10^{-12} W/m^2 and the pain threshold is about 1.0 W/m^2, determine the desirable range of sound in decibels.

Solution:

The lower threshold in decibels corresponds to the lower intensity; therefore, since $I_0 = 1.0 \times 10^{-12}\ \text{W/m}^2$,

$$\beta_{\text{lower}} = 10\log\left(\frac{I}{I_0}\right) = 10\log\left(\frac{I_0}{I_0}\right) = 10\log(1) = 0\ \text{dB}$$

The upper part of the desirable range is

$$\beta_{\text{upper}} = 10\log\left(\frac{I}{I_0}\right) = 10\log\left(\frac{1.0\ \text{W/m}^2}{1.0 \times 10^{-12}\ \text{W/m}^2}\right) = 10\log(10^{12}) = 120\log(10) = 120\ \text{dB}$$

Notice here that by using intensity level β instead of intensities I, the range of values you have to work with is much much smaller without any loss of information. This is one of the nice uses of the logarithm.

Practice Quiz

5. The intensity of a sound 2.66 m from an isotropic source is 8.48×10^{-8} W/m^2. What will the intensity be 5.32 m from the source (in W/m^2)?

   (a) 4.24×10^{-8} **(b)** 1.70×10^{-7} **(c)** 3.39×10^{-7} **(d)** 2.12×10^{-8} **(e)** none of the above

6. If the intensity of a sound becomes a factor of 10 greater, describe what happens to the intensity level.

   (a) The intensity level increases by 1 dB.

   (b) The intensity level decreases by 1 dB.

   (c) The intensity level increases by a factor of 10.

   (d) The intensity level increases by a factor of log(10).

   (e) None of the above

14–6 The Doppler Effect

When there is relative motion between the source of a sound and the observer (or receiver) of the sound the pitch of the sound changes. The pitch gets higher if the source and the observer are moving closer to each other, and it gets lower if they are moving farther apart. The phenomenon just described is called the **Doppler effect**. This effect has analogs for all other types of waves in addition to sound.

Since the pitch of a sound is directly associated with its frequency, we treat the Doppler effect by finding the frequency perceived by the observer, f', as compared with the frequency emitted by the source f. We take the speed of sound in air to be constant, v = 343 m/s, and call u_o the speed of the observer and u_s the speed of the source. With these definitions, and the provision that both u_o and u_s be less than v, we represent the Doppler effect for sound by the equation

$$f' = \left(\frac{1 \pm u_o/v}{1 \mp u_s/v}\right) f$$

In the numerator, the upper sign is used if the motion of the observer is toward the source, and the lower sign is used if it is away from the source. Similarly, in the denominator, the upper sign is used if the motion of the source is toward the observer, and the lower sign is used if it is away. Hence, upper signs are for "toward" and lower signs are for "away." There are four possibilities, as indicated in the following table.

Motion	Signs
u_s ← [S] [O] → u_o	$\left(\frac{-}{+}\right)$
u_s ← [S] u_o ← [O]	$\left(\frac{+}{+}\right)$
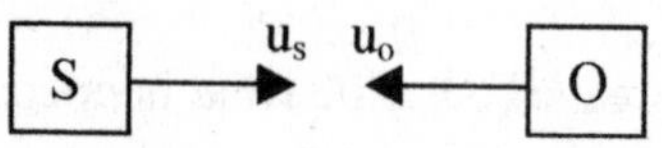 [S] → u_s u_o ← [O]	$\left(\frac{+}{-}\right)$
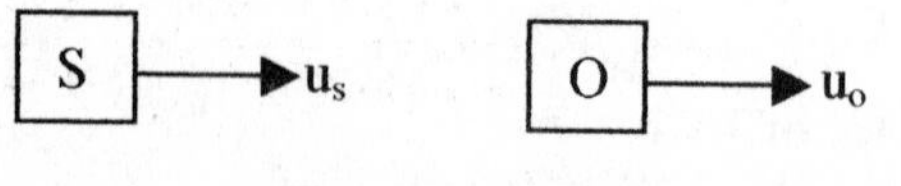 [S] → u_s [O] → u_o	$\left(\frac{-}{-}\right)$

Example 14–5 Doppler Effect **(a)** A car moves with a speed of 45.0 mi/h toward a stationary observer as its horn is blown, emitting a frequency of 445 Hz. What frequency is heard by the observer? **(b)** Two cars move toward each other, each with a speed of 22.5 mi/h. If one car's horn is blown with a frequency of 445 Hz, what frequency is heard by observers in the other car?

Picture the Problem The sketch shows (a) the source car moving toward the stationary observer car, and (b) source and observer cars moving toward each other.

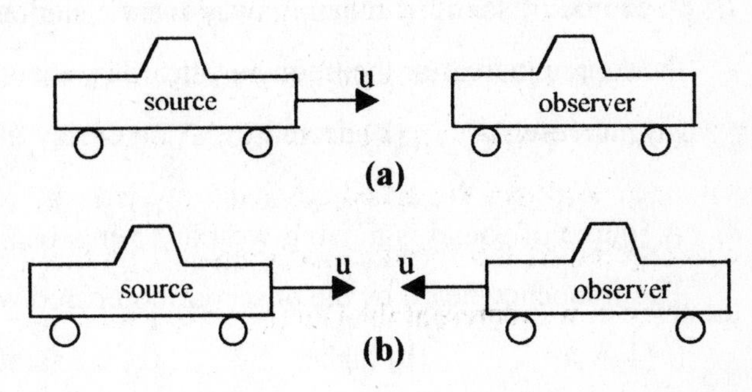

Strategy We apply the expression for the Doppler effect in both parts (a) and (b), making sure to use the proper signs to describe the given situations.

Solution

Part (a)

1. Since the source is moving toward the observer we use the minus sign in the denominator, with $u_o = 0$:

$$f' = \left(\frac{1 \pm u_o/v}{1 \mp u_s/v}\right) f \Rightarrow f' = \left(\frac{1}{1 - u_s/v}\right) f$$

2. Convert the speed to m/s:

$$u_s = 45.0 \text{ mi/h}\left(\frac{0.447 \text{ m/s}}{\text{mi/h}}\right) = 20.115 \text{ m/s}$$

3. Solve for the numerical answer:

$$f' = \left(\frac{1}{1 - u_s/v}\right) f = \left(1 - \frac{20.115 \ \frac{\text{m}}{\text{s}}}{343 \ \frac{\text{m}}{\text{s}}}\right)^{-1} (445 \text{ Hz}) = 473 \text{ Hz}$$

Part (b)

4. Convert the speeds to m/s:

$$u_s = u_o = 22.5 \text{ mi/h}\left(\frac{0.447 \text{ m/s}}{\text{mi/h}}\right) = 10.058 \text{ m/s}$$

5. For this case we have a + in the numerator and a – in the denominator:

$$f' = \left(\frac{1 + u_o/v}{1 - u_s/v}\right) f$$

$$\therefore f' = \left(\frac{1 + (10.058 \text{ m/s})/(343 \text{ m/s})}{1 - (10.058 \text{ m/s})/(343 \text{ m/s})}\right)(445 \text{ Hz})$$

$$= 472 \text{ Hz}$$

Insight Even though the relative speeds between the two cars are the same in both parts (a) and (b), the frequencies heard are slightly different. Why?

Practice Quiz

7. A source of sound is moving away from a stationary observer at 15 m/s. If the frequency emitted by the source increases, the frequency heard by the observer will
   (a) increase. **(b)** decrease. **(c)** stay the same. **(d)** none of the above

8. A source of sound is moving west at 15 m/s behind an observer who is also moving west at 12 m/s. The frequency heard by the observer, compared with the frequency emitted by the source, will be
   (a) lower **(b)** higher **(c)** the same **(d)** none of the above

14–7 – 14–8 Superposition, Interference, and Standing Waves

When more than one wave passes through the same medium simultaneously the resulting wave is called a **superposition** of the individual waves. The superposition of two waves, $y_1(x,t)$ and $y_2(x,t)$, results from the addition of the individual displacements to form a single wave, $y(x,t) = y_1(x,t) + y_2(x,t)$. After the interaction, the individual waves continue on their way unaltered. The wave that results from the superposition of other waves is said to produce an **interference pattern** of the individual waves.

Characteristic of wave interference patterns are regions where two special cases can be identified. One of these special cases occurs when the individual waves add in such a way that their maxima and/or their minima are at the same place at the same time. The result of this occurrence is that the amplitude of the resultant wave equals the sum of the amplitudes of the individual waves (the resultant wave is, in a sense, enhanced relative to the individual waves). This effect is called **constructive interference**. In short, constructive interference between two waves occurs at places where the waves are **in phase**. The other special case, **destructive interference**, occurs when the maximum of one wave meets the minimum of another. In this latter case the amplitude of the resultant wave equals the difference of the amplitudes of the individual waves (the resultant wave is therefore diminished relative to the individual waves). When destructive interference occurs the waves are said to have **opposite phase**.

Example 14–6 Two Sources with Opposite Phases Each of two point sources a distance s = 3.0 m apart emits sound waves of equal wavelength with opposite phase. If a detector placed 5.2 m from the line joining the sources, and one quarter of the way from one to the other, detects constructive interference between the two waves, what is the maximum possible wavelength of the sound wave?

Picture the Problem The diagram shows the two sources (top and bottom). The detector is indicated as being a distance L from the line joining the sources and 1/4 of the way between them from the top source.

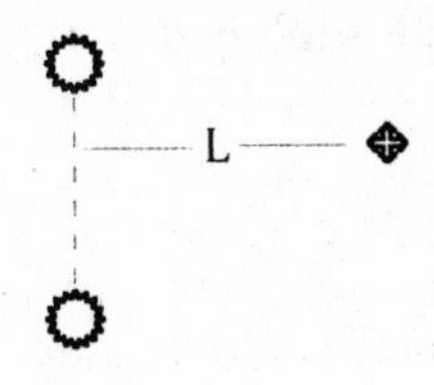

Strategy We need to determine the path difference between waves from each source in terms of the distance L. We must then correctly identify how this path difference relates to the wavelength to produce constructive interference.

Solution

1. The path length of the top source is the hypotenuse of the right triangle with legs L and s/4:

$$d_T = \sqrt{L^2 + (s/4)^2} = \sqrt{(5.2\text{ m})^2 + \left(3.0\text{ m}/4\right)^2} = 5.254\text{ m}$$

2. The path length of the bottom source is the hypotenuse of the right triangle with legs L and $\frac{3}{4}s$:

$$d_B = \sqrt{L^2 + \left(\frac{3s}{4}\right)^2} = \sqrt{(5.2\text{ m})^2 + \left(\frac{9.0\text{ m}}{4}\right)^2} = 5.666\text{ m}$$

3. The path difference between the two sources is:

$$\Delta d = d_B - d_T = 5.67\text{ m} - 5.25\text{ m} = 0.412\text{ m}$$

4. For constructive interference the path difference must be an integral multiple of $\frac{1}{2}\lambda$ since the waves start with opposite phases. Thus, the maximum possible wavelength is:

$$\Delta d = \lambda/2 \Rightarrow \lambda = 2(\Delta d) = 2(0.412\text{ m}) = 0.82\text{ m}$$

Insight The answer gives only one possible wavelength. What are some others?

One of the most important consequences of the superposition of waves occurs when a wave and its reflection travels through a medium. Under the right conditions, these two waves, traveling in opposite directions, combine to produce what is called a **standing wave**. A standing wave is a wave that oscillates in time but is fixed in its spatial location. Standing waves contain positions (or regions), called **nodes,**

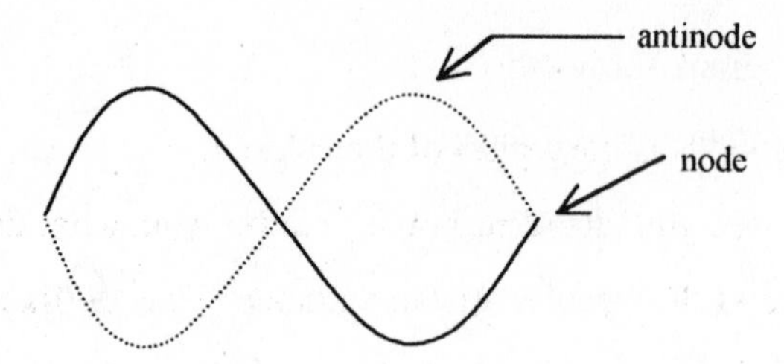

where no oscillation occurs and other positions where the oscillation of the medium has maximum amplitude, called **antinodes**.

Waves on a String

To establish a standing wave on a string with fixed ends, such as a guitar string, each fixed end must be a node. Therefore, the only standing waves that can be established on such a string are those that meet the condition of having nodes at the fixed end. The longest wavelength of a standing wave that will satisfy these conditions is called the **fundamental mode** (in terms of wavelength) or the **first harmonic** (in terms of frequency). Try to convince yourself by drawing a picture that for a string of length L the wavelength of the fundamental mode must be $\lambda = 2L$. The corresponding frequency of the first harmonic is found from $v = \lambda f_1$ to be $f_1 = v/2L$. If we continue to examine the different modes possible on strings with fixed ends, we see that the wavelengths and frequencies can be determined by

$$\lambda_n = 2L/n \quad \text{and} \quad f_n = nf_1$$

where n = 1, 2, 3, ...

Vibrating Columns of Air

Standing waves in air columns are understood in ways very similar to those for standing waves on strings; however, it is important to recognize that in air columns we are talking about longitudinal sound waves rather than transverse vibrations. Two types of air columns are in common use for generating sounds: one type has one end of the column closed off (with the other end open), and the other type has both ends open.

Standing waves in air columns with both ends open display the same characteristics as waves on a string with both ends fixed:

$$f_n = nf_1 \quad \text{and} \quad \lambda_n = 2L/n$$

for n = 1, 2, 3, ... The fundamental frequency is given by

$$f_1 = \frac{v}{2L}$$

where v is the speed of sound, and L is the length of the column.

In air columns with one closed end, standing waves exhibit somewhat different characteristics. In this case the closed end is a node, and the open end is an antinode. Thus, unlike in the two previous cases the ends of the standing wave are different. The longest wavelength of a standing wave (the fundamental mode) that can be set up in a column of length L is $\lambda_1 = 4L$. The other wavelengths are

$$\lambda_n = \frac{4L}{n} \quad \text{with} \quad n = 1,3,5,\ldots$$

For the frequencies only odd harmonics (odd multiples of the fundamental frequency) are present:

$$f_n = nf_1 \quad \text{with} \quad n = 1,3,5,\ldots$$

where the fundamental frequency is

$$f_1 = \frac{v}{4L}$$

All the above information can be used to explain about the generation of sound, especially music.

Example 14–7 Guitar String A certain guitar string of length 0.90 m has a linear density of 0.0075 kg/m. When properly tuned, the string has a 4th harmonic of 1024 Hz. What is the tension for properly tuning this string?

Picture the Problem The sketch shows the 4th harmonic of a guitar string with fixed ends.

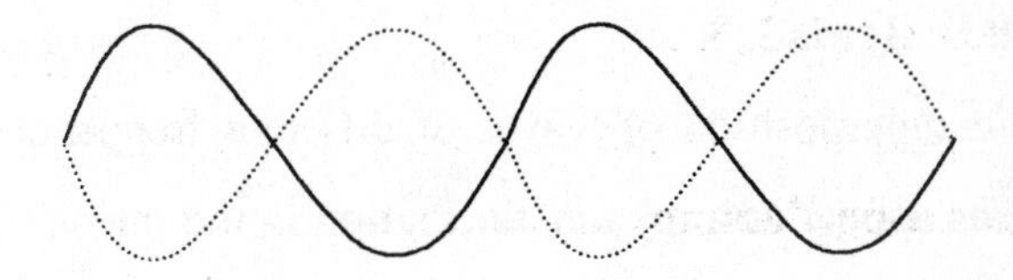

Strategy Since we know how tension relates to the speed of the wave, we can express the speed in terms of the wavelength and frequency and relate these to the tension.

Solution

1. From the diagram we see that we have two wavelengths for the fourth harmonic: $L = 2\lambda \;\therefore\; \lambda = L/2 = (0.90 \text{ m})/2 = 0.45 \text{ m}$

2. The relationship between λ, f, and F is: $\lambda f = \sqrt{\frac{F}{\mu}} \;\therefore\; F = \mu\lambda^2 f^2$

3. Obtain the numerical result:

$$F = \mu\lambda^2 f^2 = (0.0075 \text{ kg/m})(0.45 \text{ m})^2 (1024 \text{ Hz})^2$$
$$= 1600 \text{ N}$$

Insight Practice drawing different modes and harmonics of standing waves both on strings and in air columns.

Practice Quiz

9. Suppose two sources of identical waves emit them in phase. When these waves superimpose in the surrounding space, which of the following is a condition on the path difference, Δd, between the waves that will produce destructive interference?

   (a) $\Delta d = \lambda$ **(b)** $\Delta d = \lambda/2$ **(c)** $\Delta d = \lambda/3$ **(d)** $\Delta d = \lambda/4$ **(e)** none of the above

10. What is the wavelength of the third harmonic for standing waves on a string of length 1.35 m with fixed ends?

    (a) 1.5 m **(b)** 1.35 m **(c)** 0.900 m **(d)** 4.05 m **(e)** 0.45 m

11. What is the wavelength of the third harmonic for standing waves in an air column of length 1.35 m with only one open end?

    (a) 5.4 m **(b)** 1.35 m **(c)** 1.80 m **(d)** 1.08 m **(e)** 0.45 m

14–9 Beats

The superposition of waves of different frequencies gives rise to the phenomenon called **beats**. These beats appear as a regular fluctuation in the intensity of the wave that results from the superposition. This variation in intensity results from a variation in the amplitude of the resultant wave. The frequency of the successive intensity maxima is called the **beat frequency**. Two waves of frequencies f_1 and f_2 will produce a beat frequency equal to the difference between them:

$$f_{beat} = |f_1 - f_2|$$

Beats are often heard when two guitar strings are played at the same time and are often used to tune musical instruments to the desired frequency.

14–10 Matter Waves

Waves play a much more fundamental role in the physical behavior of matter than is obvious in everyday life. We now know that even objects that are traditionally viewed as *matter particles* (such as the electrons and protons that make up atoms) exhibit wavelike behavior; We call these **matter waves**. If we consider a particle of mass m moving with speed v, this particle has a linear momentum of magnitude p = mv. It turns out that we must also associate with this particle-like behavior a wave whose wavelength, called its *deBroglie wavelength* λ, is related to its momentum according to

$$\lambda = \frac{h}{p}$$

where h is a fundamental constant of nature called *Planck's constant*. This constant has the value

$$h = 6.63 \times 10^{-34}\ \text{J} \cdot \text{s}$$

Notice that the particle-like nature (linear momentum) and the wavelike nature (wavelength) are intimately connected — you can't have one without the other (at a fundamental level). This understanding was the key that unlocked the door to the modern technological revolution.

Exercise 14–8 A Macroscopic Particle How fast must a particle of mass 1.00 g move to have a deBroglie wavelength of 1.00 mm?

Solution: We are given the following information.

Given: $m = 1.00$ g, $\lambda = 1.00$ mm; **Find**: v

The deBroglie wavelength is given by $\lambda = h/p$, where p = mv. Therefore,

$$\lambda = \frac{h}{mv} \quad \therefore \quad v = \frac{h}{m\lambda} = \frac{6.63 \times 10^{-34}\ \text{J} \cdot \text{s}}{\left(1.00 \times 10^{-3}\ \text{kg}\right)\left(1.00 \times 10^{-3}\ \text{m}\right)} = 6.63 \times 10^{-28}\ \text{m/s}$$

This velocity is essentially zero. At this speed, it would take an object an amount of time equal to the age of the universe to move a distance the size of an atom. What does this result say about seeing the effects of matter waves for macroscopic objects?

Practice Quiz

12. As the speed of a particle increases, its deBroglie wavelength

(a) decreases. (b) increases. (c) stays the same. (d) none of the above

Reference Tools and Resources

I. Key Terms and Phrases

wave results from the connection of a series of oscillators

harmonic waves waves for which the oscillators move with simple harmonic motion

transverse waves waves for which the oscillation is perpendicular to the direction of propagation

longitudinal waves waves for which the oscillation is along the direction of propagation

period (for waves) the minimum amount of time it takes for a wave to repeat

frequency the number of cycles of a wave's oscillations per unit of time

wavelength the minimum repeat length of a wave

intensity the amount of energy per unit area per unit time

intensity level a measure of a sound's loudness relative to a standard reference

Doppler effect the perceived shift in frequency due to relative motion between the source and the observer

superposition the addition of two or more waves

interference pattern the wave pattern that results from the superposition of two or more waves

in phase: when the crests and/or troughs of different waves occur at the same time.

opposite phase when the crests of a wave occur at the same time as the troughs of another wave

constructive interference when waves superimpose in phase resulting in a wave of larger amplitude

destructive interference when waves superimpose out of phase resulting in a wave of smaller amplitude

standing wave a stationary wave resulting from the superposition of two waves traveling in opposite directions

node positions on a standing wave that do not oscillate

anti-node positions on a standing wave that oscillate with maximum amplitude

fundamental mode (first harmonic) the longest wavelength standing wave

beats the variation in the intensity of a wave that results from the superposition of waves of different frequency

beat frequency the frequency of successive intensity maxima of a wave that exhibits beats

matter wave the wavelike behavior of particles in accordance with the deBroglie relation

II. Important Equations

Name/Topic	Equation	Explanation
wave speed	$v = \frac{\lambda}{T} = \lambda f$	The speed of a traveling wave.
wave speed	$v = \sqrt{\frac{F}{\mu}}$	The speed of a wave on a string of tension F and linear density μ.
harmonic waves	$y(x,t) = A\cos\left(\frac{2\pi}{\lambda}x - \frac{2\pi}{T}t\right)$	The functional form for a harmonic traveling wave.
sound intensity	$I = \frac{P}{4\pi r^2}$	The intensity of a sound from a point source that emits isotropically falls off as the inverse square of the distance from the source.
intensity level	$\beta = 10\log\left(\frac{I}{I_0}\right)$	The intensity level (in decibels) for the loudness of a sound.
Doppler effect	$f' = \left(\frac{1 \pm u_o/v}{1 \mp u_s/v}\right)f$	The 4 possible relationships between the frequency heard f′ and the frequency emitted f due to the Doppler effect.
standing waves	$\lambda_n = 2L/n$ and $f_n = nf_1$	The modes and harmonics for standing waves on strings with fixed ends and in air columns with both ends open. The possible values of n are 1, 2, 3,...
standing waves	$\lambda_n = \frac{4L}{n}$ and $f_n = nf_1$	The modes and harmonics for standing waves in air columns with only one open end. The possible values of n are 1, 3, 5,...
beats	$f_{beat} = \lvert f_1 - f_2 \rvert$	The frequency of beats produced by the superposition of two waves.
matter waves	$\lambda = \frac{h}{p}$	The wavelength associated with a particle's fundamental wave nature.

III. Know Your Units

Quantity	Dimension	SI Unit
period (T)	[T]	s
frequency (f)	$[T^{-1}]$	Hz
wavelength (λ)	[L]	m
linear density (μ)	[M]/[L]	kg/m
intensity (I)	$[M]/[T^3]$	W/m^2
intensity level (β)	dimensionless	dB
Planck's constant (h)	$[M]\cdot[L^2]/[T]$	J·s

Practice Problems

1. A harmonic wave is given by y(x, t) = 4cos(4.75x – 6.83t). To the nearest hundredth of a meter what is the wavelength?
2. In problem 1 what is its period to the nearest thousandth of a second?
3. What is the velocity of the wave in problem 1, to the nearest hundredth of a m/s?
4. What will be the value of y, to two decimal places, in problem 1 if x = 1.4 and t = 0.7?
5. A point source of a wave emits 154 watts. What is its intensity at a distance of 8.14 meters, to 2 decimal places?
6. Two speakers emit sounds, one with an intensity of 4.9×10^7 W/m^2 and the other 7.18×10^7 W/m^2. What is the difference in their intensity levels, to the nearest tenth of a dB?
7. Taking right to be positive, if a speaker is moving to the right at a velocity of 5 m/s while emitting a sound of frequency 1815 Hz, you will hear a frequency of...
8. If the situation is reversed in problem 7, and you are the one that is moving (to the left), what frequency will you hear? (Remember, right is positive.)
9. A vibrating stretched string has a length of 137 cm, a mass of 15 grams and is under a tension of 43 newtons. What is the frequency, to the nearest Hz of its 3rd harmonic?
10. If the source of the wave in problem 9 was an open organ pipe of the same length as the wire, what would be the frequency of the 3rd harmonic?

Puzzle

WHEN BOBBING MET SALLY

(This puzzle authored by Prof. Evelyn Patterson, United States Air Force Academy. Used by permission)

Little Sally is floating peacefully in her small plastic boat, at the position marked A in the figure below. The lake is calm, and the day is sunny. A motorboat speeds by, about 30 m offshore, and the waves reach the shore about half a minute later. Suddenly, little Sally's older brother and sister decide to give Sally some fun. Each sibling is in a rowboat firmly tied to moorings out in the lake, as shown in the figure. They both jump up and down in their respective rowboats, bobbing up and down in synch together, about 15 times every half a minute. Soon, large waves are emanating from each of the bobbing boats, and the older brother and sister are giggling fiendishly.

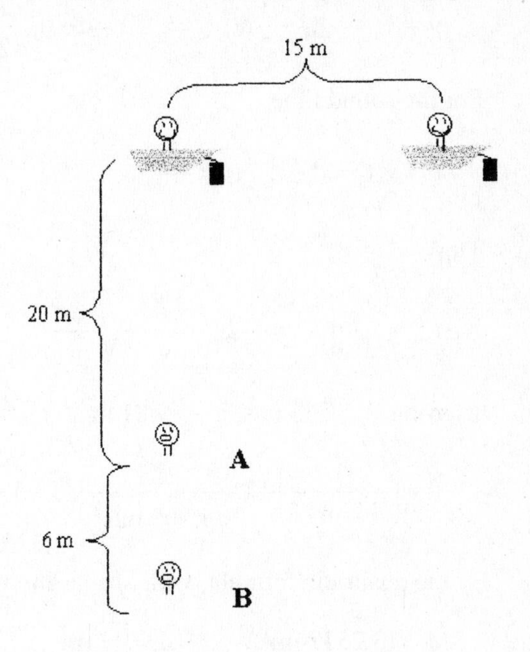

Sally's father, seeing the situation, and having taken the equivalent of honors physics, does some quick thinking and moves Sally to the position marked B in the figure. Sally likes to float in calm water. Does Sally's father do Sally a favor by moving her?

Selected Solutions

11. The tension, F, is related to the speed, v, by $F = \mu v^2$.

(a) The speed can be determined from the length of the wire and the time, t:

$$F = \mu v^2 = \frac{m}{L}\left(\frac{L}{t}\right)^2 = \frac{mL}{t^2} = \frac{(0.085\text{ kg})(7.3\text{ m})}{(0.94\text{ s})^2} = \boxed{0.70\text{ N}}$$

(b) $F \propto m$ for a given v and L. So increased mass means $\boxed{\text{increased}}$ tension.

(c) For a different value of m the tension is given by the same formula:

$$F = \frac{mL}{t^2} = \frac{(100.0\text{ kg})(7.3\text{ m})}{(0.94\text{ s})^2} = \boxed{0.83\text{ N}}$$

23. (a) total time = fall time + sound travel time, so $t_{tot} = t_f + t_s$

For the fall time:

$$y = y_0 + v_0 t_f + \frac{1}{2} a t_f^2 \quad \Rightarrow \quad 0 = d + 0 + \frac{1}{2} a t_f^2 \quad \therefore \quad t_f = \sqrt{\frac{-2d}{a}}$$

For the sound time:

$$d = v_s t_s = d \quad \therefore \quad t_s = \frac{d}{v_s}$$

Thus,

$$t_{tot} = t_s + t_f \quad \Rightarrow \quad t_{tot} = \frac{d}{v_s} + \sqrt{\frac{-2d}{a}}$$

Inserting $v_s = 343$ m/s, $a = -9.81$ m/s^2, $t_{tot} = 1.5$ s, and rearranging, we obtain

$$0 = \left(\frac{1}{343 \text{ m/s}}\right) d + \frac{1}{\sqrt{4.905 \text{ m/s}^2}} \sqrt{d} - 1.5 \text{ s}$$

By the quadratic formula, with $\sqrt{d}$ as the variable, we get

$$\sqrt{d} = (3.2537 \text{ m})^{1/2} \quad \therefore \quad d = \boxed{11 \text{ m}}$$

(b) $\boxed{\text{Less}}$, because although the sound travel time would double, the fall time would less than double (increasing only by a factor of $\sqrt{2}$).

43. (a) Since they are approaching each other,

$$f' = \left(\frac{1 + \frac{u_o}{v}}{1 - \frac{u_s}{v}}\right) f = \left(\frac{1 + \frac{8.5 \text{ m/s}}{343 \text{ m/s}}}{1 - \frac{8.5 \text{ m/s}}{343 \text{ m/s}}}\right)(300.0 \text{ Hz}) = \boxed{320 \text{ Hz}}$$

(b) $\boxed{\text{Bicyclist A speeding up}}$. For equal changes in speed, moving-source effects are greater than moving-observer effects.

55. Since the observer moves from hearing destructive interference *directly* to hearing constructive interference, the difference in distances from the two speakers must equal half a wavelength.

$$d_1 = \sqrt{(5.0 \text{ m})^2 + \left(\frac{3.5}{2} \text{ m} + 0.84 \text{ m}\right)^2} = 5.631 \text{ m}$$

$$d_2 = \sqrt{(5.0 \text{ m})^2 + \left(\frac{3.5}{2} \text{ m} - 0.84 \text{ m}\right)^2} = 5.082 \text{ m}$$

Since $d_1 - d_2 = \lambda/2$ we have that

$$f = \frac{v}{\lambda} = \frac{v}{2(d_1 - d_2)} = \frac{343 \text{ m/s}}{2(5.631 \text{ m} - 5.082 \text{ m})} = \boxed{0.31 \text{ kHz}}$$

63. (a) The wave shown represents the 3rd harmonic, which corresponds to $f_3 = 3f_1$. Therefore,

$$f_3 = 3\left(\frac{v}{4L}\right) = \frac{3(343 \text{ m/s})}{4(2.5 \text{ m})} = \boxed{0.10 \text{ kHz}}$$

(b) $f_1 = \frac{v}{4L} = \frac{343 \text{ m/s}}{4(2.5 \text{ m})} = \boxed{34 \text{ Hz}}$

Answers to Practice Quiz

1. (b) **2.** (d) **3.** (a) **4.** (c) **5.** (d) **6.** (e) **7.** (a) **8.** (b) **9.** (b) **10.** (c) **11.** (c) **12** (a)

Answers to Practice Problems

1. 1.32 m	**2.** 0.920 s
3. 1.43 m/s	**4.** −1.23 m
5. 0.18 W/m^2	**6.** 1.7 dB
7. 1842 Hz	**8.** 1789 Hz
9. 69 Hz	**10.** 376 Hz

CHAPTER 15

FLUIDS

Chapter Objectives

After studying this chapter, you should:

1. understand the essential characteristics of fluid behavior.

2. know the difference between atmospheric and gauge pressure.

3. understand how pressure depends on depth in a static fluid.

4. understand, and be able to apply, Pascal's principle.

5. understand, and be able to apply, Archimedes' principle.

6. understand the behavior described by the equation of continuity.

7. understand, and be able to apply, Bernoulli's equation.

8. understand the basic reasons for viscosity and surface tension.

Warm-Ups

1. The atmospheric pressure at sea level is 1.013×10^5 N/m^2. The unit of pressure N/m^2 is also called a pascal (Pa). The surface area of an average adult body is about 2 m^2. How much crushing force does the atmosphere exert on people? Why don't we get crushed? Calculate this value in pounds using the fact that the weight of a 1-kilogram mass is 2.2 pounds.
2. The atmospheric pressure decreases with altitude exponentially according to the relation $p = p_o e^{-(0.00012)h}$ where h is the altitude above sea level in meters, and $p_o = 1.013 \times 10^5$ N/m^2. The average atmospheric pressure in Florida is close to p_o. Estimate the average atmospheric pressure in Denver (The elevation at Denver is about 1500 m.)
3. Water is pumped up a pipeline as shown. The water pours out at the top and to the ground. The pump is running at constant speed. Compare the water speed at the three points A, B, and C in the pipeline. *State your reasons for all the answers.*

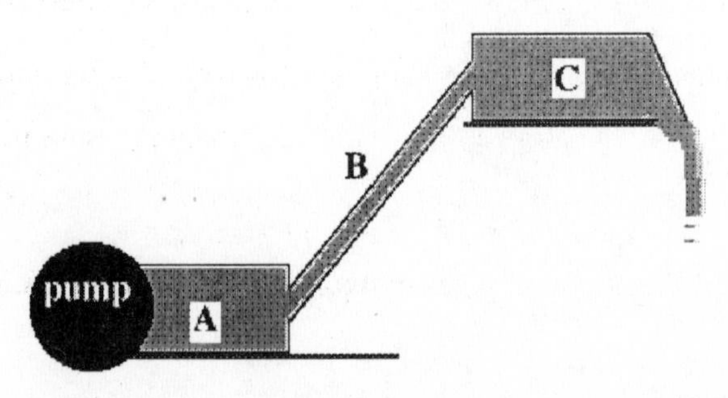

4. Estimate the force on a 1.5-m^2 windshield as a 60 mi/h (30 m/s) wind passes near its surface. The air in the car is stationary, and the car windows are shut tight.

Chapter Review

In this chapter we study **fluids**. Fluids are characterized by their ability to flow; both liquids and gases are considered to be fluids. An understanding of fluid behavior is essential to life, and applications of this understanding are essential to many of the conveniences of modern living.

15–1 – 15–2 Density and Pressure

One of the most convenient properties used to describe a fluid is its **density**, ρ. The density of a substance is a measure of how compact the substance is; that is, how much mass is packed into a volume of the substance. On average, the density of a substance is the amount of mass M divided by the volume V taken up by that mass

$$\rho = \frac{M}{V}$$

Another important quantity in the study of fluids is **pressure**, P. On average, the pressure that is applied to an object is the amount of force F (normal to the surface) divided by the area A over which the force spreads

$$P = \frac{F}{A}$$

As the preceding expressions indicate, the unit of pressure is that of force divided by area; in SI units this is called the pascal (Pa): 1 Pa = 1 N/m^2. An important property of the pressure in a fluid is that it is equally applied in all directions (at a given depth) and applies forces that are perpendicular to any surface in the fluid.

When considering the pressure on or within a fluid it is important to recognize that for many applications we must account for a constant atmospheric pressure ($P_{at} = 1.01 \times 10^5$ Pa). Because of the constant presence of the atmosphere we are often interested only in the pressure above and beyond that applied by the atmosphere. This additional pressure is called the **gauge pressure**, P_g

$$P_g = P - P_{at}$$

where P is the total pressure applied (sometimes called the *absolute pressure*).

Exercise 15–1 The Density of a Fluid If 1.00 gallons of a certain fluid weighs 3.22 lb, what is its density?

Solution: We are given the following information:

Given: V = 1.00 gal, W = 3.22 lb; **Find**: ρ

Since the density is given by $\rho = m/V$ we need to find the mass from the weight and convert everything to SI units. To get the mass we use

$$m = \frac{W}{g} = \frac{3.22\text{ lb}\left(\frac{4.45\text{ N}}{\text{lb}}\right)}{9.81\text{ m/s}^2} = 1.461\text{ kg}$$

We can immediately convert the volume can be to give

$$V = 1.00\text{ gal}\left(\frac{3.785\times10^{-3}\text{m}^3}{\text{gal}}\right) = 3.785\times10^{-3}\text{m}^3$$

We are now ready to calculate the density:

$$\rho = \frac{m}{V} = \frac{1.461\text{ kg}}{3.785\times10^{-3}\text{m}^3} = 386\text{ kg/m}^3$$

Example 15–2 Gauge Pressure of Water A uniform cylindrical container has a radius of 7.8 cm and a height of 13.2 cm. If this container is completely filled with water, what gauge pressure does the water apply to the bottom of the container?

Picture the Problem The sketch shows a cylindrical container filled with water.

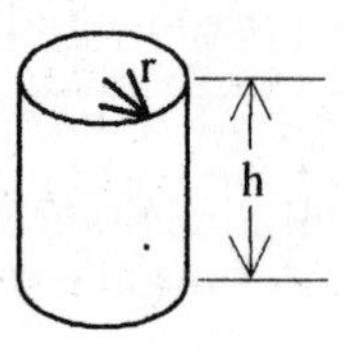

Strategy The pressure applied by the water is the force that the water applies (equal to its weight) divided by the bottom area of the cylinder.

Solution

1. The volume of water equals the volume of the cylinder: $V = Ah = \pi r^2 h = \pi(0.078\text{m})^2(0.132\text{m}) = 2.523\times10^{-3}\text{m}^3$

2. Determine mass of the water: $m = \rho V = (1000\text{ kg/m}^3)(2.523\times10^{-3}\text{ m}^3) = 2.523\text{ kg}$

3. Determine the weight of the water: $W = mg = (2.523 \text{ kg})(9.81 \text{ m/s}^2) = 24.75 \text{ N}$

4. Obtain the gauge pressure: $P = \frac{F}{A} = \frac{mg}{\pi r^2} = \frac{24.75 \text{ N}}{\pi(0.078 \text{ m})^2} = 1.3 \times 10^3 \text{ Pa}$

Insight The answer is the gauge pressure because we ignored atmospheric pressure.

Practice Quiz

1. If it takes twice as much volume of fluid 1 to weigh the same as fluid 2, how do their densities compare?

   (a) Fluid 1 is twice as dense as fluid 2.

   (b) Fluid 1 is half as dense as fluid 2.

   (c) Fluid 1 and fluid 2 have equal densities.

   (d) Fluid 1 is four times less dense than fluid 2.

   (e) none of the above

2. If a uniform cylinder of height 0.850 m and radius 0.250 m is completely filled with water, what is the total pressure on the bottom of the cylinder?

   (a) 8.34 kPa **(b)** 133 kPa **(c)** 2.08 kPa **(d)** 109 kPa **(e)** 46.2 kPa

15–3 Static Equilibrium in Fluids: Pressure and Depth

In this and the next several sections we shall consider the properties of static fluids, that is, fluids that do not flow. In the case of static fluids, every part of the fluid and every object within the fluid is in static equilibrium. One of the basic properties of fluids that is very important to our ability to understand fluid behavior is known as **Pascal's principle**:

External pressure applied to an enclosed fluid is transmitted unchanged throughout the fluid.

Pascal's principle is important in determining the dependence of pressure on the depth within a fluid. Without any external pressure applied to the outer surface of a fluid, the pressure measured at a depth h beneath the surface arises from the weight of the fluid above the given level. The amount of this increase in pressure is given by ρgh. If there is external pressure on the fluid, such as from the atmosphere and/or any other source, this pressure is transmitted undiminished to every point in the fluid and it must be added to the pressure due to the weight of the fluid. Thus, for the dependence of pressure on depth we have

$$P_2 = P_1 + \rho gh$$

where P_2 is the pressure at a given level within the fluid and P_1 is the pressure at a height h above that level.

Example 15–3 Blood Pressure with Depth Human blood has a density of approximately 1.05×10^3 kg/m^3. Use this information to estimate the difference in blood pressure between the brain and the feet in a person who is approximately 6 feet tall.

Picture the Problem The sketch shows a person approximately 6 feet in height.

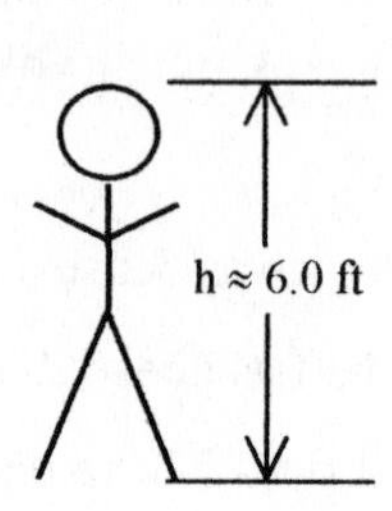

Strategy We attempt this approximation to two significant figures by applying the above result for the dependence of pressure on depth using blood as the fluid.

Solution

1. Convert the height to meters: $h = 6.0 \text{ ft}\left(\dfrac{\text{m}}{3.28 \text{ ft}}\right) = 1.83 \text{ m}$

2. The difference in pressure is given by: $P_2 - P_1 = \rho gh$

3. Obtain the numerical result: $P_2 - P_1 = \left(1.05 \times 10^3 \text{ kg/m}^3\right)\left(9.81 \text{ m/s}^2\right)\left(1.83 \text{ m}\right) = 19 \text{ kPa}$

Insight This is only an estimate for many reasons. The blood in the body is not a static fluid; it flows, and between the head and the feet is a pump (the heart) that will affect the result. If you plan to study medicine, see if you can find out the difference in blood pressure between the head and the feet.

Pascal's principle is also key to understanding the hydraulic lift. This device uses fluid pressure to convert a small input force into a large output force. The input force F_1 is applied to a fluid over a comparatively small area A_1 giving rise to a pressure change of F_1/A_1. Since this pressure change is transmitted undiminished throughout the fluid, $F_1/A_1 = F_2/A_2$, we must get a larger output force $F_2 > F_1$ if it is spread over a larger area $A_2 > A_1$; therefore,

$$F_2 = F_1\left(\frac{A_2}{A_1}\right)$$

The consequence of getting this larger output force is that the distance through which this force can move an object at the output, d_2, is smaller than the distance at the input, d_1. Since the same volume of fluid moves at the input and output, $A_1d_1 = A_2d_2$, the output distance is given by

$$d_2 = d_1\left(\frac{A_1}{A_2}\right)$$

Exercise 15–4 Hydraulic Lift Your job at Dave's Manufacturing Company is to design a hydraulic lift for a client. This client typically needs to raise objects through a height of 0.500 meters. The system should easily be used by people of average height, so the input distance should not exceed 5.00 ft. What would be a good ratio of output area to input area to consider?

Solution: We are given the following information.

Given: $d_1 = 5.00$ ft , $d_2 = 0.500$ m; **Find**: A_2/A_1

From the information given, we know that the ratio of the input distance to the output distance is directly proportional to the ratio we seek:

$$\frac{d_1}{d_2} = \frac{A_2}{A_1}$$

Therefore, the ratio is

$$\frac{A_2}{A_1} = \frac{5.00\ \text{ft}\left(\frac{\text{m}}{3.28\ \text{ft}}\right)}{0.500\ \text{m}} = 3.05$$

The output area should be at least 3.05 times greater than the input area. Of course, in a more realistic situation you'd want more information from the client. What additional information might you want?

Practice Quiz

3. What is the gauge pressure 3.4 m below the surface of a container filled with a fluid of density 550 kg/m^3?

   (a) 33 kPa **(b)** 1900 Pa **(c)** 18 kPa **(d)** 100 kPa **(e)** 550 Pa

4. The gauge pressure at a particular location in a fluid of density 870 kg/m^3 is 120 kPa. What is the gauge pressure in the fluid 5.9 m above this location?

   (a) 70 kPa **(b)** 50 kPa **(c)** 62 kPa **(d)** 58 kPa **(e)** 20 kPa

15–4 – 15–5 Archimedes' Principle and Buoyancy

When an object is submerged in a fluid, the volume taken up by the object displaces an equal volume of the fluid. The pressure applied by the fluid onto the object results in an upward force on the object; this phenomenon is known as **buoyancy**. This phenomenon is governed by **Archimedes' principle**:

An object immersed in a fluid experiences an upward force equal to the weight of the fluid displaced by the object.

The weight of the fluid displaced by the object equals the mass of this fluid times the acceleration due to gravity, mg. When dealing with buoyancy it is usually more convenient to express the mass in terms of the density, $m = \rho V$; therefore, for an object submerged in a fluid, the buoyant force on it is

$$F_b = \rho g V$$

Archimedes' principle explains the phenomenon of **flotation**, which occurs when the bouyant force acting on an object equals its weight. Often, floating objects are not completely submerged in the fluid. The amount of volume submerged, V_{sub}, for a solid object of volume V_s floating in a fluid of density ρ_f is given by

$$V_{sub} = V_s\left(\frac{\rho_s}{\rho_f}\right)$$

where ρ_s is the density of the solid object.

Exercise 15–5 The Secret of Magic Many magic tricks are based on physical principles. In order to fool her audience a magician uses an object that sinks in the freshwater made available to the audience, but floats in the seawater that she uses on stage. What is the maximum percentage of the object's volume that will float above the seawater?

Picture the Problem The sketch shows a floating object partially submerged in seawater.

Strategy According to the above discussion, more of an object will be submerged if its density approaches that of the fluid, so we get the maximum above-surface float for the smallest possible object density. Since it must sink in fresh water, the smallest object density is 1000 kg/m^3.

Solution

1. The minimum fraction of volume submerged is: $\frac{V_{sub}}{V} = \frac{\rho_{obj}}{\rho_{sea}} = \frac{1000\ \text{kg/m}^3}{1025\ \text{kg/m}^3} = 0.976$

2. Find the maximum amount of volume floating above the surface is: $V_{above} = (1 - 0.976)V = 0.024V$

3. Calculate the percentage: $\frac{V_{above}}{V} = 0.024 = 2.4\%$

Insight What could the magician do to the seawater to make a larger percentage of the object float?

Practice Quiz

5. An object of density 750 kg/m^3 is half submerged in a fluid. What is the density of this fluid?

 (a) 1500 kg/m^3 (b) 188 kg/m^3 (c) 375 kg/m^3 (d) 2250 kg/m^3 (e) 750 kg/m^3

6. If the volume of the object in question 5 is 0.33 m^3, what is the buoyant force on this object?

 (a) 9810 N (b) 1210 N (c) 3240 N (d) 1000 N (e) 2430 N

15–6 Fluid Flow and Continuity

In this section we begin to discuss properties of fluid flow. During the smooth flow of a constrained fluid (e.g., through a pipe) we can assume that the same amount of mass passes through each cross section of pipe in a given amount of time. This smooth-flow condition leads to what is known as the **equation of continuity**, which says that the mass m_1 flowing through an area A_1 in a given time equals the mass m_2 flowing through area A_2 in that same amount of time. The amount of mass per unit time of a fluid of density ρ flowing through an area A at speed v is ρAv. Therefore, the equation of continuity is

$$\rho_1 A_1 v_1 = \rho_2 A_2 v_2$$

Usually, liquids are considered to be incompressible because the density of the liquid hardly changes as it flows from one place to another. In such cases the densities in the equation of continuity are equal, $\rho_1 = \rho_2$, and we can write the equation as

$$A_1 v_1 = A_2 v_2$$

The quantity Av equals the *volume flow rate* of the fluid. Thus, the preceding equation says that this volume flow rate is constant for an incompressible fluid.

Example 15–6 Continuity Plastic bottles that are used to hold water for athletes often have a long slender nozzle out of which the water emerges. If the end of the nozzle has a diameter of 1.0 cm, and you determine the water to emerge at 25 cm/s for a typical squeeze of the bottle, what is the initial speed of the water in the neck of the bottle if its diameter is 6.0 cm?

Picture the Problem The sketch shows a squeeze water bottle with a thin nozzle on the end.

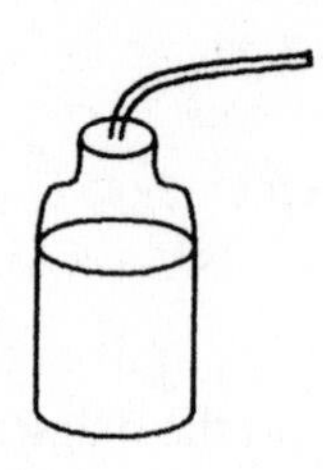

Strategy Since we don't expect the density of the water to change when flowing from inside the bottle to outside, we need only to use the fact that the volume flow rate is constant.

Solution

1. Use the equation of continuity: $A_1v_1 = A_2v_2 \Rightarrow \frac{\pi d_1^2}{4}v_1 = \frac{\pi d_2^2}{4}v_2 \Rightarrow d_1^2 v_1 = d_2^2 v_2$

2. Solve for v_1: $v_1 = v_2\left(\frac{d_2}{d_1}\right)^2$

3. Obtain the numerical result: $v_1 = (25\text{ cm/s})\left(\frac{1.0\text{ cm}}{6.0\text{ cm}}\right)^2 = 0.69\text{ cm/s}$

Insight The fact that a fluid flows more rapidly when squeezed is used in many different applications. Can you think of some others?

Practice Quiz

7. If the area through which an incompressible fluid flows decreases, the speed of flow will

   (a) decrease **(b)** increase **(c)** stay the same

8. If the area through which an incompressible fluid flows is cut in half, the speed of flow will...

   (a) also be cut in half

   (b) double

   (c) decrease to ¼ its speed

   (d) triple

   (e) none of the above

15–7 – 15–8 Bernoulli's Equation

In general, the same concepts that we use to describe the dynamics of particles also apply to fluid dynamics. With fluids it is often more convenient to express these concepts in terms of density and pressure rather than mass and force. One such example comes from the application of the work-energy theorem to fluids. The result is a mathematical relation known as **Bernoulli's equation**.

With a fluid we can replace the concept of a particle with a small region of the fluid called a *fluid element* of density ρ moving at speed v while sweeping out a volume ΔV under the action of a differential fluid pressure ΔP. Bernoulli's equation can be obtained by applying the work-energy theorem ($W_{net} = \Delta K$) to this fluid element. The work done on a fluid element due to a change in the fluid pressure is $\Delta W_{pressure} = (P_1 - P_2)(\Delta V)$. The work done by gravity as the fluid element changes vertical level from a height y_1 to y_2 is $\Delta W_{gravity} = -\rho \Delta V (y_2 - y_1)$. The change in the kinetic energy of the fluid element can be written as $\Delta K = \left(\frac{1}{2}\rho v_2^2 - \frac{1}{2}\rho v_1^2\right)\Delta V$. These three quantities combine to give Bernoulli's equation

$$P_1 + \tfrac{1}{2}\rho v_1^2 + \rho g y_1 = P_2 + \tfrac{1}{2}\rho v_2^2 + \rho g y_2$$

This expression holds in the absence of frictional losses.

Another way of stating Bernoulli's equation is

$$P + \tfrac{1}{2}\rho v^2 + \rho g y = \text{constant}$$

This form of the equation helps make the physical consequences a little more clear because you can see, for example, that for a fluid flowing at a constant vertical level an increase in the speed of flow must be accompanied by a decrease in pressure (and vice versa). This effect, often called *Bernoulli's principle*, is important in understanding the consequence of air flow in many applications.

You should also notice that Bernoulli's equation is consistent with the dependence of pressure on depth that was discussed above. Examining this dependence leads to **Torricelli's law** for the speed of a fluid flowing from an aperture in a container placed a depth h below the surface of the fluid. If, for example, both the surface of the fluid and the aperture are open to the air, then the pressure at both locations is atmospheric pressure, so that $P_1 = P_2$ in Bernoulli's equation. Assuming that the fluid is essentially static at the surface ($v_1 = 0$) we find

$$v_2 = \sqrt{2gh}$$

where $h = y_1 - y_2$.

Example 15–7 Water Flow at Constant Pressure Water flows in a horizontal segment of pipe at a pressure of 85 kPa with a speed of 2.6 m/s. The pipe widens, so that its area becomes larger by 35%. If the flow is to be at constant pressure, how far above the initial horizontal level should the pipe divert the water?

Picture the Problem The diagram shows a pipe carrying water first horizontally, then uphill, and horizontally again.

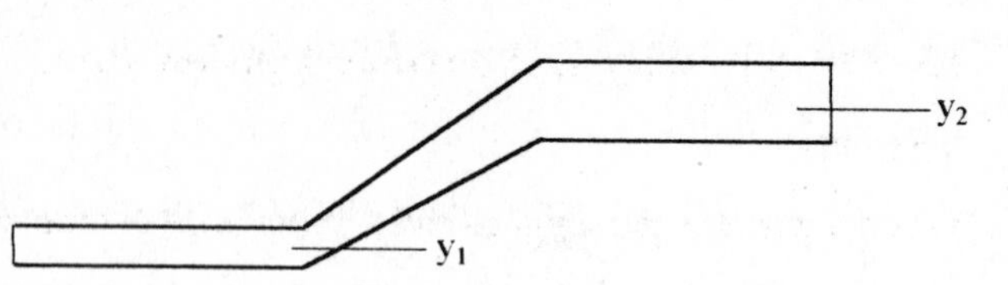

Strategy Bernoulli's equation relates the pressure to the speed of flow and the change in vertical level. However, the changing width of the pipe can be handled by the equation of continuity, so both expressions should be used.

Solution

1. Using the fact that the pressure is constant, $P_1 = P_2$, Bernoulli's equation becomes: $\frac{1}{2}\rho v_1^2 + \rho g y_1 = \frac{1}{2}\rho v_2^2 + \rho g y_2$

2. Solve for the difference in horizontal level: $(y_2 - y_1) = \frac{1}{2g}(v_1^2 - v_2^2)$

3. Use the continuity equation to solve for v_2 in terms of v_1: $v_2 = \frac{A_1 v_1}{A_2} = \frac{A_1 v_1}{A_1 + 0.35A_1} = \frac{v_1}{1.35}$

4. Substitute v_2 into the expression for the change in vertical level: $(y_2 - y_1) = \frac{1}{2g}\left[v_1^2 - (v_1/1.35)^2\right] = (0.451)\frac{v_1^2}{2g}$

5. Obtain the numerical result: $(y_2 - y_1) = (0.451)\frac{(2.6\text{ m/s})^2}{2(9.81\text{ m/s}^2)} = 0.16\text{ m}$

Insight As this example illustrates, you should keep in mind that it is often useful to use both Bernoulli's equation and the equation of continuity to perform a complete analysis.

Practice Quiz

9. If the speed of a horizontally flowing fluid decreases, the pressure in the fluid will

   (a) increase. **(b)** decrease. **(c)** stay the same.

10. If, for a constant speed of flow, a fluid begins to flow downhill, the pressure in the fluid begins to

(a) increase. **(b)** decrease. **(c)** stay the same.

15–9* Viscosity and Surface Tension

Viscosity

When a particle moves it usually experiences some sort of frictional resistance to its motion. The same is true for fluid flow. With fluids this resistance is called **viscosity**. All the previous discussion assumed an ideal fluid which, in part, means that we assumed that there was no viscosity. In this section we will take a glimpse at the effect of including this unavoidable phenomenon.

For a fluid flowing through a tube of length L and cross-sectional area A, the flow results from a pressure differential across the length of the tube, $P_1 - P_2$. What are some of the physical quantities related to this pressure difference? It works out that the pressure difference is directly proportional to the speed, v, at which the fluid flows (it requires a greater pressure difference to get faster flow). We also find that $P_1 - P_2$ is directly proportional to the length of the tube (it requires a greater pressure difference to sustain the flow of the larger amount of fluid contained in a longer tube). Finally, the pressure difference is inversely proportional to the cross-sectional area of the tube (a wider tube provides more room for the fluid to flow more freely requiring less pressure). Thus, we have that $(P_1 - P_2) \propto vL/A$. The constant of proportionality turns out to be $8\pi\eta$, where η is called the **coefficient of viscosity**. Therefore, we have the following relation

$$P_1 - P_2 = 8\pi\eta\frac{vL}{A}$$

We can see from this equation that larger values of η means the fluid is more viscous because more pressure is needed for the fluid to flow at a certain speed. The SI unit of η is $N{\cdot}s/m^2$; however, a commonly used unit of η is the *poise*, which is defined as

$$1 \text{ poise} = 1 \text{ dyne}\cdot\text{s/cm}^2 = 0.1 \text{ N}\cdot\text{s/m}^2$$

As mentioned previously, fluid flow is often characterized by the volume flow rate, $\Delta V/\Delta t = Av$. When the above expression for viscous flow is rewritten in terms of the volume flow rate for a tube of circular cross section ($A = \pi r^2$), we get Poiseuille's equation:

$$\frac{\Delta V}{\Delta t} = \frac{(P_1 - P_2)\pi r^4}{8\eta L}$$

which shows that the volume flow rate increases as the fourth power of the radius.

Surface Tension

The surface of a fluid, especially a liquid, is observed to behave in a way similar to that of an elastic membrane when objects are placed on the surface. This effect is due to the **surface tension** of the fluid. Surface tension results from internal forces between the molecules of a fluid that collectively resist the deformation of the fluid's surface from its equilibrium configuration. This effect is analogous to what happens in a spring and is responsible for the ability of insects to walk on water.

Exercise 15–8 Viscosity Using the result of Example 15–3, estimate the volume flow rate of blood from the head to the feet of the six-foot-tall person. (Assume an effective radius of 23 cm.)

Solution: We know the following information:

Given: $P_1 - P_2 = 18.8$ kPa, $L = 1.83$ m, $r = 0.23$ m, $\eta = 0.0027\ \text{N·s/m}^2$; **Find**: $\Delta V/\Delta t$

Using the given information, together with the viscosity of blood (from Table 15-2 in the text), we can directly use Poiseuille's equation:

$$\frac{\Delta V}{\Delta t} = \frac{(P_1 - P_2)\pi r^4}{8\eta L} = \frac{\pi(18.8\times10^3\ \text{Pa})(0.23\ \text{m})^4}{8(0.0027\ \text{N}\cdot\text{s/m}^2)(1.83\ \text{m})} = 4.2\times10^3\ \text{m}^3/\text{s}$$

Notice that we used three significant figures for the pressure difference. We did this because the final result of Example 15–3 is now just an intermediate result for this example.

Practice Quiz

11. In general, we expect the volume flow rate of a fluid with a large coefficient of viscosity to be...

(a) large **(b)** small **(c)** η has no relevance to the volume flow rate.

Reference Tools and Resources

I. Key Terms and Phrases

fluid a liquid or a gas

density a measure of the compactness of an object or substance, given by its mass per unit volume

pressure the normal force per unit area acting on an object or within a fluid

gauge pressure the measure of pressure that excludes the atmospheric pressure

Pascal's principle the principle that an external pressure is transmitted undiminished throughout a fluid

buoyancy the phenomenon that fluid pressure applies an upward force on immersed objects

Archimedes' principle that the buoyant force equals the weight of the fluid displaced by an immersed object

flotation occurs when the buoyant force equals an object's weight

equation of continuity the equation that expresses the constant mass flow rate within a fluid

Bernoulli's equation the application of the work-energy theorem to fluid flow

Torricelli's law the equation that determines the speed with which a fluid will flow from an aperture below the surface of a fluid in an open container

viscosity resistance to fluid flow

surface tension the phenomenon that the surfaces of fluids often behave in a way similar to an elastic membrane

II. Important Equations

Name/Topic	Equation	Explanation
density	$\rho = \frac{M}{V}$	The density of a fluid is the mass of the fluid divided by its volume.
pressure	$P = \frac{F}{A}$	The pressure in a fluid is the normal force per unit area.
pressure with depth	$P_2 = P_1 + \rho g h$	The pressure in a fluid increases with depth, h.
buoyancy	$F_b = \rho g V$	The buoyant force on an object when a volume V is submerged in a fluid of density ρ.
equation of continuity	$A_1 v_1 = A_2 v_2$	The equation of continuity for an incompressible fluid.
Bernoulli's equation	$P_1 + \frac{1}{2}\rho v_1^2 + \rho g y_1 = P_2 + \frac{1}{2}\rho v_2^2 + \rho g y_2$	The work-energy theorem applied to fluid dynamics.

viscosity	$P_1 - P_2 = 8\pi\eta\frac{vL}{A}$	The pressure difference required to cause a viscous fluid to flow with speed v through a pipe of length L and cross-sectional area A.

III. Know Your Units

Quantity	Dimension	SI Unit
density (ρ)	$[M]/[L^3]$	kg/m^3
pressure (P)	$[M][L^{-1}][T^{-2}]$	Pa
coefficient of viscosity (η)	$[M][L^{-1}][T^{-1}]$	$N{\cdot}s/m^2$

Practice Problems

1. In the figure if $h_1 = 0.5$ cm and $h_2 = 4.65$ cm, which is the height of an unknown liquid. The remaining liquid in the U-tube is water (1000 kg/m³). What is the density of the unknown liquid?

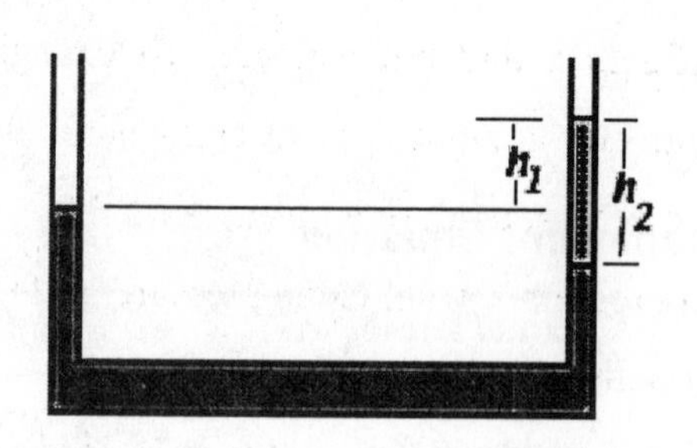

2. In the hydraulic lift shown the piston beneath W_1 has a cross-sectional area of 0.87 m², and the piston on which F_1 is applied has a cross-sectional area of 0.08 m², and the force $F_1 = 100$ N, what weight, to the nearest newton, can this configuration lift?

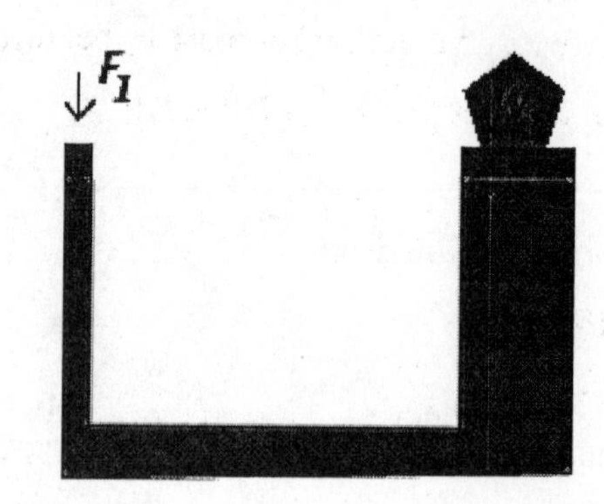

3. In problem 2 if the piston on the left is depressed 0.63 m, how far up will the piston on the right move to the nearest thousandth of a m?

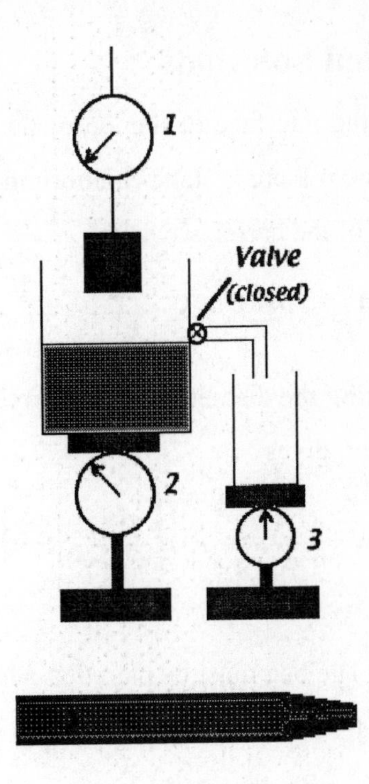

4. Water is in the big beaker in the figure on the left. Scale 1 reads 76 newtons, scale 2 reads 506 newtons and scale 3 reads 0 newtons. The hanging block has a density of 10×10^3 kg/m^3. To the nearest tenth of a newton what does scale 1 read after the block is fully lowered into the beaker of water?
5. In problem 4 to the nearest tenth of a newton what is the new reading on scale 2?
6. The experiment in problem 4 is repeated with the valve opened. What is the new reading on scale 3?
7. In problem 6, what is the new reading on scale 2?
8. The figure shows a nozzle on a hose carrying water. If the pressure at point 1 is 260×10^3 Pa, the velocity 0.58 m/s, and the diameter 10 cm, given that the diameter at point 2 is 2.5 cm, what is the velocity to the nearest tenth of a m/s at point 2?
9. In problem 8 what is the pressure, to the nearest kPa at point 2?
10. For a pipe carrying water in a building, if h = 245 meters, v_1= 2.68 m/s, and the cross-sectional area at 1 is twice that at 2, what must P_1 be, to the nearest kPa, in order that P_2= 101000 Pa?

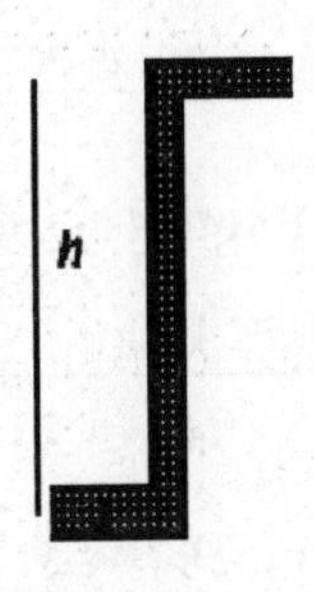

Puzzle

SINK OR SWIM

A rectangular block of wood floats submerged, 70% in water 30% in oil, as shown in the sketch. What happens if you add some more oil? What happens if you add some more water? What happens if you siphon off the oil? Do the percentages change? Which way?

Selected Solutions

17. At the interface of the top of the barrel and the bottom of the tube, the pressure can have only one value if there is static equilibrium. If W is the weight of the water in the tube and F is the force on the top of the barrel, we have

$$P_{tube} = P_{barrel} \quad \Rightarrow \quad \frac{W}{A_{tube}} = \frac{F}{A_{barrel}}$$

Taking the diameters of the barrel and tube to be D and d, respectively, and solving for the weight of water, gives

$$W = \left(\frac{A_{tube}}{A_{barrel}}\right)F = \left[\frac{\pi\left(\frac{d}{2}\right)^2}{\pi\left(\frac{D}{2}\right)^2}\right]F = \left(\frac{d}{D}\right)^2 F = \left(\frac{0.01\text{ m}}{0.75\text{ m}}\right)^2 (6430\text{ N}) = \boxed{1.1\text{ N}}$$

25. (a) The bag must be placed at a height such that the depth of the fluid will increase the fluid pressure from atmospheric pressure to the needed value of 109 kPa:

$$P = P_{at} + \rho g h$$

$$\therefore \quad h = \frac{P - P_{at}}{\rho g} = \frac{109\text{ kPa} - 101.3\text{ kPa}}{\left(1020\text{ kg/m}^3\right)\left(9.81\text{ m/s}^2\right)} = \boxed{0.770\text{ m}}$$

(b) $\boxed{\text{Increased}}$, to compensate for the reduction in pressure resulting from the lower density; the gauge pressure in a fluid is directly proportional to its density.

35. (a) Use Siri's formula to solve for the person's density.

$$x_f = \frac{4950\text{ kg/m}^3}{\rho_p} - 4.50 \quad \Rightarrow \quad \rho_p = \frac{4950\text{ kg/m}^3}{x_f + 4.50} = \frac{4950\text{ kg/m}^3}{0.184 + 4.50} = \boxed{1060\text{ kg/m}^3}$$

(b)
$$V_p = \frac{m}{\rho_p} = \frac{1}{\rho_p}\left(\frac{W}{g}\right) = \frac{768\text{ N}}{\left(1057\text{ kg/m}^3\right)\left(9.81\text{ m/s}^2\right)} = \boxed{0.0741\text{ m}^3}$$

(c) The apparent weight in water is the weight in air minus the buoyant force

$$W_a = W - F_b = W - \rho_w V_p g$$
$$= 768\text{ N} - \left(1000\text{ kg/m}^3\right)\left(0.0741\text{ m}^3\right)\left(9.81\text{ m/s}^2\right) = \boxed{41\text{ N}}$$

45. We can make direct use of the equation of continuity where the area through which the water flows equals the width of the river times its depth.

$$A_1v_1 = A_2v_2 \Rightarrow w_1d_1v_1 = w_2d_2v_2$$

$$\therefore \quad v_2 = \frac{w_1d_1v_1}{w_2d_2} = \frac{(12\,\text{m})(2.7\ \text{m})(2.2\ \text{m/s})}{(4.0\ \text{m})(0.85\ \text{m})} = \boxed{21\ \text{m/s}}$$

55. From Exercise 15–4, we know that for this situation

$$\Delta P = \tfrac{1}{2}\rho v^2 = \tfrac{1}{2}\left(1.29\frac{\text{kg}}{\text{m}^3}\right)\left(45.2\frac{\text{m}}{\text{s}}\right)^2 = 1.32\,\text{kPa}$$

$$\therefore \quad \Delta F = (\Delta P)A = (1.32\times10^3\ \text{Pa})(578\,\text{m}^2) = \boxed{762\,\text{kN}}$$

The force is directed upward because the outside wind reduces the outside pressure.

Answers to Practice Quiz

1. (b) **2.** (d) **3.** (c) **4.** (a) **5.** (a) **6.** (e) **7.** (b) **8.** (b) **9.** (a) **10.** (a) **11.** (b)

Answers to Practice Problems

1. 892 kg/m^3 2. 1088 N

3. 0.058 m 4. 68.4 N

5. 513.6 N 6. 7.6 N

7. 506 N 8. 9.3 m/s

9. 217 kPa 10. 2515 kPa

CHAPTER 16

TEMPERATURE AND HEAT

Chapter Objectives

After studying this chapter, you should:

1. understand the meaning of thermal equilibrium and the role of temperature.
2. know how to convert among the Celsius, Fahrenheit, and Kelvin temperature scales.
3. understand the basic phenomenology of thermal expansion.
4. understand the relationship between heat and work.
5. understand the difference between specific heat and heat capacity.
6. be able to perform basic calorimetry calculations.
7. know the 3 main processes for heat transfer.
8. understand the basic phenomenology of heat transfer by conduction and radiation.

Warm-Ups

1. A piece of wood at 130 degrees C can be picked up comfortably, but a piece of aluminum at the same temperature will give a painful burn. Why is this?
2. Estimate the amount of heat necessary to raise your body temperature by 1 degree Fahrenheit.
3. A metal plate has a circular hole cut in it. If the temperature of the plate increases, will the diameter of the hole increase, decrease, or remain unchanged?

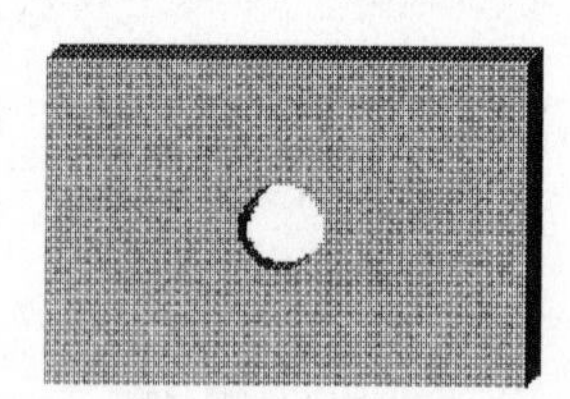

4. In an attempt to open a new jar of peanut butter, you run very hot tap water over the (steel) lid. Estimate the change in the diameter of the lid.

Chapter Review

This is the first of three chapters on thermodynamics, which can loosely be described as the study of heat and the physical processes associated with heat transfer. In this chapter we focus on the concepts of temperature and heat as well as a few physical consequences of heat transfer.

16–1 Temperature and the Zeroth Law of Thermodynamics

In this section we seek to develop working definitions of **temperature** and **heat**. The two concepts are intimately related and must be defined together. Heat is a form of energy; it is energy that flows between two systems. Systems are said to be in **thermal contact** when it is possible for heat to flow between them; however, heat does not always flow between systems that are in thermal contact. The property of systems that determines whether heat flow will occur is called the *temperature*. If there is a temperature difference between two systems that are in thermal contact, heat will flow from the system with higher temperature to the system with lower temperature. Two systems are said to be in **thermal equilibrium** if when they are brought into thermal contact, no heat transfer occurs. In this latter case the systems must have equal temperatures. These statements are embodied in the **zeroth law of thermodynamics**:

> *If system A is in thermal equilibrium with system B, and system C is also in thermal equilibrium with system B, then systems A and C will be in thermal equilibrium if brought into thermal contact.*

We thus obtain the following working definitions:

- *Heat* is the energy that is transferred between systems because of a temperature difference.
- *Temperature* is the property of systems that determines the existence and direction of the heat flow between them when they are in thermal contact.

Temperature is a new fundamental quantity, not defined in terms of length, mass, and time. It is assigned the dimensional symbol [K].

Practice Quiz

1. Which of the following statements is most accurate?
   (a) If there is a temperature difference between systems A and B, heat will flow from system A to system B.
   (b) If there is a temperature difference between systems A and B, heat will flow from system B to system A.
   (c) If there is a temperature difference between two systems, heat will flow from one system to the other.

(d) If there is a temperature difference between two systems, heat will flow from the system with higher temperature to the system with lower temperature.

(e) If there is a temperature difference between two systems, heat will flow from the system with higher temperature to the system with lower temperature if they are brought into thermal contact.

16–2 Temperature Scales

The primary temperature scales in common use are the Celsius, Fahrenheit, and Kelvin scales. Each scale is based on different choices for setting values for two convenient fixed points. The Celsius scale takes the freezing temperature of water to be 0 °C and the boiling temperature of water to be 100 °C. On the Fahrenheit scale, water freezes at 32 °F and boils at 212 °F. In addition to having different settings for the freezing and boiling points of water, the scales of these two systems are different in that a temperature change of one degree on the Fahrenheit scale corresponds to a change of only 5/9 degrees on the Celsius scale. Conversions between the Celsius and Fahrenheit scales can be accomplished using the following formulas:

$$T_F = \left(\frac{9}{5}\,^\circ F/^\circ C\right)T_C + 32\,^\circ F$$

$$T_C = \left(\frac{5}{9}\,^\circ C/^\circ F\right)\left(T_F - 32\,^\circ F\right)$$

Exercise 16–1 A Change in Temperature On a winter day in Michigan the temperature rose from a morning low of –8.0 °F to an afternoon high of 22 °F. By how many Celsius degrees did the temperature rise?

Solution: We are given the following information:

Given: $T_i = -8.0$ °F, $T_f = 22$ °F; **Find:** ΔT in C°

Since we know how to convert from Fahrenheit to Celsius let's find an expression for ΔT_C:

$$\Delta T_C = T_{C,f} - T_{C,i} = \left(\frac{5}{9}\,^\circ C/^\circ F\right)\left(T_{F,f} - 32\,^\circ F\right) - \left(\frac{5}{9}\,^\circ C/^\circ F\right)\left(T_{F,i} - 32\,^\circ F\right) = \left(\frac{5}{9}\,^\circ C/^\circ F\right)\left(T_{F,f} - T_{F,i}\right)$$

Therefore, we have

$$\Delta T_C = \left(\frac{5}{9}\,^\circ C/^\circ F\right)\left[22\,^\circ F - \left(-8.0\,^\circ F\right)\right] = 17 C^\circ$$

The Kelvin temperature scale is based on the existence of a lowest temperature below which (even *to* which) it is physically impossible to cool any system. This temperature is called **absolute zero**. The Kelvin scale is essentially identical to the Celsius scale except that the zero mark is shifted to correspond to absolute zero (−273.15 °C). The conversion between temperatures on the Kelvin scale and the Celsius scale is the following:

$$T = T_C + 273.15$$

Quoting Temperatures

You should be aware of the following with regard to quoting temperatures and temperature differences:

- When quoting a temperature in Celsius or Fahrenheit the degree symbol is both written and spoken first. That is, 10 °C ("10 degrees Celsius"), for example.

- When quoting temperature differences in Celsius or Fahrenheit the degree symbol is both written and spoken last. That is, 10 C° ("10 Celsius degrees"), for example.

- The Celsius scale is no longer alternatively referred to as the centigrade scale.

- No degree symbol is written or spoken with the Kelvin scale. That is, both temperatures and temperature differences of 10 are denoted by 10 K ("10 kelvin").

Exercise 16–2 A Temperature Conversion The normal body temperature of 98.6 °F corresponds to what temperature on the Kelvin scale?

Solution: We are given the following information:

Given: $T_F = 98.6$ °F; **Find**: T

Since we know how to convert from Fahrenheit to Celsius and from Celsius to Kelvin, we can combine the two to get a conversion from Fahrenheit to Kelvin. To get T_C we use

$$T_C = \left(\frac{5}{9}\ {}^\circ\mathrm{C}/{}^\circ\mathrm{F}\right)\left(T_F - 32\ {}^\circ\mathrm{F}\right)$$

and substitute this expression into the conversion for T to get

$$T = T_C + 273.15 = \left(\frac{5}{9}\ {}^\circ\mathrm{C}/{}^\circ\mathrm{F}\right)\left(T_F - 32\ {}^\circ\mathrm{F}\right) + 273.15$$

Therefore,

$$T = \left(\frac{5}{9}\ ^{\circ}C/^{\circ}F\right)\left(98.6\ ^{\circ}F - 32\ ^{\circ}F\right) + 273.15 = 310\ K$$

Practice Quiz

2. A temperature change of 1 kelvin corresponds to how much of a change in Fahrenheit degrees?

(a) 0.55 **(b)** 1.8 **(c)** 5/9 **(d)** 32 **(e)** 273.15

3. What temperature corresponds to absolute zero on the Fahrenheit scale?

(a) 0 °F **(b)** –273.15 °F **(c)** – 459.67 °F **(d)** – 212 °F **(e)** – 100 °F

16–3 Thermal Expansion

Most substances expand when heated and contract when cooled. The amount of expansion is different for different substances. We can identify some aspects of this behavior that are the same for nearly all substances. It is found that the amount that a substance will expand, that is, its change in length (ΔL), area (ΔA), or volume (ΔV), is directly proportional to the temperature change (ΔT) that drives the expansion. Furthermore, these changes in size are also directly proportional to the original size (L_0, A, or V) of the object being heated or cooled.

Thus, for the linear expansion of an object we have $\Delta L \propto L_0 \Delta T$. The constant of proportionality depends on the substance and is called the **coefficient of linear expansion**, α, and, an expression that describes the linear expansion of an object due to a change in its temperature is

$$\Delta L = \alpha L_0 \Delta T$$

As you can determine by dimensional analysis, the SI unit of α is K^{-1} [or $(C^{\circ})^{-1}$]. Values of the coefficient of linear expansion for different substances can be found in Table 16–1 of the text.

For the expansion of areas, the description is very much like that linear expansion. To a very high approximation, the coefficient of area expansion works out to be twice that of the coefficient of linear expansion, leading to the expression

$$\Delta A = (2\alpha)A(\Delta T)$$

Therefore, the same table of values (Table 16–1) can be used to determine area expansions. For volume expansions we have $\Delta V \propto V\Delta T$, with a proportionality constant β called the **coefficient of volume expansion**. Values of β for different substances can are listed in Table 16–1. For substances not listed in the table under volume expansion a good approximation is to take $\beta \approx 3\alpha$; therefore, we have

$$\Delta V = \beta V(\Delta T)$$

Example 16–3 The Expansion of Iron An iron cube has an edge length of 2.300 cm. If its temperature is raised from –12.00 °C to 22.00 °C, what is its volume at the higher temperature?

Picture the Problem The sketch shows exaggerated views of an iron cube **(a)** before and **(b)** after the temperature has been raised.

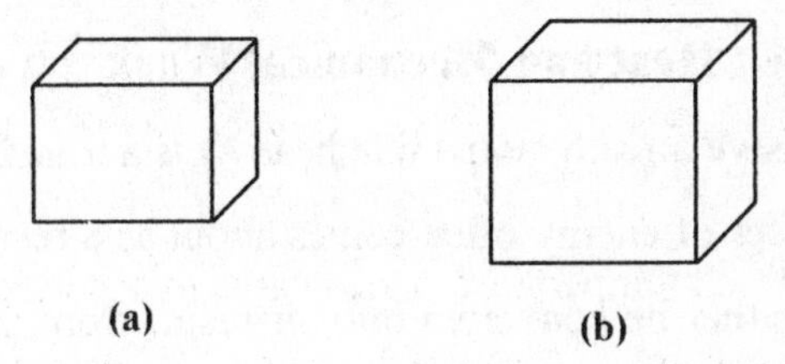

Strategy We apply the relation that describes volume expansion to determine ΔV, then add this change to the original volume to get the final result.

Solution

1. Calling the length of one edge L, determine the original volume of the cube: $V = L^3 = (2.300 \text{ cm})^3 = 12.167 \text{ cm}^3$

2. Since no value of β is given for iron in Table 16–1, make use of the result for α: $\beta = 3\alpha = 3\,(12 \times 10^{-6} \text{ K}^{-1}) = 36 \times 10^{-6} \text{ K}^{-1}$

3. Find the temperature difference: $\Delta T = 22.00 \text{ °C} - (-12.00 \text{ °C}) = 34.00 \text{ C°}$

4. Find the change in volume: $\Delta V = \beta V(\Delta T) = (36 \times 10^{-6} \text{ K}^{-1})(12.167 \text{ cm}^3)(34.00 \text{ K})$
$= 0.014892 \text{ cm}^3$

5. Calculate the final volume then: $V_f = V + \Delta V = 12.167 \text{ cm}^3 + 0.014892 \text{ cm}^3$
$= 12.18 \text{ cm}^3$

Insight In order to show a meaningful change in the volume we kept four significant figures despite the fact that we were given only two figures for the value of α. Also notice that we calculated the temperature change in Celsius degrees but used the change in Kelvin. This latter practice is permissible only when dealing with temperature differences on these scales.

Practice Quiz

4. What is the coefficient of volume expansion for lead?

(a) $29 \times 10^{-6} \text{ K}^{-1}$ **(b)** $58 \times 10^{-6} \text{ K}^{-1}$ **(c)** $87 \times 10^{-6} \text{ K}^{-1}$ **(d)** $44 \times 10^{-6} \text{ K}^{-1}$ **(e)** $99 \times 10^{-6} \text{ K}^{-1}$

5. If a 12.6-cm copper rod is cooled from 53.2 °C to –10.8 °C, by how much will its length change?

   (a) 1.4×10^{-2} cm **(b)** 9.1×10^{-3} cm **(c)** 1.1×10^{-2} cm **(d)** 2.3×10^{-3} cm **(e)** 1.7×10^{-5} cm

16–4 Heat and Mechanical Work

We have already noted that heat, Q is a transfer of energy. From previous chapters we also know that the transfer of energy often comes about as a result of mechanical work being done. It follows that heat flow can either be converted into, or result from, mechanical work. Two specialized units have been adopted for dealing with the mechanical work associated with heat flow.

One unit of heat in common use is the **calorie** (cal). One calorie is the amount of heat needed to raise the temperature of 1 gram of water by 1 Celsius degree. The equivalent amount of energy in joules is called the **mechanical equivalent of heat**:

$$1 \text{ cal} = 4.186 \text{ J}$$

Also in common use is the kilocalorie (kcal), sometimes called a "food calorie" because it is used when quoting the energy content of foods. It is important to realize that the most common notation for the kilocalories is *Cal* with a capital *C*, so, whenever energy in calories is being discussed you must note whether or not the *C* is capitalized.

Another unit of heat energy in common use is the **British Thermal Unit** (Btu). One Btu is the amount of energy needed to raise the temperature of one pound of water by 1 Fahrenheit degree. In terms of the other units of heat we have

$$1 \text{ Btu} = 0.252 \text{ kcal} = 1055 \text{ J}$$

Practice Quiz

6. If you eat 2200 Calories of food per day, what is your energy intake in units of Btu?

   (a) 8.7 **(b)** 4786 **(c)** 2200 **(d)** 8700 **(e)** 1055

16–5 Specific Heats

We have already pointed out that heat flow is associated with a temperature difference between two systems. It is also true that heat flow can result in the change in temperature of a given system. The amount of heat needed to change the temperature of an *object* is directly proportional to the change in temperature required; that is, $Q \propto \Delta T$. The proportionality constant between the two quantities depends on the type (and mass) of substance of which the object is made. The proportionality constant C is called the **heat capacity** of the object, so we have

$$Q = C\,\Delta T$$

The SI unit of heat capacity is J/K (or J/C°).

The concept of heat capacity refers to specific objects because you need to know how much of the substance you are heating up or cooling down (and the heat capacity contains that information). It is often more useful to have a quantity that is independent of the amount of substance and depends only on the nature of the substance. Such a quantity is called the **specific heat** (c), which is basically the heat capacity per unit mass of the substance: $c = C/m$. Therefore, in terms of specific heat, the amount of heat needed to change the temperature of a substance is given by

$$Q = mc\,\Delta T$$

where m is the mass of the substance into or out of which the heat flows. The SI unit of specific heat is J/(kg·K). Table 16–2 in the text contains values of the specific heats of several substances.

Example 16–4 Calorimetry An insulated container of negligible heat capacity contains 2.11 kg of water at 22.0 °C. A hot aluminum ball of mass 0.435 kg and temperature 90.0 °C is placed into the water. What will be the final equilibrium temperature of the ball and the water?

Picture the Problem The sketch shows an aluminum ball sitting in a container of water.

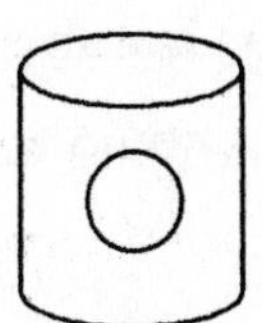

Strategy Since the container has negligible heat capacity we can ignore any heat transfer to or from it. Heat will transfer from the hot ball to the cooler water. Energy conservation requires that the net heat flow into or out of the system be zero.

Solution

1. By conservation of energy: $Q_w + Q_a = 0$

2. The heat gained by the water is: $Q_w = m_w c_w (T - T_{0w})$

3. Heat is lost by the aluminum: $Q_a = m_a c_a (T - T_{0a})$

4. The conservation of energy equation becomes: $m_w c_w (T - T_{0w}) + m_a c_a (T - T_{0a}) = 0$

5. Solve for the equilibrium temperature, T: $T = \dfrac{m_w c_w T_{0w} + m_a c_a T_{0a}}{m_w c_w + m_a c_a}$

6. Obtain the numerical result using Table 16–2 in the text to get the specific heats:

$$T = \left[(2.11\text{ kg})\left(4186\tfrac{\text{J}}{\text{kg}\cdot\text{C}^\circ}\right) + (0.435\text{ kg})\left(900\tfrac{\text{J}}{\text{kg}\cdot\text{C}^\circ}\right)\right]^{-1} \times \left\{ \begin{array}{l} (2.11\text{ kg})\left(4186\tfrac{\text{J}}{\text{kg}\cdot\text{C}^\circ}\right)(22.0\ ^\circ\text{C}) \\ +(0.435\text{ kg})\left(900\tfrac{\text{J}}{\text{kg}\cdot\text{C}^\circ}\right)(90.0\ ^\circ\text{C}) \end{array} \right\}$$

$$= 24.9\ ^\circ\text{C}$$

Insight Notice that in step 3 the heat flow is negative (because the equilibrium temperature is less than the initial temperature of the aluminum), which is indicative of heat lost as opposed to heat gained.

Practice Quiz

7. If an object made out of a certain substance has a heat capacity C, a similar object made of the same substance but having twice the mass will have a heat capacity of

(a) C/2 **(b)** C^2 **(c)** C **(d)** 2C **(e)** $\sqrt{C}$

8. An object is made out of a certain substance that has a specific heat c. To do calorimetry with a similar object made of the same substance but having twice the mass we must use a specific heat of

(a) c/2 **(b)** c^2 **(c)** c **(d)** 2c **(e)** $\sqrt{c}$

16–6 Conduction, Convection, and Radiation

Conduction

There are three main processes by which heat is transferred. One of these processes is called **conduction.** Heat conduction occurs when heat flows directly through a material object because of a temperature difference ΔT across the material. Consider an object in the form of a cylinder of circular cross-sectional area A and length L. If the ends of this cylinder have different temperatures T_2 and T_1 (with $T_2 > T_1$), then heat will conduct through the cylinder until the ends are at the same temperature. The average rate at which heat will flow, Q/t (where t is time), is found to depend on three clearly identifiable quantities:

- The heat flow rate is directly proportional to the temperature difference: $Q/t \propto \Delta T$ ($\Delta T = T_2 - T_1$).
- The heat flow rate is directly proportional to the area through which it flows: $Q/t \propto A$.
- The heat flow rate is inversely proportional to the distance through which it flows: $Q/t \propto L^{-1}$.

Combining these considerations we get $Q/t \propto A(\Delta T)/L$. The constant of proportionality k is called the **thermal conductivity** of the substance. The SI unit of thermal conductivity is W/(m·K). Table 16–3 in your text contains values of thermal conductivity for several substances. The final result is an equation that is commonly written in two ways:

$$\frac{Q}{t} = kA\frac{\Delta T}{L} \quad \text{or} \quad Q = kA\left(\frac{\Delta T}{L}\right)t$$

Exercise 16–5 Heat Loss through a Window If a 3.0-ft x 4.0-ft glass window, 3.0 mm thick, has an inner surface temperature of 30.0 °F and an outer surface temperature 20.0 °F, at what rate is energy lost through the window?

Picture the Problem The sketch shows a rectangular window and indicates the inside and outside surfaces.

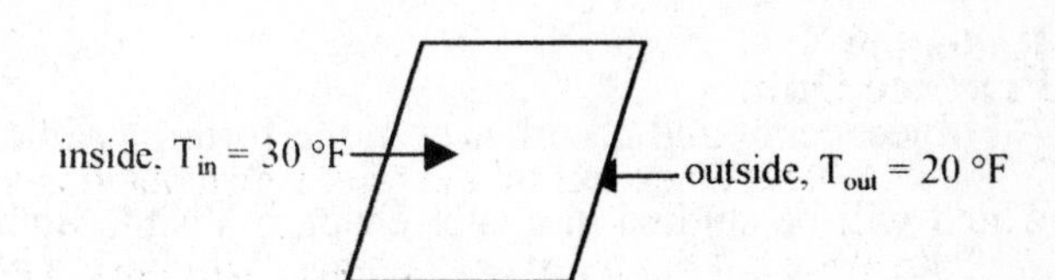

Strategy We need to apply the expression for the rate of thermal conduction. Before we use this expression we should convert quantities to SI units.

Solution

1. Using the result of Exercise 16–1, convert the temperature difference to Celsius degrees:

$$\Delta T_C = \left(\frac{5}{9}\ \text{C}^\circ/\text{F}^\circ\right)\Delta T_F = \left(\frac{5}{9}\ \text{C}^\circ/\text{F}^\circ\right)(10\ \text{F}^\circ) = 5.56\ \text{C}^\circ$$

2. Calculate the area in SI units:

$$A = 3.0\ \text{ft} \times 4.0\ \text{ft} = 12.0\ \text{ft}^2\left(\frac{\text{m}}{3.28\text{ft}}\right)^2 = 1.12\ \text{m}^2$$

3. Use Table 16–3 for the thermal conductivity of glass to solve for the rate of heat conduction:

$$\frac{Q}{t} = kA\frac{\Delta T}{L} = \left(0.84\ \tfrac{\text{W}}{\text{m}\cdot\text{C}^\circ}\right)(1.12\ \text{m}^2)\left(\frac{5.56\ \text{C}^\circ}{3.0\times10^{-3}\ \text{m}}\right) = 1700\ \text{W}$$

Insight This is a pretty high rate of energy loss. The prospect of such high energy loss is why the energy efficiency of windows is a large and profitable part of the home improvement business.

Practice Quiz

9. Connecting two identical rods end to end has what effect on the rate of heat flow across the two rods compared with the rate at which heat would conduct across just one of them?

(a) Heat will conduct at twice the rate for the connected rods.

(b) Heat will conduct at half the rate for the connected rods.

(c) Heat will conduct at the same rate for both the connected rods and the single rod.

(d) There will be no heat conduction for the connected rods.

(e) None of the above

Convection

Heat transfer also occurs by the movement of matter from one place to another, such as the rising of hot air in a room. This type of heat transfer is called **convection**. Convection is an important process in fluids where material is free to move.

Radiation

All objects emit and absorb heat in the form of **radiation**. By radiation we mean electromagnetic waves, which will be studied in a later chapter. Visible light, infrared, microwaves, X-rays, gamma rays, radio and television waves, and ultraviolet light are all forms of electromagnetic radiation. Infrared and visible light are the two that are most applicable to heat transfer.

It has been determined that the radiant power of an object is directly proportional to both the surface area of the object from which the radiation flows and the fourth power of the Kelvin temperature: $P \propto AT^4$. The proportionality factor for this expression is written as $e\sigma$, where σ is a fundamental constant called the Stefan-Boltzmann constant

$$\sigma = 5.67 \times 10^{-8}\ \mathrm{W/\left(m^2 \cdot K^4\right)}$$

and e is the emissivity, which depends on the nature of the surface. The emissivity is a dimensionless quantity whose value lies between 0 and 1; it is a measure of how effectively an object radiates heat. If $e = 1$ the object is said to be a perfect radiator. The power radiated by an object is therefore given by

$$P = e\sigma AT^4$$

Objects also absorb radiant energy according to this same relation, except that the temperature is the temperature of the environment instead of the temperature of the object. Keep in mind when using the above relation that the temperature should be in Kelvin. Because we are not dealing with a temperature difference or temperature change the Kelvin and the Celsius scales are not interchangeable.

Example 16–6 Net Radiation A metal ball of emissivity 0.65 and surface area 0.66 m^2 is heated to a temperature of 92 °C and placed into a room in which the air temperature is 22 °C. **(a)** Does heat flow into or out of the metal ball? **(b)** At what net rate does radiant heat flow between the ball and the air?

Solution We are given the following information:

Given: $e = 0.65$, $T_{ball} = 92$ °C, $T_{air} = 22$ °C; **Find**: **(a)** the direction of heat flow, **(b)** P_{net}

(a) Since the ball is at a higher temperature than the air, and heat flows from higher temperatures to lower temperatures, the heat must flow out of the metal ball.

(b) The rate at which the metal ball radiates energy away is given by

$$P_{out} = e\sigma A T_{ball}^4$$

The rate at which the metal ball absorbs energy from the air is given by

$$P_{in} = e\sigma A T_{air}^4$$

Therefore, the net rate at which the ball loses energy is

$$P_{net} = e\sigma A\left(T_{ball}^4 - T_{air}^4\right).$$

Before using this result for P_{net}, let's first convert the temperatures to Kelvin:

$$T_{ball,K} = T_{ball,C} + 273.15 = 92 + 273.15 = 365.15 \text{ K}$$

$$T_{air,K} = T_{air,C} + 273.15 = 22 + 273.15 = 295.15 \text{ K}$$

We can now calculate the net power radiated by the ball as

$$P_{net} = (0.65)\left(5.67\times10^{-8}\ \frac{\text{W}}{\text{m}^2\cdot\text{K}^4}\right)\left(0.66\text{ m}^2\right)\left[(365.15\text{ K})^4 - (295.15\text{ K})^4\right] = 250\text{ W}$$

Insight The net power is the difference between what is emitted and absorbed. Whether this net energy is emitted or absorbed is determined by the temperature difference between the object and its environment.

Practice Quiz

10. Which process of heat transfer is least important when cooking on an electric range?

(a) conduction **(b)** convection **(c)** radiation **(d)** They are all equally important

11. At what rate does an object of emissivity 0.55 absorb radiation if it has a surface area of 0.33 m^2 and sits in an environment of ambient temperature 290 K?

(a) 73 J/s **(b)** 3.0×10^{-6} J/s **(c)** 290 J/s **(d)** insufficient information to determine

Reference Tools and Resources

I. Key Terms and Phrases

temperature the property of systems that determines the existence and direction of the heat flow between them when they are in thermal contact

heat the energy that is transferred between systems because of a temperature difference

thermal contact exists between systems when it is possible for heat to flow between them

thermal equilibrium exists when systems are brought into thermal contact, and no heat transfer occurs

zeroth law of thermodynamics the fundamental law that allows a working definition of temperature

absolute zero the lower limit on physically attainable temperatures (the zero temperature on the Kelvin scale)

coefficient of expansion the substance-dependent proportionality factor that determines how much an object will expand as the result of a temperature change

calorie a unit of energy equal to the heat needed to raise the temperature of 1 gram of water by 1 Celsius degree

British Thermal Unit a unit of energy equal to the heat needed to raise the temperature of 1 pound of water by 1 Fahrenheit degree

heat capacity a quantity that determines how much heat is needed to change the temperature of an object by a certain amount

specific heat a quantity that determines how much heat is needed to change the temperature of a unit mass of an object by a certain amount

conduction heat transfer by direct flow through an object due to a temperature difference across it

thermal conductivity a quantity that determines the rate at which heat conducts through a given object

convection heat transfer by direct movement of matter from one place to another

radiation heat transfer by emission or absorption of electromagnetic waves

Stefan-Boltzmann constant a fundamental constant that determines the rate at which an object emits or absorbs radiation

emissivity a dimensionless quantity between 0 and 1 that measures how effectively an object radiates heat

II. Important Equations

Name/Topic	Equation	Explanation
temperature scales	$T_C = \left(\frac{5}{9}\,^\circ C/^\circ F\right)\left(T_F - 32\,^\circ F\right)$ $T = T_C + 273.15$	The conversions between Celsius and Fahrenheit and between Kelvin and Celsius.
thermal expansion	$\Delta L = \alpha L_0 \Delta T$ $\Delta V = \beta V(\Delta T)$	The change in size of an object or substance due to a temperature change.
specific heat	$Q = mc\,\Delta T$	The heat needed to change the temperature of an amount of mass m of a substance by ΔT.
conduction	$\frac{Q}{t} = kA\left(\frac{\Delta T}{L}\right)t$	The heat that conducts through an object of length L and cross-sectional area A in time t.
radiation	$P = e\sigma AT^4$	The rate of heat transfer by electromagnetic radiation.

III. Know Your Units

Quantity	Dimension	SI Unit
temperature (T)	[K]	K
heat (Q)	$[M][L^2][T^{-2}]$	J
coefficient of expansion (α, β)	$[K^{-1}]$	K^{-1}
heat capacity (C)	$[M][L^2][T^{-2}][K^{-1}]$	J/K
specific heat (c)	$[L^2][T^{-2}][K^{-1}]$	J/(kg·K)
thermal conductivity (k)	$[M][L][T^{-3}][K^{-1}]$	W/(m·K)

Stefan-Boltzmann constant (σ)	$[M][T^{-3}][K^{-4}]$	$W/(m^2 \cdot K^4)$
emissivity (e)	dimensionless	—

Practice Problems

1. The temperature on one thermometer is 11 °C and on another it is 50 °F, to the nearest C°, what is the temperature difference?
2. What would be the answer to problem 1 if the temperature difference was in F°?
3. A constant-volume (ideal) gas thermometer has a pressure of 50 kPa at 100 °C. What is its pressure at 171 °C, to the nearest kPa?
4. If the pressure in the thermometer in problem 3 was 118 kPa, to the nearest degree Celsius what would be the temperature?
5. A steel ($\alpha = 12 \times 10^{-6}$ C°) container with a volume of 525 cm^3 is filled with oil ($\beta = 0.7 \times 10^{-3}$ C°), if the temperature is increased by 30.8 C° how many cubic centimeters overflow, to the nearest tenth of a cm^3?
6. A metal ball of mass 0.53 kg with initial temperature of 30 °C is dropped into a container of 1.2 kg of water at 20 °C. If the final temperature is 20.4 °C, what is the specific heat of the metal, to the nearest tenth of a J/kg·K?
7. A wall is composed of two materials. Material 1 (inside) has a thickness of 3.76 cm and a thermal conductivity of 0.1, and material 2 (outside) has a thickness of 4.39 cm and a conductivity of 1. If the temperature difference between the inside and outside is 25 C°, and the wall has an area of 10 m^2, what is the energy loss per second, to the nearest watt?
8. In problem 7, if the inside temperature is 29.6 °C, what is the temperature of the interface, to the nearest degree Celsius?
9. In the figure (not to scale) a special wall has three metal studs with conductivity of 226 W/m·K, the wall otherwise has a conductivity of 1.07 W/m·K; what is the energy loss per second, to the nearest watt, if the temperature difference across the wall is 25 C°?

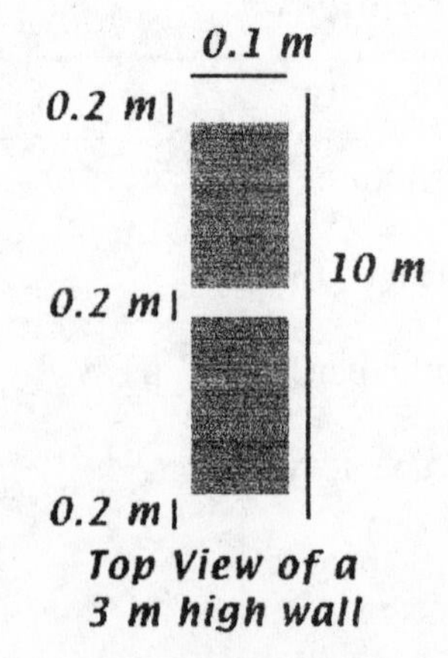

10. A spherical "person" of radius 0.55 m has a body temperature of 303.15 K and is in a room surrounded by 14.2 °C walls. If the emissivity of the person's surface is 1, what is the energy loss per second, to the nearest tenth of a watt?

Puzzle

POT PHYSICS

Advertisements for some expensive cookware have diagrams that look much like the figure. The diagram is labeled to indicate that the inner and outer surfaces of the pot are made of stainless steel, and the inner layer is made of another metal such as aluminum or copper. Please explain why pots are made this way. Discuss why the different metals are chosen, and why using a layered structure is better than using one metal and the same total thickness.

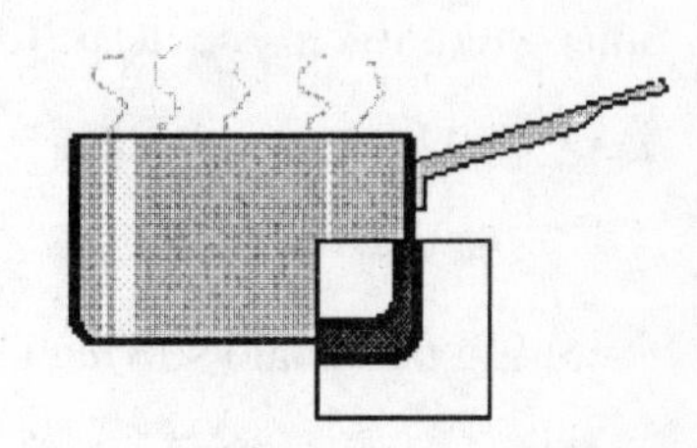

Selected Solutions

17. (a) Aluminum has the larger coefficient of volumetric expansion; therefore, the $\boxed{\text{aluminum}}$ cube will enclose a greater volume.

(b) Since the cubes initially enclose equal volumes, their difference in volume after the temperature change equals the difference between the changes in temperature.

$$\Delta V_{Al} = \beta_{Al} V_0 \Delta T$$
$$\Delta V_{Cu} = \beta_{Cu} V_0 \Delta T$$

$$\Delta V_{Al} - \Delta V_{Cu} = (\beta_{Al} - \beta_{Cu}) V_0 \Delta T$$
$$= \left[7.2\times10^{-5}(\text{C}°)^{-1} - 5.1\times10^{-5}(\text{C}°)^{-1}\right] V_0 (95\ °\text{C} - 23\ °\text{C}) = \boxed{0.0015 V_0}$$

33. Since the environment is insulated, there is zero net heat flow into or out of the system. Therefore,

$$Q_{ob} + Q_w + Q_{Al} = 0 \quad \therefore$$
$$0 = m_{ob} c_{ob} \Delta T_{ob} + m_w c_w \Delta T_w + m_{Al} c_{Al} \Delta T_{Al}$$

Solving for the specific heat of the object gives

$$c_{ob} = \frac{m_w c_w (T_w - T) + m_{Al} c_{Al} (T_{Al} - T)}{m_{Ob}(T - T_{Ob})}$$

$$= \frac{(0.103\text{ kg})\left(4186\ \frac{\text{J}}{\text{kg}\cdot{}^\circ\text{C}}\right)(20.0\ {}^\circ\text{C} - 22.0\ {}^\circ\text{C}) + (0.155\text{ kg})\left(900\ \frac{\text{J}}{\text{kg}\cdot{}^\circ\text{C}}\right)(20.0\ {}^\circ\text{C} - 22.0\ {}^\circ\text{C})}{(0.0380\text{ kg})(22.0\ {}^\circ\text{C} - 100\ {}^\circ\text{C})}$$

$$= \boxed{385\text{ J/kg}\cdot{}^\circ\text{C}}$$

The object is made of copper.

35. Since there is zero net heat flow:

$$Q_{cup} + Q_{cof} + Q_{crm} = 0 \quad \therefore$$

$$0 = m_{cup} c_{cup} (T - T_{cup}) + m_{cof} c_w (T - T_{cof}) + m_{crm} c_w (T - T_{crm})$$

Factoring out T and solving gives

$$0 = T[m_{cup} c_{cup} + (m_{cof} + m_{crm}) c_w] - [m_{cup} c_{cup} T_{cup} + (m_{cof} T_{cof} + m_{crm} T_{crm}) c_w]$$

$$T = \frac{m_{cup} c_{cup} T_{cup} + (m_{cof} T_{cof} + m_{crm} T_{crm}) c_w}{m_{cup} c_{cup} + (m_{cof} + m_{crm}) c_w}$$

$$= \frac{(0.116\text{ kg})\left(1090\ \frac{\text{J}}{\text{kg}\cdot{}^\circ\text{C}}\right)(24.0\ {}^\circ\text{C}) + [(0.225\text{ kg})(80.3\ {}^\circ\text{C}) + (0.0122\text{ kg})(5.00\ {}^\circ\text{C})]\left(4186\ \frac{\text{J}}{\text{kg}\cdot{}^\circ\text{C}}\right)}{(0.116\text{ kg})\left(1090\ \frac{\text{J}}{\text{kg}\cdot{}^\circ\text{C}}\right) + (0.225\text{ kg} + 0.0122\text{ kg})\left(4186\ \frac{\text{J}}{\text{kg}\cdot{}^\circ\text{C}}\right)}$$

$$= \boxed{70.5\ {}^\circ\text{C}}$$

41. (a) The temperature of the junction is greater than 55 °C. Since aluminum has a smaller thermal conductivity than copper, it must have a greater temperature difference across it to have the same heat flow.

(b) At the interface, the heat flow must be the same for both rods. Set $Q_{Cu} = Q_{Al}$.

$$k_{Cu} A\left(\frac{110\ {}^\circ\text{C} - T}{L}\right) = k_{Al} A\left(\frac{T - 0.0\ {}^\circ\text{C}}{L}\right) \quad \Rightarrow \quad k_{Cu}(110\ {}^\circ\text{C} - T) = k_{Al}(T - 0.0\ {}^\circ\text{C})$$

Collecting the terms involving T and solving gives

$$T(k_{Cu} + k_{Al}) = k_{Cu}(110\ {}^\circ\text{C}) \quad \Rightarrow \quad T = \frac{k_{Cu}(110\ {}^\circ\text{C})}{k_{Cu} + k_{Al}}$$

$$\therefore \quad T = \frac{\left(395\ \frac{\text{W}}{\text{m}\cdot{}^\circ\text{C}}\right)(110\ {}^\circ\text{C})}{395\ \frac{\text{W}}{\text{m}\cdot{}^\circ\text{C}} + 217\ \frac{\text{W}}{\text{m}\cdot{}^\circ\text{C}}} = \boxed{71.0\ {}^\circ\text{C}}$$

43. We can use the fact that the heat flows at an equal rate throughout the entire rod.

$$\frac{Q}{t} = \frac{Q_{23}}{t} \quad \Rightarrow \quad kA\left(\frac{87\ °C - 24\ °C}{95\ cm}\right) = kA\left(\frac{T_{23} - 24\ °C}{23\ cm}\right)$$

$$\therefore \quad T_{23} - 24\ °C = \left(\frac{23\ cm}{95\ cm}\right)(63\ °C) = 15\ °C$$

$$\therefore \quad T_{23} = 24\ °C + 15\ °C = \boxed{39\ °C}$$

Answers to Practice Quiz

1. (e) **2**. (b) **3**. (c) **4**. (c) **5**. (a) **6**. (d) **7**. (d) **8**. (c) **9**. (b) **10**. (b) **11**. (a)

Answers to Practice Problems

1. 1 C° **2**. 2 F°

3. 60 kPa **4**. 607 °C

5. 10.7 cm **6**. 394.9 J/kg·K

7. 595 W **8**. 7 °C

9. 364,145 W **10**. 64.5 W

CHAPTER 17

PHASES AND PHASE CHANGES

Chapter Objectives

After studying this chapter, you should:

1. understand the basic properties of an ideal gas.
2. understand how the kinetic theory of gases relates microscopic and macroscopic properties.
3. be able to work with stress and strain in the cases of elastic stretching (or compression), shear, and volume deformations of solids.
4. know the three phases of matter and be able to understand phase diagrams.
5. be able to calculate the heat required to change the phase of a substance.
6. know how to use the conservation of energy to keep track of heat transfer within a system.

Warm-Ups

1. According to the kinetic theory model, what is the relation between the temperature of a gas and the mechanical properties of the moving gas particles?
2. Room temperature is taken to be 70 degrees Fahrenheit. Estimate how much the temperature would rise if the kinetic energy of every gas particle in the room doubled.
3. Why is it important for the kinetic theory derivation of the gas law that the particles interact only elastically?
4. According to kinetic theory, why does the gas law begin to fail as a gas approaches the boiling point temperature?

Chapter Review

This is the second of three chapters on thermodynamics. In this chapter we use the kinetic theory of gases to see how the microscopic properties of a system give rise to the macroscopic properties with which we are more familiar. We also explore the phenomenology of phase changes. The energy needed to bring about a phase change, and application of energy conservation to phase changes and other heat transfer processes within a system are also treated.

17–1 Ideal Gases

An **ideal gas** is a gas in which the gas particles (atoms or molecules) do not interact; that is, they move freely and independently of each other. The behavior of ideal gases is a close approximation to the behavior of most real gases. That state of the system of particles making up the gas is determined by the pressure P, temperature T, number of particles N, and volume V of the gas. An equation that shows how these quantities depend on one another is called an **equation of state**. The detailed study of gases has shown that the equation of state for an ideal gas is

$$PV = NkT$$

where k is a fundamental constant called the Boltzmann constant, which has the value

$$k = 1.38 \times 10^{-23}\ \text{J/K}$$

An alternative way to write the equation of state for an ideal gas uses the concept of the **mole** (mol). A mole is the amount of a substance that contains 6.022×10^{23} entities; this value is called **Avogadro's number**, N_A. The masses of atoms (and molecules) are often quoted by stating the mass of one mole of atoms (or molecules), called the **atomic mass** (or molecular mass), of the substance. When moles are used in the equation of state, the number of particles in a gas is written as $N = nN_A$, where n is the number of moles in the gas. The product of the two constants N_Ak is another constant called the gas constant R, which has the value

$$R = 8.31\ \text{J/(mol} \cdot \text{K)}$$

Putting these factors together in the equation of state gives

$$PV = nRT$$

as a commonly used alternative version of the equation of state for an ideal gas; this latter version is often called the *ideal gas law*.

The ideal-gas equation of state contains all the information about how the relevant quantities relate. One relation is **Boyle's law**, which states that for *fixed N and T* the product of pressure and volume is constant:

$$P_iV_i = P_fV_f$$

Curves that are plotted under the condition of fixed temperature are called **isotherms**. Similarly, the ideal gas law also incorporates **Charles' law**, which states that for *fixed N and P* the ratio V/T is constant:

$$\frac{V_i}{T_i} = \frac{V_f}{T_f}$$

Notice that fixed N also implies fixed n.

Exercise 17–1 Inflating a Tire An automotive worker needs to pump up an empty tire that has an inner volume of 0.0192 m^3. If the temperature in the manufacturing plant is 28.5 °C, what will be the gauge pressure in tire, in psi, if 2.70 moles of air is pumped into it?

Solution: We are given the following information:

Given: $V = 0.0192\ \text{m}^3$, $T = 28.5\ °\text{C}$, $n = 2.70\ \text{mol}$; **Find**: P in psi

We can solve this problem by using the ideal-gas equation of state. Since we are given the number of moles of air, we shall use the form that contains n. Solving this equation for pressure we obtain

$$P = \frac{nRT}{V}$$

Since this expression assumes that T is in Kelvin, we must remember to apply the conversion. Doing this conversion gives a pressure of

$$P = \frac{(2.70\ \text{mol})(8.31\ \text{J/mol}\cdot\text{K})\left[(28.5\ \text{C}^\circ + 273.15)\text{K}\right]}{0.0192\ \text{m}^3} = 3.525 \times 10^5\ \text{Pa}$$

The above result gives the absolute pressure in the tire. What we really want is the gauge pressure which is obtained by subtracting atmospheric pressure. Thus,

$$P_g = P - P_{at} = 3.525 \times 10^5\ \text{Pa} - 1.013 \times 10^5\ \text{Pa} = 2.512 \times 10^5\ \text{Pa}\left(\frac{1.450 \times 10^{-4}\ \text{psi}}{1\ \text{Pa}}\right) = 36.4\ \text{psi}$$

which is us the final result in pounds per square inch.

Example 17–2 Compressed Air If you must transfer air from a 3.75-m^3 container at atmospheric pressure to a 1.35-m^3 container at the same temperature, to what pressure must you compress the air?

Picture the Problem The sketch shows the larger container connected to the smaller container by a tube through which the air will pass. [The mechanism that compresses the air is not shown.]

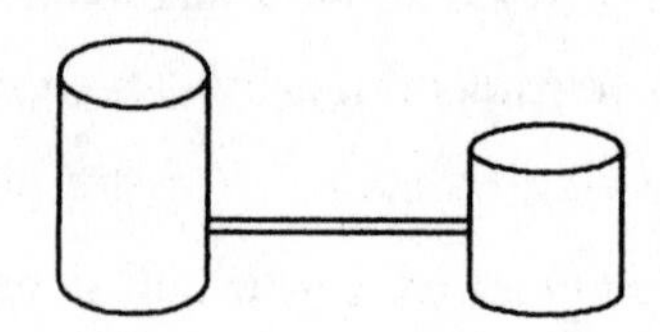

Strategy Both the amount of gas, N, and the temperature, T, remain fixed in this problem, so we can use Boyle's law to determine the final pressure.

Solution

1. Use Boyle's law to solve for the final pressure: $P_i V_i = P_f V_f \Rightarrow P_f = \frac{P_i V_i}{V_f}$

2. Obtain the numerical result: $P_f = \frac{(1.013\times10^5\ \text{Pa})(3.75\ \text{m}^3)}{1.35\ \text{m}^3} = 2.81\times10^5\ \text{Pa}$

Insight The fact that the pressure will increase if you squeeze the same gas into a smaller container should also appeal to your intuition. If it doesn't, think about it.

Practice Quiz

1. For an ideal gas confined to a constant volume, if the temperature is increased by a factor of 2, what happens to the pressure?

   (a) The pressure becomes a factor of 2 smaller.
   (b) The pressure becomes a factor of 2 larger.
   (c) There is no change in the pressure.
   (d) The pressure goes to zero.
   (e) None of the above

2. How many molecules are contained in 5.70 moles of a gas?

   (a) 6.02×10^{23} **(b)** 6 **(c)** 3.43×10^{24} **(d)** 22 **(e)** none of the above

17–2 Kinetic Theory

The **kinetic theory of gases** uses the motion of the microscopic gas particles (atoms or molecules) to explain the macroscopic properties of gases. Specifically, we shall use kinetic theory to gain deeper insight into the origin of the pressure and temperature of a gas. In its simplest form the kinetic theory of gases makes several reasonable assumptions: (a) a large number N of identical gas particles are in an impenetrable rigid container, (b) the particles move randomly within the container, and (c) the only interactions particles are elastic collisions with each other and with the walls of the container.

Kinetic theory shows that the pressure exerted by a gas on a container of volume V is due to the collisions of the gas particles with the walls of the container. The speed of the gas particles (of mass m) are not all the same but rather are distributed over a wide range of speeds according to the Maxwell speed distribution (see Figure 17-8 in the text). If you consider the change in momentum, Δp, of a typical gas particle on collision with a wall of the container together with the amount of time between collisions with this wall, Δt, the force that this particle exerts on the wall, $F = \Delta p/\Delta t$, can be determined. Upon dividing this force by the area of the wall for all N particles in the gas, the pressure works out to be

$$P = \frac{1}{3}\left(\frac{N}{V}\right)m\left(v^2\right)_{av}$$

This expression, in terms of the average translational kinetic energy of the particles, $K_{av} = \left(\frac{1}{2}mv^2\right)_{av}$, gives

$$P = \frac{2}{3}\left(\frac{N}{V}\right)K_{av}$$

This latter expression clearly shows that the pressure of a gas is directly proportional to the average translational kinetic energy of a gas particle.

If we compare the above results with the ideal-gas equation of state, $PV = NkT$ we also find that the Kelvin temperature of a gas is directly proportional to the average kinetic energy of a gas particle. Specifically,

$$K_{av} = \tfrac{3}{2}kT$$

This result not only serves to give us a better idea of the physical meaning of the concept of temperature but also shows why the Kelvin temperature scale is a physically more appropriate scale to use in scientific work — it has the most direct connection to the energy of the particles in a system. Because the average kinetic energy is an important quantity, the speed associated with K_{av} is also a useful measure of the behavior of the gas. This speed is the square root of $(v^2)_{av}$, which is called the root-mean-square speed (or rms speed):

$$v_{rms} = \sqrt{\left(v^2\right)_{av}} = \sqrt{\frac{3kT}{m}}$$

such that $K_{av} = \frac{1}{2}mv_{rms}^2$.

The total internal energy, U, of a gas whose particles consist of a single type of atom (a monatomic gas) is given by the average kinetic energy of a single particle times the number of particles:

$$U = \tfrac{3}{2}NkT = \tfrac{3}{2}nRT$$

where n is the number of moles of the gas. If the gas particles consist of molecules of two or more atoms, other contributions must be added to the internal energy. This latter case will not be treated in this chapter.

Exercise 17–3 Kinetic Energy If 15 moles of a gas are contained in a volume of 0.10 m^3 at atmospheric pressure, what is the average kinetic energy of a gas particle?

Solution: We are given the following information:

Given: $n = 15$ mol, $V = 0.10$ m^3, $P = 101$ kPa; **Find**: K_{av}

The pressure is directly related to the average kinetic energy of a gas particle by kinetic theory. Solving this expression for the kinetic energy gives

$$K_{av} = \frac{3PV}{2N}$$

The only quantity we don't directly know at this point is the number of particles, N. However, we know that N relates to the number of moles n according to $N = nN_A$. Therefore,

$$K_{av} = \frac{3PV}{2nN_A} = \frac{3(1.01\times10^5\ \text{Pa})(0.10\ \text{m}^3)}{2(15\ \text{mol})(6.022\times10^{23}\ \text{mol}^{-1})} = 1.7\times10^{-21}\ \text{J}$$

You might look at this value and think that something must be wrong because it seems ridiculously small. Keep in mind, however, that most gas particles have very small mass, so this value is actually typical.

Practice Quiz

3. If you pump energy into a gas such that the rms speed of the molecules triples, what happens to the temperature of the gas?

   (a) The temperature of the gas triples.

   (b) The temperature is reduced to one-third its previous value.

   (c) The temperature is reduced to one-ninth its previous value.

   (d) The temperature becomes nine times its previous value.

   (e) None of the above

4. What is the average kinetic energy of a molecule in a monatomic gas if the temperature of the gas is 20 °C?

   (a) 4.1×10^{-22} J **(b)** 2.8×10^{-22} J **(c)** 4.0×10^{-21} J **(d)** 7.3×10^{-21} J **(e)** 6.1×10^{-21} J

17–3 Solids and Elastic Deformation

As with gases, the behavior of solids at the microscopic level help us understand some the macroscopic properties of solids. The intermolecular forces at work within a solid give rise the "springlike" behavior of solids when they are deformed by external forces. In this section we treat the three cases of changing the length, shape, and volume of a solid.

When a force is applied perpendicularly to a side of an object of area A to stretch (or compress) the object, the amount that the length of the object increases (or decreases), ΔL, is found to be directly proportional to the pressure, F/A, causing the deformation. Furthermore, the change in length of the object, for a given applied force per unit area, increases in direct proportion to the original length, L_0, of the object; therefore, we have $\Delta L \propto (F/A)L_0$. The proportionality constant that relates these quantities is called **Young's modulus**, Y. The final equality for the change in length of a solid is

$$F = Y\left(\frac{\Delta L}{L_0}\right)A$$

Dimensional analysis of the above equation shows that the SI unit of Young's modulus is N/m^2. While the above expression applies to both stretching and compression, you should be aware that some materials have different values of Y for these two processes. Values of Young's modulus for different materials are listed in Table 17–1 in your textbook.

Another form of deformation (a *shear* deformation) occurs when the applied force is directed along the surface of the area over which it is applied; let's call it the top surface for convenience. In this case, if the opposite side of the object, the bottom surface, is held fixed, the shape of the object will change because the top surface will extend a distance Δx beyond the bottom surface. If the distance between the top and bottom surfaces is L_0, then

$$F = S\left(\frac{\Delta x}{L_0}\right)A$$

where S is called the **shear modulus**. This quantity plays the same role in shear deformation as Young's modulus plays in stretching. Values of S for different materials are listed in Table 17–2 in your text.

Similarly, deformations that change the entire volume of an object, such as when fluid pressure is applied perpendicularly over the entire surface of a completely submerged object, obey a similar relation. If the pressure acting on an object changes by an amount ΔP, the volume of the object will change in response by an amount ΔV. The relationship between these two quantities is

$$\Delta P = -B\left(\frac{\Delta V}{V_0}\right)$$

where V_0 is the original volume of the object. The quantity B is called the **bulk modulus**, which serves the same purpose as Y and S (and is also in the same units, N/m^2). Values of B are listed in Table 17–3 in your text.

Each of the preceding three cases can be written such that we have an applied force per unit area that equals a substance-dependent factor (the modulus) times a quantity that represents how the material deforms. In general, the applied force per area is called a **stress** on the the material, and the resulting deformation ($\Delta L/L_0$, $\Delta x/L_0$, $\Delta V/V_0$) is called the **strain** in the material. The preceding three cases all show that stress $\propto$ strain, which is a direct result of the springlike behavior of the material, that is in accordance with Hooke's law, to a good approximation. However, this behavior is true only within a limited range of stresses and strains known as the **elastic limit**. If a stress deforms an object beyond its elastic limit, the object no longer behaves like a spring. Instead, it becomes permanently deformed, and the preceding relations no longer hold.

Example 17–4 Temperature-Induced Stress An aluminum cube has an edge length of 4.70 cm. If its temperature is raised from –2.00 °C to 26.0 °C, causing its volume to expand, what is the magnitude of the stress applied to the aluminum cube associated with this expansion?

Picture the Problem The diagram shows exaggerated views of an iron cube **(a)** before and **(b)** after the temperature has been raised.

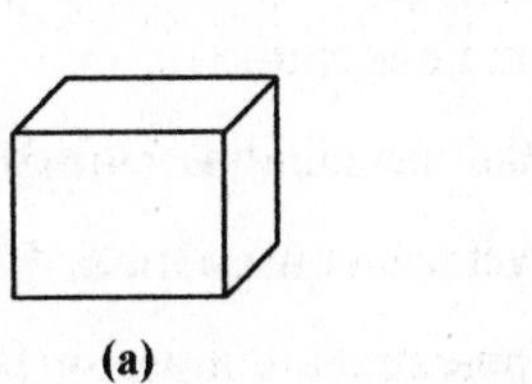
(a)

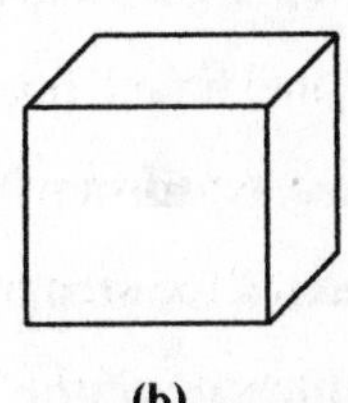
(b)

Strategy We can determine the stress, $|\Delta P|$, for this case of volume expansion by finding ΔV using the equation for volumetric thermal expansion.

Solution

1. Substitute ΔV from thermal expansion into the expression for the stress in volume expansion:
$$|\Delta P| = B\left(\frac{\beta V_0(\Delta T)}{V_0}\right) = B\beta(\Delta T)$$

2. Since no value of β is given for aluminum in Table 16–1 of the text, use the result for α:
$$\beta = 3\alpha = 3\left(24\times10^{-6}\ \text{K}^{-1}\right) = 72\times10^{-6}\ \text{K}^{-1}$$

3. Take the value of the bulk modulus for aluminum from Table 17–3 of the text, and calculate the final result:
$$|\Delta P| = B\beta(\Delta T) = \left(7.0\times10^{10}\ \text{N/m}^2\right)\times\left(72\times10^{-6}\ \text{K}^{-1}\right)\left[26.0\ ^\circ\text{C} - \left(-2.00\ ^\circ\text{C}\right)\right] = 1.4\times10^{8}\ \text{Pa}$$

Insight Here we calculated the stress as a magnitude because the pressure comes from within the solid rather than externally, as assumed by the minus sign in the expression. Also notice that the temperatures were not converted to Kelvin because we needed only the temperature difference, which is the same in Kelvin and Celsius.

Practice Quiz

5. A certain pressure differential ΔP decreases the volume of an object by an amount ΔV. If the object's volume was twice as much, its change in volume would be

   (a) $\Delta V/2$ **(b)** $2(\Delta V)$ **(c)** $4(\Delta V)$ **(d)** $\Delta V/4$ **(e)** ΔV

17–4 Phase Equilibrium and Evaporation

Evaporation is the release of molecules from the liquid phase to the gaseous phase. As with gases, molecules in a liquid have speeds that are distributed throughout a range of values. Some of the higher-speed molecules move sufficiently fast to spontaneously escape the liquid and enter the gaseous phase. If this liquid-gas mixture is in a closed container, then some of the gas molecules can also spontaneously enter the liquid phase. The result is that molecules flow between the two phases. When these flows become equal, so that the number of molecules in each phase remains constant, the molecules have reached a state of **phase equilibrium**. The pressure of the gas in phase equilibrium is called the **equilibrium vapor pressure**. A liquid boils at the temperature at which the equilibrium vapor pressure equals the external pressure on the liquid.

A plot of the equilibrium vapor pressure versus temperature produces the *vapor pressure curve*; on one side of this curve are the conditions (of temperature and pressure) that produce a gas, and on the other side are the conditions that produce a liquid. An analogous curve for the transition between the solid and liquid phases is the **fusion** curve, and for the transition between the solid and gas phases is the **sublimation** curve. When these three curves are plotted on the same diagram we have a **phase diagram** of the substance in question. This diagram provides graphical information on what properties of a substance correspond to what phase of matter.

Practice Quiz

6. Besides fusion, a more common term for the phase transition from liquid to solid is

   (a) condensation **(b)** boiling **(c)** freezing **(d)** melting **(e)** liquefaction

17–5 – 17–6 Latent Heats, Phase Changes, and Energy Conservation

Generally, heat flow is required to change the phase of a substance. During a phase change, there is no change in the temperature of the system undergoing the transformation, as all the heat is used to bring about the phase change. The amount of heat that is required to completely convert one kilogram of a substance from one phase to another is called the **latent heat**, L. Therefore, the amount of heat required to convert a mass m of a substance from one phase to another is given by

$$Q = mL$$

The SI unit of latent heat is J/kg. The value of L depends on the type of phase change being considered. For phase changes between the liquid and gas phases we use the latent heat of vaporization, L_v, between the liquid and solid phases we use the latent heat of fusion, L_f. Values of L_v and L_f are listed in Table 17–4 of your text.

Often, instead of an external agent adding or removing heat to or from a system, we have heat exchanges within a given system. In these cases, it is useful to look at the energy balance between the parts of the system that lose energy and the parts that gain energy. In other words, by energy conservation, we set the magnitude of heat lost by one part of the system equal to the magnitude of heat gained by another part. When you perform this type of analysis, you must treat each step of the energy transfer separately (temperature changes should be handled separately from phase changes) and to recognize when a phase change will take place.

Example 17–5 Boiling Ice How much heat is required to convert 2.88 kg of ice at –3.00 °C to steam at 100 °C if the mixture is housed in an insulated container?

Picture the Problem In the sketch the container on the left first has ice inside it, then it holds only liquid water, and finally it holds only steam.

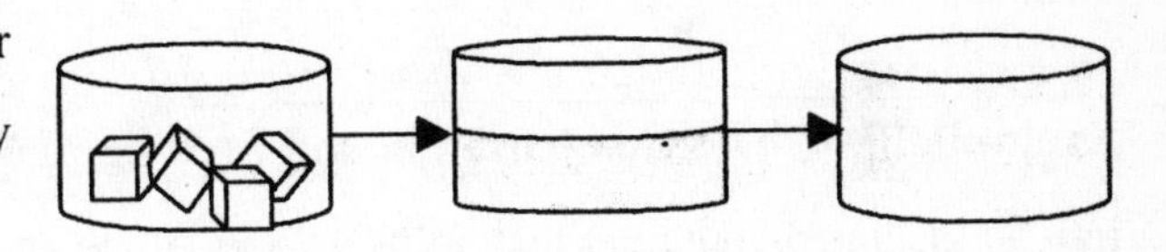

Strategy We divide the process into several steps. First, the ice warms to the melting point, then it melts into liquid water, then the liquid warms to the boiling point, and then it boils to become steam. In all phases, the mass remains equal to the mass of the ice.

Solution

1. The heat needed to warm the ice to the melting temperature 0 °C is: $Q_1 = m_{ice}c_{ice}\left(0\ ^\circ\mathrm{C} - T_{0,ice}\right)$

2. The heat needed to melt the ice at 0 °C is: $Q_2 = m_{ice}L_f$

3. The heat needed to warm the liquid water from $0\,^\circ\mathrm{C}$ to the boiling temperature, $100\,^\circ\mathrm{C}$ is:

$$Q_3 = m_{ice} c_w \left(100\,^\circ\mathrm{C} - 0\,^\circ\mathrm{C}\right)$$

4. The heat needed to convert the liquid to steam at $100\,^\circ\mathrm{C}$ is:

$$Q_4 = m_{ice} L_v$$

5. Taking values of specific and latent heats from the appropriate tables in the textbook we can calculate the total heat required:

$$Q = m_{ice} c_{ice} \left(0\,^\circ\mathrm{C} - T_{0,ice}\right) + m_{ice} L_f + m_{ice} c_w \left(100\,^\circ\mathrm{C} - 0\,^\circ\mathrm{C}\right) + m_{ice} L_v$$

$$\therefore$$

$$Q = m_{ice}\left[c_{ice}\left(3.00\,^\circ\mathrm{C}\right) + L_f + c_w\left(100\,^\circ\mathrm{C}\right) + L_v\right]$$

$$= (2.88\text{ kg})\begin{bmatrix}\left(2090\text{ J/kg}\cdot\mathrm{C}^\circ\right)\left(3.00\,^\circ\mathrm{C}\right) + 33.5\times10^4\text{ J/kg} \\ +\left(4186\text{ J/kg}\cdot\mathrm{C}^\circ\right)\left(100\,^\circ\mathrm{C}\right) + 22.6\times10^5\text{ J/kg}\end{bmatrix}$$

$$= 8.70\times10^6\text{ J}$$

Insight Approaching problems of these types in the step-by-step manner shown here is the best way to avoid mistakes. Also, notice that the units of specific heat are written with C° instead of K; remember, it doesn't matter which you use.

Example 17–6 The Condensation and Melting of Water How much steam at 100 °C must be placed into an insulated container that holds 0.75 kg of ice at –11 °C to finally produce liquid water at 12 °C?

Picture the Problem The sketch shows an ice–steam mixture that later becomes all liquid water.

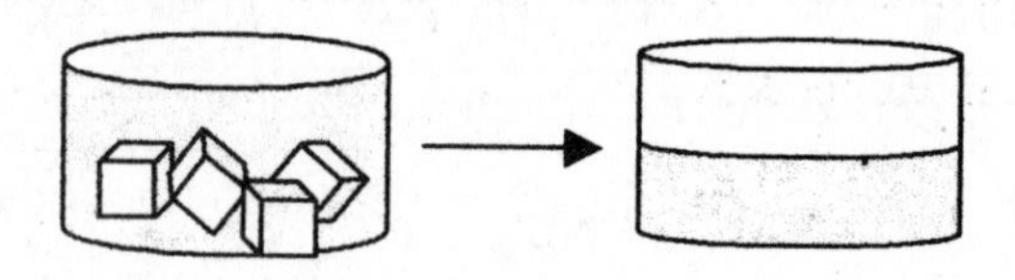

Strategy There will be heat lost by the water that starts as steam, and heat gained by the water that starts as ice. We can follow all the temperature changes and phase changes step by step and set the heat lost equal to the heat gained.

Solution

1. Heat is lost by the steam in condensing to liquid water at 100 °C:

$$Q_{lost,1} = m_{steam} L_v$$

2. Heat is subsequently lost by this liquid water in cooling to the equilibrium temperature: $Q_{lost,2} = m_{steam}c_w(100\,^\circ C - T_{equil})$

3. Heat is gained by the ice to warm it to its melting temperature, 0 °C: $Q_{gain,1} = m_{ice}c_{ice}(0\,^\circ C - T_{0,ice})$

4. Heat is gained by the ice to melt it at 0 °C: $Q_{gain,2} = m_{ice}L_f$

5. Heat is subsequently gained by this liquid water in warming to the equilibrium temperature: $Q_{gain,3} = m_{ice}c_w(T_{equil} - 0\,^\circ C)$

6. Setting the heat lost equal to the heat gained:

$$m_{steam}L_v + m_{steam}c_w(100\,^\circ C - T_{equil}) = m_{ice}c_{ice}(0\,^\circ C - T_{0,ice}) + m_{ice}L_f + m_{ice}c_w(T_{equil} - 0\,^\circ C)$$

7. Solving for the amount of steam gives:

$$m_{steam} = \frac{m_{ice}\left[c_{ice}(0\,^\circ C - T_{0,ice}) + L_f + c_w(T_{equil} - 0\,^\circ C)\right]}{L_v + c_w(100\,^\circ C - T_{equil})}$$

8. Obtain the numerical result:

$$m_{steam} = \frac{(0.75\text{ kg})\left[\left(2090\ \tfrac{J}{kg\cdot K}\right)(11\,C^\circ) + 33.5\times10^4\ \tfrac{J}{kg} + (4186\text{ J/kg}\cdot\text{K})(12\,C^\circ)\right]}{22.6\times10^5\text{ J/kg} + (4186\text{ J/kg}\cdot\text{K})(88\,C^\circ)}$$

$$= 0.12\text{ kg}$$

Insight All the individual calculations of heat lost were calculated as positive quantities. Therefore, ΔT represents the magnitude of the temperature difference and is *not* always the final temperature minus the initial temperature.

Practice Quiz

7. Does it require more or less heat transfer to fuse a certain amount of liquid water into ice than to convert the same amount of ice into liquid water? (Assume everything takes place at 0 °C.)

   (a) more **(b)** less **(c)** the same amount **(d)** The answer cannot be determined

8. When water vapor is transformed into liquid water

   (a) heat is lost by the vapor.

   (b) heat is gained by the vapor.

   (c) no heat transfer is required.

(d) the liquid water immediately freezes.

(e) none of the above

Reference Tools and Resources

I. Key Terms and Phrases

ideal gas a gas in which the gas particles do not interact except for elastic collisions

equation of state (for ideal gases) an equation that relates the temperature, pressure, volume, and number of particles of the gas

mole the amount of a substance that contains 6.022×10^{23} entities

Avogadro's number the number of entities in a mole, equal to 6.022×10^{23}

atomic (molecular) mass the mass of one mole of atoms (molecules)

isotherms pressure-versus-volume curves plotted under the condition of fixed temperature (and number of particles)

kinetic theory the theory that relates the motion of the microscopic particles making up a system to its macroscopic properties

stress the applied force per unit area that deforms a substance

strain the deformation that results from a stress applied to an object

evaporation the release of molecules from the liquid phase to the gaseous phase

fusion the freezing of a liquid to a solid

sublimation the direct transformation between the solid and gas phases

phase diagram a graph which shows the conditions (often of temperature and pressure) under which a substance will exist in different phases

latent heat the amount of heat required to completely convert one kilogram of a substance from one phase to another

II. Important Equations

Name/Topic	Equation	Explanation
equation of state	$PV = NkT = nRT$	Two forms of the equation of state for an ideal gas.
kinetic theory	$P = \frac{2}{3}\left(\frac{N}{V}\right)K_{av}$ $K_{av} = \frac{3}{2}kT$	Kinetic theory shows that both the pressure and Kelvin temperature of a gas are directly proportional to the average translational kinetic energy of a gas particle.
elastic deformation	$F = Y\left(\frac{\Delta L}{L_0}\right)A$ $F = S\left(\frac{\Delta x}{L_0}\right)A$ $\Delta P = -B\left(\frac{\Delta V}{V_0}\right)$	Expressions describing the elastic deformation of solids for stretching, shearing, and volumetric stresses.
latent heat	$Q = mL$	The expression for the heat required to change the phase of an amount of mass m of a substance.

III. Know Your Units

Quantity	Dimension	SI Unit
the Boltzmann constant (k)	$[M][L^2][T^{-2}][K^{-1}]$	J/K
mole	dimensionless	mol
Avogadro's number (N_A)	dimensionless	molecules/mol
gas constant (R)	$[M][L^2][T^{-2}][K^{-1}]$	J/(mol·K)
Young's, shear, and bulk moduli (Y, S, B)	$[M][L^{-1}][T^{-2}]$	N/m^2
stress	$[M][L^{-1}][T^{-2}]$	Pa (or N/m^2)
strain	dimensionless	—
latent heat (L)	$[L^2][T^{-2}]$	J/kg

Practice Problems

1. You inflate your car tires to a gauge pressure of 1.66 atmospheres at a temperature of 24 °C. After driving a couple of miles, the temperature of the tire increases by 30 C°. What is the new gauge pressure, to the nearest hundredth of an atmosphere?
2. What fraction (to three decimal places) of the air would you have to let out of the tire in order for the pressure to return to its original value?
3. A 1-cm^3 air bubble at a depth of 202 meters and at a temperature of 4 °C rises to the surface of the lake, where the temperature is 17.8 °C. To the nearest tenth of a cm^3, what is its new volume?
4. How many micromoles (to one decimal place) of air are in the bubble in problem 3?
5. What is the average kinetic energy of gas molecules at a temperature of 277 K in units of 10^{-21} J, to two decimal places?
6. If the molecules in problem 5 have a mole mass of 25 g, to the nearest m/s what is their rms velocity?
7. To the nearest joule, what is the total internal energy of 12 moles of a monatomic ideal gas at a temperature of 319 K?
8. One kilogram of ice is taken from a temperature of –54 °C into steam at 135 °C. To the nearest kJ, what is the energy used?
9. Two kilograms of ice at 0 °C is dropped into 2.48 kg of water at 22 °C. To two decimal places, what fraction of the ice melts?
10. One kilogram of ice at –10 °C is dropped into 25 kg of water at 44 °C. To the nearest tenth of a celsius degree, what is the final temperature of mixture?

Puzzle

VACUUM SUCKS!

A vacuum chamber designed to be pumped down to a pressure of 10^{-4} Pa has stainless steel walls that are 3 mm thick. You must design a new chamber that can be pumped down to a pressure of 10^{-5} Pa. How thick must the walls of the new chamber be?

a. 0.3 mm

b. 3 mm

c. 1 cm

d. 3 cm

Selected Solutions

13. Since the gas is at constant temperature we know that

$$P_iV_i = P_fV_f$$

Plugging in for the volumes gives

$$P_i(\pi r^2 h_i) = P_f(\pi r^2 h_f) \quad \therefore \quad P_f = \frac{P_i h_i}{h_f} = \frac{(137\text{ kPa})(23.4\text{ cm})}{20.0\text{ cm}} = \boxed{160\text{ kPa}}$$

21. (a) The rms speed of H_2O is greater than the rms speed of O_2, since H_2O is less massive but has the same average kinetic energy.

(b) First, we use O_2 to find the temperature:

$$v_{rms} = \sqrt{\frac{3RT}{M}} \Rightarrow T = \frac{v_{rms}^2 M}{3R} = \frac{\left(1550\ \frac{m}{s}\right)^2 \left(0.0320\ \frac{kg}{mol}\right)}{3\left(8.31\ \frac{J}{mol \cdot K}\right)} = 3084\text{ K}$$

Now, for H_2O:

$$v_{rms} = \sqrt{\frac{3RT}{M}} = \sqrt{\frac{3\left(8.31\ \frac{J}{mol \cdot K}\right)(3084\text{ K})}{\left[15.9994\ \frac{g}{mol} + 2\left(1.00794\ \frac{g}{mol}\right)\right]\left(\frac{1\text{ kg}}{1000\text{ g}}\right)}} = 2.07 \times 10^3\text{ m/s} = \boxed{2.07\text{ km/s}}$$

33. For each rod, the expression for the compressive force is

$$F = Y\left(\frac{\Delta L}{L_0}\right)\left(\frac{\pi D^2}{4}\right) \Rightarrow \Delta L = \frac{4L_0 F}{Y\pi D^2}$$

The total compression is given by $\Delta L_{total} = \Delta L_{Al} + \Delta L_{Br}$; therefore,

$$\Delta L_{total} = \frac{4L_0F}{\pi D^2}\left(\frac{1}{Y_{Al}} + \frac{1}{Y_{Br}}\right) = \frac{4(0.25\text{ m})(7600\text{ N})}{\pi(0.0075\text{ m})^2}\left(\frac{1}{6.9 \times 10^{10}\ \frac{N}{m^2}} + \frac{1}{9.0 \times 10^{10}\ \frac{N}{m^2}}\right) = \boxed{1.1\text{ mm}}$$

49. The amount of heat used to raise the temperature of the ice to the melting temperature of 0 °C is

$$Q_1 = mc\Delta T = (1.1\text{ kg})\left(2090\ \frac{J}{kg \cdot C°}\right)(5.0°C) = 1.15 \times 10^4\text{ J}$$

The amount of heat available for other processes is

$$Q_2 = Q_{total} - Q_1 = 2.6 \times 10^5\text{ J} - 1.15 \times 10^4\text{ J} = 2.49 \times 10^5\text{ J}$$

The amount of heat needed to melt all the ice is

$$Q_f = mL_f = (1.1\text{ kg})\left(33.5 \times 10^4\text{ J/kg}\right) = 3.69 \times 10^5\text{ J}$$

Since $Q_f > Q_2$, not all the ice melts, so the final temperature of the water is 0 °C.

The mass that melts is

$$m_f = \frac{Q_2}{L_f} = \frac{2.49\times10^5\ \text{J}}{33.5\times10^4\ \frac{\text{J}}{\text{kg}}} = 0.74\ \text{kg}$$

Therefore, the mass that remains is

$$m_{ice} = 1.1\ \text{kg} - m_f = 1.1\ \text{kg} - 0.74\ \text{kg} = \boxed{0.36\ \text{kg}}$$

53. Let us first assume that $T_f = 100$ °C. Then, the heat lost by iron is

$$Q_{Fe} = -m_{Fe}c_{Fe}(T_f - T_{Fe,i}) = -(0.825\ \text{kg})\left(560\ \frac{\text{J}}{\text{kg}\cdot\text{C}^\circ}\right)(100\ ^\circ\text{C} - 352\ ^\circ\text{C}) = 116\ \text{kJ}$$

The heat needed to raise the water to the boiling temperature of 100 °C is

$$Q_w = m_w c_{water}\left(T_f - T_{w,i}\right) = (0.0400\ \text{kg})\left(4186\ \frac{\text{J}}{\text{kg}\cdot\text{C}^\circ}\right)(100\ ^\circ\text{C} - 20.0\ ^\circ\text{C}) = 13.4\ \text{kJ}$$

Since $|Q_{Fe}| > Q_w$, we know that water will boil and that $T_f \geq 100\ ^\circ\text{C}$. The heat needed to boil the water is

$$Q_v = m_w L_v = (0.0400\ \text{kg})\left(22.6\times10^5\ \text{J/kg}\right) = 90.4\ \text{kJ}$$

Since $Q_{Fe} > Q_w + Q_v$, all the water vaporizes. Therefore, the steam gains an amount of heat equal to $Q_s = Q_{Fe} - Q_w - Q_v$ while the temperature rises to the equilibrium temperature. Therefore,

$$Q_s = m_w c_{steam}(T_f - 100\ ^\circ\text{C}) = m_{Fe}c_{Fe}(T_{Fe,i} - T_f) - Q_w - Q_v$$

Solving this expression for T_f gives

$$T_f = \frac{m_{Fe}c_{Fe}T_{Fe,i} + m_w c_{steam}(100\ ^\circ\text{C}) - Q_w - Q_v}{m_{Fe}c_{Fe} + m_w c_{steam}}$$

$$= \frac{(0.825\ \text{kg})\left(560\ \frac{\text{J}}{\text{kg}\cdot\text{K}}\right)(352\ ^\circ\text{C}) + (0.0400\ \text{kg})\left(2010\ \frac{\text{J}}{\text{kg}\cdot\text{K}}\right)(100\ ^\circ\text{C}) - 13.4\times10^3\ \text{J} - 90.4\times10^3\ \text{J}}{(0.825\ \text{kg})\left(560\ \frac{\text{J}}{\text{kg}\cdot\text{K}}\right) + (0.0400\ \text{kg})\left(2010\ \frac{\text{J}}{\text{kg}\cdot\text{K}}\right)}$$

$$= \boxed{123\ ^\circ\text{C}}$$

Answers to Practice Quiz

1. (b) **2.** (c) **3.** (d) **4.** (e) **5.** (b) **6.** (c) **7.** (c) **8.** (a)

Answers to Practice Problems

1. 1.83 atm 2. 0.092

3. 21.6 cm^3 4. 905.4 μmol

5. 5.73 **6**. 526 m/s

7. 47,716 J **8**. 3,192 kJ

9. 0.34 **10**. 39.0 °C

CHAPTER 18

THE LAWS OF THERMODYNAMICS

Chapter Objectives

After studying this chapter, you should:

1. understand the relationship between the first law of thermodynamics and the conservation of energy.
2. understand the behavior of gases that undergo constant-pressure, constant-volume, isothermal, and adiabatic processes.
3. know the difference between the specific heat at constant volume and the specific heat at constant pressure and how the two relate for a monatomic ideal gas.
4. know the basic functions of heat engines, refrigerators, and similar devices and how they function within the laws of thermodynamics.
5. develop a basic understanding of the concept of entropy and how it relates to order and disorder.
6. know the four laws of thermodynamics.

Warm-Ups

1. Explain how it is possible to compress a sample of gas to half its original volume without changing the pressure.
2. Estimate the change in temperature when one cubic meter of air is compressed to half its original volume adiabatically (start at room temperature and pressure).
3. Is melting an ice cube an irreversible process? How about freezing one?
4. Estimate the total work by a system taken around the cycle shown in the figure. Does the system have to be an ideal gas?

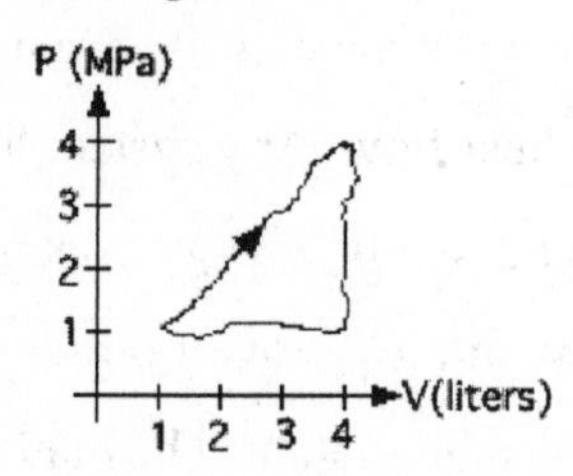

5. In your textbook, an illustration like this one is used to represent a heat engine. If this sketch is to represent your car, what parts of your car correspond to the parts of the diagram?

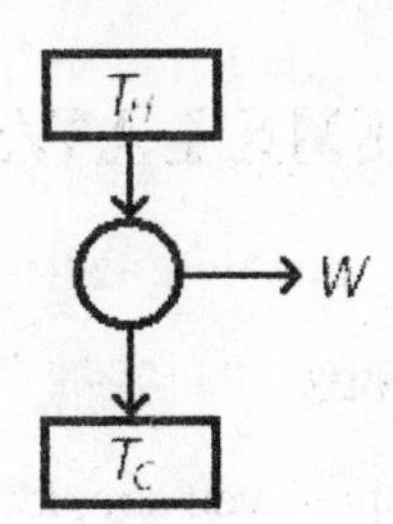

Chapter Review

In this final chapter on thermodynamics we bring all four laws together, hoping to provide a basic understanding of the importance of thermodynamics and many of its applications.

18–1 The Zeroth Law of Thermodynamics

The zeroth law of thermodynamics was introduced in Chapter 16; we review this law here for the sake of completeness. The zeroth law of thermodynamics lays important groundwork for all of thermodynamics, especially for providing a working definition of the concept of temperature. The zeroth law says the following:

> *If system A is in thermal equilibrium with system B, and system C is also in thermal equilibrium with system B, then systems A and C will be in thermal equilibrium if brought into thermal contact.*

The property that determines whether two systems will be in thermal equilibrium is temperature. When two systems are at the same temperature they will be in thermal equilibrium if brought into thermal contact.

18–2 The First Law of Thermodynamics

The first law of thermodynamics is essentially just an application of the conservation of energy to situations involving heat flow. Any system has a certain amount of internal energy U; this energy consists of all the potential and kinetic energy contained within the system. Changes in the internal energy result from either heat flow into (positive Q) or out of (negative Q) the system, and/or work done by (positive W) or on (negative W) the system. The mathematical expression of this law is

$$\Delta U = Q - W$$

Notice that this expression is consistent with the intuitive notion that the internal energy will increase when either heat is added to the system and/or work is done on the system.

An important distinction between internal energy, heat, and work is that changes in internal energy depend only on the initial and final states of the system (which are determined by pressure, temperature, and volume), and U is therefore called a **state function**. Both heat and work depend not only on the states involved but also on the process by which a system is changed from one state to another.

Example 18–1 The First Law If 4530 J of work is done on a 0.750-kg piece of copper while its temperature rises from 18.2 °C to 31.2 °C, what is the change in internal energy of the piece of copper? (Assume a constant pressure of 1 atm.)

Picture the Problem The sketch shows the piece of copper in question.

Strategy Since we are given the amount of work W, if we can determine the heat flow Q, we can use the first law to determine ΔU.

Solution

1. Use the specific heat, Table 16–2 in the text, to determine Q:

$$Q = mc\Delta T = (0.750 \text{ kg})(390 \text{ J/kg}\cdot\text{C}^{\circ})(13.0 \text{ C}^{\circ}) = 3802.5 \text{ J}$$

2. The work is done on the copper so:

$$W = -4530 \text{ J}$$

3. According to the first law of thermodynamics:

$$\Delta U = Q - W = 3802.5 \text{ J} - (-4530 \text{ J}) = 8330 \text{ J}$$

Insight Remember when using the first law of thermodynamics that it is important to keep track of whether the heat flow is into or out of the system and if the work is done on or by the system.

Practice Quiz

1. The first law of thermodynamics is most closely related to

   (a) the conservation of energy

   (b) the conservation of linear momentum

   (c) the conservation of angular momentum.

   (d) Newton's second law

   (e) none of the above

2. If 13 J of work is done on a system to remove 9.0 J of heat, what is the change in the internal energy of the system?

   (a) – 4.0 J **(b)** 22 J **(c)** – 22 J **(d)** 4.0 J **(e)** None of the above

18–3 Thermal Processes

As mentioned at the end of the previous section, some quantities, such as Q and W, depend on the process by which a system is altered. Because of this fact, we need to understand the basic types of processes used in the study and application of thermodynamics.

In this chapter we consider only processes that are quasi-static and free from dissipative forces. Quasistatic means that a process takes place so slowly that we can consider the system to be in thermal equilibrium with its surroundings throughout the process. These conditions allow processes to be **reversible**, which means that both the system and its environment can be returned to their precise states at the beginning of the process. Processes for which these conditions do not apply are called **irreversible**.

Constant Pressure and Constant Volume

For a gas that expands or contracts during a process while held at constant pressure (sometimes called an *isobaric* process), the work done by the gas during the process is found to be

$$W = P(\Delta V) \qquad \textit{(for constant pressure)}$$

where $\Delta V = V_f - V_i$ is the change in the volume of the gas. The above result can be interpreted graphically as the area under the curve of a pressure-versus-volume plot (see below). In fact, the area under the curve of a pressure-versus-volume plot for an expanding (or contracting) gas equals the work done by (or on) the gas for any process, not just processes at constant pressure.

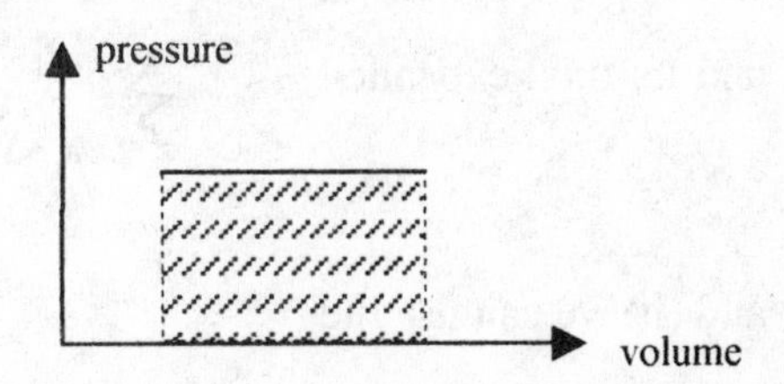

At constant volume (an *isochoric* process) no work is done by the gas during a reversible process. This fact is consistent with the preceding expression because $\Delta V = 0$ when the volume is held constant. Also, intuition suggests that there should be no work done because work results when a force acts through distance, and if the gas does not expand or contract through any distance, the net work done by the gas is zero. Thus,

$$W = 0 \qquad \textit{(for constant volume)}$$

Isothermal Processes

An **isothermal process** is one that takes place at constant temperature. For an ideal gas, the relationship between the pressure and the volume during an isothermal process is

where the constant is NkT. The work done by the gas during an isothermal process can be derived from the ideal-gas equation of state to be

Adiabatic Processes

If no heat flows into or out of a system during a process, the process is called **adiabatic**. This type of process occurs when a system is well insulated or when the process takes place so rapidly that heat doesn't have time to flow. During adiabatic processes, the pressure, volume, and temperature may all change. The pressure-versus-volume curve for a system undergoing an adiabatic process is called an *adiabat*.

Example 18–2 Expanding Gas A monatomic ideal gas consisting of 6.32 moles of atoms expands from a volume of 14.1 m^3 to 27.6 m^3. How much work is done by the gas if the expansion is **(a)** at a constant pressure of 133 kPa and **(b)** isothermal at $T = 303$ K?

Picture the Problem The sketch shows the gas at its initial volume on the left and its final expanded volume on the right.

Strategy For both parts (a) and (b) we can use the above expressions to calculate the work done.

Solution

Part (a)

1. Use the expression for the work done at constant pressure:

Part (b)

2. Use the expression for the work done during an isothermal processes:

$$W = nRT\ln\left(\frac{V_f}{V_i}\right)$$

$$= (6.32\text{ mol})(8.31\text{ J/mol}\cdot\text{K})(303\text{ K})\ln\left(\frac{27.6\text{ m}^3}{14.1\text{ m}^3}\right)$$

$$= 1.07\times10^4\text{ J}$$

Insight Recall that the natural logarithm, ln, is the same as log to the base e.

Practice Quiz

3. During an isothermal process, a gas expands to twice its previous volume. The pressure in the gas...

   (a) doubles

   (b) triples

   (c) becomes larger by a factor of $\sqrt{2}$

   (d) stays the same

   (e) none of the above

4. The graph shown on the right most likely represents what type of process?

   (a) constant pressure **(b)** constant volume

   (c) isothermal **(d)** irreversible

   (e) none of the above

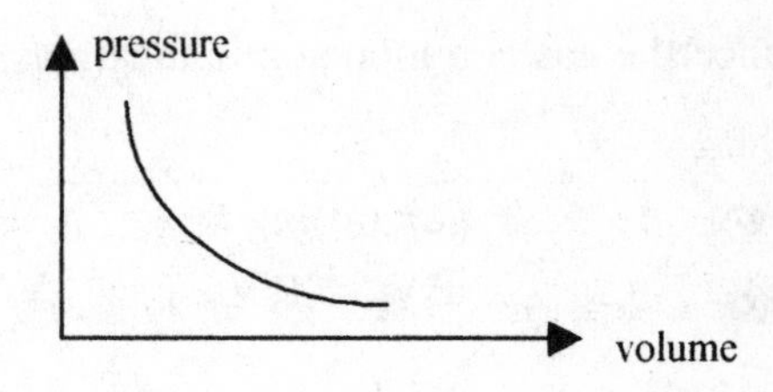

18–4 Specific Heats for an Ideal Gas: Constant Pressure, Constant Volume

It was mentioned previously that the amount of heat flow depends on the nature of the process; that is, heat is not a state function. This means that the specific heat is different for different processes. The relationship between the specific heats for constant-pressure and constant-volume processes is particularly useful. For constant volume, the specific heat c_v (lowercase c) is defined by

$$Q_v = mc_v\Delta T = nC_v\Delta T$$

where C_v is called the **molar specific heat** (with a capital C) representing the heat needed to change the temperature of 1 mole of a substance by 1 Celsius degree. Similarly, the molar specific heat for constant pressure is determined by

$$Q_p = nC_p\Delta T$$

For a monatomic ideal gas the molar specific heats are

$$C_v = \tfrac{3}{2}R \quad \text{and} \quad C_p = \tfrac{5}{2}R$$

which further implies that $C_p - C_V = R$. This last relation is true for all ideal gases, not just monatomic ones. Given this information we are now ready to write down a pressure-volume relation for adiabatic processes, which we could not do in the previous section. If we define the quantity $\gamma = C_p/C_V$, then for an ideal gas undergoing an adiabatic process we have

$$PV^{\gamma} = \text{constant} \qquad \textit{(for adiabatic processes)}$$

For a monatomic gas $\gamma = 5/3$.

Exercise 18–3 An Adiabatic Process During an adiabatic process of a monatomic ideal gas containing 100 moles, the pressure in the gas increases from 105 kPa to 115 kPa. If the original volume of the gas was 2.25 m^3, **(a)** what is the new volume of the gas and **(b)** what is its final temperature?

Solution: We are given the following information:

Given: $n = 100$ mol, $P_i = 105$ kPa, $P_f = 115$ kPa, $V_i = 2.25$ m^3; **Find**: (a) V_f, (b) T_f

Part (a)

Since the gas is a monatomic ideal gas, we know that $\gamma = 5/3$ and we can use the result PV^{γ} = constant. Thus,

$$P_i V_i^{5/3} = P_f V_f^{5/3} \quad \Rightarrow \quad V_f = V_i \left(\frac{P_i}{P_f}\right)^{3/5} = \left(2.25\ \text{m}^3\right)\left(\frac{105\ \text{kPa}}{115\ \text{kPa}}\right)^{3/5} = 2.13\ \text{m}^3$$

Part (b)

Now that we know the final pressure and volume, we can use the equation of state of an ideal gas to determine the temperature:

$$PV = nRT \quad \Rightarrow \quad T_f = \frac{P_f V_f}{nR} = \frac{\left(1.15\times10^5\ \text{Pa}\right)\left(2.130\ \text{m}^3\right)}{\left(100\ \text{mol}\right)\left(8.31\ \text{J/mol}\cdot\text{K}\right)} = 295\ \text{K}$$

Practice Quiz

5. If the volume of a monatomic ideal gas expands adiabatically such that its volume triples, what factor times the initial pressure equals the final pressure of the gas?

   (a) 6.2 **(b)** 0.16 **(c)** 3 **(d)** 0.33 **(e)** 1.9

18–5 The Second Law of Thermodynamics

In Chapter 16 we noted that temperature is the quantity that indicates the existence and direction of the flow of heat from one system to another when the systems are in thermal contact. The second law of thermodynamics states this direction explicitly:

> *When two systems at different temperatures are brought into thermal contact, the spontaneous flow of heat is always from the system at the higher temperature to the system at the lower temperature.*

18–6 – 18–7 Heat Engines, Carnot's Theorem, Refrigerators, Air Conditioners, and Heat Pumps

A device that converts heat to work is called a **heat engine**. Heat engines require a high-temperature region to supply energy to the system, a low-temperature region to exhaust wasted heat, an engine that operates cyclically, and a working substance. If we take the *magnitude* of the heat supplied to the engine as Q_h and the *magnitude* of the heat exhausted to the low temperature region as Q_c, the work done by the heat engine is given by

$$W = Q_h - Q_c$$

Heat engines are often characterized by their **efficiency**, e, which is the fraction of the heat supplied that appears as work:

$$e = \frac{W}{Q_h} = 1 - \frac{Q_c}{Q_h}$$

The maximum possible efficiency of a heat engine that operates from a single hot reservoir at temperature T_h and a single cold reservoir at T_c results when all processes in the cycle are reversible — this statement is known as **Carnot's theorem**. Analysis of engines of this type gives the maximum efficiency to be

$$e_{max} = 1 - \frac{T_c}{T_h}$$

Notice that the maximum possible efficiency depends only on the temperatures of the reservoirs and not on any details of engine. Furthermore, this theorem shows that no engine can be perfectly efficient. Thus, we can see that the maximum work that a particular heat engine can perform is

$$W_{max} = e_{max} Q_h = \left(1 - \frac{T_c}{T_h}\right) Q_h$$

Although it may not be obvious, Carnot's theorem is equivalent to the statement of the second law of thermodynamics given previously and is often considered to be an alternative statement of it.

Example 18–4 The Efficiency of an Engine A heat engine generates 4.1 kJ of work at 35% efficiency. How much heat is exhausted in the cold reservoir?

Picture the Problem The diagram shows a schematic of the heat engine.

HOT → Q_h → engine → Q_c → COLD; engine → W

Strategy Since both the work done and the efficiency depend on Q_h, we can combine those two expressions to eliminate Q_h and solve for Q_c.

Solution

1. The expressions for W and e are: $W = Q_h - Q_c$ and $e = \frac{W}{Q_h} \Rightarrow Q_h = \frac{W}{e}$

2. Substitute Q_h into the expression for the work: $W = \frac{W}{e} - Q_c$

3. Solve this equation for Q_c: $Q_c = \frac{W}{e} - W = W\left(\frac{1-e}{e}\right)$

4. Obtain the numerical result: $Q_c = \left(4.1\times10^3\ \text{J}\right)\left(\frac{1-0.35}{0.35}\right) = 7.6\ \text{kJ}$

Insight Notice here that the efficiency is used as a decimal number in the solution even though it is quoted as a percentage in the problem. It may seem strange that more heat is exhausted than is converted to work, but that should be expected for an efficiency of less than 50%.

Recall that a heat engine uses the natural tendency for heat to flow from hot to cold as a means of generating work. Refrigerators, air conditioners, and heat pumps use work to generate a flow of heat *against* its natural tendency, that is, from cold to hot. A **refrigerator** is almost the precise reverse of a heat engine. In a refrigerator work is the input that forces heat to flow from a cooler region (the refrigerator) to a warmer region (the room). The effectiveness of a refrigerator is indicated by its **coefficient of performance**, COP, which compares the heat forced from the cold reservoir, Q_c, to the amount of work required to do it, W:

$$\text{COP}_{ref} = \frac{Q_c}{W}$$

An **air conditioner** is the same as a refrigerator except that the room being cooled is the cold reservoir, and the outdoor air is the warm region.

A **heat pump** is the reverse of an air conditioner. This device consumes work to remove heat from the cold reservoir of outdoor air and pump it into the warm reservoir of the room being heated. In an ideal (reversible) heat pump the amount of work required to pump an amount of heat Q_h into a room is given by the same relationship as for a Carnot heat engine: $W = Q_h(1 - T_c / T_h)$. The COP for a heat pump is:

$$COP_{hp} = \frac{Q_h}{W}$$

Practice Quiz

6. If a heat engine requires 4500 J of heat to produce 1200 J of work, what is its efficiency?
   (a) 0.27 **(b)** 100% **(c)** 0.73 **(d)** 0.36 **(e)** 0%

7. In terms of process and function which pair of devices are most alike?
   (a) heat engine and refrigerator
   (b) heat engine and heat pump
   (c) refrigerator and air conditioner
   (d) air conditioner and heat pump
   (e) air conditioner and heat engine

8. If heat flow to the hot reservoir increases over time for a refrigerator (assuming Q_c remains constant), its COP
   (a) increases
   (b) stays the same
   (c) decreases

18–8 – 18–9 Entropy, Order, and Disorder

The fact that heat flows naturally only from systems of higher temperature to systems of lower temperature is only one part of a larger "directionality" to the laws of thermodynamics. The basic quantity in physics that encompasses that directionality is called **entropy**, S. The entropy of a system is a state function, and the change in entropy is defined to be

$$\Delta S = \frac{Q}{T}$$

where it is understood that the heat Q is transferred reversibly at a fixed absolute temperature T.

The concept of entropy becomes particularly important when we consider the total change in entropy of a system *and its surroundings* during a process. It is found that although the entropy of an individual system can decrease, this total entropy never decreases. For reversible processes the change in the total entropy is zero, and for all other processes the total entropy increases. Remembering that reversibility is an idealization and that all processes are irreversible, we see that in reality *the total entropy of the universe is always increasing*. This last point is the sense in which entropy provides us with directionality to the laws of physics — an arrow of time. The statement that the entropy of the universe is increasing is yet another equivalent way to state the second law of thermodynamics. In fact, the second law of thermodynamics is very commonly called "the law of increase of entropy."

One can gain some intuitive insight into the physical meaning of entropy by thinking of it as a measure of the amount of disorder in the universe. When processes occur, the universe always comes away more disordered than it was before. An example is the natural flow of heat from a hot reservoir to a cold reservoir. In the cold reservoir the molecules are moving more slowly and scattering off each other at a slower rate. If the system is cold enough they may even clump together and solidify, forming a crystal lattice. When heat from the hot reservoir comes in, that heat increases the speed of the molecules causing them to move about more randomly and increasing the disorder (and therefore the entropy) of the system. Of course, the disorder (entropy) of the hot reservoir decreases, but it can be shown that the increase in entropy of the cold reservoir is always greater in magnitude than the decrease in entropy of the hot reservoir. The total disorder always increases.

Example 18–5 Ice and Steam A mass of 2.88 kg of ice at 0.00 °C is mixed with 0.500 kg of steam at 100 °C. Estimate the change in entropy only up to the point when all the ice has melted and all the steam has condensed.

Picture the Problem The sketch on the left shows an initial mixture of ice and steam that ultimately becomes all liquid water.

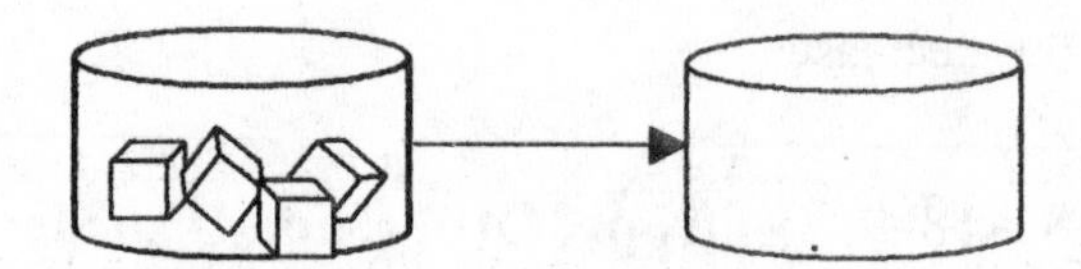

Strategy To estimate the total change in entropy we need to calculate the heat flow for all the processes that take place until we reach the point of having all liquid water.

Solution

1. Use Table 17–4 of the text to obtain the latent heat of fusion for water, the heat needed to melt the ice at 0 °C:

$$Q_i = m_i L_f = (2.88\text{ kg})(33.5\times10^4\text{ J/kg}) = 9.648\times10^5\text{ J}$$

2. Also using Table 17–4, find the heat needed to condense the steam: $Q_s = m_s L_v = (0.500\text{ kg})(22.6\times10^5\text{ J/kg}) = 1.130\times10^6\text{ J}$

3. Since $Q_s > Q_i$, the melted ice warms up; the amount of heat ΔQ that warms up the liquid is: $\Delta Q_w = Q_s - Q_i = 1.130\times10^6\text{ J} - 9.648\times10^5\text{ J} = 1.652\times10^5\text{ J}$

4. Using Table 16–2 for the specific heat, the change in temperature of the melted ice is:

$$\Delta Q_w = m_i c_w \Delta T_{wat} \Rightarrow \Delta T_w = \Delta Q_w / m_i c_w$$

$$\therefore\ \Delta T_w = \frac{1.652\times10^5\text{ J}}{(2.88\text{ kg}\times4186\text{ J/kg}\cdot\text{K})} = 13.7\text{ K}$$

5. The change in entropy for the melting ice is: $\Delta S_i = \dfrac{Q_i}{T_i} = \dfrac{9.648\times10^5\text{ J}}{273.15\text{ K}} = 3532\text{ J/K}$

6. The change in entropy to condense the steam is: $\Delta S_s = \dfrac{-Q_s}{T_i} = \dfrac{-1.130\times10^6\text{ J}}{373.15\text{ K}} = -3028\text{ J/K}$

7. The approximate change in entropy to warm the water is:

$$\Delta S_w \approx \frac{\Delta Q_w}{T_{w,av}} = \frac{\Delta Q_w}{\left[T_i + (T_i + \Delta T_w)\right]/2}$$

$$= \frac{1.652\times10^5\text{ J}}{\left[(273.15+286.85)/2\right]\text{K}} = 590\text{ J/K}$$

8. The total change in entropy is:

$$\Delta S_{tot} \approx \Delta S_i + \Delta S_s + \Delta S_w$$

$$= (3532 - 3028 + 590)\text{ J/K} = 1090\text{ J/K}$$

Insight This result is only an estimate because we had to estimate ΔS_W due to the fact that there was a temperature change. To handle this change in temperature we used the average temperature during the change. As long as the change is comparatively small, this estimate is reasonable. Note also that the temperatures were converted to Kelvin because the change in entropy is defined in terms of the Kelvin temperature. Also keep in mind that, as the problem requests, this result gives the total entropy change only up to the point when the system is all liquid and does not include the change in entropy for the system to reach an equilibrium temperature.

Practice Quiz

9. Can the entropy of a system ever decrease?

 (a) nó, because entropy always increases

 (b) yes, but only in some reversible processes

 (c) no, because the change in entropy is zero for any system

 (d) yes, because entropy always decreases

 (e) yes, if there's a net heat flow out of the system

10. When the total entropy increases during a process the systems involved become

 (a) more orderly

 (b) less energetic

 (c) crystallized

 (d) more disordered

 (e) more energetic

18–10 The Third Law of Thermodynamics

This chapter ends with the statement of the third law of thermodynamics. This law states the following:

It is impossible to lower the temperature of a system to absolute zero in a finite number of steps.

Perhaps the simplest way to accept the impossibility of absolute zero is to try to find a process that would produce the first system to be cooled all the way to this temperature. It would require either bringing the system into thermal contact with a cooler system (which would have to reach absolute zero first) or pumping heat out of the system, which would have to be isolated from everything warmer (including the pump) to prevent thermal conduction from increasing the temperature. Again, this is a heuristic discussion designed to appeal to your intuition — think about it.

Reference Tools and Resources

I. Key Terms and Phrases

state function: a quantity that only depends on the thermodynamic state of a system (determined by P, V, and T).

reversible process a process that allows a system to return precisely to a previous state

irreversible process a process that is not reversible

isothermal process a process that takes place at constant temperature

adiabatic process a process during which no heat is transferred

molar specific heat the heat needed to change the temperature of one mole of a substance by one Celsius degree

heat engine a device that converts heat into work

Carnot's theorem states the conditions that give the maximum efficiency of a heat engine

refrigerator a device that uses work to cause heat to flow from a cooler region to a warmer region

entropy the ratio of the heat that flows at a fixed temperature to that temperature for a reversible process. It measures the amount of disorder in a system

II. Important Equations

Name/Topic	Equation	Explanation
the first law of thermodynamics	$\Delta U = Q - W$	The conservation of energy for systems with heat flow.
constant pressure processes	$W = P(\Delta V)$	The work done by a gas expanding under constant pressure.
constant temperature processes	$W = NkT \ln\left(\frac{V_f}{V_i}\right) = nRT \ln\left(\frac{V_f}{V_i}\right)$	The work done by a gas expanding under constant temperature.
specific heats	$C_v = \frac{3}{2}R, \quad C_p = \frac{5}{2}R$	The molar specific heat at constant volume and at constant pressure.
adiabatic processes	$PV^{\gamma} = \text{constant}$	The behavior of an ideal gas during an adiabatic process.
heat engines	$e = \frac{W}{Q_h} = 1 - \frac{Q_c}{Q_h}$	The efficiency of a heat engine.
entropy	$\Delta S = \frac{Q}{T}$	The change in entropy of a system at fixed temperature under a reversible process.

III. Know Your Units

Quantity	Dimension	SI Unit
molar specific heat (C)	$[M][L^2][T^{-2}][K^{-1}]$	J/(mol·K)
efficiency (e)	dimensionless	—
coefficient of performance (COP)	dimensionless	—
entropy (S)	$[M][L^2][T^{-2}][K^{-1}]$	J/K

Practice Problems

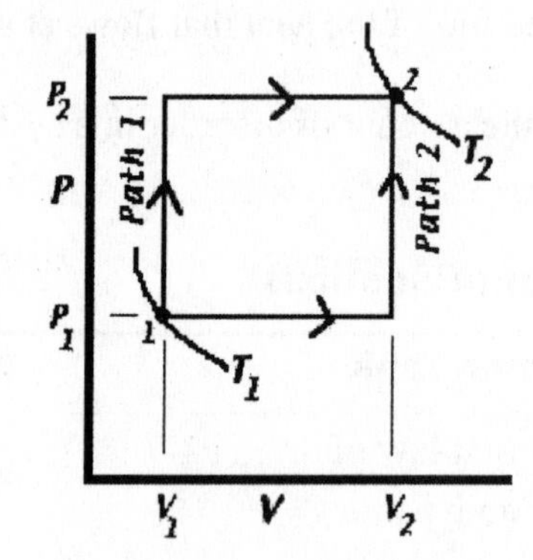

1. In the figure if 10 mol of an ideal gas is taken from point 1, T_1 = 261, to point 2, T_2 = 335, what is its change in internal energy, to the nearest tenth of a joule?
2. If P_1= 173 kPa, V_1= 1995 cm³, and P_2= 200 kPa, V_2= 10000 cm³, what is the heat absorbed (+) or liberated (–) to the nearest tenth of a Joule in Path 1?
3. What would be the answer to problem 2 if Path 2 was followed?
4. Ten moles of a gas in thermal contact with an oil bath at temperature 300 K is compressed isothermally from a volume of 8828 cm³ to a volume of 1892 cm³. To the nearest tenth of a joule what is the work done by the piston?
5. What would be the answer to problem 4 if the gas was compressed adiabatically? (It is a monatomic ideal gas.)

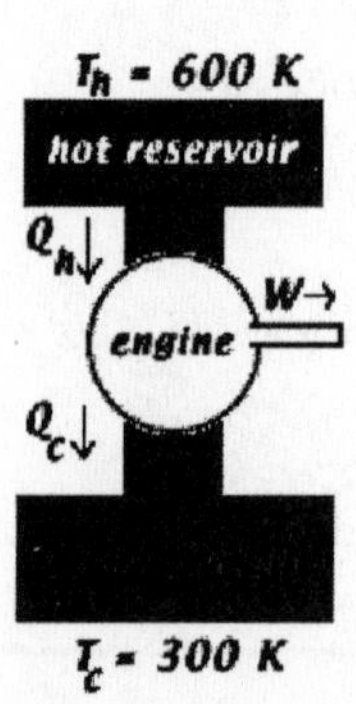

6. The heat engine in the figure absorbs Q_h = 3712 joules of heat and ejects Q_c = 1504 joules of heat. To two decimal places, what is its efficiency?
7. If the temperature of the lower reservoir is 266 K, and the engine in problem 6 is a Carnot engine, what is the temperature of the upper reservoir to, the nearest kelvin.
8. In the engine of problem 6 Q_h= 1646 and Q_c= 1194. What is the change in entropy of the upper reservoir, to two decimal places?

9. In problem 8 what is the change in entropy of the universe?

10. If 0.75 kg of ice at temperature T'= 0 °C is melted into water at the same temperature by placing it in contact with a reservoir at temperature T = 22.2 °C, what is the change in the entropy of the universe, to two decimal places?

Puzzle

PROCESS PROBLEMS

The figure shows five different processes (labeled a-e). Each leads from an initial state (P = 3 atm., V = 1 m^3) to a final state (P = 1 atm., V = 3 m^3). Please answer each of the following questions. For which process is W the largest? The smallest? For which process is Q the largest? The smallest?

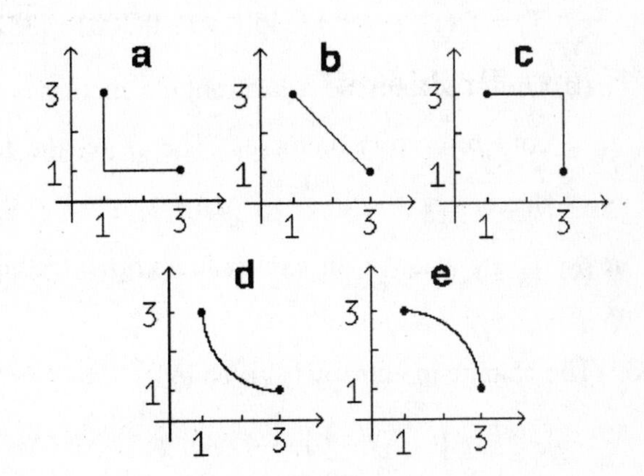

Selected Solutions

25. **(a)** At constant pressure the work done by the gas is W = P(ΔV):

$$W = P(V_f - V_i) = P(2V_i - V_i) = PV_i = (160\text{ kPa})(0.83\text{ m}^3) = \boxed{130\text{ kJ}}$$

$$\textbf{(b)}\ W = P(V_f - V_i) = P\left(\frac{V_i}{3} - V_i\right) = -\frac{2}{3}PV_i = -\frac{2}{3}(160\text{ kPa})(0.83\text{ m}^3) = \boxed{-89\text{ kJ}}$$

37. **(a)** Since the system is thermally insulated, the process is adiabatic; therefore

$$P_iV_i^{\gamma} = P_fV_f^{\gamma} \Rightarrow V_f = \left(\frac{P_i}{P_f}\right)^{\frac{1}{\gamma}} V_i = \left(\frac{115\text{ kPa}}{145\text{ kPa}}\right)^{3/5}(0.0750\text{ m}^3) = \boxed{0.0653\text{ m}^3}$$

(b) From the ideal gas equation of state we have

$$\frac{P_iV_i}{T_i} = \frac{P_fV_f}{T_f}$$

Therefore,

$$\frac{T_f}{T_i}=\left(\frac{P_f}{P_i}\right)\left(\frac{V_f}{V_i}\right)=\left(\frac{V_f}{V_i}\right)^{-\gamma}\left(\frac{V_f}{V_i}\right)=\left(\frac{V_f}{V_i}\right)^{1-\gamma}$$

$$\therefore\quad V_f=\left(\frac{T_f}{T_i}\right)^{1/(1-\gamma)}V_i=\left(\frac{295\text{ K}}{325\text{ K}}\right)^{1/(1-\frac{5}{3})}(0.0750\text{ m}^3)=\boxed{0.0867\text{ m}^3}$$

49. **(a)** For a Carnot engine

$$T_c=T_h(1-e_{max})=(545\text{ K})(1-0.300)=\boxed{382\text{ K}}$$

(b) The efficiency of a heat engine increases as the difference in temperature between the hot and cold reservoirs increases. Therefore, the temperature of the low-temperature reservoir must be decreased.

(c) $T_c=T_h(1-e_{max})=545\text{ K}(1-0.400)=\boxed{327\text{ K}}$

63. The change in entropy is given by

$$\Delta S=\left(\frac{Q}{T}\right)_{inside}+\left(\frac{Q}{T}\right)_{outside}$$

The rate, therefore, is

$$\frac{\Delta S}{t}=\left(\frac{Q/t}{T}\right)_{inside}+\left(\frac{Q/t}{T}\right)_{outside}=\frac{-20.0\text{ kW}}{(273.15+22)\text{ K}}+\frac{20.0\text{ kW}}{[273.15+(-14.5)]\text{ K}}=\boxed{9.6\text{ W/K}}$$

79. From Problem 37, for the adiabatic process,

$$T_hV_2^{\gamma-1}=T_cV_3^{\gamma-1}\quad\text{and}\quad T_hV_1^{\gamma-1}=T_cV_4^{\gamma-1}$$

Dividing these equations gives $\frac{V_2}{V_1}=\frac{V_3}{V_4}$

For the isothermal processes,

$$Q_h=W_1=nRT_h\ln\left(\frac{V_2}{V_1}\right)\quad\text{and}\quad Q_c=W_3=nRT_c\ln\left(\frac{V_3}{V_4}\right)=nRT_c\ln\left(\frac{V_2}{V_1}\right)$$

Dividing these equations gives $\frac{Q_h}{Q_c}=\frac{T_h}{T_c}$

This leads directly to

$$\boxed{e=1-\frac{Q_c}{Q_h}=1-\frac{T_c}{T_h}}$$

Answers to Practice Quiz

1. (a) **2**. (d) **3**. (e) **4**. (c) **5**. (b) **6**. (a) **7**. (c) **8**. (c) **9**. (e) **10**. (d)

Answers to Practice Problems

1. 9224.1 J	**2**. 7623.1 J
3. 7839.2 J	**4**. 38399.5 J
5. 67022.8 J	**6**. 0.59
7. 657 K	**8**. −2.74 J/K
9. 1.24 J/K	**10**. 69.14 J/K

CHAPTER 19

ELECTRIC CHARGES, FORCES, and FIELDS

Chapter Objectives

After studying this chapter, you should:

1. know the different types of electric charge and the magnitude of the smallest available charge.
2. know the difference between insulators and conductors.
3. be able to use Coulomb's law and apply it to situations involving many charges.
4. know the definition of, and basic uses for, the electric field.
5. be able to sketch electric field lines.
6. know how electrostatic shielding works.
7. be able to use Gauss's law to find the electric field in certain situations.

Warm-Ups

1. Can there be an electric field at a point where there is no charge? Can there be a charge at a place where there is no field? Please write a one- or two-sentence answer to each of these questions.
2. Let's say you are holding two tennis balls (one in each hand), and let's say these balls each have a charge Q. Estimate the maximum value of Q so that the balls do not repel each other so hard that you can't hold on to them.
3. Is it possible for two electric field lines to cross? If so, under what conditions? Do electric field lines ever end? If so, under what conditions?
4. Near its surface, the earth has an electric field that points straight down and has a magnitude of about 150 N/C. Estimate the charge that you would have to place on a basketball so that the electric force on the ball would balance its weight.

Chapter Review

This chapter begins our study of electricity and magnetism. Electric and magnetic forces are due to a phenomenon that we have not studied in previous chapters, namely, electric charge. The existence of electric charge and its consequences creates a completely new branch of physics that adds to the concepts and phenomena that we studied in previous chapters.

19–1 – 19–2 Electric Charge, Insulators, and Conductors

Through a series of experiments over many years we have now come to understand that we are all made up of smaller objects (atoms) that contain **electric charge**. To the best of our present knowledge, electric charge is a fundamental property of nature that comes in two types, called **positive charge** and **negative charge**. Most bulk materials contain an equal number of positive and negative charges and are said to be electrically **neutral** (or to have zero net charge).

Atoms consist of a small central nucleus that contains positively charged particles called **protons**. The nucleus is surrounded by an equal number of negatively charged particles called **electrons**. The electric charge of protons and electrons has the same magnitude

$$e = 1.60 \times 10^{-19}\ \mathrm{C}$$

An electron has a charge of –e, and protons carry a charge of +e. The SI unit of electron charge is called the **coulomb** (C). This is a new unit that is not derived from just [L], [M], and [T]. However, the coulomb is officially considered to be a derived unit; we will see its derivation in a later chapter. For now we will denote the dimension associated with electric charge as [C].

One of the properties of electric charge is that it is **quantized**. This fact means that charge comes only in discrete units. The smallest available charge is that of a proton or electron, e. Another property of electric charge is that it is conserved. Therefore, *in any physical process electric charge is never created or destroyed; the total electric charge of the universe remains constant.*

The positive and negative charges in an object can become separated, usually by the movement of electrons, so that one side of the object contains more of the negative charge, and the other side is left with more of the positive charge. Such objects are said to be **polarized**. In atoms (and molecules — bound groups of atoms) electrons can be completely removed, or extra electrons can be added. An atom with one or more electrons removed will have a net positive charge and is called a **positive ion**; an atom with extra electrons will have a net negative charge and is called a **negative ion**.

Important to our ability to make practical use of electricity is the fact that in some materials, called **conductors**, electrons are relatively free to move, while in other materials, called **insulators**, electrons are not very free to move. Metals are typically good conductors of electricity. Examples of commonly used insulators are rubber, plastic, and wood. There are also materials, called **semiconductors**, whose behavior is not clearly conducting or insulating. These materials can be manipulated to be more conducting in some situations and more insulating in others. The ability to manipulate semiconductors to control the flow of electricity has been a major triumph of modern technology.

Practice Quiz

1. Which of the following values is *not* a possible charge on an ion?

   (a) 1.60×10^{-19} C **(b)** 3.20×10^{-19} C **(c)** -1.60×10^{-19} C

   (d) 4.80×10^{-19} C **(e)** -2.80×10^{-19} C

19–3 Coulomb's Law

Electric charges exert forces one another. The rule for determining this force is called **Coulomb's law**. This law states that for two stationary point charges of magnitudes q_1 and q_2, the magnitude of the electrostatic force between them is proportional to the product $q_1 \cdot q_2$ and inversely proportional to the square of the distance r between them. Therefore, the magnitude of the force is given by

$$F = k\frac{q_1 q_2}{r^2}$$

where k is a proportionality that has the value

$$k = 8.99 \times 10^9 \, N \cdot m^2/C^2$$

Two additional rules allow us to determine the direction of the force. One rule is that *the force is directed along the line joining the charges*. The second rule is that *like charges repel each other and opposite charges attract each other*. Note that there are actually two forces, one exerted on q_1 by q_2 ($\mathbf{F}_{12}$) and one exerted on q_2 by q_1 ($\mathbf{F}_{21}$). These forces have equal magnitudes, given by F, and point in opposite directions in accordance with Newton's law of action and reaction. When more than two charges are involved, the force on any one charge can be determined using the principle of **superposition**. This principle states that the force on any charge is the vector sum of the forces that each of the other charges exerts on it individually.

In addition to individual point charges, it is also common to deal with continuous distributions of charge. An interesting example is that of an amount of charge Q distributed uniformly over the surface of a sphere. The force experienced by a point charge q outside the sphere works out, by superposition, to look just like Coulomb's law:

$$F = k\frac{qQ}{r^2}$$

where r, in this case, is the distance of q from the center of the sphere. When dealing with continuous distributions of charge it is often convenient to work with charge densities. In the case at hand we have a

surface charge density, σ, which is the charge per area ($\sigma = Q/A$) over the sphere. Thus, for an amount of charge Q spread over the surface of a sphere of radius R the surface charge density is given by

$$\sigma = \frac{Q}{4\pi R^2}$$

and its SI unit is C/m^2.

Example 19–1 Coulomb's Law and Superposition Given a configuration of three charged particles such that charge $q_1 = 5.3$ nC has (x, y) coordinates of (1.5 m, 0.0 m), charge $q_2 = 9.2$ nC is located at (0.0 m, 2.0 m), and charge $q_3 = -2.4$ nC is located at the origin, determine the net force on charge q_3.

Picture the Problem The diagram shows the charge configuration of q_1, q_2, and q_3.

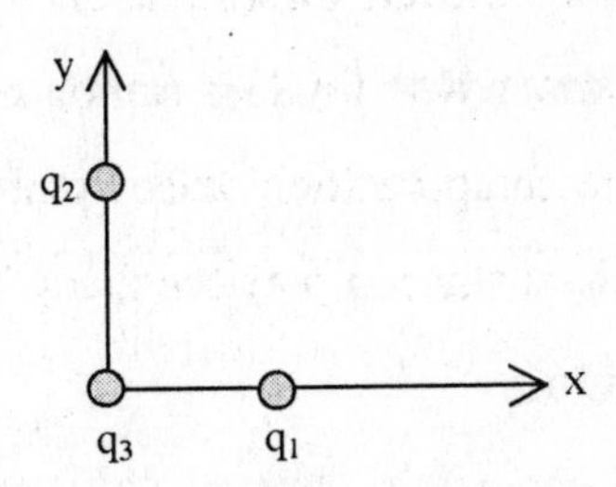

Strategy Our basic strategy, in keeping with the principle of superposition, is to determine the force on q_3 due to q_1 and q_2 separately using Coulomb's law, then calculate the net force on q_3 as the vector sum of the two.

Solution

1. From the given coordinates, the distance between q_1 and q_3 is: $r_{13} = 1.5$ m

2. Using Coulomb's law, the magnitude of the force on q_3 due to q_1 is:

$$F_{31} = \frac{k|q_1 q_3|}{r_{13}^2}$$

$$= \frac{(8.99\times 10^9\ \text{N}\cdot\text{m}^2/\text{C}^2)\left|(5.3\times 10^{-9}\ \text{C})(-2.4\times 10^{-9}\ \text{C})\right|}{(1.5\,\text{m})^2}$$

$$= 5.1\times 10^{-8}\ \text{N}$$

3. Since the charges have opposite signs, $\mathbf{F}_{31}$ must point toward q_1: $\mathbf{F}_{31} = (5.1\times 10^{-8}\ \text{N})\hat{\mathbf{x}}$

4. From the given coordinates, the distance between q_2 and q_3 is: $r_{23} = 2.0$ m

5. Using Coulomb's law, the magnitude of the force on q_3 due to q_2 is:

$$F_{32} = \frac{k|q_2q_3|}{r_{23}^2}$$

$$= \frac{(8.99\times10^9\ \text{N}\cdot\text{m}^2/\text{C}^2)|(9.2\times10^{-9}\ \text{C})(-2.4\times10^{-9}\ \text{C})|}{(2.0\text{m})^2}$$

$$= 5.0\times10^{-8}\ \text{N}$$

6. Since the charges have opposite signs $\mathbf{F}_{31}$ must point toward q_2:

$$\mathbf{F}_{32} = (5.0\times10^{-8}\ \text{N})\hat{\mathbf{y}}$$

7. The net force on q_3 is the sum of the two forces:

$$\mathbf{F}_3 = \mathbf{F}_{31} + \mathbf{F}_{32} = (5.1\times10^{-8}\ \text{N})\hat{\mathbf{x}} + (5.0\times10^{-8}\ \text{N})\hat{\mathbf{y}}$$

Insight Notice that because q_3 is a negative charge, we used the absolute value of the product of the charges in each application of Coulomb's law.

Practice Quiz

2. If two oppositely charged particles are placed side by side, which charge experiences the greater magnitude of force?
   (a) the positive charge
   (b) the negative charge
   (c) Neither charge experiences any force.
   (d) They experience forces of equal magnitude.
   (e) None of the above

3. If two oppositely charged particles are placed side by side with the positive charge to the left of the negative charge, what is the direction of the force on the positive charge?
   (a) to the left
   (b) to the right
   (c) perpendicular to the line joining them.
   (d) There is no force on the positive charge.
   (e) None of the above

4. A point charge of +Q placed a distance d away from the origin on the positive x-axis applies a force **F** on a charge q at the origin. If another charge of –Q is placed a distance d away from the origin on the negative x-axis, what will be the net force on q?
   (a) – **F**   **(b) F**   **(c)** – 2**F**   **(d)** 2**F**   **(e)** 0

19–4 – 19–5 The Electric Field and Electric Field Lines

In considering Coulomb's law you may have noticed that the two charges apply forces to each other even though they are not in physical contact. It is useful, and ultimately very important, to define an intermediary for the electrostatic force. This intermediary is called the **electric field**. We say that every charge creates an electric field in the space around it and that this electric field exerts a force on other charges placed within it. The electric field **E** is a vector quantity defined as the force per unit of positive charge at a given location:

$$\mathbf{E} = \mathbf{F}/q_0$$

Above, q_0 (imagined to be placed at the given location) represents a *test charge*, which is a positive charge so small in magnitude that its presence does not significantly alter the electric field in the region. The electric field is independent of q_0. The SI unit of electric field is N/C. Notice from the above definition that the electric field is defined in terms of the force on a positive charge. Therefore, a positive charge experiences a force in the direction of **E**, and a negative charge experiences a force in the opposite direction of **E**.

The electric field due to a point charge can easily be determined using Coulomb's law with one charge q as the source of the field and the other charge as a test charge. This configuration reveals that the magnitude of the electric field a distance r away from a point charge is given by

$$E = k\frac{q}{r^2}$$

To determine the electric field due to more than one charge we can use the principle of superposition. Each point charge contributes to the total electric field vector according to the above equation for the magnitude and the rule that electric fields points in the direction of the force on a positive charge. The total electric field is found by taking the vector sum of the electric fields due to each of the point charges.

Example 19–2 Electric Fields and Superposition A configuration of three charges whose values are $q_1 = q_2 = -q_3 = 3.88\ \mu C$ is shown below. Charge q_1 is located at (–3.00 m, –3.00 m), charge q_2 is located at (–3.00 m, 3.00 m), and charge q_3 is located at (3.00 m, 3.00 m). **(a)** Determine the magnitude and direction of the electric field at the origin. **(b)** Determine the magnitude and direction of the force on a –1.00 nC charge placed at the origin.

Picture the Problem The sketch shows of the charge configuration.

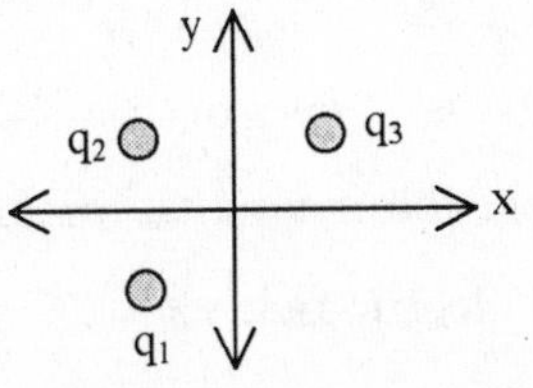

Strategy For part (a) we can use superposition to determine the electric field by adding the electric fields due to each charge. For part (b) we can use the definition of electric field to determine the force.

Solution

Part (a)

1. From the given coordinates, the squared distances of the charges from the origin are:

$$r_1^2 = (-3.00\,\text{m})^2 + (-3.00\,\text{m})^2 = 18\,\text{m}^2 = r_2^2 = r_3^2$$

2. Since all the charges have equal magnitudes and are at equal distances from the origin, the magnitude of the field due to each charge is:

$$E_1 = E_2 = E_3 = \frac{kq}{r^2}$$
$$= \frac{(8.99\times10^9\,\text{N}\cdot\text{m}^2/\text{C}^2)(3.88\times10^{-6}\,\text{C})}{18\,\text{m}^2} = 1938\ \text{N/C}$$

3. Each charge is an equal distance from both the x- and y-axes. Since electric fields point radially away from positive charges and radially toward negative charges, each field makes a 45° angle with the x- and y-axes:

$$\sin(45^\circ) = \cos(45^\circ) = 1/\sqrt{2}\ \therefore$$
$$\mathbf{E}_1 = \frac{1938\,\text{N/C}}{\sqrt{2}}[\hat{\mathbf{x}} + \hat{\mathbf{y}}]$$
$$\mathbf{E}_2 = \frac{1938\,\text{N/C}}{\sqrt{2}}[\hat{\mathbf{x}} - \hat{\mathbf{y}}]$$
$$\mathbf{E}_3 = \frac{1938\,\text{N/C}}{\sqrt{2}}[\hat{\mathbf{x}} + \hat{\mathbf{y}}]$$

4. The net electric field then is:

$$\mathbf{E} = \mathbf{E}_1 + \mathbf{E}_2 + \mathbf{E}_3$$
$$= \frac{1938\,\text{N/C}}{\sqrt{2}}(3\hat{\mathbf{x}} + \hat{\mathbf{y}}) = (4111\,\text{N/C})\hat{\mathbf{x}} + (1370\,\text{N/C})\hat{\mathbf{y}}$$

5. The magnitude of the electric field is:

$$E = \left(E_x^2 + E_y^2\right)^{1/2} = \left[(4111\,\text{N/C})^2 + (1370\,\text{N/C})^2\right]^{1/2}$$
$$= 4330\ \text{N/C}$$

6. The direction of the electric field is:

$$\theta_E = \tan^{-1}\left(\frac{E_y}{E_x}\right) = \tan^{-1}\left(\frac{1370\ \text{N/C}}{4111\ \text{N/C}}\right) = 18.4^\circ$$

Part (b)

7. By definition, the magnitude of the force on a charge in an electric field is:

$$F = |q|E = |-1.00\times10^{-9}\,\text{C}|(4333\ \text{N/C}) = 4.33\times10^{-6}\ \text{N}$$

8. Since the charge in question is negative, the direction of the force on it must be opposite that of the electric field: $\theta_F = 18.4^\circ + 180^\circ = 198^\circ$

Insight Be absolutely sure you understand how the result in step 3 relates to the position of the charges and the fact that the fields of point charges are radial.

A visual representation of electric fields can be obtained by drawing **electric field lines**. For an example, see Figure 19–16 in the text. The electric field lines can be drawn by following four rules:

1. Electric field lines are tangent to the electric field at every point;
2. Electric field lines start on positive charge (or at infinity);
3. Electric field lines end on negative charge (or at infinity);
4. The number of electric field lines is proportional to the magnitude of the source charge.

Exercise 19–3 Electric Field Lines **(a)** Sketch the electric field lines of a point charge of positive charge Q. **(b)** Sketch the electric field lines for a point charge of charge –2Q.

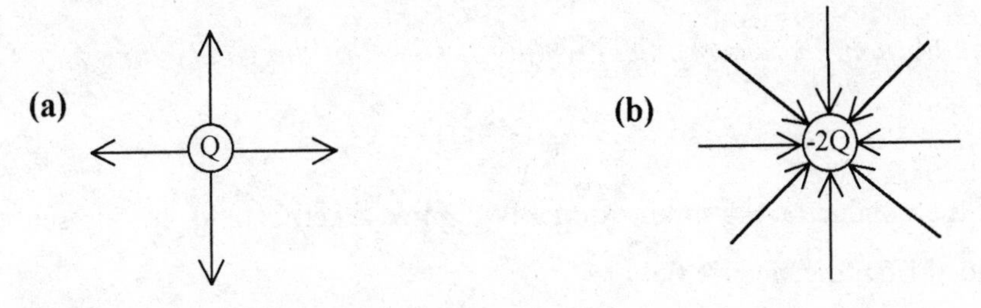

There are only two significant difference between the two sketches. (1) The field lines point away from the positive charge and toward the negative charge. (2) Twice as many lines are drawn for the charge with twice the magnitude.

Practice Quiz

5. The magnitude of the electric field a distance of 5.28×10^{-11} m away from a proton is

(a) 8.28×10^{-8} N/C

(b) 27.3 N/C

(c) 57.5 N/C

(d) 5.16×10^{11} N/C

(e) none of the above.

6. If, in a scaled drawing, 4 electric field lines are drawn around a point charge q_1 and 12 electric field lines are drawn around a different point charge q_2, this indicates that

   (a) the magnitude of q_1 is 12 times that of q_2

   (b) the magnitude of q_2 is 1/4 that of q_1

   (c) the magnitude of q_2 is 3 times that of q_1

   (d) q_1 and q_2 have equal magnitudes.

   (e) No statement can be made from this information.

19–6 Shielding and Charging by Induction

This section collects several useful facts related to the electrostatics of conductors. The first of these facts states what happens to excess charge on a conductor:

> *In electrostatic equilibrium, any excess charge on a conductor, whether positive or negative, must lie on the surface of the conductor.*

The next fact dealing with conductors has to do with the phenomenon called **electrostatic shielding**:

> *In electrostatic equilibrium, the electric field within the body of the conducting material must be zero.*

This fact remains true even if the conductor sits in an externally applied electric field. In this sense the inside of the conductor is shielded from the electric field.

The fact that charges are relatively free to move in conductors makes it possible for conductors to have a net positive or negative charge through a process called **charging by induction**. In this process a charged object (with charge Q) is brought near a neutral conductor. Free electrons in the conductor will be attracted to the side of the conductor nearer to, or pushed to the side farther from, the charged object (depending on the sign of Q). If the conductor is then **grounded** (that is, connected to an object like Earth that can give or receive a large supply of electrons) free electrons can either leave or flow onto the conductor, leaving it with a net electric charge once the conductor is disconnected from the ground. The sign of the net charge on the conductor will be opposite that of the object used to induce the charge.

Practice Quiz

7. If a conductor is placed in an external electric field, the electric field inside the body of the conductor

   (a) always equals zero

(b) never equals zero

(c) is non-zero only when electrostatic equilibrium is reached

(d) is zero only if it carries excess charge

(e) none of the above

19–7 Electric Flux and Gauss's Law

In this section we learn what some consider to be the fundamental law of electrostatic, **Gauss's law**. (We'll see later that this law is also important for electrodynamics as well.) Important to understanding this law is the concept of **electric flux**, which we review first.

In any region of space you can imagine a surface (or a membrane) of area A. An electric field **E**, assumed to be uniform in that region, may flow through this surface. The flow of the electric field through a surface is represented by the electric flux, Φ, of **E** through this surface. The extent to which **E** flows through the surface depends on the angle θ between the electric field vector and the direction normal to the surface. Specifically, we have that

$$\Phi = EA\cos\theta$$

Notice that the flux is not a vector quantity and that its SI unit is $N{\cdot}m^2/C$. If the surface is closed, such as the surface of a sphere, then the following sign convention for the flux Φ must be used:

- Φ is positive for electric field lines that leave the enclosed volume.
- Φ is negative for electric field lines that enter the enclosed volume.

Example 19–4 Electric Flux A uniform electric field of magnitude 250 N/C makes an angle of 35° with the normal to a circular aperture of radius 65 cm. If the aperture lies entirely in the electric field, what is the electric flux through the aperture?

Picture the Problem The perspective sketch shows a circular aperture, the electric field vector, and a vector **n̂** representing the direction normal to the surface.

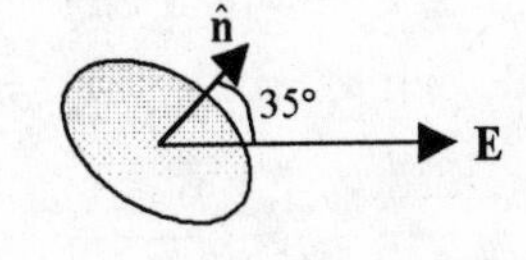

Strategy The expression for the electric flux through an area A can be used directly for this calculation. We need only to first calculate the area of the aperture.

Solution

1. The area of the circular aperture is: $A = \pi r^2$

2. The electric flux then is: $\Phi = E(\pi r^2)\cos\theta = (250\ \text{N/C})\pi(0.65\ \text{m})^2\cos(35°)$

$= 270\ \text{N}\cdot\text{m}^2/\text{C}$

Insight The main part of this problem is getting the picture correct. Once you correctly identify the situation being described the mathematical calculation is straightforward.

Charged objects create electric fields. Any surface that encloses a charged object will, therefore, have an electric flux through it. Gauss's law relates the electric flux through the surface to enclosed charge in the following way:

If a charge q is enclosed by an arbitrary surface, the electric flux through the surface is $\Phi = q/\varepsilon_0$.

The constant ε_0 is called the **permittivity of free space**; its value is

$$\varepsilon_0 = 8.85 \times 10^{-12} \frac{C^2}{N \cdot m^2}$$

and is related to the constant k in Coulomb's law by $k = 1/4\pi\varepsilon_0$.

Gauss's law can be very useful for calculating electric fields in certain cases where the source of the field is highly symmetric. When we enclose a charge with an imaginary surface for the purpose of using Gauss' law, we call this surface a **Gaussian surface**. For point charges and spherically symmetric charge distributions take the Gaussian surface to be a spherical surface centered on the charge (or center of the charge distribution) in question. For a flat sheet of charge take the Gaussian surface to be a cylindrical surface with its axis perpendicular to the sheet (the only nonzero flux will be through the top and bottom edges). For linear or cylindrical charge distributions also take the Gaussian surface to be a cylinder (the only nonzero flux will be through the sides).

Example 19–5 Gauss' Law An infinite thin sheet of charge has a uniform surface charge density of $-7.2\ \text{nC/m}^2$. Charge configurations of this type create uniform electric fields on each side of the sheet. Determine the electric field created by this sheet of charge.

Picture the Problem The sketch shows a section of the infinite sheet of charge, some of the electric

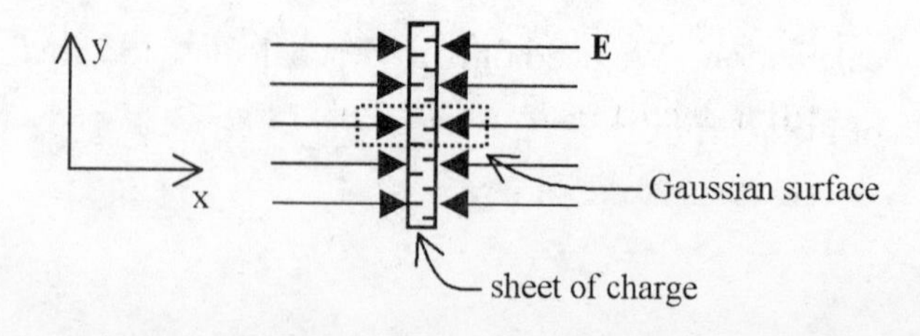

field lines, and a cylindrical Gaussian surface that will be used to solve this problem.

Strategy We apply Gauss's law to the cylindrical Gaussian surface of radius r to determine the electric field.

Solution

1. The only nonzero contributions to the electric flux through the Gaussian cylinder come from the two circular ends: $\Phi = -2EA = -2E\pi r^2$

2. Determine the charge enclosed by the Gaussian surface from the surface charge density: $\sigma = q/A \Rightarrow q = \sigma A = \sigma\pi r^2$

3. From Gauss's law we have: $\Phi = q/\varepsilon_0 \Rightarrow -2E\pi r^2 = \sigma\pi r^2 \Rightarrow E = -\sigma/2\varepsilon_0$

$$\therefore \quad E = -\frac{-7.2\times10^{-9}\ \text{C/m}^2}{2\left(8.85\times10^{-12}\ \text{C}^2/\text{N}\cdot\text{m}^2\right)} = 410\ \text{N/C}$$

4. The electric field on the right side is: $\mathbf{E}_{\text{right}} = -(410\ \text{N/C})\hat{\mathbf{x}}$

5. The electric field on the left side of the sheet is: $\mathbf{E}_{\text{left}} = (410\ \text{N/C})\hat{\mathbf{x}}$

Insight Notice that the electric field is discontinuous at the surface. You can see this same effect in Active Example 19–3 in the text. Can you explain why the results are different for this example and the one in the text?

Practice Quiz

8. An electric field is directed along the normal to a surface that lies in the field. If the area of the surface is doubled, what happens to the electric flux through the surface?

   (a) It doubles.

   (b) It's cut in half.

   (c) It increases by a factor of $\sqrt{2}$.

   (d) It decreases by a factor of $\sqrt{2}$.

   (e) It stays the same.

9. An electric field is directed at 90° to the normal to a surface that lies in the field. If the electric field is doubled, what happens to the electric flux through the surface?

   (a) It doubles.

   (b) It's cut in half.

   (c) It increases by a factor of $\sqrt{2}$.

   (d) It decreases by a factor of $\sqrt{2}$.

   (e) It stays the same.

10. A positive point charge is enclosed by a spherical surface of radius r producing an electric flux Φ through this surface. If this same charge is enclosed by a spherical surface with a radius of 2r, the electric flux through this larger surface will be

    (a) Φ **(b)** 2Φ **(c)** 4Φ **(d)** Φ^2 **(e)** zero

Reference Tools and Resources

I. Key Terms and Phrases

electric charge a property of particles that is important to the structure of atoms and molecule and acts as the source of a fundamental force of nature

protons positively charged particles, in atomic nuclei, that posses the smallest measurable electric charge

electrons negatively charged particles found in atoms that posses the smallest measurable electric charge

conductors materials, such as metals, in which electrons easily flow

insulators materials, such as rubber, in which electrons do not readily flow

semiconductors materials with electrical properties intermediate between conductors and insulators

Coulomb's law the force law between two point charges

superposition the principle that electrical forces and fields that result from multiple sources are obtained by vector addition of the results from the individual sources

charge density a measure of the compactness of electric charge that is especially useful with continuous charge distributions

electric field the electric force per unit of positive charge in space

electric field lines a pictoral representation of an electric field

electrostatic shielding the phenomenon that the electric field inside a conductor is zero, in static equilibrium, even when the conductor is exposed to an external electric field

electric flux a measure of the extent to which an electric field flows through an area

Gauss's law the law that relates the electric flux through a closed surface to the net charge it encloses

permittivity of free space the fundamental constant of electrostatics

gaussian surface an imaginary surface used for applying Gauss's law

II. Important Equations

Name/Topic	Equation	Explanation
Coulomb's law	$F = k\frac{q_1 q_2}{r^2}$	The force between two point charges.
electric field	$E = F/q_0$	The electric field is the force per unit positive charge.
electric flux	$\Phi = EA\cos\theta$	Electric flux measures the amount of electric field flowing through an area.
Gauss's law	$\Phi = q/\varepsilon_0$	Gauss's law relates electric charge to the flux generated by its electric field.

III. Know Your Units

Quantity	Dimension	SI Unit
electric charge (q)	[C]	C
electric field (E)	$[M][L][T^{-2}][C^{-1}]$	N/C
electric flux (Φ)	$[M][L^3][T^{-2}][C^{-1}]$	$\frac{N\cdot m^2}{C}$
permittivity of free space (ε_0)	$[C^2][T^2][M^{-1}][L^{-3}]$	$\frac{C^2}{N\cdot m^2}$

Practice Problems

1. Two charges are separated by a distance 0.73 meters, each having a charge 18.4 mC. To the nearest newton, what is the force on each?

2. In the figure the charge on the lower left is 7 mC, the charge on the lower right is 7 mC, and the upper charge is 4 mC. All distances are in meters. To the nearest newton, what is the force on the upper charge?

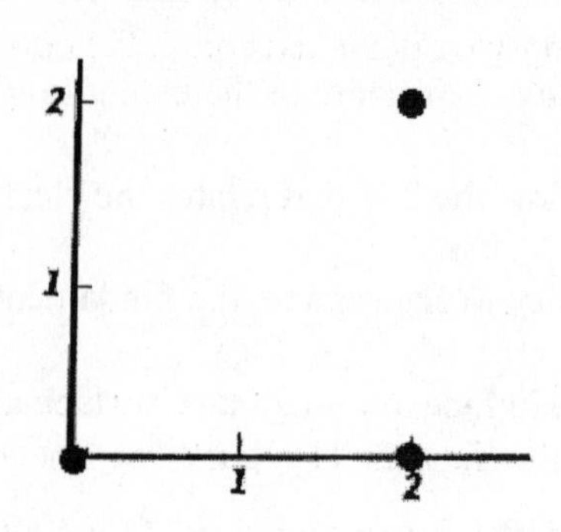

3. Using angles measured counterclockwise from east, to the nearest degree, what is the angle of the force in problem 2?

4. A negative charge of 7 μC and mass 5 μkg orbits a massive positive charge of 19 μC in a circular orbit of radius 5 meters. To the nearest tenth of a m/s, what is its speed?

5. A conducting sphere of radius R = 0.19 meters rests on a insulated column and has a negative uniform charge density of 5.9 μC/m^2. What is the force, to the nearest tenth of a mN on a charge q = 0.71 μC at point p a distance of r = 0.37 meters?

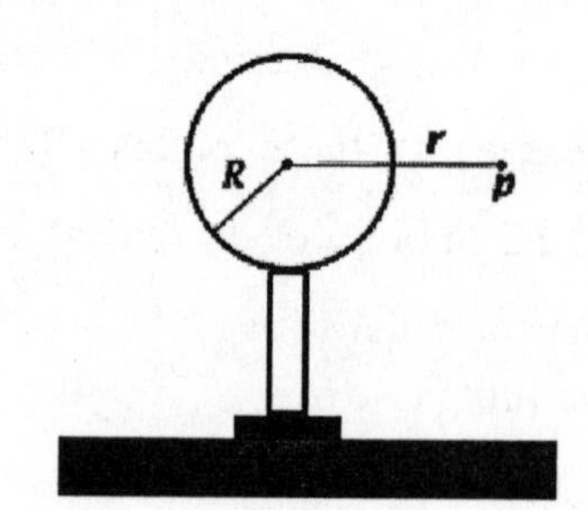

6. In problem 5 what is the magnitude of the electric field, to the nearest N/C?

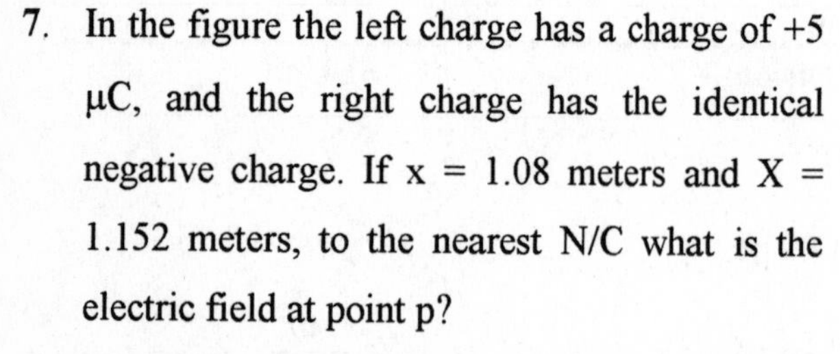

7. In the figure the left charge has a charge of +5 μC, and the right charge has the identical negative charge. If x = 1.08 meters and X = 1.152 meters, to the nearest N/C what is the electric field at point p?

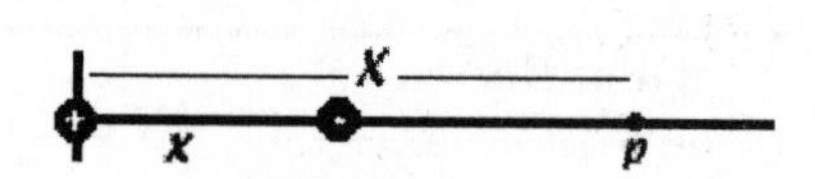

8. In problem 7 if the charge on the left charge is five times that on the right charge, what must be the value of X, to the nearest mm, for the electric field to be zero?

9. A capacitor has an electric field of 18 x 10^5 N/C. To four digits what is the charge per square mm?

10. A 3.1-μC charge hangs from a string in the capacitor and makes an angle of 11.5 degrees with the vertical. To the nearest thousandth of a newton, what is the tension in the cord?

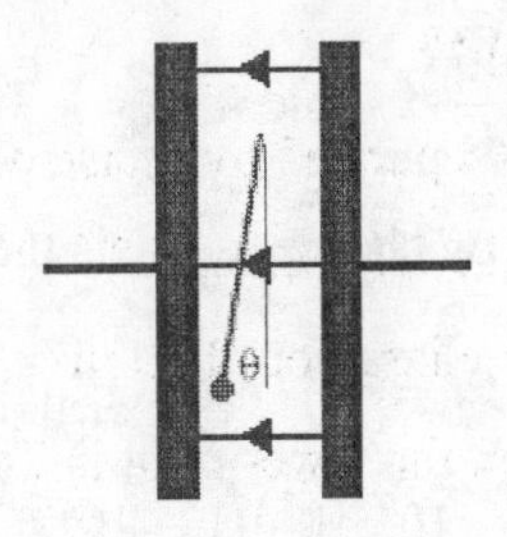

Puzzle

SYMMETRY

Consider the figure. The positively charged balls are located at (-1,1,0) and (1,-1,0) and each carries a charge +Q. The negatively charged balls are located at (1,1,0) and (-1,-1,0), and each carries a charge -Q. There are three specially marked points: O is the origin, P1 is (0,0,2), and P2 is (2,0,0). There are no charges at these points. Rank the points O, P1, and P2 in order of decreasing electric field strength. The correct answer is:

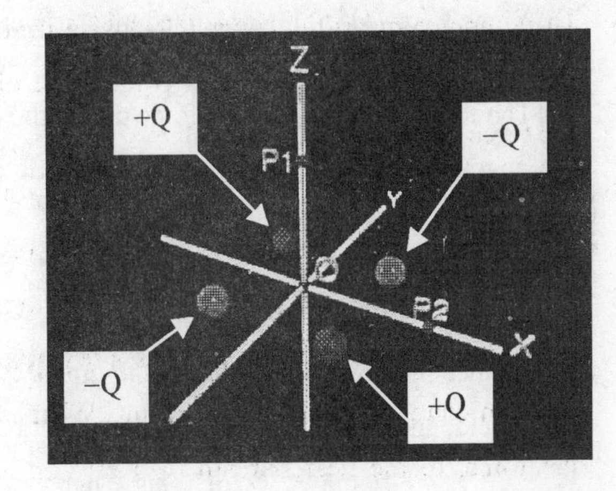

1. P1 > P2 > O
2. O > P1 > P2
3. P1 > O >P2
4. O > P2 > P1
5. O > P2 = P1
6. P2 > P1 = O
7. P2 > P1 > O
8. P1 = P2 > O
9. The fields are all equal.
10. None of the above

Selected Solutions

13. (a) Let the x-axis be along the line of the three charges with the positive direction pointing from q_2 to q_3. Then, because opposite charges attract, we have that F_{21} points in the negative x direction and F_{23} points in the positive x direction. When combined with Coulomb's law for the magnitudes, this condition gives

$$\mathbf{F}_{21} = -k\frac{q_2 q_1}{d^2}\hat{\mathbf{x}} \quad \text{and} \quad \mathbf{F}_{23} = k\frac{q_2 q_3}{d^2}\hat{\mathbf{x}}$$

The net force on charge q_2, $\mathbf{F}_2$, is the vector sum of these two forces:

$$\mathbf{F}_2 = \mathbf{F}_{21} + \mathbf{F}_{23} = \frac{k}{d^2}[-q_2 q_1 + q_2 q_3]\hat{\mathbf{x}} = \frac{k}{d^2}[-(2.0q)q + (2.0q)(3.0q)]\hat{\mathbf{x}}$$

$$= \frac{\left(8.99\times10^9\ \frac{\text{N}\cdot\text{m}^2}{\text{C}^2}\right)(12\times10^{-6}\ \text{C})^2}{(0.16\text{m})^2}(-2.0+6.0)\hat{\mathbf{x}} = (200\ \text{N})\hat{\mathbf{x}}$$

Therefore, the net electrostatic force exerted on q_2 is $\boxed{\text{200 N toward } q_3}$.

(b) Since $F_2 \propto 1/d^2$, if d was tripled, the answer to part (a) would $\boxed{\text{decrease by a factor of } 1/3^2\text{, or } 1/9}$.

31. **(a)** The forces balance. So, the force due to the electric field must be opposite that due to gravity, and since the charge is negative, the electric field must be directed downward.

$$qE = mg \quad \therefore \quad E = \frac{mg}{q} \quad \Rightarrow \quad \mathbf{E} = \frac{mg}{q}(-\hat{\mathbf{y}}) = -\frac{(0.012\,\text{kg})\left(9.81\,\frac{\text{m}}{\text{s}^2}\right)}{3.6\times10^{-6}\,\text{C}}\hat{\mathbf{y}} = \boxed{(-3.3\times10^4\ \text{N/C})\hat{\mathbf{y}}}$$

(b) Since the downward force due to gravity was balanced by the upward force due to the electric field, and since the charge on the object has now increased, the acceleration will be upward.

$$F_q - F_g = ma \quad \therefore \quad 2qE - mg = ma$$

$$2q\left(\frac{mg}{q}\right) - mg = ma \quad \therefore \quad mg = ma \quad \Rightarrow \quad a = g$$

Thus, the acceleration is $\boxed{\left(9.81\ \text{m/s}^2\right)\hat{\mathbf{y}}}$.

37. **(a)** The electric field lines begin at q_1 and q_3 and end at q_2. Also, electric field lines start at positive charges or at infinity and end at negative charges or at infinity. So, since q_2 is negative, q_1 and q_3 must be $\boxed{\text{positive}}$.

(b) q_1 has 8 lines leaving it. q_2 has 16 lines entering it. Since 8 is half of 16, and since the number of lines entering or leaving a charge is proportional to the magnitude of the charge, the magnitude of q_1 is half of q_2, or $\boxed{5.00\ \mu\text{C}}$.

(c) By the reasoning of part (b), the magnitude of q_3 is $\boxed{5.00\ \mu\text{C}}$.

39.

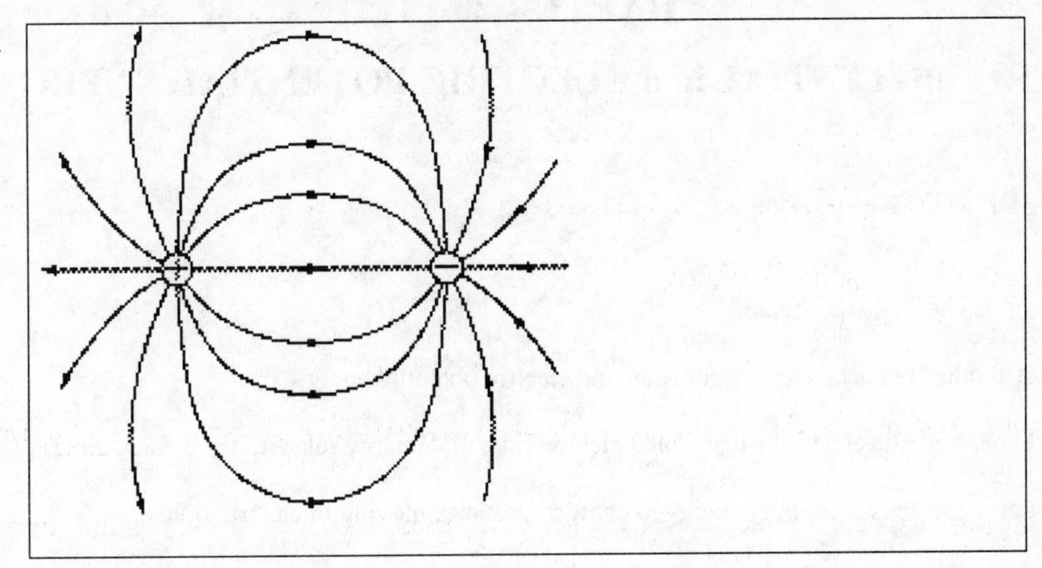

47. According to Gauss's law $\Phi = q/\varepsilon_0$, where $\Phi = \Phi_1 + \Phi_2 + \Phi_3 + \Phi_4 + \Phi_5 + \Phi_6$; therefore,

$$q = \varepsilon_0 \Phi = \left(8.85 \times 10^{-12} \frac{C^2}{N \cdot m^2}\right)(150.0 + 250.0 - 350.0 + 175.0 - 100.0 + 450.0)\frac{N \cdot m^2}{C} = \boxed{5.09 \times 10^{-9}\, C}.$$

Answers to Practice Quiz

1. (e) **2**. (d) **3**. (b) **4**. (d) **5**. (d) **6**. (c) **7**. (e) **8**. (a) **9**. (e) **10**. (a)

Answers to Practice Problems

1. 5711493 N
2. 88037 N
3. 75°
4. 218.7 m/s
5. 124.8 mN
6. 175,762 N/C
7. –8,637,040 N/C
8. 1954 mm
9. 15.93 C/mm^2
10. 27.988 N

CHAPTER 20

ELECTRIC POTENTIAL and ELECTRIC POTENTIAL ENERGY

Chapter Objectives

After studying this chapter, you should:

1. know the difference between electric potential and electric potential energy.
2. understand how both the electric potential and electric potential energy relate to the electric field.
3. be able to apply the conservation of energy to charged particles moving in electric fields.
4. be able to determine the electric potential of a configuration of point charges.
5. understand the relationship between equipotential surfaces and electric fields.
6. be able to calculate the capacitance of, electric field of, and energy stored in a parallel-plate capacitor with and without a dielectric.

Warm-Ups

1. Suppose you are given three capacitors consisting of aluminum disks configured as in the figure. You put equal amounts of positive charge on the top disks of each of the three sets and corresponding amounts of negative charge on the bottom disks. If you were to measure the electric fields between the disks, which set would give you the highest value? Which one the lowest? How about the voltages?

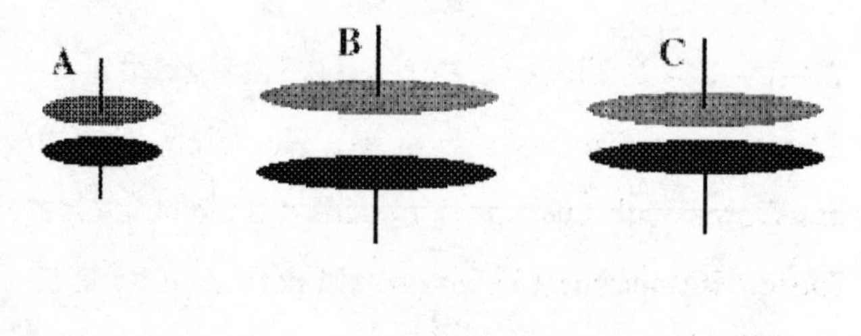

2. Electronic flashes in cameras flash when a capacitor is discharged through a material that produces light energy when electric charge passes through it. This energy comes from a battery inside the flash unit. Estimate the capacitance of the capacitor used to store the energy. (Hint: Assume that the flash gives you the same amount of light as a 500-watt bulb that is turned on for a 60th of a second).

3. A fresh "D cell" battery can provide about 5000 J of electrical energy before it must be discarded. Estimate the number of coulombs that must pass through it during its lifetime.
4. The voltage across a capacitor is given by the formula V=Q/C, where Q is usually called "the charge on the capacitor." Where is this charge in a capacitor? Does the capacitor really have a net charge? If two capacitors are connected together, is the charge on one the same as on the other?
5. You and a close friend stand facing each other. You are as close as you can get without actually touching. If a wire is attached to each of you, you can act as the two conductors in a capacitor. Estimate the capacitance of this "human capacitor."
6. Contour maps like the one to the right show equally spaced lines of constant elevation at Earth's surface. Explain in what sense they are "equally spaced." Also, explain in what sense these lines could be called gravitational equipotentials

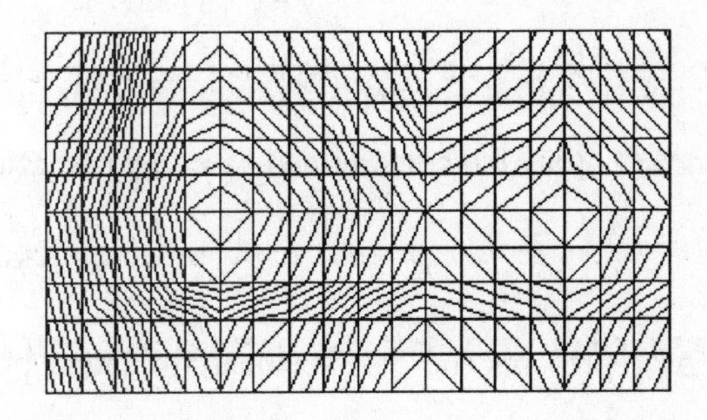

7. Estimate the total charge that passes through a lightbulb in one second.

Chapter Review

In this chapter we study the potential energy associated with the electric force and the electric field. This chapter also includes our first detailed treatment of an important electrical device called a capacitor.

20–1 – 20–2 Electric Potential Energy, Electric Potential, and Energy Conservation

The electric force is a conservative force. This means that it is useful to define a potential energy associated with this force; we call it the **electric potential energy**. As always, only changes in potential energy are measurable, and so we define only the change in potential energy, ΔU. The change in electric potential energy is defined as the negative of the work done by the electric force

$$\Delta U = -W_E$$

Since work done depends on the force applied, and the electric force, in turn, depends on the charge on which it is applied, the change in potential energy must depend on the charge. Recall that the concept of the electric field (force per unit charge) gives us a way to handle information about the electric force without reference to any specific charge; in a similar way we can define a quantity, the **electric potential**, to handle information about the energy without reference to a specific charge. The electric potential difference ΔV is defined as

$$\Delta V = \frac{\Delta U}{q_0}$$

where q_0 is a test charge that moves from one point to another. The SI unit of electric potential is J/C and is called a **volt** (V): 1V = 1 J/C. Often, instead of electric potential, the quantity V is called the *voltage*. The above definition also implies that $\Delta U = q\,\Delta V$; a commonly used unit of energy based on this relation is the **electron-volt** (eV). An eV is the energy change experienced by an electron or proton when it accelerates through a potential difference of 1 V:

$$1\,\text{eV} = \left(1.60\times10^{-19}\,\text{C}\right)\left(1\text{V}\right) = 1.60\times10^{-19}\,\text{J}$$

The above definitions imply a direct relationship between ΔV and the electric field E. It works out that the electric field can be determined from the rate at which the electric potential changes with position,

$$E = -\frac{\Delta V}{\Delta s}$$

where Δs represents the displacement from one position to another. The minus sign indicates that the electric field points in the direction of decreasing electric potential. This means that positive charges accelerate in the direction of decreasing electric potential and negative charges accelerate in the direction of increasing electric potential. The preceding relation also shows that an alternative unit for electric field could be V/m instead of N/C. In fact, both units of E are in common use; 1 N/C = 1 V/m.

That we can define an electric potential energy also means that we can extend the conservation of mechanical energy to the motion of charged particles in electric fields; therefore, we can write that the sum of the kinetic and electric potential energies must be constant: $K_i + U_i = K_f + U_f$. At a given location U = qV, therefore, the above result can also be written in terms of the electric potential as

$$K_i + qV_i = K_f + qV_f$$

Example 20–1 Motion in a Constant Electric Field A particle of charge 22.4 mC and mass 57.2 mg is initially held at rest in a constant electric field of 133 V/m. If the particle is released and accelerates through a distance of 14.9 m, **(a)** determine its change in potential energy, **(b)** determine the change in electric potential it experiences, and **(c)** use energy conservation to determine its final speed.

Picture the Problem The diagram shows the electric field lines and the charged particle.

Strategy The given information allows us to calculate the work on the particle. By knowing the work we can use the definitions of electric potential energy and electric potential to determine the answers.

Solution

Part (a)

1. Since the force and displacement are in the same direction, the work done by the electric force is: $W_E = F_E d = qEd$

2. By definition, the change in potential energy must be: $\Delta U = -W_E = -qEd = -(22.4\times10^{-3}\text{ C})(133\ \tfrac{N}{C})(14.9\text{ m}) = -44.4\text{ J}$

Part (b)

3. From the definition of electric potential, we obtain ΔV: $\Delta V = \dfrac{\Delta U}{q} = \dfrac{-44.39\text{ J}}{0.0224\text{ C}} = -1.98\text{ kV}$

Part (c)

4. According to energy conservation: $\Delta K = -\Delta U \Rightarrow K_f - 0 = -\Delta U \ \therefore\ \frac{1}{2}mv_f^2 = -\Delta U$

5. Solving for the final speed gives: $v_f = \sqrt{\dfrac{-2(\Delta U)}{m}} = \sqrt{-\dfrac{2(-44.39\text{ J})}{57.2\times10^{-6}\text{ kg}}} = 1.25\times10^3\text{ m/s}$

Insight For convenience I switched the unit of electric field from V/m to N/C; remember V/m = N/C.

Practice Quiz

1. An electric force does 25.0 J of work in accelerating a particle of charge −15.0 μC. What is the change in the electric potential energy of the charge?

   (a) 15.0 J **(b)** 25.0 J **(c)** −15.0 J **(d)** −25.0 J **(e)** 0 J

2. An electric force does 25.0 J of work in accelerating a particle of charge −15.0 μC. What change in electric potential does the charge experience?

   (a) 1.67 MV **(b)** −1.67 MV **(c)** −25.0 MV **(d)** 25.0 MV **(e)** 0 MV

3. An electric force does 25.0 J of work in accelerating a particle of charge −15.0 μC. What is the change in the kinetic energy of the charge?

   (a) 15.0 J **(b)** 25.0 J **(c)** −15.0 J **(d)** −25.0 J **(e)** 0 J

20–3 The Electric Potential of Point Charges

A detailed treatment of point charges using the definitions of the previous section reveals that for a test charge moving from point B to A relative to a charge q fixed at the origin, the change in electric potential energy is

$$\Delta U = kqq_0\left(\frac{1}{r_A} - \frac{1}{r_B}\right)$$

In order to assign a value of potential energy to a given charge configuration a reference configuration must be chosen at which we set U = 0. For point charges, this reference is typically chosen to be infinitely far away from the charge. With this choice of reference understood, we can say that the value of the potential energy of two point charges q and q_0 separated by a distance r is

$$U = \frac{kqq_0}{r}$$

Applying the definition of the electric potential to the above results for the potential energy suggests that the value of the electric potential a distance r away from a point charge q can be taken as

$$V = \frac{kq}{r}$$

This result is for the electric potential of a single point charge. For a system of two or more point charges we can use the fact that the electric potential obeys a principle of superposition appropriate for a scalar quantity:

The electric potential due to two or more point charges equals the algebraic sum of the potentials due to each point charge individually.

The electric potential energy of a system of three or more point charges obeys a similiar superposition principle:

The electric potential energy of a system of three or more point charges equals the algebraic sum of the potential energy of each pair of charges.

Example 20–2 Potential and Potential Energy A configuration of three charged particles is such that charge q_1 = 5.3 nC has (x, y) coordinates of (1.5 m, 0.0 m), charge q_2 = 9.2 nC is located at (0.0 m, 2.0 m), and charge q_3 = –2.4 nC is located at the origin. **(a)** Determine the electric potential at position P =

(1.5 m, 2.0 m). **(b)** Determine the change in potential energy if a 7.4-nC charge is placed at P from infinitely far away.

Picture the Problem The diagram shows the charge configuration of q_1, q_2, and q_3. The point P is indicated by the open circle.

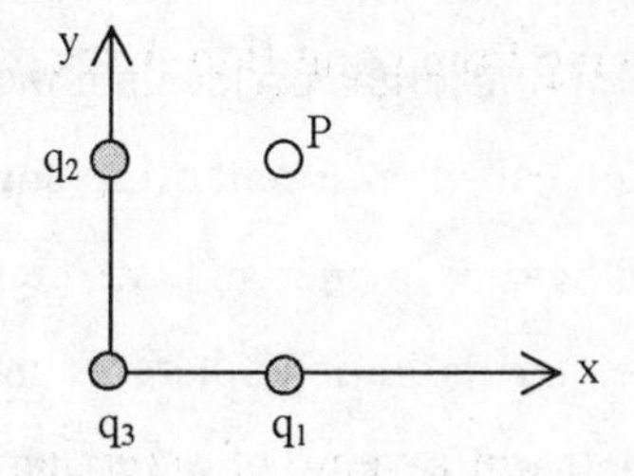

Strategy We can solve this problem by using the result for the potential energy due to a point charge together with the principle of superposition.

Solution

Part (a)

1. The potential at P is the sum of the potentials due to each charge: $V_P = V_1 + V_2 + V_3 = k\left(\frac{q_1}{r_1} + \frac{q_2}{r_2} + \frac{q_3}{r_3}\right)$

2. The distances can be determined from the coordinates: $r_1 = 2.0 \text{ m}, \quad r_2 = 1.5 \text{ m}, \quad r_3 = \sqrt{(1.5 \text{ m})^2 + (2.0 \text{ m})^2}$

3. The potential at P then is: $V_P = \left(8.99 \times 10^9 \frac{\text{N}\cdot\text{m}^2}{\text{C}^2}\right)\left(\frac{5.3 \text{ nC}}{2.0 \text{ m}} + \frac{9.2 \text{ nC}}{1.5 \text{ m}} - \frac{2.4 \text{ nC}}{2.5 \text{ m}}\right)$ $= 70 \text{ V}$

Part (b)

4. We can use the above result to get the change in potential energy if we bring in the fourth charge: $\Delta U = q(\Delta V) = (7.4 \text{ nC})(70.3 \text{ V}) = 5.2 \times 10^{-7} \text{ J}$

Insight In part (b) most of the information for getting ΔU was already determined when we calculated V at P. The result for V_P equaled ΔV for the fourth charge because at infinity $V = 0$.

Practice Quiz

4. Two point charges, $q_1 = 23\ \mu\text{C}$ and $q_2 = -75\ \mu\text{C}$, are moved such that the distance between them changes from 17.3 m to 4.77 m. What is the change in the potential energy of this system of charges?

   (a) –2.4 J (b) 2.4 J (c) – 3.3 J (d) – 0.90 J (e) 0.90 J

5. Two point charges, $q_1 = 23\ \mu\text{C}$ and $q_2 = -75\ \mu\text{C}$, are separated by a distance of 13 m. What is the electric potential at the midpoint between them?

   (a) 0 V (b) –100 kV (c) –32 kV (d) – 36 kV (e) –72 kV

20–4 Equipotential Surfaces and the Electric Field

Recall that to get a visual feel for the electric field of a configuration of charges we can draw electric field lines. In a similar way, we can obtain a visual representation of the electric potential due to a configuration of charges by locating the surfaces on which the potential will have the same value. These surfaces are called **equipotential surfaces**. In two-dimensional drawings these surfaces are often represented by simple lines and are often called *equipotential lines*, or just *equipotentials*.

Because, by definition, there is no change in electric potential when a charge moves along an equipotential surface, we know that no work is done on this particle by the electric force; therefore, the electric field must not have any component along an equipotential surface. This result leads to the useful fact that

the electric field is always perpendicular to the equipotential surfaces.

That charged particles on the surfaces of conductors configure themselves such that they are free to move indicates that conducting surfaces are also equipotential surfaces. As a result, of this fact, charge becomes more concentrated near sharp edges on a conductor than near blunt ends. The above discussion implies that when a conducting surface is in an electric field, the electric field must be perpendicular to the surface in static equilibrium.

Exercise 20–3 Estimating the Electric Field Two equipotential surfaces of 28.2 V and 29.9 V are separated by 3.36 mm. Estimate the magnitude of the average electric field between the surfaces.

Solution We are given the following information:

Given: $V_1 = 28.2$ V, $V_2 = 29.9$ V, $\Delta x = 3.36$ mm; **Find**: E

Here we can make direct use of the relationship between the electric field and the electric potential, however, since we only want the magnitude of the electric field, we can ignore minus signs:

$$E = \frac{\Delta V}{\Delta x} = \frac{V_2 - V_1}{\Delta x} = \frac{29.9\text{ V} - 28.2\text{ V}}{3.36 \times 10^{-3}\text{ m}} = 506\text{ V/m}$$

Practice Quiz

6. A particle of charge −15.0 μC is moved a distance of 2.00 m along an equipotential surface of value −24.0 V. What is the change in the potential energy of the charge?

   (a) 3.60×10^{-4} J **(b)** 0 J **(c)** 7.20×10^{-4} J **(d)** 12.0 J **(e)** none of the above

20–5 – 20–6 Capacitors, Dielectrics, and Electrical Energy Storage

A **capacitor** — an important device used in many electrical applications — consists of two conductors separated by some distance. The conductors that make up a capacitor are generically called *plates*. Capacitors store charge and electrical energy when a potential difference V is applied across the plates. Equal and opposite amounts of charge is stored on each plate. The magnitude of charge Q that can be stored is characterized by its **capacitance**, C, which is defined as

$$C = \frac{Q}{V}$$

The charge Q and the potential difference V are usually related to each other such that the capacitance typically does not depend on either. Rather, the capacitance is determined by the size, shape, and distance between the conductors. As can be seen from the preceding expression, the SI unit of capacitance is C/V, which is called a farad (F): 1 F = 1 C/V. In practice, 1 F is a large amount of capacitance; therefore, we usually work with capacitances in the range between picofarad (1 pF = 10^{-12} F) and microfarad (1 μF = 10^{-6} F).

A prototype for studying capacitance is the parallel-plate capacitor, which consists of two flat, parallel plates, each of area A, separated by a distance d. Two useful facts about parallel-plate capacitors are (a) the electric field between the plates is uniform (except near the edges), and its magnitude is given by

$$E = \frac{Q}{\varepsilon_0 A}$$

and (b) the capacitance of a parallel-plate capacitor is given by

$$C = \frac{\varepsilon_0 A}{d}$$

These results assume that there is vacuum between the plates of the capacitor. A common practice that increases the capacitance of a capacitor is to insert an insulating material, called a *dielectric*, between the plates. When a dielectric is present in a capacitor, polarization of the molecules that make up the dielectric gives rise to a reduced electric field between the plates:

$$E = \frac{E_0}{\kappa}$$

where E_0 is the electric field in vacuum, and κ is a dimensionless quantity greater than 1, called the **dielectric constant**, that depends on the material being used. Similarly, in the presence of a dielectric the potential difference between the plates is reduced from the value V_0 in vacuum by a factor of κ:

$$V = \frac{V_0}{\kappa}$$

These results imply that the capacitance is increased in the presence of a dielectric by a factor of κ:

$$C = \kappa C_0$$

For the particular case of a paralle-plate capacitor containing a dielectric we have

$$C = \frac{\kappa \varepsilon_0 A}{d}$$

It takes energy to store an amount of charge Q on an initially uncharged capacitor; this energy is stored as electric potential energy in the capacitor. The amount of energy U (called E in the text) stored in a capacitor that holds a charge Q for a potential difference V across its plates can be determined by three equivalent expressions:

$$U = \tfrac{1}{2}QV = \tfrac{1}{2}CV^2 = \frac{Q^2}{2C}$$

This energy is stored in the electric field that exists between the plates of the capacitor. The relationship between the stored energy and the electric field is most conveniently written in terms of the energy density (energy per volume) of the electric field:

$$\text{energy density} = \tfrac{1}{2}\varepsilon_0 E^2$$

Example 20–4 A Parallel-Plate Capacitor A certain parallel-plate capacitor has plates of area 13.7 cm^2 that are separated by a distance of 1.55 cm. If a material with a dielectric constant of 2.0 exists between the plates, and a potential difference of 2.50 V is applied across them, **(a)** how much charge will the plates hold, and **(b)** how much energy is stored in the capacitor?

Picture the Problem The sketch shows a side view of the plates of a parallel-plate capacitor containing a dielectric.

Strategy Because it is a parallel-plate capacitor we know how to calculate the capacitance; since we are given the potential difference, we can determine the charge and energy stored.

Solution

Part (a)

1. The capacitance is given by: $C = \frac{\kappa \varepsilon_0 A}{d}$

2. Therefore charge stored is:

$$Q = VC = \frac{V\kappa\varepsilon_0 A}{d}$$

$$= \frac{(2.50\text{ V})2.0\left(8.85\times10^{-12}\,\tfrac{C^2}{N\cdot m^2}\right)\left(13.7\times10^{-4}\,m^2\right)}{1.55\times10^{-2}\text{ m}}$$

$$= 3.91\text{ pC}$$

Part (b)

3. The energy stored is:

$$U = \tfrac{1}{2}QV = \tfrac{1}{2}\left(3.911\times10^{-12}\text{ C}\right)\left(2.50\text{ V}\right) = 4.9\text{ pJ}$$

Insight We did not need a numerical calculation of the capacitance, so to eliminate the possibility of round-off error, we did not do one.

Practice Quiz

7. A 2.0-μF capacitor is connected across a 4.0-V battery. How much charge is stored on the negative plate?

   (a) −8.0 μC (b) −0.50 μC (c) −6.0 μC (d) −2.0 μC (e) −1.0 μC

8. Which of the following statements is *not* correct for an isolated charged capacitor?

   (a) Inserting a dielectric into a previously empty capacitor decreases the potential difference across it.
   (b) Inserting a dielectric into a previously empty capacitor decreases the electric field across it.
   (c) Inserting a dielectric into a previously empty capacitor increases the capacitance.
   (d) Inserting a dielectric into a previously empty capacitor allows it to store more charge.
   (e) None of the above

9. A certain parallel-plate capacitor stores an amount of energy U between its plates. If the potential difference across it is doubled, the energy stored equals

   (a) U (b) 2U (c) 4U (d) U/2 (e) U/4

10. A certain parallel-plate capacitor stores an amount of energy U between its plates. If the charge stored on it is doubled, the energy stored equals

    (a) U (b) 2U (c) 4U (d) U/2 (e) U/4

Reference Tools and Resources

I. Key Terms and Phrases

electric potential energy the potential energy associated with the electric force

electric potential the electric potential energy per unit charge

volt the SI unit of electric potential

electron-volt a unit of energy equal to 1.60×10^{-19} J

equipotential surface a surface on which every point has the same value of electric potential

capacitor an electrical device, used to store charge and energy, that consists of a two conductors separated by a finite distance

capacitance a measure of a capacitor's ability to store charge

dielectric an insulating material often used between the plates of a capacitor

energy density energy per unit volume

II. Important Equations

Name/Topic	Equation	Explanation
electric potential energy	$\Delta U = -W_E$	The definition of electric potential energy.
electric potential difference	$\Delta V = \frac{\Delta U}{q_0}$	The definition of electric potential.
electric potential	$V = \frac{kq}{r}$	The electric potential of a point charge q assuming that V = 0 at r = ∞.
capacitors	$C = \frac{Q}{V}$	The definition of capacitance.
	$C = \frac{\kappa\varepsilon_0 A}{d}$	The capacitance of a parallel-plate capacitor with a dielectric.
	$U = \frac{1}{2}QV = \frac{1}{2}CV^2 = \frac{Q^2}{2C}$	The energy stored in a capacitor.

III. Know Your Units

Quantity	Dimension	SI Unit
electric potential (V)	$[M][L^2][T^{-2}][C^{-1}]$	V
capacitance (C)	$[C^2][T^2][M^{-1}][L^{-2}]$	F

Practice Problems

1. If the battery voltage in the figure is 12.48 volts and the plate separation is 0.49 cm, what is the electric field, to the nearest N/C?
2. If a charge of 5.7 mC moves from the upper plate to the lower one, what is its change in electric potential energy, to the nearest hundredth of a mJ?

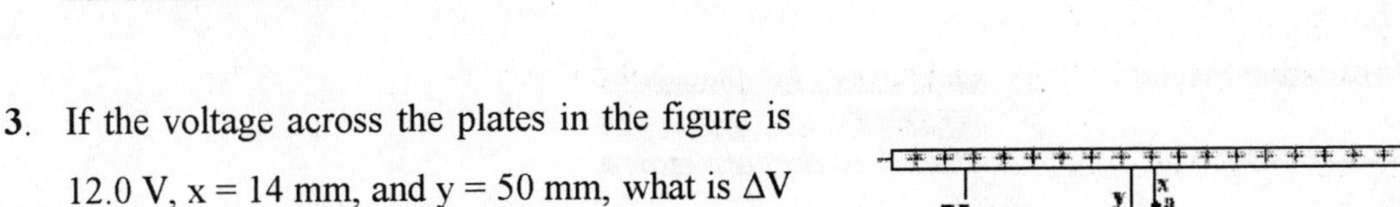

3. If the voltage across the plates in the figure is 12.0 V, x = 14 mm, and y = 50 mm, what is $\Delta V = V_B - V_A$, to the nearest hundredth of a volt?
4. In problem 3 what is the electric field at b to the nearest N/C?
5. In problem 3 if a particle with charge 11.27 mC and 11.82 g is released from rest at the upper plate what will be its velocity when it reaches the lower plate, to the nearest tenth of a m/s?

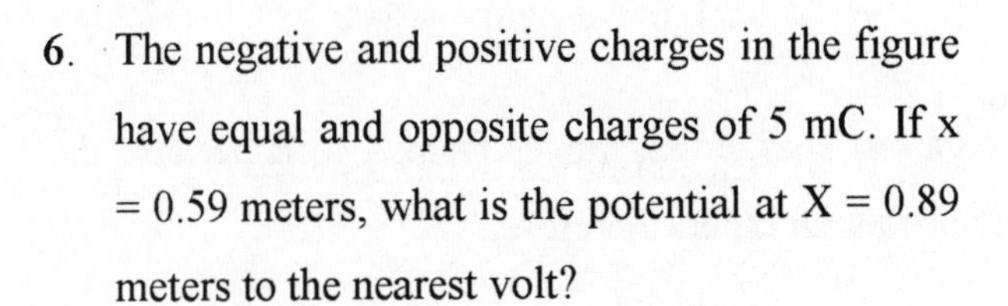

6. The negative and positive charges in the figure have equal and opposite charges of 5 mC. If x = 0.59 meters, what is the potential at X = 0.89 meters to the nearest volt?

7. The upper charge has a charge of –4 mC, the charge on the lower left is 5 mC, and the charge on the lower right is –2 mC. What is the potential energy of the configuration to the nearest tenth of a millijoule?
8. In problem 7 what is the potential at the point x = 1 and y = 1, to the nearest volt?

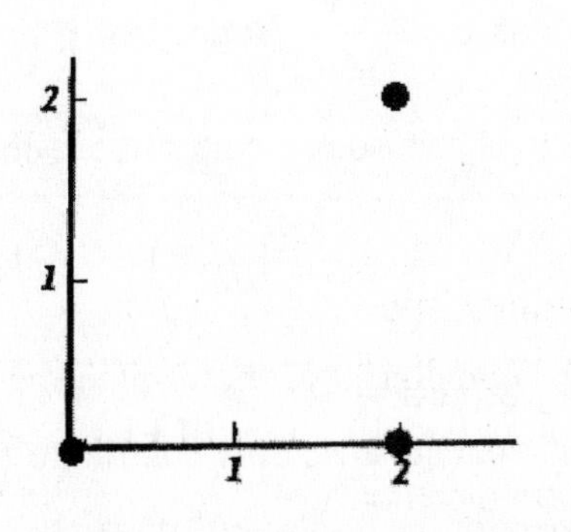

9. A capacitor with a 12-V potential difference across it has a plate area, A, of $0.05\ \text{m}^2$ and a plate separation, d, of 0.44 mm. What is its capacitance to the nearest hundredth of a nF?

10. In problem 9 what is the energy stored in the capacitor, to the nearest nJ?

Puzzle

STUFFED CAPACITOR

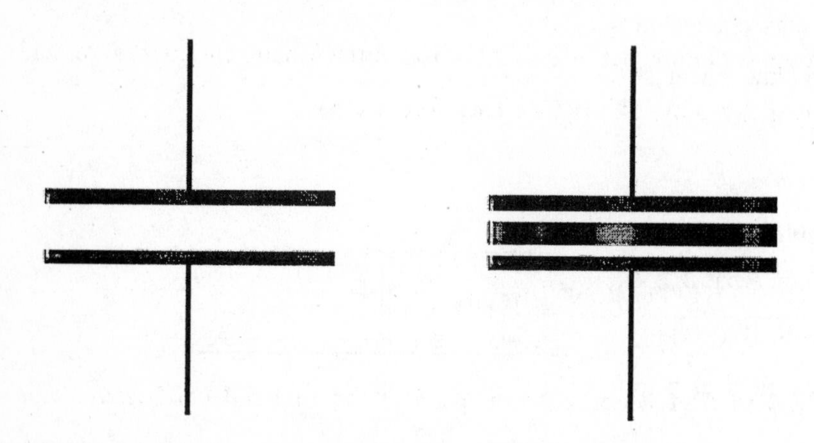

Consider the figures above. On the left, is a plain parallel-plate capacitor (area = A, gap = d). On the right is a system consisting of the same capacitor with a metal plate inserted into the gap. Assume the thickness of the plate is d/2. How will the new capacitance compare with the old? Answer this question in words, not equations, briefly explaining how you obtained your answer.

Selected Solutions

11. **(a)** No work will be done by the electric force on a test charge that moves from A to B because the force and the displacement will be perpendicular to each other; therefore,

$$V_B = V_A \quad \Rightarrow \quad \Delta V = V_B - V_A = \boxed{0}$$

(b) Recall that $E = -\dfrac{\Delta V}{\Delta s}$. Notice that **E** points in the negative x direction, and the displacement Δs, from C to B, is also in the negative x-direction. We therefore have

$$\Delta V = V_B - V_C = -(-E)(\Delta s) = -\left(-1200\frac{\text{N}}{\text{C}}\right)(-0.040\,\text{m}) = \boxed{-48\,\text{V}}$$

(c) Only the component of the displacement parallel to the electric field matters; any displacement perpendicular to **E** doesn't contribute to the change in the electric potential. Notice that the x component of the displacement is in the positive x direction. Therefore,

$$\Delta V = V_C - V_A = -(-E)(\Delta s) = -\left(-1200\frac{N}{C}\right)(0.040\,m) = \boxed{48\,V}$$

(d) $\boxed{\text{No}}$, it is not possible to determine V_A. We only know the potential differences. We need to know either V_B or V_C to determine V_A.

19. (a) The particle has a positive charge, so it will move in the direction of the electric field, which is the $\boxed{\text{negative x direction}}$.

(b) From the conservation of energy we know that $\Delta K = -\Delta U$. The initial kinetic energy is zero, and the change in potential energy is given by $\Delta U = q\Delta V$. Therefore, we have

$$\tfrac{1}{2}mv^2 = -q\Delta V = qEd$$

Solving this equation for v gives

$$v = \sqrt{\frac{2qEd}{m}} = \sqrt{\frac{2(0.045\times10^{-6}\,C)(1200\,\tfrac{N}{C})(0.050\,m)}{0.0035\,kg}} = \boxed{3.9\,cm/s}$$

(c) Its increase in speed will be $\boxed{\text{less than}}$ its increase in speed in the first 5.0 cm because v is proportional to the square root of the distance traveled. To see it another way, since E is constant, then F is constant, which means that the acceleration a is constant. For constant a we know that $\Delta v \propto \Delta t$. Because of the acceleration, the particle will be moving faster during the second 5.0 cm than it was during the first, so the second 5.0 cm will require less time than the first, and the particle will therefore experience a smaller increase in speed.

31. (a) The work required by an external force to move the +2.7-μC charge to infinity equals the work done by the electric force in bringing this charge in from infinity. By definition, this work equals the negative of the change in the electric potential energy when brought in from infinity. Let's call the charges

$$q_1 = -6.1\,\mu C,\quad q_2 = +2.7\,\mu C,\quad q_3 = -3.3\,\mu C$$

Therefore,

$$W = -\Delta U = -\left(\frac{kq_2q_1}{r_{21}} + \frac{kq_2q_3}{r_{23}}\right) = -kq_2\left(\frac{q_1}{r_{21}} + \frac{q_3}{r_{23}}\right)$$

$$\therefore\quad = -\left(8.99\times10^9\frac{N\cdot m^2}{C^2}\right)(2.7\times10^{-6}\,C)\left[\frac{-6.1\times10^{-6}\,C}{0.25\,m} + \frac{-3.3\times10^{-6}\,C}{\sqrt{(0.25\,m)^2+(0.16\,m)^2}}\right] = \boxed{0.86\,J}$$

(b) Again, this work equals $-\Delta U$:

$$W = -\Delta U = -\left(\frac{kq_1q_2}{r_{12}} + \frac{kq_1q_3}{r_{13}}\right) = -kq_1\left(\frac{q_2}{r_{12}} + \frac{q_3}{r_{13}}\right)$$

$$= -\left(8.99\times10^{9}\frac{N\cdot m^2}{C^2}\right)(-6.1\times10^{-6}\,C)\left[\frac{2.7\times10^{-6}\,C}{0.25\,m} + \frac{-3.3\times10^{-6}\,C}{0.16\,m}\right] = \boxed{-0.54\,J}$$

43. (a) From the definition of capacitance we have that $V = Q/C$. Plugging in the expression for C for a parallel-plate capacitor gives

$$V = \frac{Qd}{\kappa\varepsilon_0 A} = \frac{(4.7\times10^{-6}\,C)(0.88\times10^{-3}\,m)}{2.0\left(8.85\times10^{-12}\,\frac{C^2}{N\cdot m^2}\right)(0.012\,m^2)} = \boxed{19\,kV}$$

(b) The answer to part (a) will $\boxed{\text{decrease}}$ because V is inversely proportional to κ

(c) Using the same expression as above, gives

$$V = \frac{Qd}{\kappa\varepsilon_0 A} = \frac{(4.7\times10^{-6}\,C)(0.88\times10^{-3}\,m)}{4.0\left(8.85\times10^{-12}\,\frac{C^2}{N\cdot m^2}\right)(0.012\,m^2)} = \boxed{9.7\,kV}$$

57. (a) $Q = CV = (850\times10^{-6}\,F)(330\,V) = \boxed{0.28\,C}$

(b) $U = \frac{1}{2}CV^2 = \frac{1}{2}(850\times10^{-6}\,F)(330\,V)^2 = \boxed{46\,J}$

Answers to Practice Quiz

1. (d) 2. (a) 3. (b) 4. (a) 5. (e) 6. (b) 7. (a) 8. (e) 9. (c) 10. (c)

Answers to Practice Problems

1. 2547 N/C 2. −71.14 mJ

3. −5.76 V 4. 160 N/C

5. 4.8 m/s 6. −99328 V

7. −72.6 mJ 8. −6357 V

9. 1.01 nF 10. 72 nJ

CHAPTER 21

ELECTRIC CURRENT AND DIRECT CURRENT CIRCUITS

Chapter Objectives

After studying this chapter, you should:

1. understand the meaning of electric current.
2. be able to apply Ohm's law to basic circuits involving resistors.
3. know how to determine power consumption in electric circuits.
4. be able to correctly combine resistors and capacitors in series and parallel.
5. be able to apply Kirchhoff's rules to electric circuits.
6. understand the behavior of RC circuits.
7. know how ammeters and voltmeters should be properly connected into circuits.

Warm-Ups

1. In the circuit shown, the battery provides a current I = 0.3 A to the resistor. How much current is returned by the resistor to the negative terminal of the battery? How much current flows inside the battery from the negative to the positive terminal?

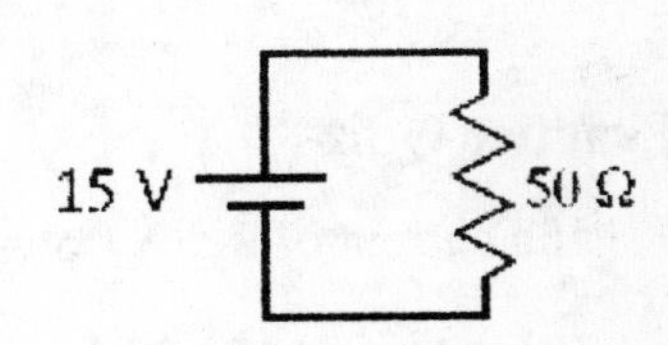

2. Estimate the resistance of a typical electrical cord in your home, say, one that attaches a lamp to the wall socket.
3. According to your textbook, the power dissipated by a resistor is given by $P = V^2/R$. However, your textbook also says that the power dissipated by a resistor is given by $P = I^2R$. Is the power really proportional to R or is it really proportional to 1/R? Can both be correct?
4. Estimate the resistance of a typical lightbulb. (Hint: You can use "typical" values for the voltage and power.)

Chapter Review

In this chapter we study direct current circuits. The important concepts of Ohm's law and Kirchhoff's rules are covered. Capacitors as circuit elements are also introduced.

21–1 Electric Current

The flow of electric charge constitutes an **electric current**, I. The current is determined by the net magnitude of charge, ΔQ, that flows past a point in time Δt. Specifically,

$$I = \frac{\Delta Q}{\Delta t}$$

As the above definition suggests, the unit of current, called an ampere, A, can be written in terms of charge as 1 A = 1 C/s. However, for technical reasons the ampere is considered to be the fundamental SI unit. From now on, we will adopt this officially correct practice and take the fundamental dimension used for electricity to be [A] to represent current.

In this chapter we consider currents that flow through closed paths called **electric circuits**. Since the flow of current that we study here will be in one direction around the circuit, these are called **direct current circuits**. A current flows in a circuit when energy is supplied by an **electric battery**. Batteries have two ends called *terminals* across which a potential difference exists. This potential difference is called the **electromotive force**, $\mathcal{E}$, or emf, of the battery. The direction of the flow of current in a circuit is always taken to be the direction in which a positive charge would move. This last statement is true even if the actual charge that flows is negative, as is the case in almost all electrical devices.

Practice Quiz

1. If a current of 2.5 A flows through a wire for 3.0 seconds, how much charge passes through the wire?

   (a) 0.5 C **(b)** 7.5 C **(c)** 5.5 C **(d)** 1.2 C **(e)** 0.83 C

21–2 Resistance and Ohm's Law

Typically, current does not flow through a circuit unimpeded. The wires through which the current flows offer some **resistance** to this flow. The more resistance R in the wire, the smaller the current I that will flow for a given potential difference V across the wire. The relationship between these three quantities is referred to as **Ohm's law**:

$$V = IR$$

The resistance R has the SI unit V/A; this unit is called an **ohm** (Ω): 1 Ω = 1 V/A.

The amount of resistance in a particular piece of wire depends on the size and shape of the wire as well as the type of material out of which the wire is made. Resistance increases in direct proportion to the length L of the wire and decreases in inverse proportion to the cross-sectional area A of the wire. The dependence on the type of material is contained within a measured quantity called the **resistivity**, ρ. Putting this information together gives

$$R = \rho \frac{L}{A}$$

The resistivity of a material has the SI unit Ω·m.

Example 21–1 Flashlight Assume that the filament in the lightbulb of a flashlight is made of a small piece of metal alloy. If this piece of metal has a radius of 0.500 mm, a length of 1.00 cm, and resistivity of 4.39×10^{-4} Ω·m, what current runs through this filament if the flashlight operates on two 1.50-volt batteries?

Picture the Problem The diagram shows the circuit for the flashlight.

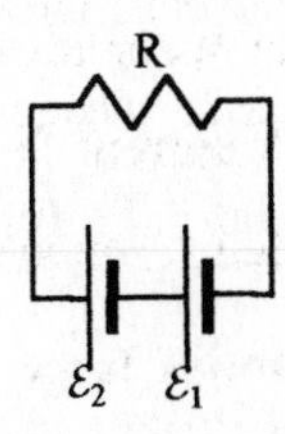

Strategy The current can be determined from Ohm's law once the resistance is known. The resistance can be calculated from the given information.

Solution

1. The resistance can be determined from the resistivity: $R = \rho \frac{L}{A} = \rho \frac{L}{\pi r^2}$

2. The net potential difference across the resistor: $\mathcal{E} = \mathcal{E}_1 + \mathcal{E}_2$

3. By Ohm's law we have: $I = \frac{\mathcal{E}}{R} = \frac{\pi r^2 (\mathcal{E}_1 + \mathcal{E}_2)}{\rho L}$

4. The numerical result is: $I = \frac{\pi (0.500 \times 10^{-3}\ \text{m})^2 (3.00\ \text{V})}{(4.39 \times 10^{-4}\ \Omega \cdot \text{m})(1.00 \times 10^{-2}\ \text{m})} = 0.537\ \text{A}$

Insight Notice that the two batteries oriented with the same polarity provide an emf equal to the sum of the individual emfs.

Practice Quiz

2. A potential difference V is applied across a resistor of resistance R producing a current I through the resistor. If a potential difference of 2 V is applied across the same resistor, what current will flow through it?

(a) I/4 **(b)** I/2 **(c)** I **(d)** 2I **(e)** 4I

3. A potential difference is applied across a resistor of resistance R, producing a current I through the resistor. If the resistor is replaced by one of resistance 2R, what current will flow through it?

(a) I/4 **(b)** I/2 **(c)** I **(d)** 2I **(e)** 4I

4. A potential difference V is applied across a resistor, made from a copper wire of length L and radius r, producing a current I through the resistor. If the resistor is replaced by another copper wire of equal length and radius 2r, what current will flow through it?

(a) I/4 **(b)** I/2 **(c)** I **(d)** 2I **(e)** 4I

21–3 Energy and Power in Electric Circuits

From the definition of electric potential (potential energy per unit charge) we know that energy is transferred when an amount of charge moves through a potential difference. Batteries transfer this energy to the charge, and resistors dissipate this energy in the form of heat. The rate at which this energy transfer takes place is the power that is either generated or dissipated by a device in the circuit. In general, when a current I flows as a result of a potential difference V the electrical power used is given by

$$P = IV$$

Recall that the SI unit of power is the watt: 1 W = 1 A·V = 1 J/s. When this power is being dissipated through a resistance R, we can apply Ohm's law (V = IR) to find other useful relationships for the power. Specifically, it is straightforward to show that

$$P = IV = I^2R = V^2/R$$

Since power is energy divided by time, then power times time will give energy. A unit of energy that is commonly used to track the household use of electricity is the kilowatt-hour, kWh. This quantity equals the amount of energy used if 1 kW of power is consumed for one hour. Its equivalent energy in joules is:

$$1\text{ kWh} = 3.6 \times 10^6\text{ J}$$

Example 21–2 Power (a) Determine the power requirements of the flashlight in Example 21.1. **(b)** If the flashlight is left on for a half-hour, how much energy in kilowatt-hours is used?

Picture the Problem The diagram shows the circuit for the bulb in the flashlight.

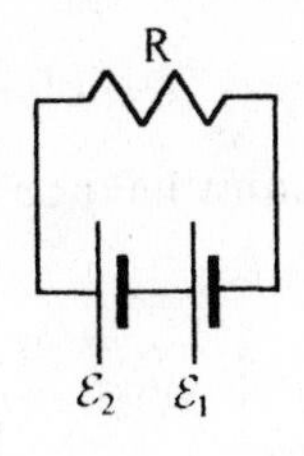

Strategy For part (a) we can use the result of Example 21.1 to obtain the power. For part (b) the energy can be found as the power times the time.

Solution

Part (a)

1. Use the current from Example 21–1 to caclulate the power: $P = IV = (0.5367\text{ A})(3.00\text{ V}) = 1.61\text{ W}$

Part (b)

2. The energy in kilowatt-hours is: $E = Pt = (0.001610\text{ kW})[0.500\text{ h}] = 8.05 \times 10^{-4}\text{ kWh}$

Insight In part (b) we converted the power to kilowatts and used the time in hours to directly get the result in the desired units.

Practice Quiz

5. What is the resistance of a device that consumes 100 W of power when connected to a 120-V source?

   (a) 144 Ω **(b)** 12,000 Ω **(c)** 1.2 Ω **(d)** 0.83 Ω **(e)** 20 Ω

6. How much energy is consumed if a 75-W light bulb is left on for 12 hours?

   (a) 900 kWh **(b)** 6.3 kWh **(c)** 0.90 kWh **(d)** 160 kWh **(e)** 0.16 kWh

21–4 Resistors in Series and Parallel

A resistor in a circuit, indicated by a jagged line, represents a circuit element (such as a lightbulb or a heater) that contains resistance. Circuit elements can be connected in different ways, and the overall resistance, or **equivalent resistance**, of the combination depends on how they are connected. In this section we learn how to determine the equivalent resistance for combinations of resistors connected in **series** and in **parallel**.

Resistors in series are connected one after the other (or end to end). Connecting resistors in series has the same effect as making one resistor longer. It is important to remember that *the same current flows through resistors that are in series with each other.*

The result is that the equivalent resistance, R_{eq}, of N resistors connected in series is the sum of the individual resistances

$$R_{eq} = R_1 + R_2 + \cdots + R_N = \sum_{i=1}^{N} R_i$$

Consequently, R_{eq} is greater than any of the individual resistances that contribute to it.

Resistors in parallel are connected across the same potential difference.

Connecting resistors in parallel has the same effect as making one resistor wider. The result is that the equivalent resistance of N resistors connected in parallel can be found by the following relation:

$$\frac{1}{R_{eq}} = \frac{1}{R_1} + \frac{1}{R_2} + \cdots + \frac{1}{R_N} = \sum_{i=1}^{N} \frac{1}{R_i}$$

Once $1/R_{eq}$ is determined, we can that the reciprocal of this result to get the value of R_{eq} in ohms. As a consequence, we can see that for resistors in parallel, R_{eq} is smaller than the smallest of the individual resistances that contribute to it.

Example 21–3 Equivalent Resistance Determine the equivalent resistance of the circuit shown, given that $R_1 = 3.3\ \Omega$, $R_2 = 4.5\ \Omega$, $R_3 = 6.1\ \Omega$, and $R_4 = 2.8\ \Omega$.

Picture the Problem The diagram shows the circuit in question.

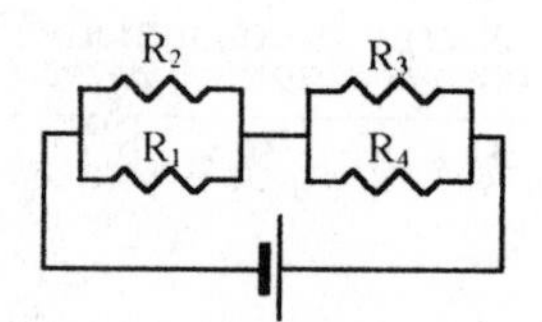

Strategy We will first get the equivalent resistance of each parallel combination, then combine those results in series.

Solution

1. The equivalent resistance of R_1 and R_2 is:

$$R_{12} = \left[\frac{1}{R_1} + \frac{1}{R_2}\right]^{-1} = \left[\frac{1}{3.3\ \Omega} + \frac{1}{4.5\ \Omega}\right]^{-1} = 1.904\ \Omega$$

2. The equivalent resistance of R_3 and R_4 is:

$$R_{34} = \left[\frac{1}{R_3} + \frac{1}{R_4}\right]^{-1} = \left[\frac{1}{6.1\ \Omega} + \frac{1}{2.8\ \Omega}\right]^{-1} = 1.919\ \Omega$$

3. The overall equivalent resistance then is: $R_{eq} = R_{12} + R_{34} = 1.904\ \Omega + 1.919\ \Omega = 3.8\ \Omega$

Insight When finding the equivalent resistance of many resistors it is often useful to take an "inside-out" approach as shown here.

Practice Quiz

7. A 2.0-Ω resistor is connected in series with a 3.0-Ω resistor. If this series combination is then connected in parallel with a third resistor of resistance 2.5 Ω, what is the equivalent resistance of this three-resistor circuit?

   (a) 2.5 Ω **(b)** 5.0 Ω **(c)** 1.7 Ω **(d)** 3.9 Ω **(e)** 7.5 Ω

8. A resistor R_1 = 3.0 Ω is connected in series with a resistor R_2 = 5.0 Ω. When this combination is connected to a battery, which resistor consumes more power?

   (a) R_1 **(b)** R_2 **(c)** They consume the same amount of power.

9. A resistor R_1 = 3.0 Ω is connected in parallel with a resistor R_2 = 5.0 Ω. When this combination is connected to a battery, which resistor consumes more power?

   (a) R_1 **(b)** R_2 **(c)** They consume the same amount of power.

21–5 Kirchhoff's Rules

As circuitry becomes more complicated it becomes difficult to analyze the circuit using simple applications of Ohm's law. For these we use **Kirchhoff's rules**, which apply the conservation of electric charge and the conservation of energy to circuit analysis.

Kirchhoff's first rule (for the conservation of charge) is called the **junction rule**:

The algebraic sum of all currents meeting at any junction in a circuit must equal zero.

A **junction** is a point in a circuit where three or more wires meet so that the current in the circuit may take different paths into or out of this point. In applying the junction rule as stated above, currents *entering* the junction are given a positive sign, and current *leaving* the junction are given a negative sign. Application of the junction rule seems to imply that the directions of the currents must be known; however, it is not always obvious what the direction of every current will be. Fortunately, in these cases it is sufficient just to guess a direction and solve the equations. If the value of the current whose direction you guessed works out to be negative, then you guessed wrong and you therefore know that it actually flows in the opposite direction from what you guessed.

Kirchhoff's second rule (for the conservation of energy) is called the **loop rule**:

The algebraic sum of all potential differences around any closed loop in a circuit must equal zero.

Applying these two rules involves making several choices. These choices include the following:

1. choose the directions of the currents in different parts of the circuit
2. choose which junctions and loops you will use for your analysis
3. choose directions for traversing the loops to generate your loop rule equations

For step 3 there are several things you should remember when writing your loop rule equations.

- Crossing a battery from the negative terminal to the positive terminal is a potential increase, and crossing it from positive to negative is a potential decrease.
- Crossing a resistor in the same direction as the current you chose is a potential decrease.
- Crossing a resistor in the direction opposite the current you chose is a potential increase.

Applying these rules carefully will give you a system of equations that can be solved for currents, resistances, and potential differences in circuits.

Exercise 21–4 Kirchhoff's Rules Use Kirchhoff's rules to determine the overall current and the current through each of the resistors in the circuit shown. The relevant data is that $R_1 = 2.5\ \Omega$, $R_2 = 3.3\ \Omega$, $R_3 = 1.8\ \Omega$, $\mathcal{E}_1 = 20$ V, and $\mathcal{E}_2 = 6.0$ V.

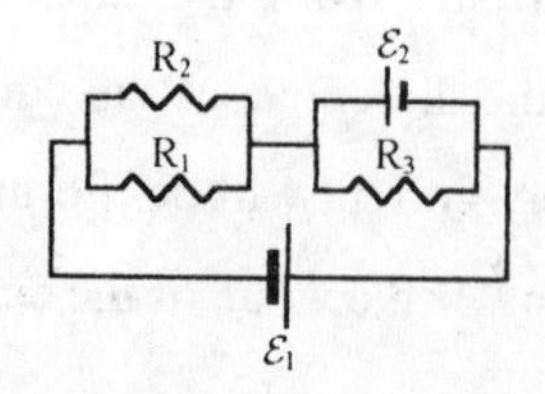

Solution

First, we choose the directions for the currents in the circuit realizing that some or all of these choices may not be correct. The figure below also labels the junctions in the circuit **A** – **D**.

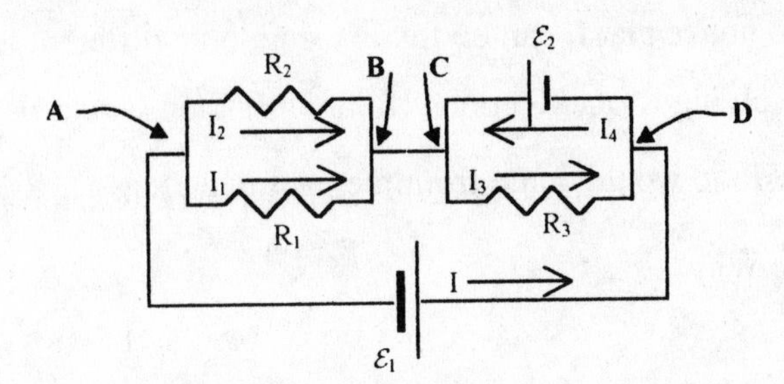

By examining the circuit, we see that I_3 can be determined by applying the loop rule to the small loop involving R_3 and $\mathcal{E}_2$ shown below.

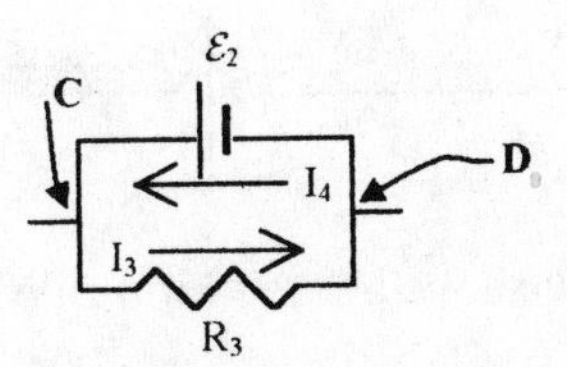

Moving around this loop counterclockwise from junction **D** back to **D** produces the loop rule equation

$$\mathcal{E}_2 - I_3R_3 = 0$$

which gives

$$I_3 = \frac{\mathcal{E}_2}{R_3} = \frac{6.0\text{ V}}{1.8\ \Omega} = 3.3\text{ A}$$

The fact that we got a positive value for I_3 shows that we chose the correct direction.

Further examination suggests that I_2 can be determined by applying the loop rule to the large loop that includes $\mathcal{E}_1$ and $\mathcal{E}_2$ and R_2.

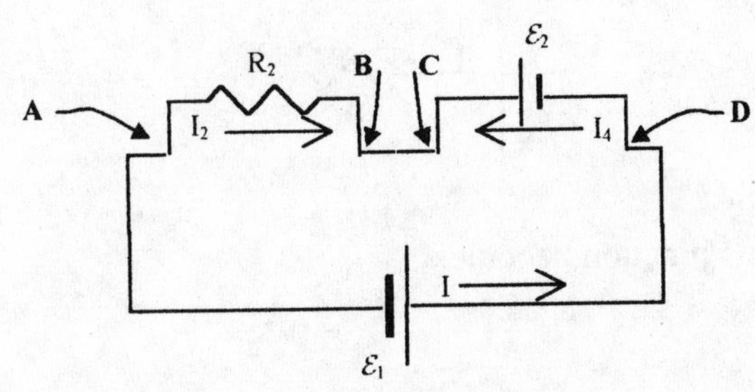

Moving completely around this loop counterclockwise produces the loop rule equation

$$\mathcal{E}_1 + \mathcal{E}_2 + I_2R_2 = 0$$

which gives

$$I_2 = -\frac{\mathcal{E}_1 + \mathcal{E}_2}{R_2} = -\frac{20\text{ V} + 6.0\text{ V}}{3.3\ \Omega} = -7.88\text{ A}$$

The negative sign means that we chose the wrong direction for I_2.

Let's pause for a moment to notice that I_1 can be found using three different loops: (a) One loop would be the parallel combination involving R_1 and R_2 only, (b) another loop would be the larger path including $\mathcal{E}_1$, $\mathcal{E}_2$, and R_1, and (c) the third loop would come from the path including $\mathcal{E}_1$, R_3, and R_1. The choice with the least chance for error is (b). Why?

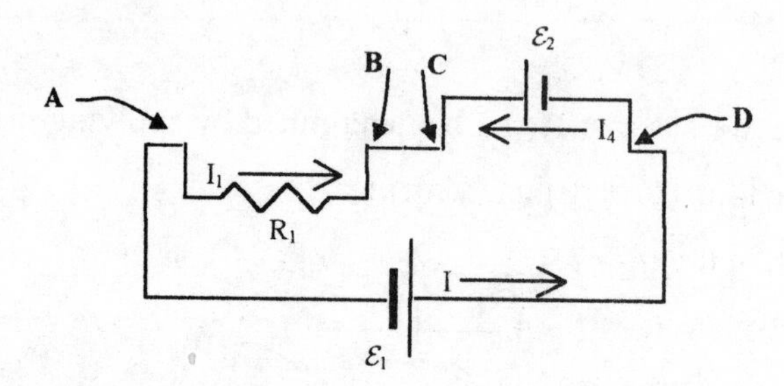

Moving around this loop counterclockwise we get the loop rule equation

$$\mathcal{E}_1 + \mathcal{E}_2 + I_1 R_1 = 0$$

which gives

$$I_1 = -\frac{\mathcal{E}_1 + \mathcal{E}_2}{R_1} = -\frac{20\text{ V} + 6.0\text{ V}}{2.5\ \Omega} = -10.4\text{ A}$$

You can see that the wrong direction was also chosen for I_1.

All that remains is to find the overall current I. Inspection of the circuit suggests that we can determine I by applying the junction rule at junction **A**.

The junction rule equation at this junction becomes

$$-I - I_1 - I_2 = 0$$

This gives

$$I = -(I_1 + I_2) = -(-10.4\text{ A} - 7.88\text{ A}) = 18\text{ A}$$

Thus, our final results are $I_1 = 10\text{ A}$, $I_2 = 7.9\text{ A}$, $I_3 = 3.3\text{ A}$, and $I = 18\text{ A}$

Make sure you understand the sign associated with each term in the loop rule and junction rule equations; it is very important that these be done consistently and correctly. Also, keep in mind that when you get a negative result for a current, that negative result must be used in any subsequent mathematical step; only drop the negative when stating the final result for that current. It is always better if you can choose the correct current directions in the beginning. In the circuit for this problem the polarities of the two emfs would immediate suggest that the directions of I_1 and I_2 should be opposite to what was chosen. Try to see this fact? The choice made for this example was done only to show you what happens when the wrong choice is made.

Practice Quiz

10. Which of the following is a correct junction rule equation for the diagram on the right?

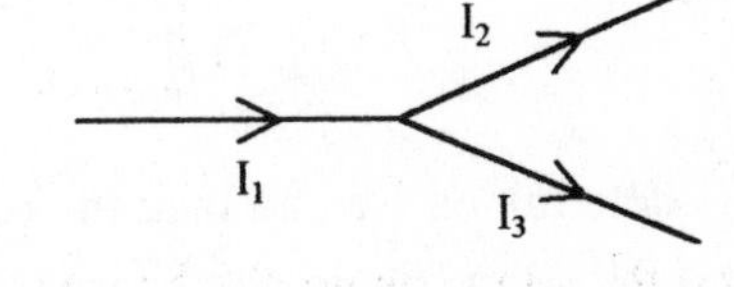

(a) $I_1 + I_2 + I_3 = 0$ **(d)** $I_1 - I_2 - I_3 = 0$

(b) $I_1 + I_2 - I_3 = 0$ **(e)** $I_1 - I_2 + I_3 = 0$

(c) $-I_1 - I_2 + I_3 = 0$

11. Which of the following is a correct loop rule equation for the diagram on the right?

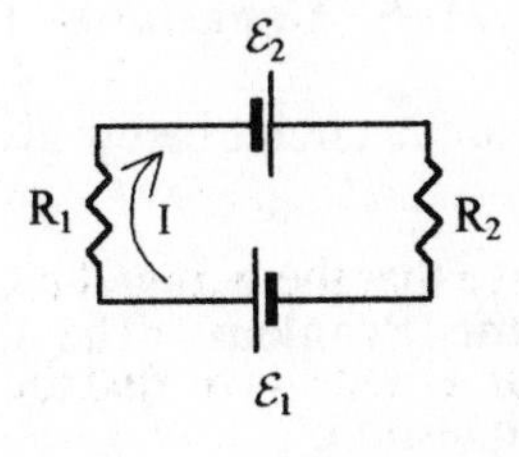

(a) $\mathcal{E}_1 + IR_1 + \mathcal{E}_2 + IR_2 = 0$ **(d)** $-\mathcal{E}_1 - IR_1 + \mathcal{E}_2 + IR_2 = 0$

(b) $\mathcal{E}_1 + IR_1 - \mathcal{E}_2 + IR_2 = 0$ **(e)** $\mathcal{E}_1 - IR_1 + \mathcal{E}_2 - IR_2 = 0$

(c) $\mathcal{E}_1 - IR_1 + \mathcal{E}_2 + IR_2 = 0$

21–6 Circuits Containing Capacitors

Capacitors are commonly used in circuits and can be combined in series and parallel to produce an overall equivalent capacitance C_{eq} that depends on the type of connection. As with resistors,

capacitors in parallel are connected across the same potential difference.

In the case of parallel-plate capacitors ($C = \varepsilon_0 A/d$), connecting capacitors in parallel is equivalent to making the plates of one capacitor wider. The result is that for N capacitors connected in parallel the equivalent capacitance is

$$C_{eq} = C_1 + C_2 + \cdots + C_N = \sum_{i=1}^{N} C_i$$

Therefore, in parallel, the equivalent capacitance is larger than any of the individual capacitances that contribute to it.

Just as resistors in series have the same current flowing through them, capacitors in series have the same current flow onto them, and thus each capacitor stores the same amount of charge Q. Therefore,

capacitors in series store equal amounts of charge.

Connecting capacitors in series has the effect of increasing the distance between the plates of one capacitor. The result for N capacitors connected in series is that the equivalent capacitance can be determined by

$$\frac{1}{C_{eq}} = \frac{1}{C_1} + \frac{1}{C_2} + \cdots + \frac{1}{C_N} = \sum_{i=1}^{N} \frac{1}{C_i}.$$

Once $1/C_{eq}$ is determined we can take the reciprocal to determine C_{eq} in farads. Therefore, in series, the equivalent capacitance is smaller than the smallest of the individual capacitances that contribute to it.

Example 21–5 Capacitors Determine both the charge on and the potential difference across each capacitor in the circuit below given that $C_1 = 12\ \mu F$, $C_2 = 5.0\ \mu F$, $C_3 = 6.5\ \mu F$, and $\mathcal{E} = 12$ V.

Picture the Problem The diagram shows the circuit in question.

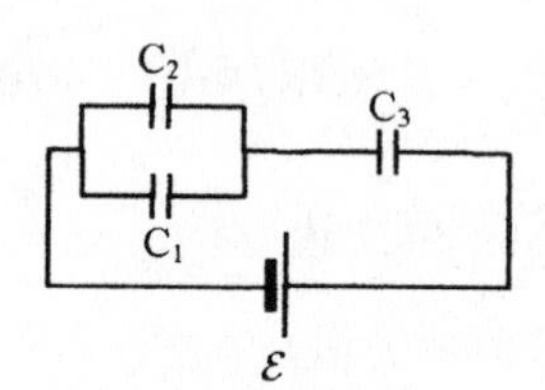

Strategy We can use the known facts about capacitors in series and parallel to obtain the required information.

Solution

1. The equivalent capacitance of C_1 and C_2 is: $C_{12} = C_1 + C_2 = 12\mu F + 5.0\mu F = 17.0\mu F$

2. The overall equivalent capacitance is: $C_{eq} = \left[\frac{1}{C_{12}} + \frac{1}{C_3}\right]^{-1} = \left[\frac{1}{17.0\mu F} + \frac{1}{6.5\mu F}\right]^{-1} = 4.702\ \mu F$

3. The total charge stored is: $Q_{tot} = C_{eq}V = (4.702\ \mu F)(12\ V) = 56.43\ \mu C$

4. Since the same charge is stored on capacitors in series the charge stored on C_3 is: $Q_3 = Q_{tot} = 56\ \mu C$

5. The potential difference across C_3 is: $V_3 = \frac{Q_3}{C_3} = \frac{56.43\ \mu C}{6.5\ \mu C} = 8.7\ V$

6. By the loop rule, the potential difference across the C_1 and C_2 combination is: $V_{12} = \mathcal{E} - V_3 = 12\ V - 8.681\ V = 3.319\ V$

7. Since capacitors in parallel are connected across the same potential difference: $V_1 = V_2 = 3.3\ V$

8. The charge stored on C_1 is: $Q_1 = C_1 V_1 = (12\ \mu F)(3.319\ V) = 40\ \mu C$

9. The charge stored on C_2 is: $Q_2 = C_2 V_2 = (5.0\ \mu F)(3.319\ V) = 17\ \mu C$

Insight Make sure you understand the results of steps 4 and 7; these steps are the keys to the solution.

Practice Quiz

12. A 6.2-μF capacitor is connected in parallel with a 2.4-μF capacitor. If this parallel combination is then connected in series with a third capacitor of capacitance 3.8 μF, what is the equivalent capacitance of this three-capacitor circuit?

(a) 5.5 μF **(b)** 1.7 μF **(c)** 8.6 μF **(d)** 12.4 μF **(e)** 2.6 μF

13. Given two capacitors of equal capacitance, which type of connection would allow them to store the most combined charge?

(a) series **(b)** parallel **(c)** Both connections would store the same amount of charge.

21–7 RC Circuits

For our purposes an **RC circuit** is one that contains resistance and capacitance in series. In these circuits the charge on the capacitor builds up (and dies off) slowly in comparison to what happens in circuits containing only resistance and no capacitance. For an RC circuit in which the capacitor is *initially uncharged*, the charge builds up on the capacitor according to the equation

$$q(t) = C\mathcal{E}\left(1 - e^{-t/\tau}\right) \qquad \text{[charging]}$$

where $\mathcal{E}$ is the source emf. The quantity τ is given by τ = RC and is called the **time constant** of the circuit. The time constant provides a characteristic amount of time for the charge to build because after an amount of time t = τ the capacitor will be most of the way (63.2%) to its maximum charge of $q_{max} = C\mathcal{E}$.

In the present case of an initially uncharged capacitor, the current starts out at its maximum value and falls off exponentially:

$$I(t) = \frac{\mathcal{E}}{R} e^{-t/\tau} \qquad \text{[charging]}$$

If the capacitor *is initially charged* with a charge Q, has a potential difference V across it, and is allowed to discharge, both the charge and current fall off exponentially:

$$q(t) = Qe^{-t/\tau}; \quad I(t) = \frac{V}{R} e^{-t/\tau} \qquad \text{[discharging]}$$

Example 21–6 An RC Circuit Determine the time constant of the RC circuit shown below.

Picture the Problem The diagram shows the circuit in question.

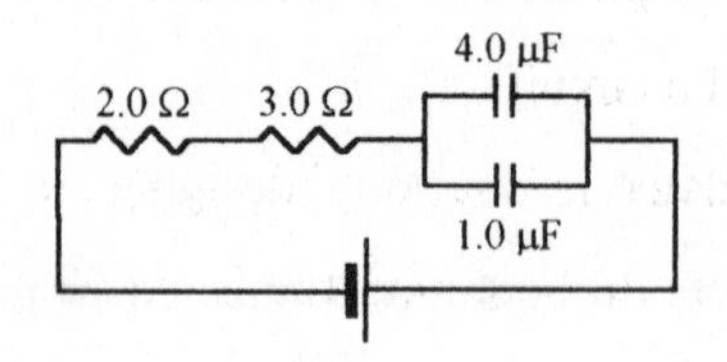

Strategy We need to determine the correct values of R_{eq} and C_{eq} and then use them to get τ.

Solution

1. The equivalent resistance is: $R_{eq} = R_1 + R_2 = 2.0\ \Omega + 3.0\ \Omega = 5.0\ \Omega$

2. The equivalent capacitance is: $C_{eq} = C_1 + C_2 = 4.0\ \mu F + 1.0\ \mu F = 5.0\ \mu F$

3. The time constant is: $\tau = R_{eq}C_{eq} = (5.0\ \Omega)(5.0\ \mu F) = 25\ \mu s$

Insight Here, despite the fact that there are two resistors and two capacitors, the resistance and capacitance are still in series.

Practice Quiz

14. Which of the following statements is true for charging a capacitor?

(a) The time constant is how long it takes the charge to build up to 36.8% of its maximum value.

(b) The time constant is how long it takes the current to build up to 63.2% of its maximum value.

(c) The time constant is how long it takes the current to build up to 36.8% of its maximum value.

(d) The time constant is how long it takes the charge to build up to 63.2% of its maximum value.

(e) The time constant is how long it takes the current to fall off to 63.2% of its maximum value.

15. Which of the following statements is true for discharging a capacitor?

(a) The time constant is how long it takes the charge to fall off to 36.8% of its maximum value.

(b) The time constant is how long it takes the current to build up to 63.2% of its maximum value.

(c) The time constant is how long it takes the current to build up to 36.8% of its maximum value.

(d) The time constant is how long it takes the charge to fall off to 63.2% of its maximum value.

(e) The time constant is how long it takes the current to fall off to 63.2% of its maximum value.

*21–8 Ammeters and Voltmeters

An **ammeter** is a device specially made for measuring currents in a circuit. An ammeters should be connected in series with the device whose current is sought. Ideally, an ammeter should have zero resistance. In practice, the resistance of an ammeter should be much less than the resistances of other devices in the circuit.

A voltmeter is specially designed to measure potential differences across devices in a circuit. A voltmeter should be connected in parallel with the device across which the potential difference is being sought. Ideally, a voltmeter should have infinite resistance. In practice, the resistance of a voltmeter should be much greater than the resistances of other devices in the circuit.

Reference Tools and Resources

I. Key Terms and Phrases

electric current results from the flow of electric charge

electric circuits closed paths containing circuit elements through which current can flow

direct current (DC) circuits circuits in which the current always flows in one direction

electric battery a device, that maintains a potential difference and is used as an energy source for electric circuits

electromotive force (emf) the potential difference across an ideal battery

resistance the opposition to the flow of charge through a wire due to the properties of the wire

ohm the SI unit of resistance

resistivity the property of a substance that partially determines the resistance of objects made of that substance

connected in series circuit elements connected one after the other such that the same current flows through them

connected in parallel circuit elements connected across the same potential difference

Kirchhoff's rules two rules that apply the conservation of charge (junction rule) and the conservation of energy (loop rule) to electric circuits

junction a point in a circuit where three or more wires meet so that the current in the circuit may take different paths into or out of this point

loop any closed path in a circuit

time constant the characteristic amount of time, $\tau = RC$, for an initially uncharged capacitor to charge up to 63.2% of its maximum value in an RC circuit

ammeter a device designed to measure the current through a circuit element

voltmeter a device designed to measure the potential difference across a circuit element

II. Important Equations

Name/Topic	Equation	Explanation
Ohm's law	$V = IR$	The relationship between potential difference, current, and resistance.
resistivity	$R = \rho \frac{L}{A}$	How the resistance of a wire depends on the properties of the wire.
electrical power	$P = IV = I^2R = V^2/R$	The electrical power transformed by a device in a circuit.
resistors in series	$R_{eq} = R_1 + R_2 + \cdots + R_N$	How resistors in series combine.
resistors in parallel	$\frac{1}{R_{eq}} = \frac{1}{R_1} + \frac{1}{R_2} + \cdots + \frac{1}{R_N}$	How resistors in parallel combine.

capacitors in parallel	$C_{eq} = C_1 + C_2 + \cdots + C_N$	How capacitors in parallel combine.
capacitors in series	$\frac{1}{C_{eq}} = \frac{1}{C_1} + \frac{1}{C_2} + \cdots + \frac{1}{C_N}$	How capacitors in series combine.

III. Know Your Units

Quantity	Dimension	SI Unit
electric current (I)	[A]	A
resistance (R)	$[M][L^2][A^{-2}][T^{-3}]$	Ω
resistivity (ρ)	$[M][L^3][A^{-2}][T^{-3}]$	Ω·m

Practice Problems

1. A battery with an emf of 1.5 volts supplies a current of 1.09 amps to a light bulb for 10 seconds. How much charge is supplied to the bulb, to the nearest tenth of a coulomb?
2. A wire with a length of 1.02 meters and a diameter of 1.45 mm caries a current of 1.7 amps. What is its resistance, to the nearest tenth of a mΩ, if its resistivity is 1.7×10^{-8} Ω·m?
3. In the circuit on the left in the diagram $\mathcal{E}$ = 34 volts, R_1 = 18 Ω, and R_2 = 15 Ω. What is the current, to the nearest tenth of an amp?
4. What would be the answer to problem 3 if the circuit was as shown on the right in the diagram?

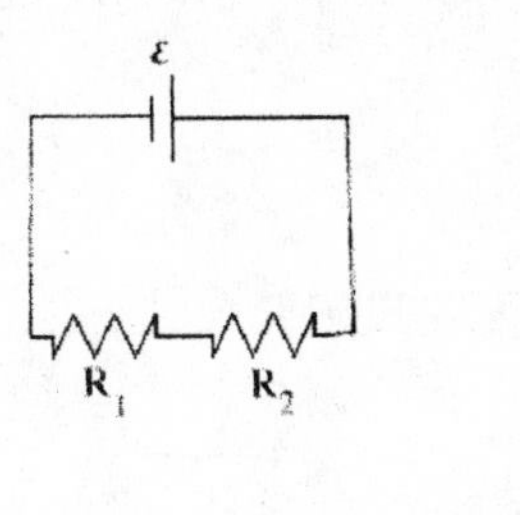

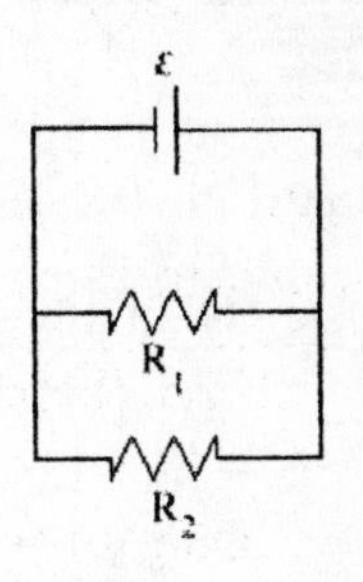

5. In the circuit R_1 = 20 Ω, R_2 = 18 Ω, and R_3 = 3 Ω. What is the current, to the nearest tenth of an amp?

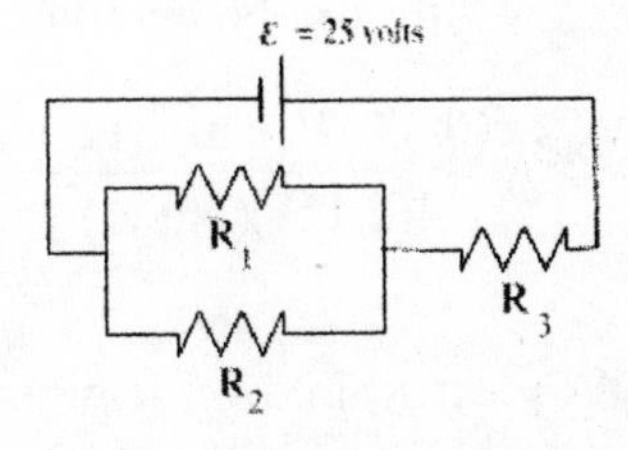

6. In this circuit $\mathcal{E}_1$= 4.83 volts and $\mathcal{E}_2$= 3.99 volts, to the nearest hundredth of an amp what is I_1?
7. What is I_3 in problem 6?

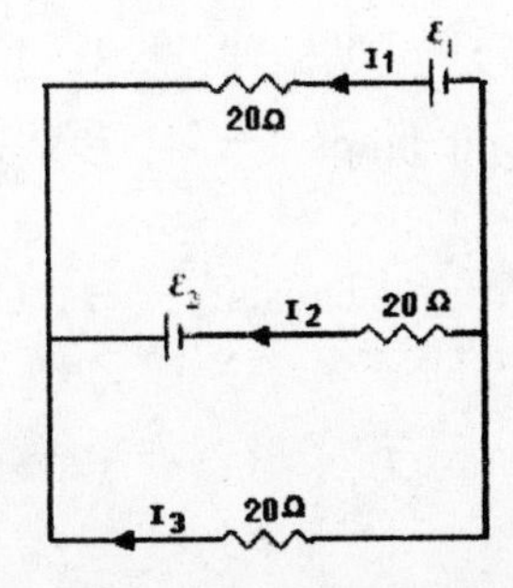

8. In the figure $C_1 = 20$ mF, $C_2 = 10$ mF, and $C_3 = 20$ mF. What is the charge on C_1, to the nearest hundredth of a mC?

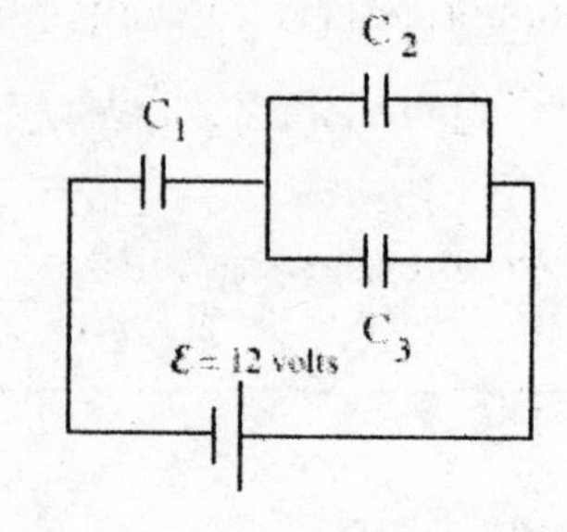

9. In problem 8 what is the charge on C_2?

10. In the circuit on the right, if R = 1884 Ω and C = 9.97 μF, what is the current, to the nearest hundredth of an amp, 1 second after the switch is closed?

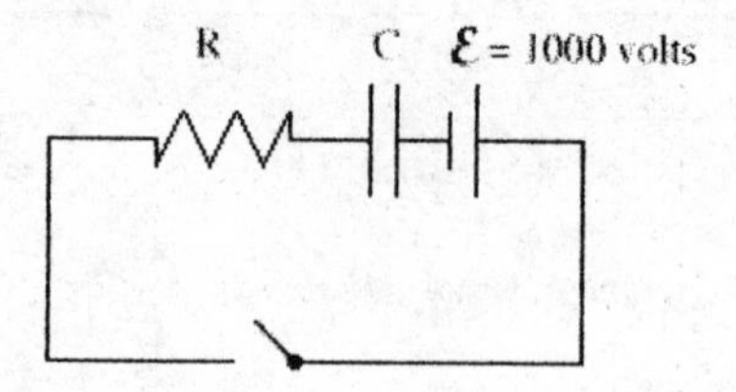

Puzzle

BULB MARKET

Houses, boats, and cars have electrical systems based on 120 V, 24 V, and 12 V respectively. Let's say we have a 60-watt bulb from each, and enough adapters to plug any bulb into any system. Which combination of bulb and system will give the most light? Which will give the least? Answer these questions in words, not equations, briefly explaining how you obtained your answers.

Selected Solutions

15. Using Ohm's law, we can write

$$V = IR = I\left(\rho\frac{L}{A}\right) \quad \Rightarrow \quad \rho = \frac{AV}{IL}.$$

Since $A = \pi d^2/4$, this can be rewritten as

$$\rho = \frac{\pi d^2 V}{4IL} = \frac{\pi(0.33\times10^{-3}\text{ m})^2(12\text{ V})}{4(2.1\text{ A})(6.9\text{ m})} = \boxed{7.1\times10^{-8}\,\Omega\cdot\text{m}}$$

25. Recall that the electrical power is given by P = IV, which gives the amount of energy (U) delivered per second. The total amount of energy delivered, then, is U = Pt = (IV)t. Therefore,

905 cranking amps $\Rightarrow$ $U = IVt = (905\text{ A})(7.2\text{ V})(30.0\text{ s}) = 2.0\times10^5\text{ J}$

155-min reserve capacity $\Rightarrow$ $U = IVt = (25\text{ A})(10.5\text{ V})(155\text{ min})\left(\frac{60\text{s}}{\text{min}}\right) = 2.4\times10^6\text{ J}$

The 155-minute reserve capacity rating represents the greater amount of energy delivered by the battery.

37. Let us label the resistors as follows:

$$R_1 = 1.5\ \Omega,\ R_2 = 2.5\ \Omega,\ R_3 = 6.3\ \Omega,\ R_4 = 4.8\ \Omega,\ R_5 = 3.3\ \Omega,\ R_6 = 8.1\ \Omega$$

The equivalent resistance of the parallel combination of R_4, R_5, and R_6 is given by

$$R_{456} = \left(\frac{1}{R_4} + \frac{1}{R_5} + \frac{1}{R_6}\right)^{-1}$$

The above parallel combination R_{456} is in series with R_3; their equivalent resistance is given by

$$R_{3456} = R_3 + R_{456} = R_3 + \left(\frac{1}{R_4} + \frac{1}{R_5} + \frac{1}{R_6}\right)^{-1}$$

Now, we see that R_{3456} is in parallel with R_1 and R_2. The overall equivalent resistance can be found from

$$\frac{1}{R_{eq}} = \frac{1}{R_1} + \frac{1}{R_2} + \frac{1}{R_{3456}}$$

$$\therefore\ R_{eq} = \left[\frac{1}{1.5\Omega} + \frac{1}{2.5\Omega} + \frac{1}{6.3\Omega + \left(\frac{1}{4.8\Omega} + \frac{1}{3.3\Omega} + \frac{1}{8.1\Omega}\right)^{-1}}\right]^{-1} = \boxed{0.84\ \Omega}$$

45. (a) Take current, I, to flow from the positive terminal of the 12 V battery. At junction A, have I split into I_1, which flows through the 1.2-Ω resistor, and I_2, which flows through the 6.7-Ω resistor. Applying the junction rule at junction A gives

$$I = I_1 + I_2 \qquad \text{(eq. 1)}$$

If we apply the loop rule to the loop on the left side of the circuit in the figure (traversing it in a clockwise direction) we get

$$0 = 12\ \text{V} - I(3.9\Omega) - I_1(1.2\Omega) - I(9.8\Omega)$$

If we solve this loop rule equation for I_1 it becomes

$$I_1 = 10\ \text{A} - \frac{13.7}{1.2}I \qquad \text{(eq. 2)}$$

Applying the loop rule (clockwise) to the loop around the outside of the circuit gives

$$0 = 12\ \text{V} - I(3.9\ \Omega) - I_2(6.7\ \Omega) - 9.0\ \text{V} - I(9.8\ \Omega)$$

Solving for I_2 gives

$$I_2 = \frac{3}{6.7}\text{ A} - \frac{13.7}{6.7}I \qquad \text{(eq. 3)}$$

Substituting eq. 2 and eq. 3 into eq. 1 gives,

$$I = 10\text{ A} - \frac{13.7}{1.2}I + \frac{3}{6.7}\text{ A} - \frac{13.7}{6.7}I \;\Rightarrow\; I = \frac{10\text{ A} + \frac{3}{6.7}\text{ A}}{1 + \frac{13.7}{1.2} + \frac{13.7}{6.7}} = 0.72\text{ A}$$

From eq. 1 we can now find I_1:

$$I_1 = 10\text{ A} - \frac{13.7}{1.2}\left(\frac{10\text{ A} + \frac{3}{6.7}\text{ A}}{1 + \frac{13.7}{1.2} + \frac{13.7}{6.7}}\right) = 1.8\text{ A}$$

From eq. 2 we can find I_2:

$$I_2 = \frac{3}{6.7}\text{ A} - \frac{13.7}{6.7}\left(\frac{10\text{ A} + \frac{3}{6.7}\text{ A}}{1 + \frac{13.7}{1.2} + \frac{13.7}{6.7}}\right) = -1.0\text{ A}$$

The currents through each resistor are as follows:

3.9 Ω, 9.8 Ω: $\boxed{0.72\text{ A}}$; 1.2 Ω: $\boxed{1.8\text{ A}}$; 6.7 Ω: $\boxed{1.0\text{ A}}$

(b) The potential at point A is $\boxed{\text{greater than}}$ that at point B because the potential has been decreased by the 1.2-Ω resistor between A and B.

(c) By Ohm's law, the potential difference between A and B will have a magnitude of $V_A - V_B = I_1(1.2\ \Omega)$.

$$V_A - V_B = \left[10\text{ A} - \frac{13.7}{1.2}\left(\frac{10\text{ A} + \frac{3}{6.7}\text{ A}}{1 + \frac{13.7}{1.2} + \frac{13.7}{6.7}}\right)\right](1.2\Omega) = \boxed{2.1\text{ V}}$$

47. Let's label the capacitors as follows: $C_1 = 15\ \mu F$, $C_2 = 8.2\ \mu F$, $C_3 = 22\ \mu F$

The equivalent capacitance of the series combination of C_2 and C_3 is given by

$$C_{23} = \left(\frac{1}{C_2} + \frac{1}{C_3}\right)^{-1}$$

This equivalent capacitance is in parallel with C_1; therefore, the overall equivalent capacitance is given by

$$C_{eq} = C_1 + C_{23} = C_1 + \left(\frac{1}{C_2} + \frac{1}{C_3}\right)^{-1} \quad \therefore$$

$$C_{eq} = 15\mu F + \left(\frac{1}{8.2\mu F} + \frac{1}{22\mu F}\right)^{-1} = \boxed{21\mu F}$$

Answers to Practice Quiz

1. (b) **2**. (d) **3**. (b) **4**. (e) **5**. (a) **6**. (c) **7**. (c) **8**. (b) **9**. (a) **10**. (d) **11**. (e) **12**. (e) **13**. (b) **14**. (d) **15**. (a)

Answers to Practice Problems

1. 10.9 C **2**. 10.5 mΩ

3. 1.0 A **4**. 4.2 A

5. 2.0 A **6**. 0.09 A

7. −0.15 A **8**. 144.00 mC

9. 48 mC **10**. 0.50 A

CHAPTER 22

MAGNETISM

Chapter Objectives

After studying this chapter, you should:

1. know the rules for how magnetic poles interact and how to draw magnetic field lines.
2. be able to determine the magnetic force on a moving charge in a magnetic field.
3. be able to describe the motion of charged particles in uniform magnetic fields.
4. be able to calculate the force on a current-carrying wire, and the torque on a current loop, in a magnetic field.
5. be able to determine the magnetic field of a long, straight current in the center of a current loop and inside a solenoid.

Warm-Ups

1. The force on a charged particle in a magnetic field is very different from the force due to an electric field. Please list as many differences as you can. Don't forget to include differences in the direction as well as in the magnitude.
2. Estimate the force of the earth's magnetic field on a 10-cm segment of a typical wire in your home. (Hint: the magnitude of Earth's magnetic field is about 5.5×10^{-5} T.)
3. A proton moving downward enters a region of space with magnetic field **B**, which points eastward. In which direction is the force on the proton?

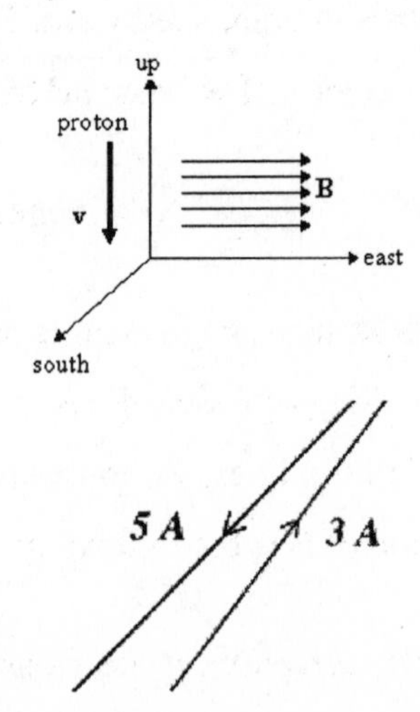

4. Two long parallel wires are separated by 0.2 meters and carry currents of 3 and 5 amps as shown in the figure. What is the direction of the magnetic force felt by the wire on the right? By the wire on the left? How will the magnitudes of these forces compare?

5. Let's say you shuffle across a carpet on a dry winter day and pick up a charge of 5 microcoulombs. What force will you feel due to the magnetic field of Earth.

6. The magnetic field can often be calculated easily by using Ampere's law. This law is similar to Gauss's Law, where we had to use a "Gaussian surface." Now, we must use an imaginary "Amperian loop." What are the essential features of an Amperian loop?

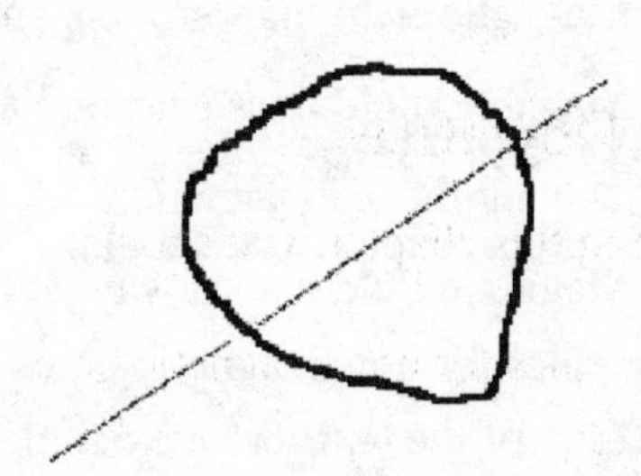

7. You can make a good approximation to a "long solenoid" by buying a 50-yd. spool of wire and winding it carefully (nice, even coils) around a cylindrical core (say a broomstick), then removing the core. Estimate the maximum magnetic field that can be produced in such a solenoid.

Chapter Review

In this chapter we study magnetism. It turns out that electricity and magnetism are interconnected. We begin to glimpse at this connection in this chapter. The full story will emerge later.

22–1 The Magnetic Field

Some materials are almost always magnetic as a direct result of their structure; objects made of these materials are called **permanent magnets**. The prototype object for discussing the behavior of these materials is the simple bar magnet. The two ends of a bar magnet behave differently. One end, the **north pole**, of the magnet tends to point northward with respect to the earth. The other end, the **south pole**, of the magnet tends to point southward. Magnets always have both a north and a south pole. There is a force between two magnets. The basic behavior of this force is reminiscent of the force between two charges:

like magnetic poles repel; opposite poles attract.

As with electricity, magnetism is associated with the presence of a **magnetic field**, for which we use the symbol **B**. Magnetic field lines can be drawn to get a visual representation of this field. In order to draw magnetic field lines, we must know how to determine the direction of the magnetic field. The rule for the direction of **B** is the following:

The direction of the magnetic field at a given location is the direction that the north pole of a compass needle would point if placed at that location.

With the above definition, we can now state the rules for drawing magnetic field lines:

(1) Magnetic field lines are tangent to the magnetic field at every point

(2) Magnetic field lines start on the north poles of a magnet and end on its south pole

(3) The number of magnetic field lines is proportional to the magnitude of the field

(4) Magnetic field lines always form closed loops.

See Figure 22–4 of the text for an example of magnetic field lines.

The fact that bar magnets interact with Earth is evidence that Earth has a magnetic field. The poles of Earth's magnetic field are near to Earth's geographic poles. However, the definition of the magnetic poles and how they behave requires that the magnetic pole of the earth that's near to Earth's north pole (called the **north magnetic pole**) is actually the *south pole* of Earth's magnetic field. Similarly, the magnetic pole of the earth that's near to Earth's south pole (called the **south magnetic pole**) is actually the *north pole* of Earth's magnetic field (see Figure 22–6 of the text).

Practice Quiz

1. If a magnetic field is directed along the negative y-axis, toward which direction would the south pole of a compass needle point if placed in this magnetic field?

   (a) +x axis **(b)** -x axis **(c)** +y axis **(d)** -y axis **(e)** none of the above

22–2 The Magnetic Force on Moving Charges

In addition to the behavior described above, magnets also apply a force to charged objects that are moving through the magnetic field. The magnitude of the force on this charge depends on the magnitude of the charge q, the magnitude of its velocity v, the magnitude of the magnetic field B, and the smallest angle θ between the vectors **B** and **v**:

$$F = qvB\sin\theta$$

Notice that if $\theta = 0°$ (or $180°$), that is, if the charge moves in the direction of the magnetic field, the force on it is zero. Therefore, a force is exerted only if the velocity of the charge has a component perpendicular to the magnetic field.

The above expression for the force on a moving charge can be used to obtain the magnitude (or strength) of the magnetic field:

$$B = \frac{F}{qv\sin\theta}$$

This relation shows that the SI unit of the magnetic field can be derived from N/(C·m/s). After some rearranging, this unit can be written as N/(A·m), which is called a **tesla** (T): 1 T = 1 N/A·m.

The direction of the force on a moving charge is found to be perpendicular to both **B** and **v**. That is, **F** is perpendicular to the plane formed by the vectors **B** and **v**. However, this statement still leaves two possible directions for **F** on either side of the **B**–**v** plane. To determine the direction of **F** more precisely we use the **magnetic force right-hand-rule**:

> *To find the direction of the magnetic force on a positive charge, point the fingers of your right hand in the direction of* **v**. *Orient your hand such that your fingers can curl toward the direction of* **B**. *Your thumb then indicates the direction of* **F**. *If the charge is negative, the force points in the opposite direction indicated by your thumb.*

Essentially, the above rule determines on which side of the **B**–**v** plane **F** points; since you know **F** must be perpendicular to this plane, then you know the precise direction of **F**.

The magnetic force on a moving charge is necessarily a three-dimensional situation. On two-dimensional sheets of paper this force involves the directions directly into the page and out of the page. These directions are indicated by using the symbol ⊗ for *into the page*, and ⊙ for *out of the page*.

Example 22–1 Force on a Moving Charge The magnetic field in a region is $\mathbf{B} = (2.5\ \mathrm{T})\hat{\mathbf{x}} + (2.5\ \mathrm{T})\hat{\mathbf{y}}$. A particle of charge –1.7 C moves into this region with a velocity of $\mathbf{v} = (4.5\ \mathrm{m/s})\hat{\mathbf{x}}$. Determine the magnitude and direction of the force on this charge.

Picture the Problem The sketch shows the magnetic field and velocity vectors.

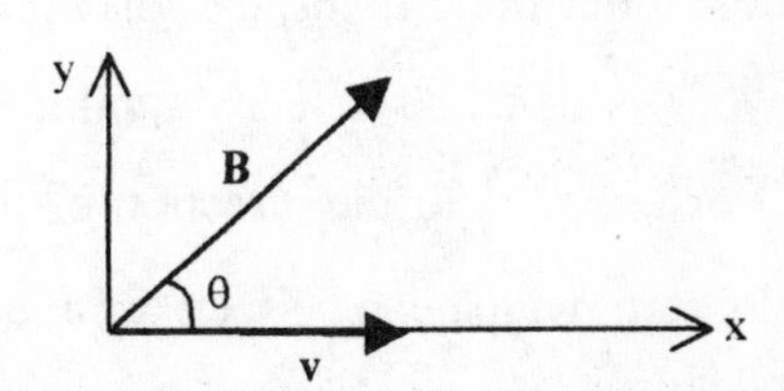

Strategy From the components given, it is clear that **B** makes a 45° angle in the 1st quadrant. With this insight we can proceed using the expression for the force.

Solution

1. The magnitude of **B** is: $B = \sqrt{B_x^2 + B_y^2} = \sqrt{2(2.5\ \mathrm{T})^2} = 3.54\ \mathrm{T}$

2. The magnitude of the force is: $F = qvB\sin\theta = (1.7\ \mathrm{C})(4.5\ \tfrac{\mathrm{m}}{\mathrm{s}})(3.54\ \tfrac{\mathrm{m}}{\mathrm{s}})\sin(45^\circ) = 19\ \mathrm{N}$

3. The right-hand rule gives the direction as: into the page, ⊗

Insight Only the magnitude of the charge was used to get the magnitude of the force as appropriate. Also, make sure you recognize that because the charge is negative, the direction of the force is opposite to the direction indicated by your thumb when you use the magnetic force right-hand rule.

Practice Quiz

2. A magnetic field is directed vertically from the bottom of this page toward the top, ↑. A positively charged particle moves with an initial velocity from the left toward the right, →. What is the direction of the force on this particle?

   (a) ↓ **(b)** ← **(c)** ⊗ **(d)** ⊙ **(e)** F = 0, so there is no direction.

3. An *electron* moves with a velocity that is into the page, ⊗. It experiences a magnetic force to the left, ←. What is the direction of the magnetic field in which the electron is moving?

   (a) ↓ **(b)** ↑ **(c)** ⊙ **(d)** → **(e)** B = 0, so there is no direction.

4. A particle of charge q moves with speed v at an angle θ with respect to a magnetic field **B**, producing a force of magnitude F on the particle. If a different particle of charge –q moves with speed v/2 at the same angle θ through this same magnetic field, the magnitude of the force on this charge will be

   (a) F/4 **(b)** F/2 **(c)** F **(d)** 2F **(e)** 4F

22–3 The Motion of Charged Particles in a Magnetic Field

For uniform magnetic fields, the force law for moving charges gives rise to three types of motion. One of these types is constant-velocity motion. This motion occurs when the velocity of the charge is either parallel or opposite to the direction of the magnetic field. For this situation, the magnetic force on the charge is zero, which can be seen from the fact that, for $\theta = 0°$ or $180°$, $\sin(\theta) = 0$.

Another type of motion that results in a uniform magnetic field is that of uniform circular motion. This motion occurs when the velocity of the charge is perpendicular to the magnetic field. Since the magnetic force is perpendicular to the velocity, so is the acceleration. When the acceleration of an object remains perpendicular to its velocity uniform circular motion is the result. For a charged particle of mass m, whose charge has magnitude q, moving with velocity **v** in a direction perpendicular to **B**, the radius of its uniform circular motion is given by

$$r = \frac{mv}{qB}$$

This radius is sometimes called the *cyclotron radius.*

The third type of motion is a combination of the above linear and circular motions. This occurs when the velocity of the charge is neither along nor perpendicular to the magnetic field lines. The component of the velocity that is perpendicular to **B**, $v_{\perp}$, contributes a circular component to its motion

$$r = \frac{mv_{\perp}}{qB}$$

and the fact that the velocity has a component along the magnetic field lines, $v_{\parallel}$, means that there is also a constant-velocity drift, with speed $v_{\parallel}$, taking place. This combination of linear and circular motion is called *helical motion* because the path of the particle is that of a helix.

Example 22–2 Motion in a Magnetic Field A uniform magnetic field of 3.75 T points in the positive x direction. Particles of mass 25.0 g and charge 6.20 mC are fired into this magnetic field at angles of 90.0°, 45.0°, and 0.00° above the positive x-axis, in the x-y plane, with a speed of 35.0 cm/s. Calculate the radius of each particle's path and describe its motion.

Picture the Problem The sketch shows the magnetic field vector together with three velocity vectors in the stated directions.

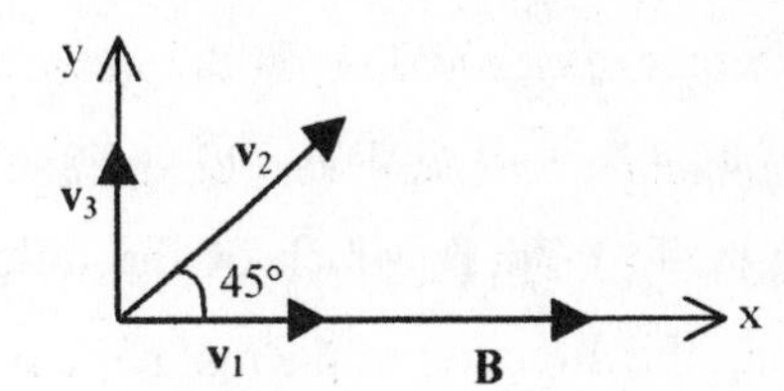

Strategy We can use the results discussed above to determine the radius in each case and then consider which type of motion we have.

Solution

1. For the case with v_1 there is no component perpendicular to B; therefore: we get constant-velocity motion with velocity $\mathbf{v}_1$.

2. For particle 2 the perpendicular component of $\mathbf{v}_2$ is given by: $v_{2\perp} = v_2 \sin(45^\circ)$

3. The radius of the motion for particle 2 is: $r_2 = \frac{mv_{2\perp}}{qB} = \frac{(0.025\ \text{kg})(0.350\ \frac{\text{m}}{\text{s}})\sin(45^\circ)}{(6.20\times10^{-3}\ \text{C})(3.75\text{T})} = 0.266\ \text{m}$

4. The parallel component of $\mathbf{v}_2$ is: $v_{2\parallel} = v_2 \cos(45^\circ) = (0.350\,\text{m/s})\cos(45^\circ) = 0.247\,\text{m/s}$

5. The motion of particle 2 is: helical motion of radius 0.266 m with a constant speed drift of 0.247 m/s in the positive x direction

6. Particle 3 moves perpendicular to B. The radius of its path is: $r_3 = \dfrac{mv_3}{qB} = \dfrac{(0.025\text{ kg})(0.350\text{ m/s})}{(6.20\times10^{-3}\text{C})(3.75\text{T})} = 0.376\text{ m}$

7. Since $\mathbf{v}_3$ has no parallel component: we have uniform circular motion with speed 35.0 cm/s and radius 37.6 cm.

Insight Each of the three types of motion is treated here.

Practice Quiz

5. For a particle of mass m and charge q moving at an angle θ with respect to the direction of a uniform magnetic field **B**, which of the following values of θ will produce the path of largest radius?

   (a) 0° **(b)** 30° **(c)** 45° **(d)** 60° **(e)** 90°

6. For a particle of mass m and charge q moving at an angle θ with respect to the direction of a uniform magnetic field **B**, which of the following values of θ will produce the fastest constant-speed drift along the direction of the magnetic field?

   (a) 0° **(b)** 30° **(c)** 45° **(d)** 60° **(e)** 90°

22–4 The Magnetic Force Exerted on a Current-Carrying Wire

Since a force is exerted on a moving charge in a magnetic field, and current consists of moving charges, then a force will be exerted on a current-carrying wire in a magnetic field. The force exerted on this wire is the resultant of the forces on the moving charges that make up the current. The magnitude of this net result works out to be

$$F = ILB\sin\theta$$

where θ is the direction between the current flow and the magnetic field. The direction of this force is given by the same magnetic force right-hand rule as for individual charges. In this case the direction of the velocity (for a positive charge) is the same as the direction of the current.

Practice Quiz

7. A uniform magnetic field is directed into the page, ⊗. A current I flows vertically from the bottom of this page toward the top, ↑. What is the direction of the force on the wire carrying this current?

   (a) ↓ **(b)** → **(c)** ⊗ **(d)** ⊙ **(e)** none of the above

8. A current I flows out of the page, ⊙. It sits in a magnetic field B that's directed into the page, ⊗. What is the direction of the force on the wire carrying this current?

   (a) ↓ **(b)** ← **(c)** ↑ **(d)** → **(e)** F = 0, so there is no direction.

22–5 Loops of Current and Magnetic Torque

When an entire loop of current sits in a magnetic field, the forces on the different parts of the loop can result in a net torque on the loop depending on how the loop is oriented with respect to the magnetic field. The magnitude of this torque is given by

$$\tau = NIAB\sin\theta$$

where N is the number of "turns," that is, the number of times the wire carrying the current I is wrapped around the loop, A is the area of the loop, and B is the magnetic field. In the case of the torque on a current loop, the angle θ refers to the angle between the magnetic field and the direction normal to the plane of the loop. The normal direction used to determine θ in the above expression is determined by its own right-hand rule

The normal direction for a current loop is in the direction of the thumb when the fingers of the right hand curl around the loop in the direction of the current.

The effect of this torque on a current loop is to rotate the loop such that its normal direction coincides with the direction of the magnetic field.

Example 22–3 Torque Conducting wire is wrapped once around a plastic rectangular slab that is 10.0 cm × 15.0 cm in size. The construction is attached to an axle so that it is free to rotate about its center. If a current of 1.50 A travels through the wire in a clockwise direction as viewed from the face of the slab, what is the magnitude and direction of a uniform magnetic field that causes the slab to experience a maximum torque of 2.50 N·m rotating it counterclockwise as viewed from above the right-hand side of the axle?

Picture the Problem The diagram shows the current loop and an eye indicating how the rotating loop is viewed.

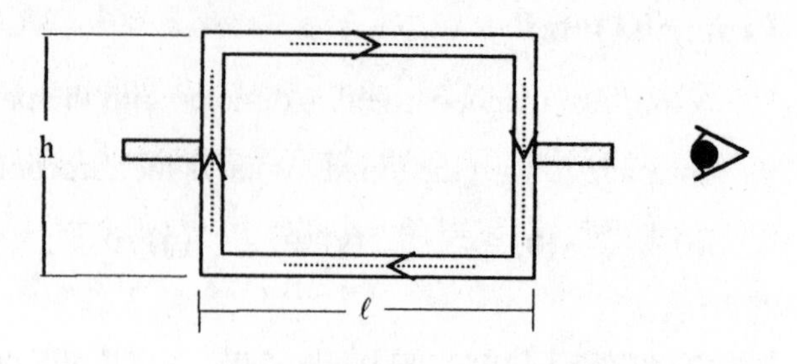

Strategy The maximum torque occurs for sin θ = 1, so we can apply the expression for torque to find the magnitude of **B**. For the direction of **B** we'll consider the needed direction of the force on a segment of wire.

Solution

1. Setting sin θ = 1 we can use the relationship between B and τ to solve for B:

$$\tau_{max} = NIAB \Rightarrow B = \frac{\tau_{max}}{NIA} = \frac{\tau_{max}}{IA}$$

2. The area is $A = \ell h$ so B is:

$$B = \frac{\tau_{max}}{I\ell h} = \frac{2.50\ \text{N}\cdot\text{m}}{(1.50\ \text{A})(0.150\ \text{m})(0.100\ \text{m})} = 111\ \text{T}$$

3. For the eye to see counterclockwise rotation: The direction of force on the upper wire is ⊙.

4. Given the above direction of the force, the magnetic force right-hand rule gives a magnetic field direction: The direction of **B** on the upper wire must be upward, ↑.

5. Thus, the magnetic field is: **B** = 111 T, pointing upward.

Insight Verify that using this direction for **B** on the other wire segments is consistent with the required torque.

Practice Quiz

9. A single-turn, circular loop of current 2.3 A has a radius of 8.9 cm and experiences a maximum torque of 0.065 N·m when placed in a magnetic field. What is the magnitude of the magnetic field?
(a) 1.1 T **(b)** 1.3 T **(c)** 0.0 T **(d)** 0.32 T **(e)** 0.10 T

22–6 Electric Currents, Magnetic Fields, and Ampere's Law

The connection between current and magnetism goes further than just the force on a current-carrying wire. It turns out that currents are sources of magnetism in that a current creates a magnetic field in the space around it. Recall that magnetic fields form closed loops, so the magnetic field created by a current loops around the current. The direction in which the magnetic field loops around the current is determined by a **magnetic field right-hand rule**:

> *If you orient the thumb of your right hand along the current-carrying wire in the direction of the current, your fingers will curl around the wire in the direction of the magnetic field.*

The closed loops formed by the magnetic field enclose the current that creates the field. The relationship between the enclosed current and the field it creates is given by **Ampere's law**:

$$\sum B_{\parallel}\Delta L = \mu_0 I_{\text{enclosed}}$$

In this expression, $B_{\parallel}$ is a component of **B** that is parallel to a little segment, ΔL, of a path that encloses the current I_{enclosed}. The factor μ_0 is a fundamental constant called the **permeability of free space**; it has the value

$$\mu_0 = 4\pi \times 10^{-7}\ \text{T}\cdot\text{m/A}$$

Ampere's law is somewhat like Gauss's law in that it is most useful for finding the magnetic field in cases with a high degree of symmetry. One such case is that of a long, straight wire carrying a current I. In this case, Ampere's law gives the result for the magnitude of the magnetic field to be

$$B = \frac{\mu_0 I}{2\pi r}$$

Now that we know that a current both creates a magnetic field and experiences a force in a magnetic field, then two parallel current-carrying wires, I_1 and I_2, will exert forces on each other. Each current "feels" the force due to the magnetic field created by the other. They will exert equal and opposite forces on each other of magnitude

$$F = \frac{\mu_0 I_1 I_2}{2\pi d} L$$

where L is the length of the wires (assumed to be equal for simplicity). The direction of the force on each wire can be determined by the magnetic force right-hand rule. By applying this rule to each wire, we find that if the currents are in the same direction, the two wires attract each other, and if the currents are in opposite directions, the wires repel each other.

Example 22–4 Force between Currents and the Ampere The same current runs through two parallel wires that are separated by 1.00 m. If the force per unit length on each wire is 2.00 x 10^{-7} N/m, what current runs through the wires?

Picture the Problem The sketch shows two parallel wires carrying the same current I.

I

I

Strategy We can solve this problem by manipulating the equation for the force between the currents.

Solution

1. The force per unit length is:

$$F = \frac{\mu_0 I_1 I_2}{2\pi d} L \quad \Rightarrow \quad \left(\frac{F}{L}\right) = \frac{\mu_0 I^2}{2\pi d}$$

2. The current is given by:

$$I = \sqrt{\frac{2\pi d (F/L)}{\mu_0}} = \sqrt{\frac{2\pi(1.00\text{ m})(2.00\times 10^{-7}\text{ N/m})}{4\pi\times 10^{-7}\text{ T}\cdot\text{m/A}}}$$

$$= 1.00\text{ A}$$

Insight This calculation is the basis for using the ampere as the fundamental quantity of electricity rather than the coulomb. It turns out to be more accurate to measure one ampere of current this way than it is to measure one coulomb of charge.

Practice Quiz

10. A long, straight current I produces a magnetic field of magnitude B a distance r away from it. What is the magnitude of the magnetic field at a distance of r/2 from the wire?

    (a) B/4 **(b)** B/2 **(c)** B **(d)** 2B **(e)** 4B

11. Two parallel wires of equal length, separated by a distance d, carry the same current I in opposite directions. If the current is reduced to I/2, the force between the wires becomes

    (a) I/4 **(b)** I/2 **(c)** I **(d)** 2I **(e)** 4I

22–7 Current Loops and Solenoids

In the previous section the result for the magnetic field of a long, straight wire was given. There are other useful configurations of current-carrying wires such as current loops and **solenoids**. For a circular loop of wire of radius R, carrying a current I, and containing N turns, the magnitude of the magnetic field at the center of the loop is given by

$$B = \frac{N\mu_0 I}{2R}$$

A solenoid is an electrical device in which a wire has been wound into the geometry of a helix. The circular loops are usually tightly packed. Solenoids are sometimes referred to as *electromagnets.* A solenoid carrying a current I produces a nearly uniform magnetic field inside the loops. Ampere's law shows that the magnitude of this field is given by

$$B = \mu_0 \frac{N}{L} I = \mu_0 n I$$

where N is the number of turns and L is the length of the solenoid. The quantity $n = N/L$ is the number of turns per unit length.

Practice Quiz

12. How many turns per centimeter are needed to produce a 1.00 T magnetic field inside a solenoid with a 10.0-A current?

(a) 10.0 **(b)** 796 **(c)** 1.00 **(d)** 79,600 **(e)** $4\pi \times 10^7$

22–8 Magnetism in Matter

The magnetic properties of matter is an important branch of applied physics. The very fact that materials have magnetic properties can be traced back to the fact that the electrons in the atoms of a substance possess little magnetic fields as part of their fundamental character. For many atoms, there is a net cancellation of these little magnetic fields. However, the structure of some substances (as with iron and nickel) is such that there is a nonzero net magnetic field. In bulk, the magnet fields of these atoms tend to align, giving rise to a permanent magnetic field. Materials that behave this way are called **ferromagnets**. Ferromagnetic materials can often lose their magnetism at high temperatures.

In some materials the tendency for the magnetic fields of the atoms to align themselves is too weak to produce a significant effect. However, an external magnetic field applied to the material will produce alignment, resulting in a magnetized material. This behavior is called **paramagnetism**. We briefly mention one more effect, called **diamagnetism**, in which an external magnetic field applied in a given direction produces an oppositely directed magnetic field in the material.

Reference Tools and Resources

I. Key Terms and Phrases

north pole the north pole of a magnetic is the end that tends to point northward with respect to the earth

south pole the south pole of a magnetic is the end that tends to point southward with respect to the earth

magnetic field the field created by a magnetic material or a current that carries its magnetic affect

magnetic field lines a pictoral representation of a magnetic field

tesla the SI unit of the magnetic field

magnetic force right-hand rule the rule for determining the direction of the magnetic force on a moving charge or current. For a positive charge, place the fingers of your right hand along **v** so that they would curl toward **B**; your thumb then gives the direction of **F**. Reverse the direction for a negative charge

magnetic field right-hand rule the rule for determining the direction of the magnetic field created by a current. Orient the thumb of your right hand along the current, your fingers then curl in the direction of **B**

Ampere's law the fundamental law relating a current to the magnetic field it creates

permeability of free space the fundamental constant of magnetism in Ampere's law

solenoid an electrical device in which a wire has been wound into the geometry of a helix

ferromagnetism the phenomenon that some materials have a spontaneously created "permanent" magnetic field

paramagnetism the phenomenon that some materials obtain a "permanent" magnetic field after exposure to an external magnetic field

diamagnetism the phenomenon that matter produces a magnetic field in the opposite direction of an external magnetic field

II. Important Equations

Name/Topic	Equation	Explanation
magnetic force on a moving charge	$F = qvB\sin\theta$	The force law for the magnitude of the force on a moving charge in a magnetic field.
magnetic force on a current-carrying wire	$F = ILB\sin\theta$	The magnitude of the magnetic force on a current-carrying wire.
electric currents and magnetic fields	$B = \frac{\mu_0 I}{2\pi r}$	The magnitude of the magnetic field due to a long, straight current I.
two parallel current-carrying wires	$F = \frac{\mu_0 I_1 I_2}{2\pi d} L$	The magnitude of the force between two parallel current-carrying wires.
solenoids	$B = \mu_0 \frac{N}{L} I = \mu_0 n I$	The magnitude of the magnetic field inside a solenoid.

III. Know Your Units

Quantity	Dimension	SI Unit
magnetic field (**B**)	$[M][A^{-1}][T^{-2}]$	T
permeability of free space (μ_0)	$[M][L][A^{-2}][T^{-2}]$	T·m/A

Practice Problems

1. A charge of 52 μC moves in a magnetic field of 1.5 T with a speed of 800 m/s, making an angle of 41 degrees with the field, to the nearest tenth of a mN (millinewton), what is the force acting on it?

2. A charge of 8 μC moves at –366 m/s (right is positive) as shown in the figure; the electric field is 3478 N/C, and the magnetic field is 3 T. To the nearest tenth of a mN, what is the net force acting on the particle (up is positive)?

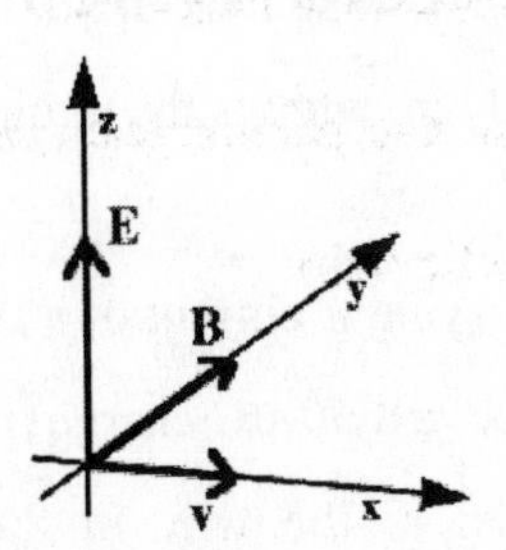

3. If the -8-μC charge is moving at 89 m/s (right is positive), B= 3.5 T, and E = 3478 N/C (up is positive), what is the magnitude of the force acting on it to the nearest tenth of a mN?

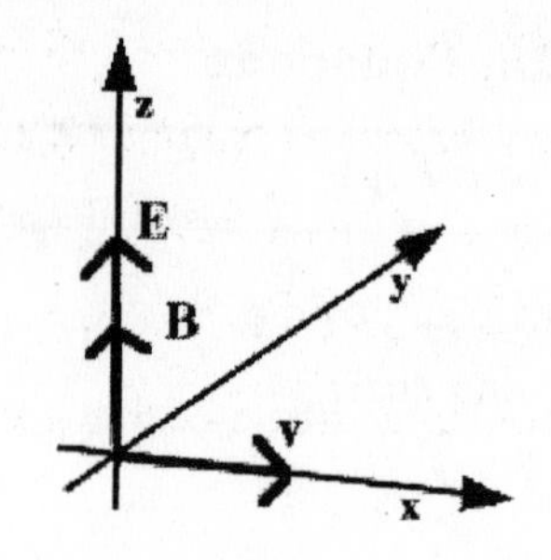

4. In problem 3 what is the angle of the force, to the nearest tenth of a degree, in the y-z plane measured counter-clockwise from the positive y-axis?

5. In the figure what is the velocity of the charged particle, to the nearest tenth of a m/s, that will allow it to travel undeflected through the crossed magnetic field if E = 8887 N/C and B = 4.5 T?

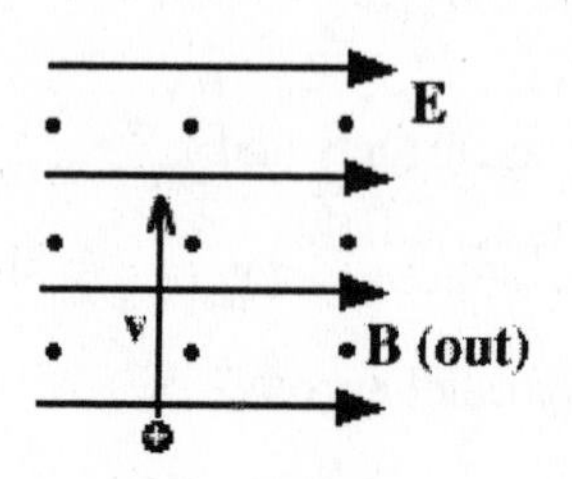

6. The left wire carries a current of I_1 = 36 amps. To the nearest hundredth of a millitesla, what is the magnitude of the magnetic field at the location of the right wire?

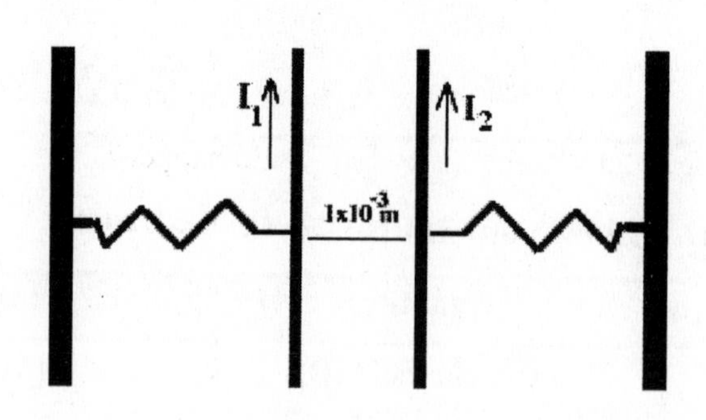

7. If the right wire carries the same current as the left one, and each is 10 meters long, and the spring constant is 38 N/m for each spring, to the nearest hundredth of a meter how far is each spring stretched?

8. The 10-turn current coil shown in the sketch has an area of 0.10 m^2 and carries a current of 16.4 amps while making an angle of 59 degrees with a 1.5-T magnetic field. To the nearest tenth of a N·m, what is the torque?

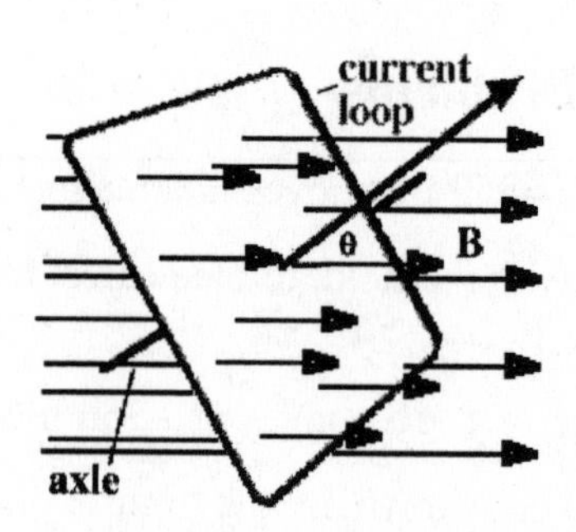

9. A wire carrying a current of -19 amps (right is positive) is directly above a 50-μC charge traveling with a speed of 700 m/s parallel to the wire in the direction shown. To the nearest tenth of a nN, what is the force (up is positive) on the charge if the distance between the charge and the wire is 0.16 meters?

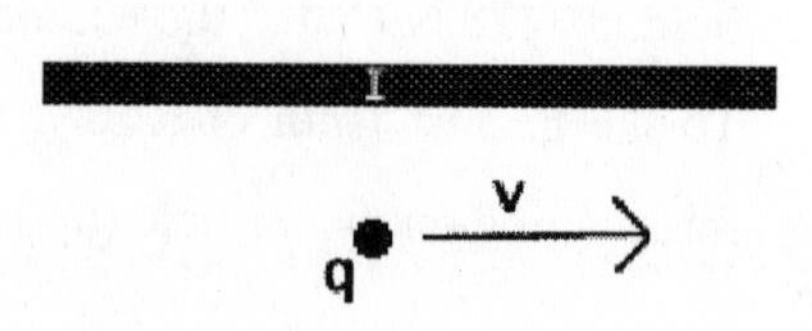

10. A 400-turn, 40 cm solenoid carries a current of 17 amps. What is the magnitude of the force, to the nearest tenth of a mN, on a 50-mC charge traveling at a speed of 1000 m/s on the axis of the solenoid at an angle of 8 degrees?

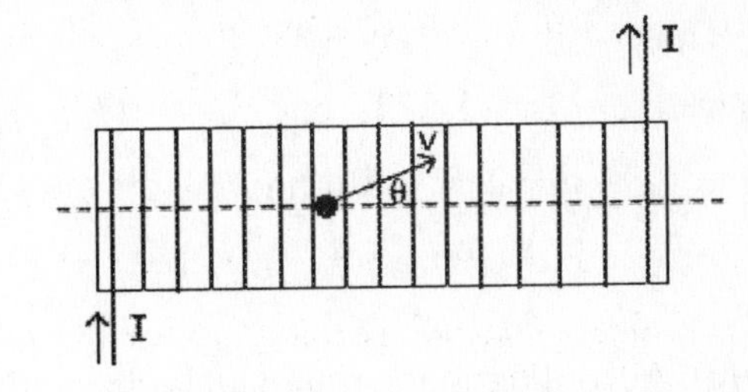

Puzzle

THE SNAKE

The figure shows a flexible wire loop lying on a frictionless tabletop. The wire can slide around and change its shape freely, but it cannot move perpendicular to the table.

Part a:

If a current were made to flow counterclockwise in the loop, which of the following would happen?

1. The wire would stretch out into a circle.
2. The wire would contract down to a tangled mess.
3. Nothing would happen.
4. There is not enough information to tell.

Part b:

If the current went clockwise, would that affect your answer to part a? Answer this in words, not equations, briefly explaining how you obtained your answer.

Selected Solutions

17. By conservation of energy, the magnitude of the electric potential energy lost by the electron, eV, is converted into kinetic energy. Therefore,

$$\frac{1}{2}mv^2 = eV \quad \Rightarrow \quad v = \sqrt{\frac{2eV}{m}}$$

We can rewrite the expression for the radius of the path to solve for B:

$$r = \frac{mv}{eB} \quad \Rightarrow \quad B = \frac{mv}{er}$$

Thus,

$$B = \frac{m}{er}\sqrt{\frac{2eV}{m}} = \frac{1}{r}\sqrt{\frac{2mV}{e}} = \frac{1}{0.17\text{ m}}\sqrt{\frac{2(9.11\times10^{-31}\text{ kg})(310\text{ V})}{1.60\times10^{-19}\text{ C}}} = \boxed{0.35\text{ mT}}$$

19. (a) According to the right-hand rule, a positive charge would experience a force to the left. Since the particle is experiencing a force to the right, it must have $\boxed{\text{negative}}$ charge.

(b) We can rewrite the expression for the radius of the path to solve for m

$$r = \frac{mv}{eB} \Rightarrow m = \frac{erB}{v}$$

Thus,

$$m = \frac{(1.60\times10^{-19}\text{ C})(0.520\text{ m})(0.180\text{ T})}{\left(1.67\times10^{-27}\ \frac{\text{kg}}{\text{u}}\right)\left(6.0\times10^{6}\ \frac{\text{m}}{\text{s}}\right)} = \boxed{1.5\text{ u}}$$

35. Since $\tau = IAB\sin\theta$, and we are given that $\tau = \frac{1}{2}\tau_{max}$, therefore, $\sin\theta = \frac{1}{2}$. Thus,

$$\theta = \sin^{-1}\left(\frac{1}{2}\right) = \boxed{30°}$$

47. (a) The force per unit length is given by

$$\frac{F}{L} = \frac{\mu_0 I_1 I_2}{2\pi d} = \frac{\left(4\pi\times10^{-7}\ \frac{\text{T}\cdot\text{m}}{\text{A}}\right)(4.33\text{ A})(1.75\text{ A})}{2\pi(0.322\text{ m})} = \boxed{4.71\times10^{-6}\text{ N/m}}$$

(b) Assuming that the lengths of each wire are the same, the force per meter exerted on the 4.33-A wire is $\boxed{\text{the same}}$ as the force per meter exerted on the 1.75-A wire because these forces form an action-reaction pair.

61. The magnetic field at each point is the vector sum of the magnetic fields due to each wire:

$$\mathbf{B}_A = \mathbf{B}_{1A} + \mathbf{B}_{2A} = \frac{\mu_0 I}{2\pi r_1}\odot + \frac{\mu_0 I}{2\pi r_2}\odot = \frac{\mu_0 I}{2\pi}\left(\frac{1}{r_1} + \frac{1}{r_2}\right)\odot$$

$$= \frac{\left(4\pi\times10^{-7}\ \frac{\text{T}\cdot\text{m}}{\text{A}}\right)(2.2\text{ A})}{2\pi}\left(\frac{1}{0.075\text{ m}} + \frac{1}{3(0.075\text{ m})}\right)\odot = \boxed{(7.8\,\mu\text{T})\odot}$$

$$\mathbf{B}_B = \mathbf{B}_{1B} + \mathbf{B}_{2B} = \frac{\mu_0 I}{2\pi}\left(\frac{1}{r_1}\otimes + \frac{1}{r_2}\odot\right) = \frac{\mu_0 I}{2\pi}\left(\frac{1}{r} - \frac{1}{r}\right)\odot = \boxed{0}$$

$$\mathbf{B}_C = \mathbf{B}_{1C} + \mathbf{B}_{2C} = \frac{\mu_0 I}{2\pi}\left(\frac{1}{r_1}\otimes + \frac{1}{r_2}\otimes\right) = \frac{\left(4\pi\times10^{-7}\ \frac{\text{T}\cdot\text{m}}{\text{A}}\right)(2.2\text{ A})}{2\pi}\left(\frac{1}{3(0.075\text{ m})} + \frac{1}{0.075\text{ m}}\right)\otimes = \boxed{(7.8\,\mu\text{T})\otimes}$$

Answers to Practice Quiz

1. (c) **2**. (d) **3**. (b) **4**. (b) **5**. (e) **6**. (a) **7**. (e) **8**. (e) **9**. (a) **10**. (d) **11**. (a) **12**. (b)

Answers to Practice Problems

1. 40.9 mN **2**. 19.0 mN

3. 27.9 mN **4**. 84.9 °

5. 1974.9 m/s **6**. 7.20 mT

7. 0.07 m **8**. 21.1 N·m

9. 831.2 nN **10**. 148.7 mN

CHAPTER 23

MAGNETIC FLUX AND FARADAY'S LAW OF INDUCTION

Chapter Objectives

After studying this chapter, you should:

1. know how to calculate the magnetic flux through a surface of area A.

2. know the relationship between induced emf and magnetic flux: Faraday's law of induction.

3. be able to use Lenz's law to determine the direction of an induced current.

4. know the relationship between a magnetic field and an induced electric field.

5. understand the basic principle of how electric generators produce alternating current.

6. understand the concept of self-inductance and the behavior of RL circuits.

7. be able to calculate the energy stored in the magnetic field of an inductor.

8. understand how step-up and step-down transformers work.

Warm-Ups

1. Lenz's law is said to be a consequence of the principle of conservation of energy. Explain this statement by describing what would happen if Lenz's law were reversed. Take the example of a wire loop with a changing magnetic flux, and describe what would happen if current were induced opposite to the way it really is.
2. Let's say you take an ordinary wire coat hanger and straighten out the hook shaped part that normally hangs over the coat rack. Now, you can spin the (roughly) triangular part around by twisting the straightened part between your fingers. Estimate the EMF that you can generate by spinning the hanger in Earth's magnetic field (about 5×10^{-5} T).

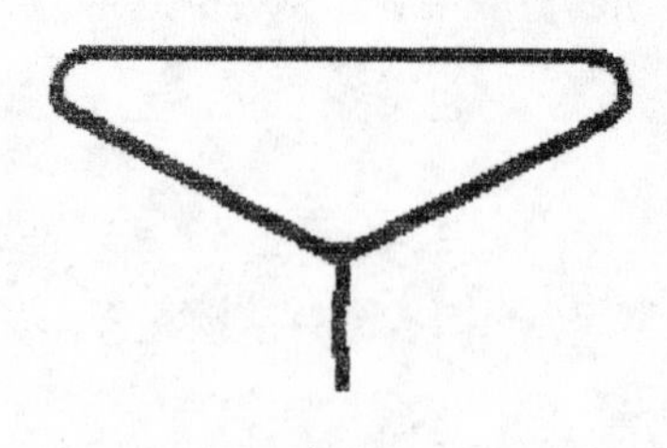

3. Here's one way of understanding a capacitor: It is a device that won't let the voltage between two points change too rapidly, because it stores up charge and has $V = Q/C$. The charge cannot be changed instantaneously, so the voltage cannot either. Describe an inductor in a similar way, that is, say what cannot be changed rapidly and why.

4. Estimate the inductance of a solenoid made by buying a typical spool of wire from hardware store and winding it carefully around a broom handle.
5. A wire hoop surrounds a long solenoid. If the current in the solenoid increases, will there be any current induced in the loop? If so, which way will the current flow?
6. A car battery has an emf of only 12 V, yet energy from the battery provides the 20,000 V spark that ignites the gasoline. How is this possible?

Chapter Review

This chapter provides more detail about the connection between electricity and magnetism. The very important concept of electromagnetic induction is introduced. Some of the most important applications of this concept, such as electric generators, motors, and transformers, are also treated.

23–1 – 23–2 Induced EMF and Magnetic Flux

The fundamental fact on which this chapter is based is that a magnetic field, when it changes, can produce an electric field. This is seen in the fact that when the magnetic field passing through a coil changes, and an **induced current** is observed in the coil. The induced current is present just as if a battery had been placed in the coil, so we say that there is an **induced emf** as a result of the changing magnetic field. The magnitude of the induced emf is directly proportional to the rate at which the magnetic field changes.

The precise magnitude of the induced emf is nicely represented in terms of the **magnetic flux** Φ through the coil. As with electric flux, magnetic flux represents the "flow" of the magnetic field through a surface. The magnetic flux through a surface of area A is calculated in the same way as the electric flux:

$$\Phi = BA\cos\theta$$

where, again, θ is the angle between the magnetic field and the direction normal to the surface. The SI unit of magnetic flux is called the **weber** (Wb): 1 Wb = 1 T·m^2.

Example 23–1 Magnetic Flux A uniform magnetic field of magnitude 0.250 T makes an angle of 35° with the normal to a circular aperture of radius 65 cm. If the aperture lies entirely in the magnetic field, what is the magnetic flux through the aperture?

Picture the Problem The sketch shows a circular aperture, the magnetic field vector and a unit vector, $\hat{\mathbf{n}}$ representing the direction normal to the surface.

$\hat{\mathbf{n}}$ 35° **B**

Strategy The expression for the magnetic flux through an area A can be used directly for this calculation. We need only to first calculate the area of the aperture.

Solution

1. The area of the circular aperture is: $A = \pi r^2$

2. The magnetic flux then is:

$$\Phi = B(\pi r^2)\cos(\theta) = (0.250\text{ T})\pi(0.65\text{ m})^2\cos(35°)$$
$$= 0.27\text{ Wb}$$

Insight This example is almost the same as Example 19–4 for electric flux. Thus, if you understood electric flux then, you understand magnetic flux now.

Practice Quiz

1. What angle between **B** and the normal to a surface produces the largest magnetic flux through that surface?

   (a) 0° **(b)** 30° **(c)** 45° **(d)** 60° **(e)** 90°

2. A magnetic field **B** passes through an area A, at angle θ, giving rise to a magnetic flux Φ. If B is doubled and A is halved, the magnetic flux through the area will be

   (a) Φ/4 **(b)** Φ/2 **(c)** Φ **(d)** 2Φ **(e)** 4Φ

23–3 – 23–4 Faraday's Law of Induction and Lenz's Law

Using the concept of magnetic flux, we can write an expression to determine the precise magnitude of the induced emf that arises in a coil when the magnetic field changes. What is observed is that the magnitude of the induced emf $\mathcal{E}$ in a single coil equals the rate of change of the magnetic flux through the coil. If the coil has N turns, the result for the induced emf is

$$\mathcal{E} = -N\frac{\Delta\Phi}{\Delta t}$$

This result is known as **Faraday's law of induction**. Often, only the magnitude of the induced emf is desired; in such cases Faraday's law is simply written as $|\mathcal{E}| = N|\Delta\Phi/\Delta t|$.

The minus sign in Faraday's law is there to account for the polarity (or direction) of the induced emf. The polarity of an induced emf is correctly determined by applying **Lenz's law**:

> *The polarity of the induced emf in a coil is such that the induced current creates a magnetic field whose flux opposes the original change in flux.*

Notice, the law says only that the flux due to the induced magnetic field *opposes* the original change in flux, but it does not cancel the original change in flux. As explained in the text, this result is the only possibility that preserves the conservation of energy.

Example 23–2 Faraday's Law A square loop of side length 25.0 cm sits in a magnetic field that makes an angle of 60.0° with the normal to the loop. If the magnetic field increases uniformly from 0.250 T to 1.25 T in 0.525 seconds, determine the magnitude and polarity of the induced emf in the loop.

Picture the Problem The sketch shows the square loop, the magnetic field vector, and a unit vector **n̂** representing the direction normal to the surface.

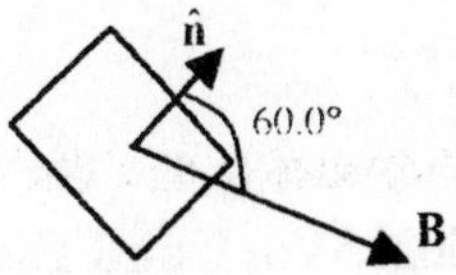

n̂
60.0°
B

Strategy Using the expression for the area of a square, we can get the change in flux using the expression for magnetic flux. Direct application of Faraday's and Lenz's laws then give the final results.

Solution

1. If we call the side length d, the area is: $A = d^2$

2. The change in flux is: $\Delta\Phi = \Phi_f - \Phi_i = B_f d^2\cos\theta - B_i d^2\cos\theta$
$= d^2\cos\theta\,(B_f - B_i)$

3. Faraday's law for the magnitude of the induced emf gives: $|\mathcal{E}| = d^2\cos\theta(B_f - B_i)/\Delta t$
$= (0.250\text{ m})^2\cos(60.0^\circ)\dfrac{[1.25\text{ T} - 0.250\text{ T}]}{0.525\text{ s}} = 0.0595\text{ V}$

4. In the diagram, the flux due to **B** is increasing outward, so: We must have an inward induced magnetic flux.

5. For an inward induced flux, the magnetic field right-hand-rule gives: a clockwise induced current and therefore a clockwise polarity for the induced emf.

Insight Reconstruct the Lenz's law argument if **B** had been decreasing instead of increasing.

Practice Quiz

3. For case 1 the current in a coil is constant, for case 2 the current in the same coil changes at a rate of +1 A/s, and for case 3 the current in this coil changes at a rate of –2 A/s. For which case will there be the largest induced emf?

   (a) case 1 **(b)** case 2 **(c)** case 3 **(d)** It cannot be determined from the given information.

4. If each of the arrows in the figure on the right represents an increasing current, which one will produce a clockwise induced emf in the circular coil?

   (a) 1 **(b)** 2 **(c)** 3 **(d)** 4 **(e)** None

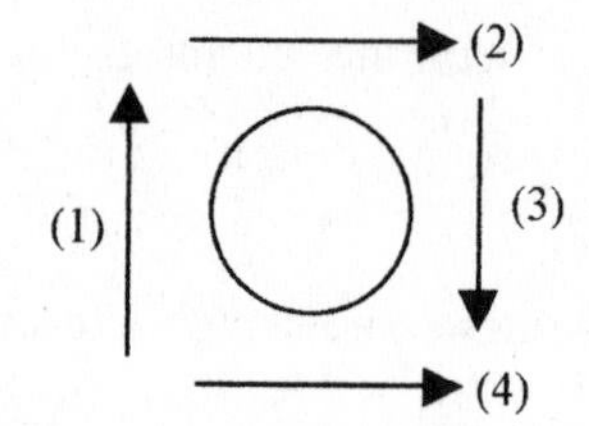

23–5 Mechanical Work and Electrical Energy

The magnetic flux through a loop can change either as a result of changing the magnetic field that passes through the loop, changing the orientation of the loop within the field, or by changing the area of the loop. One way that the area of the loop can change is by moving the conducting wire that makes up the loop; this latter possibility leads directly to the concept of **motional emf**. A motional emf is an induced emf that results from the motion of a conductor through a magnetic field **B**. Applying Faraday's law to this situation shows that the magnitude of the induced emf is given by

$$|\mathcal{E}| = \mathrm{B}\mathrm{v}\ell$$

where v is the speed of the conductor, and ℓ is its length. To be a little more precise, in the above expression, ℓ, **v**, and **B** are mutually perpendicular.

An emf, induced or otherwise, is best understood as coming from an electric field. In fact, we know that a changing magnetic flux creates an electric field. The result from the induced motional emf helps us to see the relationship between the magnetic field through which the conductor moves and the electric field induced by the changing flux:

$$\mathrm{E} = \mathrm{Bv}$$

In this simple case both E and B are constant. When we study light (or electromagnetic radiation) a relationship just like the above expression will be very important even though both the electric and magnetic fields will be changing.

If a motional emf is established in a circuit of resistance R, a current $I = Bv\ell/R$ will flow through this circuit. The magnetic force on this current will act to stop the motion of this conductor unless an equal and opposite external force is applied to maintain the constant speed v. The mechanical power supplied by this external force is precisely what supplies the electrical power consumed through the resistor at any moment. This power is

$$P_{mechanical} = P_{electrical} = \frac{(Bv\ell)^2}{R}$$

Exercise 23–3 An Induced Electric Field A metal bar 1.5 m in length moves in a direction perpendicular to a uniform magnetic field of magnitude 0.85 T with a speed of 2.2 m/s. Determine the magnitude of the induced electric field in the metal bar.

Solution: We are given the following information:

Given: $\ell = 1.5$ m, $B = 0.85$ T, $v = 2.2$ m/s; **Find**: E

The relationship between the electric and magnetic fields is $E = Bv$. Therefore,

$$E = (0.85\ \text{T})(2.2\ \text{m/s}) = 1.9\ \text{N/C}$$

Insight You should verify that the units work out correctly. As mentioned above the fact that induction relates E and B through a velocity is important in understanding electromagnetic radiation.

Practice Quiz

5. A conductor of length L moves with speed v through a magnetic field **B**, producing a motional emf $\mathcal{E}$. If the length of the conductor is doubled, the induced emf will be

   (a) $\mathcal{E}/4$ **(b)** $\mathcal{E}/2$ **(c)** $\mathcal{E}$ **(d)** $2\mathcal{E}$ **(e)** $4\mathcal{E}$

6. A conductor of length L moves with speed v through a magnetic field **B**, producing an induced electric field E. If the length of the conductor is doubled, the induced electric field will be

   (a) E/4 **(b)** E/2 **(c)** E **(d)** 2E **(e)** 4E

23–6 Generators and Motors

Faraday's law of induction has had a tremendous impact on society. This law is the basic principle behind **electric generators**. An electric generator is a device designed to convert mechanical energy to electrical energy. To accomplish this conversion, most modern generators rotate a loop of area A and N turns at an angular speed ω in a magnetic field **B**. This rotation causes a continuous change in the magnetic flux through the loop that results in an induced emf given by

$$\mathcal{E} = NBA\omega \sin \omega t$$

Notice that as time progresses the induced emf will change sign. When the induced emf changes sign the induced current changes direction and thus is an **alternating current** (AC). Generators of this type are called AC generators for this reason.

An **electric motor** has the opposite purpose of an electric generator. Motors are designed to convert electrical energy into mechanical work. The basic physics behind the workings of a motor is the torque on a current loop discussed in Chapter 22. Electrical power supplies current to a loop that sits in a magnetic field. This loop then experiences a torque, which causes the loop to rotate. The rotation of this loop can then be used to do mechanical work such as the turning of a wheel.

Practice Quiz

7. A coil of 2000 turns and area 1.5 m^2 rotates in a magnetic field of 10 T with an angular velocity of 377 rad/s. What is the maximum emf generated in this coil?

   (a) 11 MV **(b)** 1.4 MV **(c)** 0 V **(d)** 2000 V **(e)** 20 kV

23–7 Inductance

When a current flows through a loop, the magnetic field created by that current has a magnetic flux through the area of the loop. If the current changes, the magnetic field changes, and so the flux changes, giving rise to an induced emf. This phenomenon is called **self-induction** because it is the loop's own current, and not an external one, that gives rise to the induced emf. Since $\Delta\Phi/\Delta t \propto \Delta B/\Delta t \propto \Delta I/\Delta t$, Faraday's law implies that $\mathcal{E} \propto \Delta I/\Delta t$. The constant of proportionality, L, is called the **inductance** of the loop,

$$\mathcal{E} = -L\frac{\Delta I}{\Delta t}$$

The SI unit of inductance is the **henry** (H), where 1 H = 1 V·s/A.

Comparing the self-inductance form of Faraday's law with the original form, $\mathcal{E} = -N(\Delta\Phi/\Delta t)$, we see a useful way to calculate the inductance of a coil, $L = N(\Delta\Phi/\Delta I)$. Using this result shows that for a solenoid

$$L = \mu_0\left(\frac{N^2}{\ell}\right)A = \mu_0 n^2 A\ell$$

where $n = N/\ell$ is the number of turns per unit length of the solenoid.

Example 23–4 Self-Inductance in a Solenoid A solenoid of length 12.5 cm contains 250 turns and has a radius of 1.75 cm. If the current in the solenoid decreases from 1.5 A to 0.45 A in 0.0125 seconds, what is the magnitude of the induced emf in the solenoid?

Picture the Problem The sketch shows a coil representing the solenoid in a circuit with an emf.

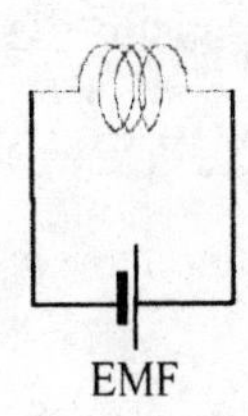

Strategy We can solve this problem using the self-inductance form of Faraday's law, but we first need to get the inductance of the solenoid.

Solution

1. The inductance of the solenoid is:

$$L = \mu_0\left(\frac{N^2}{\ell}\right)A = \mu_0\left(\frac{N^2}{\ell}\right)\pi r^2$$

$$= \left(4\pi\times10^{-7}\ \tfrac{\text{T}\cdot\text{m}}{\text{A}}\right)\left[\frac{(250)^2}{0.125\ \text{m}}\right]\pi(0.0175\ \text{m})^2 = 0.6045\ \text{mH}$$

2. Faraday's law gives:

$$|\mathcal{E}| = L\frac{|\Delta I|}{\Delta t} = L\frac{|I_f - I_i|}{\Delta t}$$

$$= \left(0.6045\times10^{-3}\ \text{H}\right)\frac{|0.45\ \text{A} - 1.5\ \text{A}|}{0.0125\ \text{s}} = 0.051\ \text{V}$$

Insight Make sure you understand the absence of the minus sign and the use of absolute values in this calculation.

Practice Quiz

8. A solenoid of length ℓ and area A consists of N turns producing an inductance L. If the number of turns is doubled, the inductance will be

   (a) L/4 **(b)** L/2 **(c)** L **(d)** 2L **(e)** 4L

23–8 RL Circuits

A coil with a finite inductance (and negligible resistance) is called an **inductor**. Inductors are often used as circuit elements; a circuit with a resistor and an inductor in series is called an **RL circuit**. Recall that induction in a coil tends to resist changes in the current. As a result, the current in a circuit that contains an inductor does not rise and fall as quickly as it otherwise would; the inductance causes it to rise and fall gradually. The larger the inductance, the more gradual the change in current.

Similar to what happens in RC circuits, when an emf is first applied (or shut off) in an RL circuit, the current increases (or decreases) exponentially with a characteristic time given by the time constant

$$\tau = \frac{L}{R}$$

For a current that is building up from zero in an RL circuit the result is

$$I = \frac{\mathcal{E}}{R}\left(1 - e^{-t/\tau}\right) \qquad \text{[building up]}$$

For a current that is falling off from the maximum value, $I_{max} = \mathcal{E}/R$ the result is

$$I = \frac{\mathcal{E}}{R}e^{-t/\tau} \qquad \text{[falling off]}$$

We can see that in the case where the current is falling off, τ repesents the amount of time it takes for the current to fall to $(e^{-1})I_{max} = (0.368)I_{max}$, that is, to 36.8% of its maximum value.

Practice Quiz

9. An RL circuit contains an inductor of inductance L and a resistor of resistance R giving a time constant of value τ. If the resistance is doubled, the time constant will be

   (a) $\tau/4$ **(b)** $\tau/2$ **(c)** τ **(d)** 2τ **(e)** 4τ

10. Which of the following statements is true concerning the current in an RL circuit?

    (a) The time constant is how long it takes the current to build up to 36.8% of its maximum value.

    (b) The time constant is how long it takes the current to build up to 63.2% of its maximum value.

    (c) The time constant is how long it takes the current to fall off to 63.2% of its maximum value.

(d) The time constant is how long it takes the current to fall to 36.8% of its maximum value from 63.2% of it.

(e) The time constant is how long it takes the current to build up to 63.2% of its maximum value from 36.8% of it.

23–9 Energy Stored in a Magnetic Field

Because the induced current in an inductor resists the buildup of current in the coil, it therefore requires energy to build up the current in an inductor against this "resistance." Once the current is established, so is the magnetic field within the coil. The energy needed to build up this current is stored in the magnetic field. For an inductor of inductance L, sustaining a current I, the amount of energy U (the text uses E) stored in the magnetic field is

$$U = \frac{1}{2}LI^2$$

Just like with the energy stored in the electric field of a capacitor, the relationship between the energy stored in an inductor and its magnetic field is nicely expressed in terms of the energy density (energy per unit volume) of the magnetic field

$$\text{energy density} = u_B = \frac{B^2}{2\mu_0}$$

This expression for the energy density applies to any magnetic field, not just that for an inductor.

Practice Quiz

11. A solenoid of inductance L sustains a current I storing an amount of energy U. If the current in the solenoid is reduced to half its previous value, the energy stored in the solenoid will be
    (a) U/4 **(b)** U/2 **(c)** U **(d)** 2U **(e)** 4U

23–10 Transformers

A very important practical devices that relies on the phenomenon of induction is the **transformer**. A transformer is a device that uses induction to increase or decrease the voltage in a circuit. A transformer consists of a *primary coil*, containing N_p turns, across which an AC voltage V_p is applied. The primary coil is bound by an iron core to a secondary coil, containing N_s turns. By induction there will be an induced potential difference V_s across the secondary coil. The relationship between the primary and secondary voltages is given by the **transformer equation**:

$$\frac{V_p}{V_s} = \frac{N_p}{N_s}$$

When $V_s > V_p$ we have a **step-up transformer** and when $V_s < V_p$ we have a **step-down transformer.**

By conservation of energy, the average power in the secondary circuit must equal the average power in the primary circuit, $I_s V_s = I_p V_p$. Therefore,

$$\frac{I_s}{I_p} = \frac{V_p}{V_s} = \frac{N_p}{N_s}$$

so that a step-up transformer steps down the current and vice versa.

Exercise 23–5 A Big Step Down Suppose you want to use one transformer to step down from a power line voltage of 50,000 V to a household voltage of 120 V. If your primary coil consists of 10,000 turns, how many turns will you need in your secondary coil?

Solution: We are given the following information:

Given: $V_p = 50{,}000$ V, $V_s = 120$ V, $N_p = 10{,}000$; **Find**: N_s

The transformer equation gives

$$\frac{V_p}{V_s} = \frac{N_p}{N_s} \quad \Rightarrow \quad N_s = N_p\left(\frac{V_s}{V_p}\right) = 10{,}000\left(\frac{120 \text{ V}}{50{,}000 \text{ V}}\right) = 24$$

Insight In practice the voltage from power lines is stepped down in several stages, not all at once.

Practice Quiz

12. The primary coil of a transformer contains 100 turns, and the secondary coil contains 200 turns. If the current in the primary coil is I, what is the current in the secondary coil?

    (a) I **(b)** 2I **(c)** I/2 **(d)** I/4 **(e)** 4I

Reference Tools and Resources

I. Key Terms and Phrases

induced emf a potential difference created by a changing magnetic flux

magnetic flux a measure of the extent to which a magnetic field “flows” through an area

weber the SI unit of magnetic flux, equal to 1 $T \cdot m^2$

Faraday's law of induction the induced emf in a coil equals the rate of change of magnetic flux through the coil

Lenz's law the polarity of an induced emf is such that the induced current gives rise to a magnetic flux that opposes the original change in flux

motional emf an induced emf that results from the motion of a conductor through a magnetic field

electric generator a device designed to convert mechanical energy to electrical energy

alternating current (AC) current that alternates in direction

electric motor a device designed to convert electrical energy into mechanical work

self-induction the creation of an induced emf in a coil due its own changing current

inductance the proportionality constant between the induced emf in a coil and the rate of change in the current in that same coil

henry the SI unit of inductance

inductor a coil with a finite inductance used as a circuit element

transformer uses induction to increase (step-up) or decrease (step-down) the voltage in a circuit

II. Important Equations

Name/Topic	Equation	Explanation
magnetic flux	$\Phi = BA\cos\theta$	Magnetic flux measures the amount of magnetic field "flowing" through an area.
Faraday's law	$\mathcal{E} = -N\frac{\Delta\Phi}{\Delta t}$	The induced emf from the rate of change in magnetic flux through a coil of N turns.
motional emf	$\lvert\mathcal{E}\rvert = Bv\ell$	The magnitude of the induced emf from the motion of a conductor through a magnetic field.
electric generators	$\mathcal{E} = NBA\omega\sin\omega t$	The emf of an AC generator.

inductance	$\mathcal{E} = -L\frac{\Delta I}{\Delta t}$	Faraday's law in terms of self-inductance.
RL circuits	$\tau = \frac{L}{R}$	The time constant for an RL circuit.
stored energy	$U = \frac{1}{2}LI^2$	The energy stored in the magnetic field of an inductor.
transformers	$\frac{V_p}{V_s} = \frac{N_p}{N_s}$	The ratio of the potential differences in a transformer equals the ratio of the turns.

III. Know Your Units

Quantity	Dimension	SI Unit
magnetic flux (Φ)	$[M][L^2][A^{-1}][T^{-2}]$	Wb
inductance (L)	$[M][L^2][T^{-2}][A^{-2}]$	H

Practice Problems

1. The circular loop in the figure has a radius of 2.6 cm and makes an angle of 19 degrees with a 2-T magnetic field. To the nearest tenth of a mWb, what is the flux through the loop?

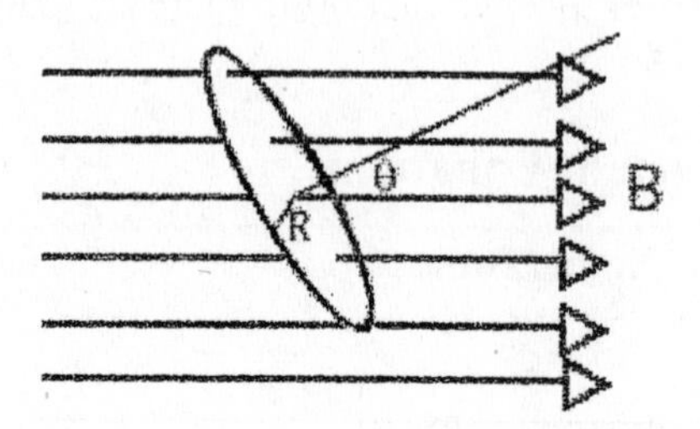

2. The 2-cm (radius) current loop shown in the figure is changing from θ_1 = 32 degrees to θ_2 = 68 degrees in a magnetic field of 2 T in 0.5 seconds. What is the magnitude of the induced emf, to the nearest hundredth of a millivolt?

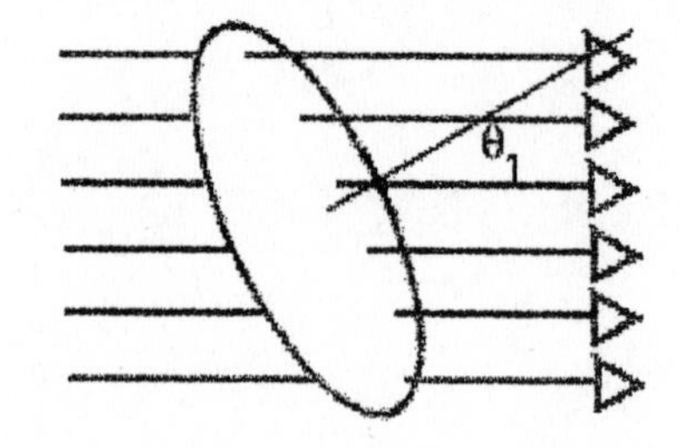

3. The wire in problem 2 has a diameter of 0.53 mm and a resistivity of 1.09 x 10^{-8} Ω·m. Find the current in the wire to the nearest tenth of a milliamp.
4. A 27-turn coil of wire with a radius of 6 centimeters is moved up to a magnet where B = 1.2 T, in 0.1 seconds. What is the induced emf to the nearest hundredth of a volt?

5. The circuit shown at the left has a resistance of 23 Ω and consumes 6.9 watts of power; the rod has a width of 1.25 m between the tracks and moves to the right at 3.5 m/s. What is the strength of the magnetic field, to the nearest hundredth of a tesla?

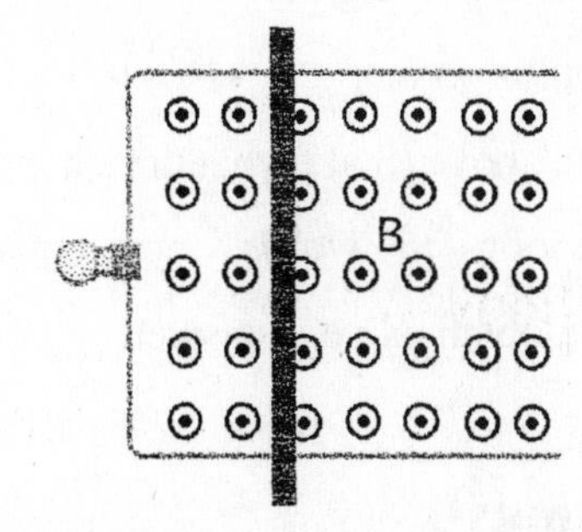

6. A coil in an electric generator rotated at 60 Hz has 147 turns with an area of 1×10^{-2} m^2. The magnetic field is 1.92 tesla. To the nearest hundredth of a volt what is the maximum voltage?
7. In problem 6 what is the voltage at t = 0.02 seconds?
8. In the circuit R = 47 Ω and L = 75 mH. What does the amp meter read 5 milliseconds after the switch is moved to position 1, to the nearest hundredth of an amp?

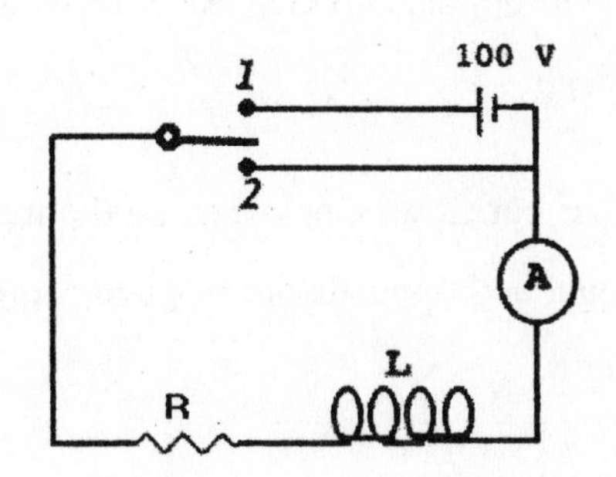

9. What is the answer to problem 8 if the switch is moved to position 2 after full current flow is established?

10. In the transformer the number of coils in the secondary is 3 times the number of turns in the primary. If R = 26 Ω, to the nearest hundredth of an amp what is the current in the secondary?

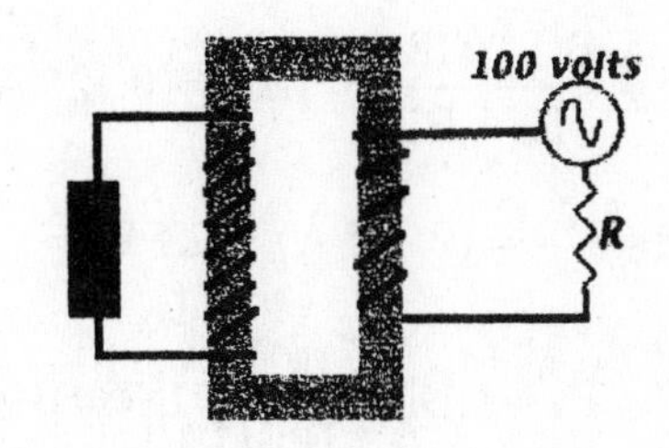

Puzzle

PLAY BULB!

The figure shows three identical lightbulbs attached to an ideal battery and an inductor. The switch is closed for a long time, then opened. Which of the following statements correctly describes what happens just after the switch is opened?

1. Bulb B stays the same, and bulbs A and C go out.
2. Bulb B stays the same and bulbs A and C get brighter
3. Bulb B goes out, and bulbs A and C get brighter
4. Bulb B gets brighter, and bulbs A and C stay the same
5. Bulb B gets brighter, and bulbs A and C go out

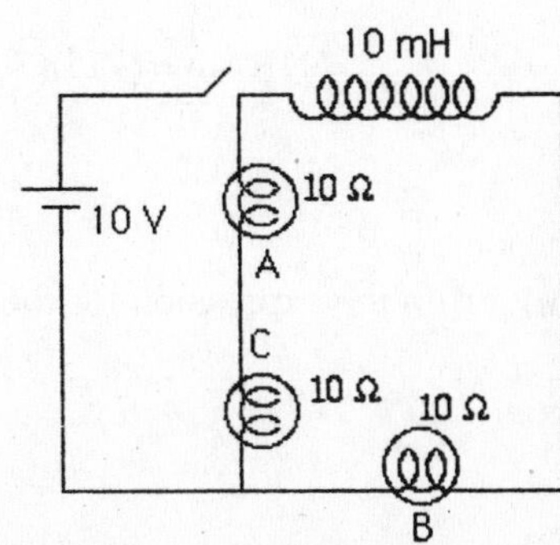

6. All three bulbs get brighter.
7. All three bulbs go out.
8. None of the above (State what does happen.)

Answer this question in words, not equations, briefly explaining how you obtained your answer.

Selected Solutions

19. If the current in the wire changes direction, the direction of the magnetic field is reversed, changing its direction from out of the page to into the page. According to Lenz's law, the current induced in the circuit will oppose this change by flowing $\boxed{\text{counterclockwise}}$, generating a field which is directed out of the page.

27. The current flows clockwise, so the magnetic force is directed to the left. The external force must be equal and opposite the magnetic force to maintain the rod's constant speed. So, it is directed to the right.

(a) The magnitude of the force is $F = \dfrac{B^2vL^2}{R}$ Now from Problem 26, $v = \dfrac{IR}{BL}$ so

$$F = \frac{B^2L^2}{R}\left(\frac{IR}{BL}\right) = IBL = (0.125\ \text{A})(0.750\ \text{T})(0.45\ \text{m}) = \boxed{42\ \text{mN}}$$

(b) $P = I^2R = (0.125\ \text{A})^2(12.5\ \Omega) = \boxed{195\ \text{mW}}$

(c) $P = Fv = (IBL)\left(\dfrac{IR}{BL}\right) = I^2R = \boxed{195\ \text{mW}}$

33. **(a)** Only the $\boxed{\text{horizontal}}$ component of the magnetic field is important. The vertical component is parallel to the plane of the circular coil at all times. Thus, it does not contribute to the flux through the coil.

(b) Since the emf is given by $\mathcal{E} = NBA\omega\sin(\omega t)$, the maximum emf is

$$\mathcal{E}_{max} = NBA\omega = (155)(3.80 \times 10^{-5}\ \text{T})\pi\left(\frac{0.220\ \text{m}}{2}\right)^2\left(1250\ \frac{\text{rev}}{\text{min}}\right)\left(\frac{2\pi\ \text{rad}}{1\ \text{rev}}\right)\left(\frac{1\ \text{min}}{60\ \text{s}}\right) = \boxed{29.3\ \text{mV}}$$

37. **(a)** From the expression for the inductance of a solenoid, we can solve for the area to get

$$A = \frac{\ell L}{\mu_0 N^2} = \frac{(0.24\text{ m})(7.3 \times 10^{-3}\text{ H})}{\left(4\pi \times 10^{-7}\ \frac{\text{T}\cdot\text{m}}{\text{A}}\right)(450)^2} = \boxed{6.9 \times 10^{-3}\text{ m}^2}$$

(b) From Faraday's law written in terms of self-inductance:

$$\mathcal{E} = -L\frac{\Delta I}{\Delta t} = -(7.3 \times 10^{-3}\text{ H})\left(\frac{-3.2\text{ A}}{55 \times 10^{-3}\text{ s}}\right) = \boxed{0.42\text{ V}}$$

55. The transformer equation, solved for the primary voltage, gives

$$V_p = V_s\left(\frac{N_p}{N_s}\right) = (4800\text{ V})\left(\frac{25}{500}\right) = \boxed{200\text{ V}}$$

In terms of the relationship between the current ratio and turns ratio we have

$$I_p = I_s\left(\frac{N_s}{N_p}\right) = (12 \times 10^{-3}\text{ A})\left(\frac{500}{25}\right) = \boxed{0.2\text{ A}}$$

Answers to Practice Quiz

1. (a) **2.** (c) **3.** (c) **4.** (d) **5.** (d) **6.** (c) **7.** (a) **8.** (e) **9.** (b) **10.** (b) **11.** (a) **12.** (c)

Answers to Practice Problems

1.	4.0	**2.**	2.37 mV
3.	3.8 mA	**4.**	3.66 V
5.	2.88 T	**6.**	1063.76 V
7.	1011.69 V	**8.**	2.03 A
9.	0.09 A	**10.**	1.28 A

CHAPTER 24
ALTERNATING CURRENT CIRCUITS

Chapter Objectives

After studying this chapter, you should:

1. know the meaning of alternating voltage and current and how to determine their root-mean-square values.

2. be able to use phasor diagrams to represent alternating voltages.

3. understand the behavior of capacitors in AC circuits and how the capacitive reactance affects the current.

4. know what the impedance in an AC circuit is and how it differs from resistance.

5. be able to calculate the phase difference between the total voltage in an AC circuit and the current.

6. understand the behavior of inductors in AC circuits and how the inductive reactance affects the current.

7. be able to calculate the total voltage, impedance, current, and phase angle for an RLC circuit.

8. understand LC oscillations and resonance in RLC circuits.

Warm-Ups

1. Home Experiment: Make a pendulum using a string 1 to 1.5 meters long (40-60 inches), and a mass similar to a lemon (go ahead, use a lemon!). This will give you a pendulum with a resonant period in the range of 2 to 2.5 seconds ($T = 2\{L/g\}^{1/2}$). In terms of frequency, $f = 1/T$, so the resonant frequency is between 0.5 and 0.4 Hz.

   Now, hold the end of the string and vibrate it back and forth by 5 cm, slowly varying the frequency of your oscillation. When you find the right frequency, the lemon will swing back and forth with quite a large amplitude. This is the resonant period for your pendulum. Note the phase between the motion of your hand and the lemon (is your hand out ahead of the lemon pulling it forward, or behind pulling it back).

Try swinging the lemon a little more slowly (longer period than at resonance). How does the amplitude change? How about the phase?

Next, try going a little more quickly than at resonance. What happens to the amplitude and phase now? Finally, compare the experiment you just did with the behavior of an RLC circuit.

2. A radio has RLC circuits inside that oscillate at the same frequencies as the radio waves they receive (88.1-107.9 MHz for FM, 540-1180 KHz for AM). Select values of L and C that could be used to make these circuits.
3. An inductor or a capacitor in an AC circuit has a (time-varying) current flowing through it and voltage across it. However, the average power $P_{AV} = 0$. Explain how this works out, recalling that power is equal to voltage times current.
4. Old radios used to tune from one frequency to another using variable capacitors. In such a device there are two rows of semicircular plates. One set of plates (all electrically connected together) are fixed to the base, and the other set (all connected together) can rotate through a half turn. At one extreme the rotating set are opposite to the fixed set, and at the other extreme the plates are fully "interleaved" with the fixed set. Assuming the whole thing must fit inside an old radio, estimate the maximum capacitance. (Hint: for ten fixed and ten rotating plates, you can treat this as 10 parallel-plate capacitors in parallel.)

Chapter Review

This chapter treats circuits with an alternating current (AC). These types of circuits are very important because AC circuits are used to provide the electricity that most of us use every day. Included with this discussion is the behavior of resistors, capacitors, and inductors in AC circuits.

24–1 Alternating Voltages and Currents

As discussed in a previous chapter, an AC generator supplies a voltage that switches (or alternates) in polarity. Typically, the voltage varies sinusoidally and can be represented by the expression

$$V = V_{max} \sin \omega t$$

where ω is the angular frequency ($\omega = 2\pi f$) of the oscillating voltage, and V_{max} is the amplitude of the varying voltage. When this voltage is connected in a circuit with resistance R, the result is a sinusoidally **alternating current**, that is, a current that alternates in direction according to

$$I = I_{max} \sin \omega t$$

where, by Ohm's law, $I_{max} = V_{max}/R$. The voltage across the resistor in the circuit and the current vary together, reaching their maxima simultaneously and equaling zero simultaneously. Because of this behavior, we say that the voltage across the resistor and the current in an AC circuit are **in phase**.

A convenient representation of alternating voltages is the use of **phasors**. A phasor is an arrow that rotates counterclockwise in an x-y coordinate system with an angular velocity ω equal to the angular frequency of the alternating voltage it represents. The length of the arrow represents, and is proportional to, the amplitude of the alternating voltage. The projection of the phasor onto the y-axis gives the instantaneous value of the voltage at any given time. The current in an AC circuit can also be represented by a phasor, and because the voltage across the resistor and current are in phase, the phasor for the current always points in the same direction as the phasor for this voltage.

Since the voltage and current in an AC circuit alternate, it is customary to characterize quantities by an average value. However, the average values of the voltage and current are zero so instead we use the **root-mean-square** (rms) values. An rms value of a quantity is the square root of the average (mean) of the squared quantity. In the case of sinusoidally varying quantities, the rms value works out to be the maximum value divided by $\sqrt{2}$:

$$V_{rms} = V_{max} / \sqrt{2}$$

$$I_{rms} = I_{max} / \sqrt{2}$$

Using rms values, we can write equations for AC circuit that mirror equations we used for DC circuits. For example, an AC version of Ohm's law for a purely resistive circuit is

$$V_{rms} = I_{rms} R$$

Also, the average power consumed by the resistance R in an AC circuit is conveniently written, in terms of rms values, by expressions that look just like the ones we used with DC circuits:

$$P_{av} = I_{rms}^2 R = I_{rms} V_{rms} = V_{rms}^2 / R$$

The instantaneous power used in the resistor is determined by the same expressions, except we replace I_{rms} with I, and V_{rms} with V. Also, any expression involving rms values can also be written for maximum values (V_{max}, I_{max}, P_{max}) simply by replacing the rms value of the quantity with its corresponding maximum value.

Example 24–1 AC A simple circuit contains an AC generator with a maximum output of 150 V operating at a frequency of 60 Hz. If the resistance in the circuit is 35 Ω, what average power is dissipated through the resistor?

Picture the Problem The diagram shows an AC generator connected to a resistor.

Strategy Of the three equivalent expressions for P_{av} we choose the one most convenient for the given information. Since we are given R and V_{max}, we will use $P_{av} = V_{rms}^2 / R$.

Solution

1. The rms voltage is: $V_{rms} = V_{max} / \sqrt{2} = (150\text{ V}) / \sqrt{2} = 106.1\text{ V}$

2. The average power then is: $P_{av} = V_{rms}^2 / R = (106.1\text{ V})^2 / 35\,\Omega = 320\text{ W}$

Insight Any of the three expressions could have been used to determine P_{av}, but this one was most convenient because it involved the fewest intermediate steps.

Practice Quiz

1. An alternating current is given by $I = (0.25\text{ A}) \sin[(377\text{ rad/s})t]$. What is the rms current?
   (a) 0.25 A **(b)** 0.18 A **(c)** 377 A **(d)** 0.12 A **(e)** 0.50 A

2. An alternating current is given by $I = (0.25\text{ A}) \sin[(377\text{ rad/s})t]$. What is the frequency f?
   (a) 377 Hz **(b)** 0.25 Hz **(c)** 60.0 Hz **(d)** 0.18 Hz **(e)** 2370 Hz

23–2 – 23–3 Capacitors in AC Circuits and RC Circuits

If you consider a circuit consisting of just an AC generator and a capacitor, with no resistance in it, the capacitor itself, because of charging and discharging, offers some opposition to the current being sent by the generator. This resistance-like opposition to the current due to the charging and discharging of the capacitor in an AC circuit is called the **capacitive reactance**, X_C; it is given by

$$X_C = \frac{1}{\omega C}$$

The capacitive reactance has the same SI unit as resistance, the ohm, Ω. Written in terms of the capacitive reactance, the rms current in the circuit is

$$I_{rms} = \frac{V_{rms}}{X_C}$$

Recall that in a DC circuit the capacitor takes time to charge and discharge; similar behavior occurs in AC circuits as well. Because of this fact, there is a phase difference between the current and the voltage across the capacitor:

the voltage across a capacitor lags the current by 90°.

Because of this lag, the phasor for the voltage across the capacitor is drawn 90° behind the phasor for the current, with both phasors rotating counterclockwise, as shown below.

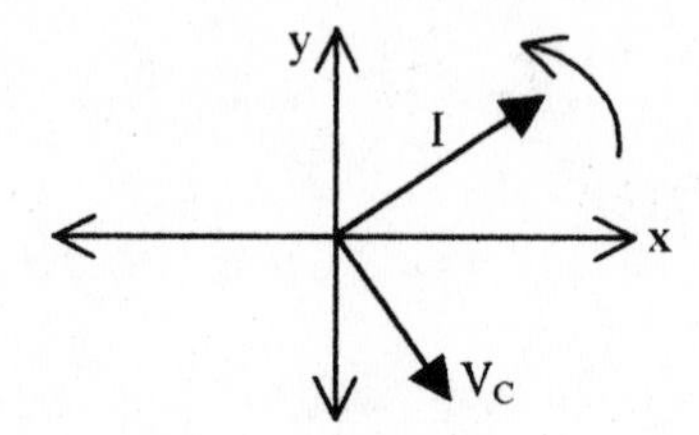

The key differences between the capacitive reactance and resistance are that (a) they result from very different types of physical processes; (b) the capacitive reactance depends on the frequency of the generator, getting smaller as ω increases; and (c) there is no net power consumption associated with X_C, as there is with R, because no net energy is used by the capacitor.

If we now consider an RC circuit run by an AC generator we must combine the considerations of Section 24–1 with the present information. The voltage across the resistor, V_R, is in phase with the current and is 90° out of phase with the voltage across the capacitor. The phasors for these two voltages can be used to form a right triangle, and the total voltage in the circuit can be determined using the Pythagorean theorem. For maximum values we have

$$V_{max} = \sqrt{V_{max,R}^2 + V_{max,C}^2}$$

This expression can also be written as $V_{max} = I_{max}\sqrt{R^2 + X_C^2}$. Making the definition

$$Z = \sqrt{R^2 + X_C^2}$$

we can write this relation in a way that look's similar to Ohm's law:

$$V_{max} = I_{max}Z$$

The quantity Z is called the **impedance** of the circuit and represents the combined resistance-like effects of the actual resistance and the capacitance. The SI unit of impedance is the ohm. The preceding expressions can also be written for rms values.

In general, the total voltage in an RC circuit will be out of phase with the current by an angle ϕ between 0 and –90°, where the minus indicates that the voltage lags the current. Using trigonometry on the phasor diagram gives

$$\tan\phi = \frac{X_C}{R} \quad \text{and} \quad \cos\phi = \frac{R}{Z}$$

these expressions can be used to find the magnitude of the angle ϕ. Because energy is only consumed by the resistor, the phase angle also comes into play in determining the average power consumed in the circuit. The result, equivalent to $P_{av} = I_{rms}^2 R$, is

$$P_{av} = I_{rms} V_{rms} \cos\phi$$

The factor $\cos\phi$ is called the **power factor**.

Example 24–2 RC Circuits An AC circuit contains a 44.3-μF capacitor in series with a 50.1-Ω resistor. The circuit is powered by a 60.0-Hz generator with an rms output of 85.0 V. Find the rms current in the circuit and the average power consumed through the resistor.

Picture the Problem The diagram shows a series RC circuit with an AC generator.

Strategy We start with the basic expression for I_{rms} and find all the necessary quantities for that expression. Then, we should also have enough information to determine P_{av}.

Solution

1. An expression for I_{rms} is:

$$I_{rms} = \frac{V_{rms}}{Z} = \frac{V_{rms}}{\sqrt{R^2 + X_C^2}}$$

2. The capacitive reactance is:

$$X_C = \frac{1}{\omega C} = \frac{1}{2\pi f C} = \frac{1}{2\pi(60.0\ \text{Hz})(44.3\ \mu\text{F})} = 59.88\ \Omega$$

3. The current then is:

$$I_{rms} = \frac{V_{rms}}{\sqrt{R^2 + X_C^2}} = \frac{85.0\ \text{V}}{\left[(50.1\ \Omega)^2 + (59.88\ \Omega)^2\right]^{\frac{1}{2}}} = 1.09\ \text{A}$$

4. The average power can be found as:

$$P_{av} = I_{rms}^2 R = (1.089\ \text{A})^2(50.1\ \Omega) = 59.4\ \text{W}$$

Insight The formula used for the rms current is just a rearrangement of the expression involving the maximum current and voltage, with the maximum values replaced by rms values.

Practice Quiz

3. An RC circuit contains a 25-Ω resistor and a 56-μF capacitor. The circuit is run by a 60-Hz AC generator with a maximum emf of 120 V. What is the impedance of this circuit?

   (a) 47 Ω **(b)** 25 Ω **(c)** 380 Ω **(d)** 54 Ω **(e)** 60 Ω

4. An RC circuit contains a 25-Ω resistor and a 56-μF capacitor. The circuit is run by a 60-Hz AC generator with a maximum emf of 120 V. What is the rms current in this circuit?

   (a) 1.6 A **(b)** 2.2 A **(c)** 3.4 A **(d)** 1.8 A **(e)** 2.5 A

24–4 Inductors in AC Circuits

For an inductor in an AC circuit, without resistance, the induced emf delays the current in reaching its maximum, which means that the inductor offers a resistance-like opposition to the current sent by the generator. This behavior is represented by what is called the **inductive reactance**, which is given by

$$X_L = \omega L$$

where L is the inductance of the inductor. With this result the rms current in the circuit becomes

$$I_{rms} = \frac{V_{rms}}{X_L}$$

Recall that the induced emf across the inductor is greatest when the current is changing most rapidly. This maximum rate occurs at the zero point of the current oscillation. Thus, the voltage across the inductor is out of phase with the current. In fact,

the voltage across the inductor V_L leads the current by 90°.

Because of this lead, the phasor for the voltage across the inductor is drawn 90° ahead of the phasor for the current, with both phasors rotating counterclockwise, as shown below.

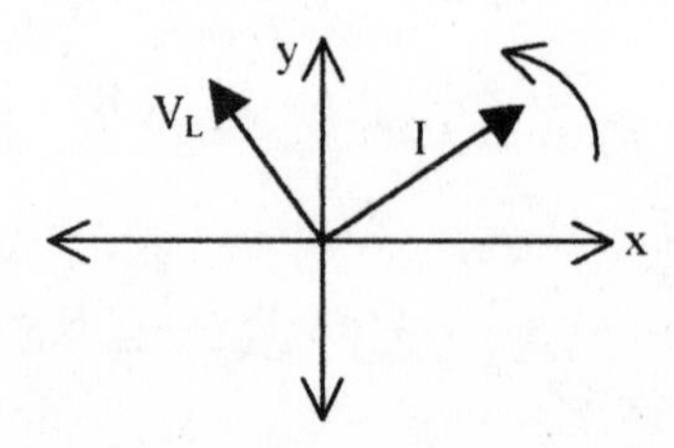

The key differences between the inductive reactance and resistance are that (a) they result from very different types of physical processes; (b) the inductive reactance depends on the frequency of the generator, getting larger as ω increases; and (c) there is no net power consumption associated with X_L, as there is with R, because no net energy is used by an ideal inductor.

If we now consider an RL circuit run by an AC generator, we notice that the voltage across the resistor lags the voltage across the inductor by 90°. The phasors for these two voltages, therefore, can be used to form a right triangle and the total voltage in the circuit is determined using the Pythagorean theorem. The result is

$$V_{max} = \sqrt{V_{max,R}^2 + V_{max,L}^2}$$

This expression can also be written as $V_{max} = I_{max}\sqrt{R^2 + X_L^2}$. Here, the impedance is given by

$$Z = \sqrt{R^2 + X_L^2}$$

and we can write this relation as

$$V_{max} = I_{max} Z$$

As with the RC circuit, in general, the total voltage in an RL circuit will be out of phase with the current. However, for the case of an RL circuit, the phase angle ϕ is between 0 and +90° because the voltage leads the current in this case. The phase angle can be determined from

$$\tan\phi = \frac{X_L}{R} \quad \text{and} \quad \cos\phi = \frac{R}{Z}$$

Exercise 24–3 An RL Circuit An AC circuit contains a 40-mH inductor in series with a 35-Ω resistor. The circuit is powered by a 25-Hz generator with a maximum output of 115 V. Find the rms voltage across the inductor.

Solution: We are given the following information:

Given: $L = 40$ mH, $R = 35\ \Omega$, $f = 25$ Hz, $V_{max} = 115$ V; **Find**: $V_{rms,L}$

The relationship for the rms voltage across the inductor is

$$V_{rms,L} = I_{rms} X_L$$

To calculate this voltage we need to find both I_{rms} and X_L. Since we know the frequency and the inductance, the inductive reactance is

$$X_L = \omega L = 2\pi f L = 2\pi(25\text{ Hz})(40\times10^{-3}\text{ H}) = 6.28\ \Omega$$

The expression for the rms current is

$$I_{rms} = \frac{V_{rms}}{Z} = \frac{V_{rms}}{\sqrt{R^2 + X_L^2}}$$

Using the given information we get

$$I_{rms} = \frac{V_{max}/\sqrt{2}}{\sqrt{R^2 + X_L^2}} = \frac{(115\text{ V})/\sqrt{2}}{\sqrt{(35\,\Omega)^2 + (6.28\,\Omega)^2}} = 2.287\text{ A}$$

Now, we are able to calculate the rms voltage across the inductor to be

$$V_{rms,L} = (2.287\text{ A})(6.28\,\Omega) = 14\text{ V}$$

This result is only the rms voltage across the inductor. It is considerably smaller than the total rms voltage because of the relatively low frequency of the generator.

Practice Quiz

5. An RL circuit contains a 25-Ω resistor and a 56-mH inductor. The circuit is run by a 60-Hz AC generator with a maximum emf of 120 V. What is the inductive reactance in this circuit?

   (a) 60 Ω **(b)** 0.35 Ω **(c)** 377 Ω **(d)** 25 Ω **(e)** 21 Ω

6. An RL circuit contains a 25-Ω resistor and a 56-mH inductor. The circuit is run by a 60-Hz AC generator with a maximum emf of 120 V. What is the maximum current in this circuit?

   (a) 4.8 A **(b)** 3.7 A **(c)** 2.6 A **(d)** 5.7 A **(e)** 3.4 A

24–5 RLC Circuits

In this section we combine all the above results to consider an AC circuit containing a resistor, an inductor, and a capacitor in series — an **RLC circuit**. The phasors for the voltage in this circuit show that V_L leads the V_R by 90°, and V_C lags V_R by 90°. Therefore, V_C lags V_L by 180°, and we can consider their combined effect as $V_L - V_C$. The total maximum voltage is given by

$$V_{max} = \sqrt{V_{max,R}^2 + \left(V_{max,L} - V_{max,C}\right)^2}$$
$$= I_{max}\sqrt{R^2 + (X_L - X_C)^2}$$

This shows that for an RLC circuit the impedance is

$$Z = \sqrt{R^2 + (X_L - X_C)^2}$$

The phase angle between the total voltage in the circuit and the current is determined by

$$\tan\phi = \frac{X_L - X_C}{R} \quad \text{and} \quad \cos\phi = \frac{R}{Z}$$

It is worth remembering that the power factor $\cos\phi$ will only give the magnitude of ϕ; it does not tell you if the voltage leads or lags the current. Expressing the phase angle in terms of the tangent, however, gives both the magnitude and sign of ϕ. If ϕ is positive, the total voltage leads the current, if it is negative the voltage lags the current.

Example 24–4 An RLC Circuit An RLC circuit has $R = 55.5\ \Omega$, $L = 15.6$ mH, and $C = 75.2\ \mu$F. If the AC generator connected to this circuit has rms output of 120.0 V and operates at a frequency of 60.0 Hz, find **(a)** the impedance of the circuit, **(b)** the rms current in the circuit, **(c)** the rms voltage across the resistor, **(d)** the rms voltage across the inductor, **(e)** the rms voltage across the capacitor, **(f)** the phase angle between the total voltage and the current, and **(g)** the power consumed through the resistor.

Picture the Problem The diagram shows an RLC circuit connected to an AC generator.

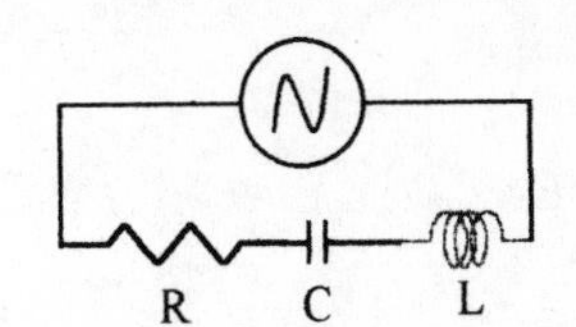

Strategy For each part we use the expression for the desired quantity as a guide for what to calculate in intermediate steps.

Solution

Part (a)

1. The expression for the impedance shows that we first need X_L and X_C:

$$Z = \sqrt{R^2 + (X_L - X_C)^2}$$

2. The inductive reactance is:

$$X_L = \omega L = 2\pi f L$$
$$= 2\pi(60.0\text{ Hz})(15.6\times10^{-3}\text{ H}) = 5.881\ \Omega$$

3. The capacitive reactance is:

$$X_C = \frac{1}{\omega C} = [2\pi f C]^{-1}$$
$$= \left[2\pi(60.0\text{ Hz})(75.2\times10^{-6}\text{ F})\right]^{-1} = 35.27\ \Omega$$

4. The impedance then becomes:

$$Z = \sqrt{(55.5\ \Omega)^2 + (5.881\ \Omega - 35.27\ \Omega)^2} = 62.8\ \Omega$$

Part (b)

5. The rms current is:

$$I_{rms} = \frac{V_{rms}}{Z} = \frac{120\text{ V}}{62.80\ \Omega} = 1.91\text{ A}$$

Part (c)

6. The rms voltage across the resistor is: $V_{rms,R} = I_{rms}R = (1.911\text{ A})(55.5\ \Omega) = 106\text{ V}$

Part (d)

7. The rms voltage across the inductor is: $V_{rms,L} = I_{rms}X_L = (1.911\text{ A})(5.881\ \Omega) = 11.2\text{ V}$

Part (e)

8. The rms voltage across the capacitor is: $V_{rms,C} = I_{rms}X_C = (1.911\text{ A})(35.27\ \Omega) = 67.4\text{ V}$

Part (f)

9. The phase angle is given by:

$$\tan\phi = \frac{X_L - X_C}{R} \Rightarrow \phi = \tan^{-1}\left(\frac{X_L - X_C}{R}\right)$$

$$\therefore\ \phi = \tan^{-1}\left(\frac{5.881\ \Omega - 35.27\ \Omega}{55.5\ \Omega}\right) = -27.9^\circ$$

Part (g)

10. The power consumed through the resistor is: $P_{av} = I_{rms}^2 R = (1.911\text{ A})^2(55.5\ \Omega) = 203\text{ W}$

Insight Once all the reactances and the impedance are known, the rest of the results are obtained rather easily. Notice that the sum of the rms voltages across the resistor, inductor, and capacitor is greater than the total rms voltage of 120 V. This fact is not a problem because these voltages are out of phase, so they reach their maxima (or rms) values at different times. The negative phase angle means that the total voltage lags the current. Can you sketch the phasor diagram?

Practice Quiz

7. An RLC circuit contains a 25-Ω resistor, a 56-mH inductor, and an 85-μF capacitor; it is run by a 60-Hz AC generator with a maximum emf of 120 V. What is the impedance of this circuit?

   (a) 25 Ω **(b)** 21 Ω **(c)** 31 Ω **(d)** 36 Ω **(e)** 27 Ω

8. An RLC circuit contains a 25-Ω resistor, a 56-mH inductor, and an 85-μF capacitor; it is run by a 60-Hz AC generator with a maximum emf of 120 V. What is the phase angle between the current and the total voltage?

   (a) +22° **(b)** +47° **(c)** −22° **(d)** −47° **(e)** 0.0°

24–6 Resonance in Electric Circuits

In a pure LC circuit, with an initially charged capacitor, no resistor, and no generator, the current will still oscillate between charging the capacitor and flowing through the inductor. This also means that the energy contained in the circuit oscillates between the electric field of the capacitor and the magnetic field of the inductor. These oscillations naturally take place at a characteristic frequency that depends on the values of the inductance and capacitance. The natural frequency of oscillation for an ideal LC circuit is given by

$$\omega = \frac{1}{\sqrt{LC}}$$

The fact that current oscillates with a natural frequency in an LC circuit leads to the phenomenon of resonance in RLC circuits. The AC generator is what drives the oscillation in an RLC circuit. When the generator drives the circuit at the natural frequency the current has its largest amplitude, and we have achieved resonance in this circuit. The current will be largest when the impedance is smallest. The smallest impedance is $Z = R$, which occurs when the frequency in the circuit is such that $X_L = X_C$. This occurs at precisely the natural frequency for the LC oscillations. These results also mean that the phase angle between the current and the total voltage is zero; that is, the current and voltage are in phase at resonance.

Example 24–5 Resonance An RLC circuit has $R = 55.5\ \Omega$, $L = 15.6$ mH, and $C = 75.2\ \mu$F. If the AC generator of this circuit has an rms output of 120.0 V and operates at a frequency of 60.0 Hz, **(a)** what is the resonant frequency of this circuit, and **(b)** what would be the average power consumed through the resistor if the generator was operated at the resonant frequency?

Picture the Problem The diagram shows an RLC circuit connected to an AC generator.

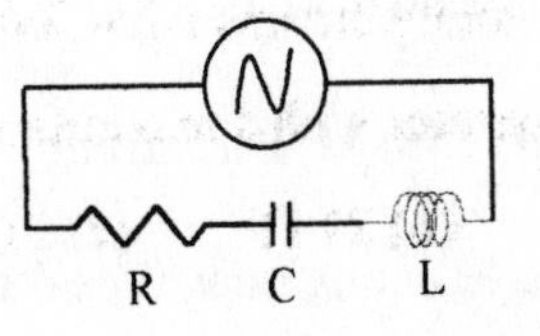

Strategy The resonant frequency is the natural frequency for the LC oscillation. We can use our knowledge of what happens at resonance to simplify the calculation of the average power.

Solution

Part (a)

1. The resonant frequency is (in rad/s):

$$\omega = 1/\sqrt{LC} = \left[\left(15.6\times10^{-3}\text{ H}\right)\left(75.2\times10^{-6}\text{ F}\right)\right]^{-1/2}$$
$$= 923 \text{ rad/s}$$

Part (b)

2. At resonance Z = R, therefore:

$$V_{rms} = I_{rms}Z = I_{rms}R = V_{rms,R}$$

3. The average power is therefore:

$$P_{av} = \frac{V_{rms,R}^2}{R} = \frac{V_{rms}^2}{R} = \frac{(120.0 \text{ V})^2}{55.5\,\Omega} = 259 \text{ W}$$

Insight There are other routes to determining the average power at resonance. Think of at least one other way, then try it.

Practice Quiz

9. An RLC circuit contains a 25-Ω resistor, a 56 mH inductor, and an 85-μF capacitor; it is run by a 60-Hz AC generator with a maximum emf of 120 V. What is the resonant frequency (in Hertz) of this circuit?

(a) 73 Hz (b) 460 Hz (c) 2.2 Hz (d) 60 Hz (e) 380 Hz

10. An RLC circuit contains a 15-Ω resistor, a 20-mH inductor, and a 75-μF capacitor; it is run by an AC generator of variable frequency. At what frequency (either ω or f) will the power output of this circuit be maximum?

(a) 60 Hz (b) 380 rad/s (c) 820 Hz (d) 130 Hz (e) 30 rad/s

Reference Tools and Resources

I. Key Terms and Phrases

alternating current (AC) a current whose direction alternates in a circuit

in phase two alternating quantities that reach their maxima and minima simultaneously oscillate in phase

phasor an arrow in an x-y coordinate system that rotates counterclockwise to represent alternating voltage or current

root-mean-square (rms) the square root of the mean of a squared quantity

capacitive reactance the resistance-like behavior of a capacitor in an AC circuit

inductive reactance the resistance-like behavior of an inductor in an AC circuit

impedance the resistance-like behavior of an AC circuit that combines the effects of the resistance, capacitive reactance and inductive reactance

RLC circuit an AC circuit containing a resistor, a capacitor, and an inductor in series

power factor a muliplicative factor, cos ϕ, that determines the power consumed through the resistor in an RLC circuit

II. Important Equations

Name/Topic	Equation	Explanation
alternating voltages and currents	$V = V_{max} \sin \omega t$ $I = I_{max} \sin \omega t$	Equations representing sinusoidally varying voltage and current.
capacitive reactance	$X_C = \frac{1}{\omega C}$	The definition of the capacitive reactance.
inductive reactance	$X_L = \omega L$	The definition of the inductive reactance.
impedance	$Z = \sqrt{R^2 + (X_L - X_C)^2}$	The impedance of an RLC circuit.
phase angle	$\tan \phi = \frac{X_L - X_C}{R}$	The tangent of the phase angle allows us to calculate ϕ for an RLC circuit.
AC voltage	$V_{max} = I_{max} Z$	The expression similar to Ohm's law for an AC circuit..
natural frequency	$\omega = \frac{1}{\sqrt{LC}}$	The natural frequency of LC oscillation that produces resonance in an RLC circuit.

III. Know Your Units

Quantity	Dimension	SI Unit
reactance and impedance (X_C, X_L, Z)	$[M][L^2][A^{-2}][T^{-3}]$	Ω

IV. Tips

There is an old mnemonic that people sometimes use to remember the phase relationships between the current and the voltages across inductors and capacitors in AC circuits. The mnemonic is based on the phrase

ELI the **ICE** man

and the associations **E** for emf (or voltage), **I** for current, **L** for inductor, and **C** for capacitor.

The word **ELI** contains the letter **L** and, therefore, refers to the inductor. Notice that the **E** in **ELI** comes before the **I** which serves to remind you that the voltage across the inductor "comes before" (or leads) the current. The word **ICE** contains the letter **C** and, therefore, refers to the capacitor. Since the **E** in **ICE** comes after the **I**, this word reminds you that the voltage across the capacitor "comes after" (or lags) the current. If you find mnemonic devices such as this useful, then this is a good one to use.

Practice Problems

1. An AC generator with a maximum voltage of 43 volts and a frequency of 60 Hz is connected to a 292-Ω resistor. What is the rms current through the resistor, to the nearest tenth of a milliamp?
2. In problem 1 what is the maximum power dissipated in the resistor, to the nearest hundredth of a watt?
3. A 110-volt AC generator is connected to a 32-Ω resistor and a 25-μF capacitor. At what frequency will the rms current be 1.2 amps to the nearest Hz?
4. By what angle, to the nearest degree, does the current lead the voltage in problem 3?
5. A 0.404-H inductor is connected in series to a 435-Ω resistor and a 30-volt (rms), 60-Hz generator. What is the rms voltage across the resistor, to the nearest tenth of a volt?
6. In problem 5 what is the rms voltage across the inductor?
7. An AC generator with a frequency of 60 Hz operates at an rms voltage of 120 volts and is connected in series with a 175-Ω resister, a 58-mH inductor, and a 17-μF capacitor. What is the phase angle, to the nearest tenth of a degree?
8. What is the rms power in the circuit in problem 7, to the nearest tenth of a watt?
9. An RLC circuit has $V_{rms} = 25$ V, C = 62 μF, and L = 97 mH. To the nearest tenth of a Hz, what is the resonant frequency?
10. The current at resonance is 2.5 A in the circuit in problem 9. What is the current at twice the resonant frequency, to the nearest thousandth of an amp?

Puzzle

UPS AND DOWNS

The figure shows a graph of the current in a series RLC circuit, $I = I_{max}\cos(t)$. There are five points marked on the graph, labeled A, B, C, D, E. State at which point (or points) each of the following quantities is a maximum, a minimum, or zero:

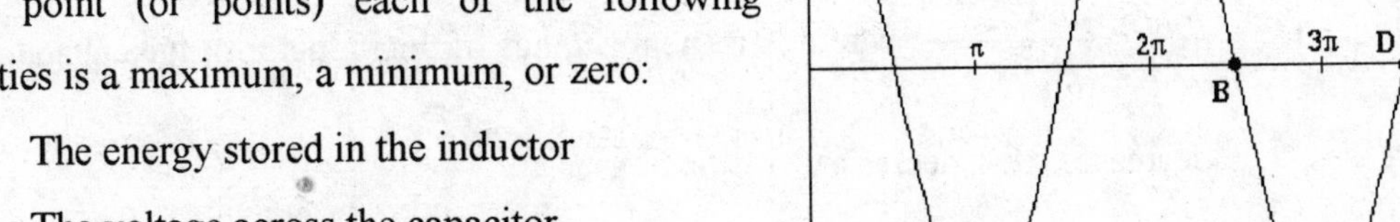

- The energy stored in the inductor
- The voltage across the capacitor
- The power input to the capacitor
- The power output by the inductor
- The charge on the capacitor
- The voltage across the resistor

Selected Solutions

31. The rms current is given by $I_{rms} = V_{rms}/Z$, where $Z = \left(R^2 - X_L^2\right)^{1/2}$; therefore,

$$I_{rms} = \frac{V_{rms}}{Z} = \frac{V_{rms}}{\sqrt{R^2 + \omega^2 L^2}} = \frac{20.0\ \text{V}}{\sqrt{(525\ \Omega)^2 + 4\pi^2(60.0\ \text{s}^{-1})^2(255\ \times\ 10^{-3}\ \text{H})^2}} = \boxed{37.5\ \text{mA}}$$

41. The impedance is given by $Z = \sqrt{R^2 + \left(X_L - X_C\right)^2}$; therefore,

$$Z = \sqrt{R^2 + \left(\omega L - \frac{1}{\omega C}\right)^2} = \sqrt{(1500\ \Omega)^2 + \left[2\pi(60.0\text{s}^{-1})(0.0500\ \text{H}) - \frac{1}{2\pi(60.0\text{s}^{-1})(125\times10^{-6}\text{F})}\right]^2} = \boxed{1.50\ \text{k}\Omega}$$

47. The inductive reactance can be found from the impedance:

$$Z = \sqrt{R^2 + \left(X_L - X_C\right)^2} \Rightarrow R^2 + (X_L - X_C)^2 = Z^2$$

$$\therefore\ (X_L - X_C)^2 = Z^2 - R^2 \Rightarrow X_L - X_C = \pm\sqrt{Z^2 - R^2}$$

$$\therefore\ X_L = X_C \pm \sqrt{Z^2 - R^2}$$

The impedance is given by $Z = \dfrac{V_{rms}}{I_{rms}}$; therefore,

$$X_L = X_C \pm \sqrt{\left(\frac{V_{rms}}{I_{rms}}\right)^2 - R^2} = 5.6\times10^3\,\Omega\ \pm\ \sqrt{\left(\frac{120\ \text{V}}{24\times10^{-3}\ \text{A}}\right)^2 - (3.3\times10^3\,\Omega)^2} = \boxed{9.4\ \text{k}\Omega\ \text{or}\ 1.8\ \text{k}\Omega}$$

55. (a) Since the resonance frequency is inversely proportional to the square root of the product of L and C, the resonance frequency will $\boxed{\text{decrease}}$ if L and C are doubled.

(b) $\omega_0 = \dfrac{1}{\sqrt{LC}} \Rightarrow \omega = \dfrac{1}{\sqrt{(2L)(2C)}} = \dfrac{1}{\sqrt{4LC}} = \dfrac{1}{2\sqrt{LC}} = \dfrac{\omega_0}{2} = \dfrac{125\text{ Hz}}{2} = \boxed{62.5\text{ Hz}}$

65. (a) Recall that $I_{rms} = \dfrac{V_{rms}}{Z} = \dfrac{V_{rms}}{\sqrt{R^2 + X^2}}$, where X is the reactance of the inductor or the capacitor. If X increases, I_{rms} decreases. If X decreases, I_{rms} increases.

* $X_L = \omega L$ increases with increasing frequency.
* $X_C = \dfrac{1}{\omega C}$ decreases with increasing frequency.

Thus, the box must contain a $\boxed{\text{capacitor}}$.

(b) The capacitance can be found from the capacitive reactance, which is part of the impedance:

$$Z = \sqrt{R^2 + X_C^2} \Rightarrow R^2 + X_C^2 = Z^2 \Rightarrow X_C = \sqrt{Z^2 - R^2}$$

$$\therefore \quad \frac{1}{\omega C} = \sqrt{Z^2 - R^2} \Rightarrow C = \frac{1}{\omega}\left(Z^2 - R^2\right)^{-1/2}$$

Since C cannot be negative, we only take the positive square root for the result. Given that Z can be determined from $Z = V_{rms}/I_{rms}$, we have

$$C = \frac{1}{\omega}\left[\left(\frac{V_{rms}}{I_{rms}}\right)^2 - R^2\right]^{-1/2} = \frac{1}{2\pi(25.0 \times 10^3\text{ s}^{-1})}\left[\left(\frac{0.750\text{ V}}{87.2 \times 10^{-3}\text{ A}}\right)^2 - (5.00\ \Omega)^2\right]^{-1/2} = \boxed{0.910\ \mu\text{F}}$$

Answers to Practice Quiz

1. (b) 2. (c) 3. (d) 4. (a) 5. (e) 6. (b) 7. (e) 8. (c) 9. (a) 10. (d)

Answers to Practice Problems

1. 104.1 mA 2. 6.33 W

3. 74 Hz 4. 70°

5. 28.3 V 6. 9.9 V

7. −37.5° 8. 51.8 W

9. 64.9 Hz 10. 0.415 A

CHAPTER 25

ELECTROMAGNETIC WAVES

Chapter Objectives

After studying this chapter, you should:

1. know what makes up an electromagnetic wave.

2. know the relationship between the directions of **E**, **B**, and the direction of propagation.

3. be able to calculate the Doppler shift for electromagnetic waves.

4. be familiar with the regions of the electromagnetic spectrum.

5. know the relationship among the wavelength, frequency, and speed of an electromagnetic wave.

6. be able to calculate the energy density, intensity, and radiation pressure of an electromagnetic wave.

7. know the relationship between the magnitudes of **E** and **B**, and the speed of light.

8. be able to use the law of Malus to determine the intensity of light passing through a polarizer.

Warm-Ups

1. In a wave on a string, the amplitude is the maximum displacement of a point on the string from its equilibrium position. What is the amplitude of an electromagnetic wave?
2. Estimate the wavelength of your favorite radio station.
3. At a given distance, say, 3 m, the light from a 200-W lightbulb is brighter than the light from a 60 W lightbulb. Does this means that light coming from the 200-W bulb has a larger electric field? A larger magnetic field? A higher frequency? A longer wavelength?
4. Light from the Sun reaches Earth's atmosphere at a rate of about 1350 W/m^2. Assuming that this light is entirely absorbed, estimate the force exerted on Earth due to radiation pressure. Is this significant compared with the force on Earth due to the Sun's gravitational attraction?

Chapter Review

In this chapter our understanding of the connection between electricity and magnetism culminates in the treatment of **electromagnetic waves**. As will be mentioned below, visible light is one example of an electromagnetic wave; these waves are sometime referred to as *light* even when they cannot be seen with the naked eye (*electromagnetic radiation* is also a commonly used term).

25–1 The Production of Electromagnetic Waves

Electromagnetic waves are generated by accelerating charges. One of the most common ways of doing this is by connecting an antenna to an AC circuit. The charges accelerating back-and-forth in the antenna produce electromagnetic waves that travel away from the antenna at the speed of light. The electromagnetic wave consists of electric and magnetic fields that are perpendicular to one another and are in phase with one another. Electromagnetic waves are transverse waves; the direction of propagation is perpendicular to both **E** and **B**. Given **E** and **B**, the direction in which the wave propagates can be found from a right-hand rule:

> *Point the fingers of your right hand in the direction of* **E** *so that they would curl toward* **B**. *Your thumb then gives the direction of propagation.*

Practice Quiz

1. Consider an electromagnetic wave for which the electric field points toward the top of this page, ↑, and the magnetic field points into this page, ⊗. What is the direction of propagation of the electromagnetic wave?

   (a) → **(b)** ⊙ **(c)** ↓ **(d)** ← **(e)** ⊗

25–2 The Propagation of Electromagnetic Waves

Unlike the other waves we have studied, electromagnetic waves do not require a medium; they can propagate in vacuum, because an electromagnetic wave is *self-sustaining* by electromagnetic induction. The changing magnetic field produces the changing electric field, and this changing electric field produces the changing magnetic field. The speed of an electromagnetic wave in vacuum is a fundamental constant of nature called the **speed of light**:

$$c = 3.00 \times 10^8 \text{ m/s}$$

From Maxwell's theory of electricity and magnetism, which predicted electromagnetic waves, we know that the speed of light in vacuum is related to the permittivity and permeability of free space,

$$c = \frac{1}{\sqrt{\varepsilon_0 \mu_0}}$$

The propagation of electromagnetic waves exhibits the Doppler effect similar to sound waves. In the case of sound waves the speed of the wave depends on the motion of the source; however, the speed of an electromagnetic wave is independent of the motion of the source. If the relative speed between the source and observer, u, is small compared to the speed of light, then the frequency received, f′, is given by

$$f' = f\left(1 \pm \frac{u}{c}\right)$$

where f is the frequency emitted by the source. The + sign is used if the source is approaching the observer, and the – sign is used if the source is receding from the observer.

Example 25–1 The Doppler Effect Some frequencies of the light from another galaxy are found to be 0.65% lower than the corresponding frequencies from stationary sources on Earth. Is this galaxy moving toward or away from us? Determine the speed at which it is moving toward or away from us.

Picture the Problem The diagram shows the galaxy moving relative to Earth, either toward or away.

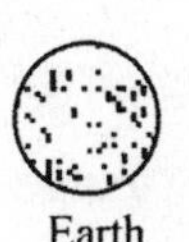

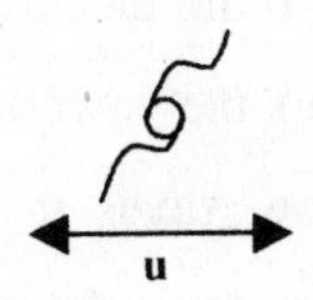

Strategy Given that the frequency is lower, we can deduce whether or not the galaxy is moving toward or away from us.

Solution

1. Since the frequency decreases, this implies: $\frac{f'}{f} < 1 \Rightarrow 1 \pm \frac{u}{c} < 1 \Rightarrow$ use – sign ∴ moving away

2. The observed frequency can be written as: $f' = f - 0.0065f = (0.9935)f$

3. The expression for the Doppler effect becomes: $\frac{f'}{f} = \left(1 - \frac{u}{c}\right) = \frac{(0.9935)f}{f} \quad \therefore \quad 1 - \frac{u}{c} = 0.9935$

4. solving for u gives: $u = (0.0065)c = (0.0065)(3.00 \times 10^8\ \mathrm{\tfrac{m}{s}}) = 2.0 \times 10^6\ \mathrm{\tfrac{m}{s}}$

Insight This may look like an unreasonably high speed, but many distant galaxies are moving away from us very rapidly.

Practice Quiz

2. If an observer determines the frequency of the light given off by a source to be 10% higher than expected, what velocity, relative to the observer, must the source of the light have (in m/s)?

(a) 3.0×10^8 **(b)** 3.0×10^7 **(c)** 3.0×10^6 **(d)** 3.0×10^5 **(e)** 3.0×10^9

25–3 The Electromagnetic Spectrum

As with other types of waves, the speed of an electromagnetic wave equals the product of its wavelength and frequency:

$$c = \lambda f$$

The difference with electromagnetic waves is that the speed is constant. The above equation suggests that any combination of λf that equals c might be a valid electromagnetic wave — this is in fact the case. The full (infinite) range of frequencies (and wavelengths) is known as the **electromagnetic spectrum**.

The finite electromagnetic spectrum that is observed is divided into several regions. **Radio waves** make up the lowest-frequency region of practical importance. This region includes both radio and television waves in the frequency range of about 10^6 Hz – 10^9 Hz. Electromagnetic waves in the frequency range from about 10^9 Hz – 10^{12} Hz are called **microwaves**. Microwaves are commonly used for long-distance communication and cooking. **Infrared waves** fall in the frequency range of about 10^{12} Hz – 10^{14} Hz. The frequencies of infrared waves are just below that of red light. Most of the heat given off by common objects is in the form of infrared waves, and most handheld remote-controls operate via infrared waves (or "IR" as it's often called). The part of the electromagnetic spectrum that we can see with our eyes is called the **visible light** region. The frequency of visible light is on the order of 10^{14} Hz. This light includes all the colors of the rainbow and the various mixtures that these colors can produce.

The frequency range just above that of visible light is **ultraviolet light**. The frequency range of this "UV" light is about 10^{15} Hz – 10^{17} Hz. Ultraviolet light from the sun can be harmful over time, but most harmful UV rays are blocked by Earth's ozone (O_3). This fact is a key reason why ozone depletion is an important current issue. **X-rays** fall in the frequency range of about 10^{17} Hz – 10^{20} Hz. These waves are very penetrating and widely used in medicine to "see" past skin and tissue. The last region of the electromagnetic spectrum is for frequencies above 10^{20} Hz; these waves are called **gamma rays**. These waves are given off by radioactive materials and are even more penetrating and damaging than X-rays.

Exercise 25–2 An Electromagnetic Wave A certain electromagnetic wave has a frequency of 5.11×10^{10} Hz. What type of radiation is it? What is its wavelength?

Solution: We are given that $f = 5.11 \times 10^{10}$ Hz.

In terms of the frequency, the wavelength is given by $\lambda = c/f$. Therefore,

$$\lambda = \frac{3.00 \times 10^{8} \text{ m/s}}{5.11 \times 10^{10} \text{ Hz}} = 5.87 \times 10^{-3} \text{ m}$$

The frequency of this wave falls within the range of microwaves.

Practice Quiz

3. A certain electromagnetic wave has a wavelength of 620 nm. What is its wavelength?

(a) 4.8×10^{14} Hz **(b)** 2.1×10^{-15} Hz **(c)** 6.2×10^{-9} Hz **(d)** 6.2×10^{-7} Hz **(e)** 3.0×10^{8} Hz

4. A certain electromagnetic wave has a wavelength of 0.1 nm. What type of radiation is it?

(a) infrared **(b)** visible light **(c)** ultraviolet **(d)** X-rays **(e)** gamma rays

25–4 Energy and Momentum in Electromagnetic Waves

You may recall that energy is stored in both electric fields and magnetic fields. Therefore, electromagnetic waves carry energy. The energy density of an electromagnetic wave is the sum of the energy densities of the electric field and the magnetic field

$$u = u_E + u_B = \frac{1}{2}\varepsilon_0 E^2 + \frac{1}{2\mu_0} B^2$$

For an electromagnetic wave $u_E = u_B$, so that we can also write

$$u = \varepsilon_0 E^2 = \frac{B^2}{\mu_0}$$

The electric and magnetic fields of an electromagnetic wave vary sinusoidally with time, just like the voltage and current in an AC circuit. Therefore, the average energy density of the wave is conveniently written in terms of the rms values of the fields:

$$u_{av} = \frac{1}{2}\varepsilon_0 E_{rms}^2 + \frac{1}{2\mu_0} B_{rms}^2 = \varepsilon_0 E_{rms}^2 = \frac{B_{rms}^2}{\mu_0}$$

where $E_{rms} = E_{max}/\sqrt{2}$, and $B_{rms} = B_{max}/\sqrt{2}$. These results can also be used to show the direct relationship between the magnitudes of **E** and **B**,

$$E = cB$$

As with sound waves, the intensity of electromagnetic waves is an important quantity. The intensity is the amount of energy delivered to a unit area in a unit time. In terms of the energy density, the intensity of electromagnetic waves is given by

$$I = uc = c\varepsilon_0 E^2 = cB^2/\mu_0$$

The average intensity is obtained by using the rms values of E and B.

In addition to carrying energy, electromagnetic waves also carry momentum. The momentum transferred by an electromagnetic wave to an area that absorbs an amount of energy U is given by

$$p = \frac{U}{c}$$

This momentum transfer results in a force applied to the area. The force per unit area is the pressure; in the case of electromagnetic waves it is called **radiation pressure**. The average radiation pressure applied by waves with an average intensity I_{av} is

$$\text{Pressure}_{av} = \frac{I_{av}}{c}$$

Example 25–3 Energy Density The electric and magnetic fields of an electromagnetic wave are oriented as shown. **(a)** What is the direction of propagation of this wave? **(b)** If the average energy density of this wave is 3.28×10^{-6} J/m^3, what are the rms values of the electric and magnetic fields?

Picture the Problem The sketch shows the directions of **E** and **B** in an electromagnetic wave.

E

B

Strategy We can use the right-hand rule to get the direction of propagation. We can obtain the rms values of the **E** and **B** field by starting with the expression for the energy density.

Solution

Part (a) The right-hand rule for the direction of propagation gives:

$\longrightarrow$ **c**

Part (b)

1. The average energy in terms of E_{rms} is:

$$u_{av} = \varepsilon_0 E_{rms}^2 \quad \Rightarrow \quad E_{rms} = \left(\frac{u_{av}}{\varepsilon_0}\right)^{1/2}$$

$$\therefore \quad E_{rms} = \left(\frac{3.28 \times 10^{-6}\ \text{J/m}^3}{8.85 \times 10^{-12}\ \text{C}^2/\text{N}\cdot\text{m}^2}\right)^{1/2} = 609\ \text{N/C}$$

2. Find the rms value of the magnetic field from E_{rms}:

$$B_{rms} = \frac{E_{rms}}{c} = \frac{608.8 \text{ N/C}}{3.00 \times 10^8 \text{ m/s}} = 2.03 \times 10^{-6} \text{ T}$$

Insight The rms magnetic field could also have been found from the energy density. Try it.

Example 25–4 Radiation Pressure For the electromagnetic wave in Example 25–3, what average force would it exert on a 1.00-cm × 1.00-cm slab?

Picture the Problem The sketch shows an electromagnetic wave impinging on a square slab.

Strategy The pressure is force per area, so we can determine the force by using the radiation pressure.

Solution

1. The average radiation pressure P_{av} is:

$$P_{av} = \frac{I_{av}}{c} = \frac{u_{av} c}{c} = u_{av} = \frac{F_{av}}{A}$$

2. Solving for the average force gives:

$$F_{av} = u_{av} A = \left(3.28 \times 10^{-6} \text{ J/m}^3\right)\left(1.00 \times 10^{-2} \text{ m}\right)^2 = 3.28 \times 10^{-10} \text{ N}$$

Insight Note that the equations work out such that the average pressure equals the average energy density. Check that this makes sense dimensionally.

Practice Quiz

5. An electromagnetic wave whose electric field has an amplitude E_{max} contains an average energy density u_{av}. Another electromagnetic wave of amplitude $2E_{max}$ would contain an average energy density of

(a) u_{av} **(b)** $\sqrt{2}\, u_{av}$ **(c)** $2u_{av}$ **(d)** $4u_{av}$ **(e)** $u_{av}/2$

6. The electric field of a certain electromagnetic wave has an rms value of 2130 N/C. What is the rms value of the associated magnetic field?

(a) 1.4×10^5 T **(b)** 6.4×10^{11} T **(c)** 7.1×10^{-6} T **(d)** 3.0×10^8 T **(e)** 2130 T

7. The electric field of an electromagnetic wave has an rms value of 950 N/C. What is the average energy density of this wave (in J/m^3)?

(a) 4.0×10^{-6} **(b)** 2.4×10^3 **(c)** 9.0×10^5 **(d)** 8.4×10^{-9} **(e)** 8.0×10^{-6}

8. An electromagnetic wave whose electric field has an amplitude E_{max} exerts an average radiation pressure P_{av}. Another electromagnetic wave of amplitude $\sqrt{2}\, E_{max}$ would exert an average radiation pressure of

(a) $\sqrt{2}P_{av}$ **(b)** P_{av} **(c)** $2P_{av}$ **(d)** $4P_{av}$ **(e)** none of the above

25–5 Polarization

In general, the electric field in an electromagnetic wave points in all directions within a plane perpendicular to the direction of propagation; however, there are processes that can cause the electric field of an electromagnetic wave to oscillate with a specific orientation. Such electromagnetic waves are said to be **polarized**. When the electric field oscillates in random directions, the waves are called **unpolarized**. These same considerations apply to the magnetic field as well; however, by tradition, the direction of polarization is taken to be direction of the electric field.

One of the ways in which electromagnetic waves are polarized is by passing through a **polarizer**. A polarizer is a material that absorbs electromagnetic waves whose electric fields are perpendicular to a certain direction called the **transmission axis**. Waves for which the electric field **E** is parallel to this axis are transmitted through the material undiminished. If the electric field makes an angle θ with the transmission axis, the component of **E** perpendicular to the axis is absorbed and the component parallel to the axis is transmitted. This leads to the **law of Malus**:

$$I = I_0 \cos^2 \theta$$

where I_0 is the intensity of initially polarized waves whose polarization direction makes an angle θ with the transmission axis of the polarizer, and I is the intensity of the waves that emerge from the polarizer. The light that emerges from the polarizer will be polarized along the direction of the transmission axis. If the initial electromagnetic wave is unpolarized, the law of Malus shows that the intensity upon passage through the polarizer is reduced by half:

$$I = \tfrac{1}{2} I_0 \qquad \text{(initially unpolarized light)}$$

Often, a second polarizer, called an **analyzer**, is used to investigate the properties of polarized light, and if it's free to rotate, it will produce light of variable intensity. (Electromagnetic waves can also be polarized by scattering and reflection.)

Example 25–5 Polarization Unpolarized light of intensity 1.20×10^3 W/m^2 is incident on a polarizer-analyzer system. If the final intensity that emerges is 504 W/m^2, what is the angle between the transmission axes of the polarizer and the analyzer?

Picture the Problem The sketch shows an unpolarized electromagnetic wave incident on a polarizer-analyzer system.

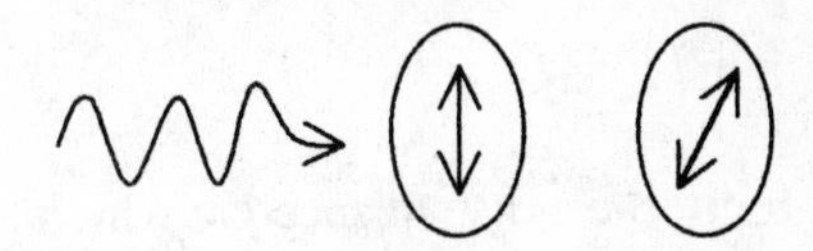

Strategy For the unpolarized light we use the result that the intensity is halved by the polarizer, then we apply the law of Malus for the light that passes through the analyzer.

Solution

1. For the incident intensity I_0 and that which emerges from the polarizer, I_1, we have: $I_1 = \frac{1}{2} I_0$

2. Call the final intensity I_2, then from Malus' law we get: $I_2 = I_1 \cos^2\theta = \frac{1}{2} I_0 \cos^2\theta$

3. solving for θ gives: $\theta = \cos^{-1}\left[\sqrt{\frac{2I_2}{I_0}}\right] = \cos^{-1}\left[\sqrt{\frac{2(504\text{ W/m}^2)}{1.20\times10^3\text{ W/m}^2}}\right] = 23.6^\circ$

Insight Here, we see that a system of polarizers is handled by applying the law of Malus to each polarizer.

Practice Quiz

9. If polarized light of intensity 750 W/m^2 is incident on a polarizer whoes transmission axis makes an angle of 20.0° with the direction of polarization, what is the intensity of the light that emerges from the polarizer?

   (a) 750 W/m^2 **(b)** 375 W/m^2 **(c)** 705 W/m^2 **(d)** 662 W/m^2 **(e)** 0 W/m^2

10. If the intensity of initially polarized light is cut in half on passing through a polarizer, what is the angle between the direction of polarization and the transmission axis of the polarizer?

   (a) 0° **(b)** 30° **(c)** 45° **(d)** 60° **(e)** 90°

Reference Tools and Resources

I. Key Terms and Phrases

electromagnetic wave a wave consisting of oscillating electric and magnetic fields that propagates at the speed of light

speed of light the constant speed at which light travels in vacuum

electromagnetic spectrum the full (infinite) range of frequencies (and wavelengths) of electromagnetic waves

radiation pressure the pressure exerted by electromagnetic waves

polarized light an electromagnetic wave whose electric field oscillates along a specific direction

polarizer an object that absorbs the component of the electric field of an electromagnetic wave that is perpendicular to its transmission axis

transmission axis the direction in a polarizer that transmits a polarized electromagnetic wave without loss of intensity if it is polarized along this direction

law of Malus the expression for the intensity that results when polarized light passes through a polarizer

analyzer a second polarizer used to analyze the polarized light that emerged from a polarizer

II. Important Equations

Name/Topic	Equation	Explanation
Doppler effect	$f' = f\left(1 \pm \frac{u}{c}\right)$	The Doppler effect for light, assuming $u \ll c$.
electromagnetic spectrum	$c = \lambda f$	The relationship between the wavelength and frequency of electromagnetic waves.
energy density	$u_{av} = \frac{1}{2}\varepsilon_0 E_{rms}^2 + \frac{1}{2\mu_0} B_{rms}^2 = \varepsilon_0 E_{rms}^2 = \frac{B_{rms}^2}{\mu_0}$	The average energy density of electromagnetic waves.

electromagnetic waves	$E = cB$	The relationship between E and B in an electromagnetic wave.
intensity	$I = uc = c\varepsilon_0 E^2 = cB^2/\mu_0$	The instantaneous intensity of electromagnetic waves.
polarization	$I = I_0 \cos^2\theta$	The law of Malus for the intensity of light emerging from a polarizer.

Practice Problems

1. In the Fizeau experiment if the wheel has 499 notches and is turning at 48 rev/s, what must be the distance, to the nearest tenth of a meter if the returning light beam just passes through the next notch? ($c = 3.00 \times 10^8$ m/s)
2. You are shot by a radar gun while traveling toward it at 39 m/s. If the gun is using a frequency of 2.76×10^9 Hz, what is the difference in frequency between the outgoing and the reflected wave?
3. At a particular time at a given instant of time in an electromagnetic field the electric field has a value of 990 N/C. To the nearest hundredth of a $\mu J/m^3$, what is the instantaneous energy density at that point?
4. To the nearest hundredth of a μT, what is the corresponding value of the magnetic field in problem 3?
5. A 68-W (average) lightbulb radiates in all directions. At a distance of 1.38 meters what is the average intensity to the nearest hundredth of a W/m^2?
6. A laser with a beam diameter of 0.73 mm has an average power of 8.9 mW. What is its intensity, to the nearest W/m^2?
7. A spaceship is pulled along by a solar sail 19,295 meters on a side. At the location of the ship the sunlight has an intensity of 1099 W/m^2. To the nearest tenth of a newton what is the force on the sail? (Assume the radiation is totally reflected.)
8. Unpolarized light of intensity 1086 W/m^2 passes through a polarizer at an angle of 63 degrees to the vertical. To the nearest tenth of a W/m^2, what is the intensity of the transmitted light?
9. The light from the filter in problem 8 passes through another filter making an angle of 80 degrees with the vertical. To the nearest tenth of a W/m^2, what is the intensity of the transmitted light?
10. Unpolarized light is incident on three polarizers, the first making an angle of 0 degrees with the vertical, the second making an angle of θ degrees with the vertical and the last one making an angle of 90 degree with the vertical. If the initial intensity is 924 W/m^2 and the final intensity is 26 W/m^2, to the nearest tenth of a degree what is the angle of the middle polarizer?

Puzzle

LIGHT DRILLS

The sketch shows the basic elements used by A. Michelson to measure the speed of light in the 1920s. A pulse of light hits the rotating 8-sided mirror and takes off on a round trip to a flat mirror some distance off. The pulse will be received by the detector if the rotating mirror is in the right orientation when the pulse returns. A moderately priced power drill can turn at about 1200 rpm . If such a drill is used to turn the mirror, how far away does the flat mirror have to be for a pulse to be observed? How about for a Dremel that turns at 30,000 rpm? (Hint: Assume the distance to the mirror (L) is large compared with the size of the 8-sided mirror.)

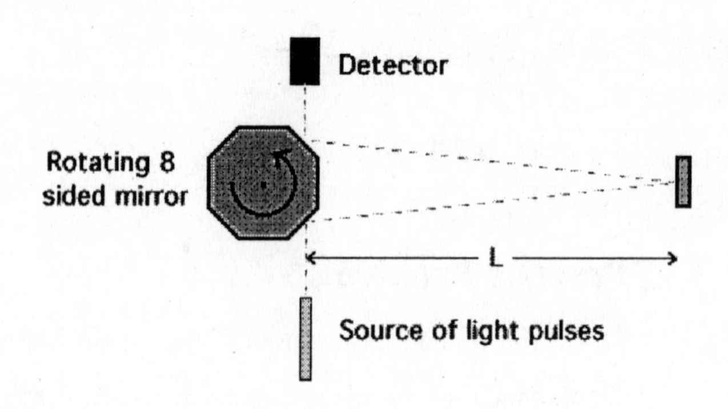

Selected Solutions

5. Using the right-hand rule we get the following:

(a) $\boxed{+z}$ **(b)** $\boxed{-z}$ **(c)** $\boxed{-x}$ **(d)** $\boxed{-x}$

17. For the case of receding objects, the Doppler effect is given by

$$f' = f\left(1-\frac{u}{c}\right) = (5.000\times10^{14}\text{ Hz})\left(1-\frac{3025\times10^{3}\text{ m/s}}{3.00\times10^{8}\text{ m/s}}\right) = \boxed{4.950\times10^{14}\text{ Hz}}$$

33. The wavelength of the waves is $\lambda = c/f$. Therefore, the length L is given by

$$L = \frac{1}{4}\lambda = \frac{c}{4f} = \frac{3.00\times10^{8}\ \frac{\text{m}}{\text{s}}}{4(890\times10^{3}\text{ Hz})} = \boxed{84\text{ m}}$$

39. **(a)** $E_{max} = cB_{max} = (3.00\times10^{8}\text{ m/s})(2.7\times10^{-6}\text{ T}) = \boxed{810\text{ V/m}}$

(b) $$I_{max} = \frac{c}{\mu_0}B_{max}^2 = \frac{3.00\times10^{8}\text{ m/s}}{4\pi\times10^{-7}\ \frac{\text{T}\cdot\text{m}}{\text{A}}}(2.7\times10^{-6}\text{ T})^2 = \boxed{1.7\text{ kW/m}^2}$$

(c) $$I_{av} = \frac{c}{\mu_0}B_{rms}{}^2 = \frac{3.00\times10^{8}\text{ m/s}}{4\pi\times10^{-7}\ \frac{\text{T}\cdot\text{m}}{\text{A}}}\left(\frac{2.7\times10^{-6}\text{ T}}{\sqrt{2}}\right)^2 = \boxed{870\text{ W/m}^2}$$

65. Using the law of Malus, with I_1 as the intensity after passing through the first polarizer and I_2 as the intensity after passing through the second, we have

$$I_1 = I_0 \cos^2 \theta_1$$

$$I_2 = I_1 \cos^2 \theta_2 = I_0 \cos^2 \theta_1 \cos^2 \theta_2$$

(a) For the three cases we have

(a) $\frac{I_2}{I_0} = \cos^2(45°)\cos^2(90° - 45°) = \cos^4(45°) = 0.25$

(b) $\frac{I_2}{I_0} = \cos^2(45°)\cos^2(0° - 45°) = \cos^4(45°) = 0.25$

(c) $\frac{I_2}{I_0} = \cos^2(45°)\cos^2(45° + 45°) = \cos^2(45°)\cos^2(90°) = 0$

case (c) is the smallest; (a) and (b) tie.

(b) Using the above expression for I_2 yields

(a) $I_2 = \left(52 \frac{W}{m^2}\right)\cos^2(45°)\cos^2(90° - 45°) = \boxed{13\ W/m^2}$

(b) $I_2 = \left(52 \frac{W}{m^2}\right)\cos^2(45°)\cos^2(0° - 45°) = \boxed{13\ W/m^2}$

(c) $I_2 = \left(52 \frac{W}{m^2}\right)\cos^2(45°)\cos^2(45° + 45°) = \boxed{0}$

Again, case (c) is the smallest; (a) and (b) tie. The results verify part (a).

Answers to Practice Quiz

1. (d) **2**. (b) **3**. (a) **4**. (d) **5**. (d) **6**. (c) **7**. (e) **8**. (c) **9**. (d) **10**. (c)

Answers to Practice Problems

1. 6262.5 m **2**. 717.6 Hz

3. 4.34 μJ/m³ **4**. 3.30 μT

5. 2.84 W/m² **6**. 21,264 W/m²

7. 2727.7 N **8**. 543.0 W/m²

9. 496.6 W/m² **10**. 14.2°

CHAPTER 26

GEOMETRICAL OPTICS

Chapter Objectives

After studying this chapter, you should:

1. know and be able to use the law of reflection.

2. be able to sketch ray diagrams to locate images formed by mirrors and lenses.

3. be able to use the mirror equation and the thin-lens equation together with their sign conventions.

4. be able to calculate the magnification of images formed by reflection and refraction.

5. be able to determine the speed of light in different media.

6. be able to use Snell's law for the refraction of light.

7. know how to determine the critical angle for total internal reflection and Brewster's angle.

Warm-Ups

1. In your own words, explain what a focal length is. Try not to use any equations or refer to any specific type of mirror or lens.
2. Estimate the focal length of a typical bathroom "magnifying mirror."
3. Images formed by spherical lenses can be real or virtual. What does it mean for an image to be virtual? Can such an image be seen?
4. Estimate the focal length of a typical magnifying glass.
5. What is the focal length of a clear, flat piece of glass, such as a normal household window?

Chapter Review

As discussed in the previous chapter, light is an electromagnetic wave; however, understanding the behavior of light does not always require a wave analysis. In this chapter and the next we study the branch of physics, called **geometrical optics**, in which conditions are such that the wave nature of light can be "glossed over" while still accurately describing its behavior.

26–1 The Reflection of Light

In the study of geometrical optics we think of light as **rays** traveling along a straight-line path. In terms of the propagating electromagnetic wave, we can track the crests of these waves (or any specific phase point). The collection of crests at a given phase can be imagined to form surfaces called **wave fronts**. Within this approximation, the light rays are perpendicular to the wave fronts and point in the direction of their propagation. As a light wave gets farther away from the source the wave fronts are so spread out that the surfaces are approximately planar; these waves are called **plane waves** and they give rise to parallel rays.

The reflection of light from a smooth boundary obeys a simple law called the **law of reflection**. This law says that the angle that the incident ray makes with the normal to the reflecting surface, called the *angle of incidence,* θ_i, is equal to the angle that the reflected ray makes with the normal to the surface, called the *angle of reflection,* θ_r. Also, the incident ray, the normal line, and the reflected ray all lie in the same plane.

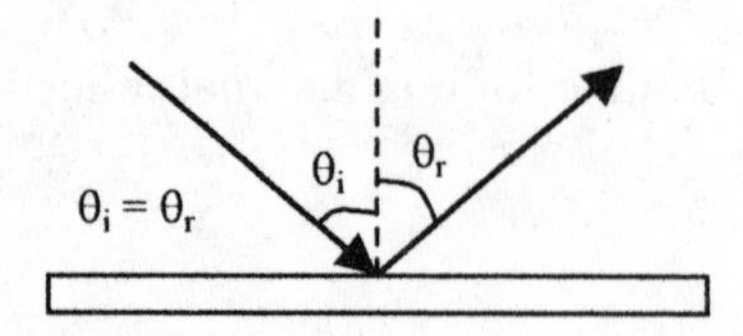

Practice Quiz

1. An incident light ray makes an angle of 33° with the normal to the surface of a plane mirror. What angle does the reflected ray make with the surface of the mirror?

   (a) 0° **(b)** 33° **(c)** 66° **(d)** 57° **(e)** none of the above

26–2 Forming Images with a Plane Mirror

A plane mirror is one that is perfectly flat. When you look at your reflection in a plane mirror what you see is called an **image** of yourself. The source that is being reflected in the mirror (you) is called the **object**. Using the law of reflection, several results about reflection with a plane mirror can be found:

* The distance between the object and the mirror, called the **object distance,** d_o, is equal to the distance between the image, on the opposite (back) side of the mirror, and the mirror, called the **image distance,** d_i.

* The image is right-side-up, referred to as **upright**.

* The image is the same size as the object.

Left and right appear to be reversed in a mirror; because the image is facing you. Just as when a person faces you, his or her right hand is on your left, and vice versa.

Practice Quiz

2. An object is 5.0 m in front of a plane mirror. How far away from the mirror is its image?

   (a) 0.0 m **(b)** 2.5 m **(c)** 5.0 m **(d)** 10.0 m **(e)** 15.0 m

26–3 Spherical Mirrors

A spherical mirror has the shape of a section of a sphere. If the reflecting surface is on the outside of this spherical section, it is called a **convex** mirror; if the reflecting surface is on the inside, it is called a **concave** mirror. The **principal axis** of the mirror is the line that passes through the center of the mirror (often called the *vertex*) perpendicular to the surface. A distance R away from the vertex along the principal axis is a point called the **center of curvature**, C, where R is the radius of the sphere from which the spherical section of mirror was taken (often called the *radius of curvature*). The point C would be located at the center of the sphere.

If light rays that are parallel to the principal axis and lie close to it (paraxial rays), are incident on a spherical mirror, they will either converge to (for a concave mirror) or diverge from (for a convex mirror) points a distance of magnitude f from the vertex, called the **focal length**. The point on the principal axis a distance of one focal length away from the vertex is called the **focal point**. For a spherical mirror the focal length is given by

$$f = \pm\tfrac{1}{2}R$$

where the + sign applies to concave mirrors, and the – sign applies to convex mirrors. The full reasoning behind this difference in sign will be discussed below. Basically, the positive sign means that the focal point is in front of the mirror, and the negative sign means that it is behind the mirror.

Practice Quiz

3. The center of curvature of a convex, spherical mirror is located 50 cm away from its vertex. What is its focal length?

   (a) 25 cm **(b)** –25 cm **(c)** 50 cm **(d)** –50 cm **(e)** none of the above

26–4 Ray Tracing and the Mirror Equation

Our main objective in image formation by reflection is to determine the size, location, and orientation of the image of a given object. There are two methods for obtaining this information, one method, **ray tracing**, is geometric, and the other method, the **mirror equation**, is algebraic. In practice, ray tracing is used qualitatively to help visualize the given situation, and the mirror equation is used to obtain accurate numerical results.

For ray tracing with spherical mirrors, there are three commonly used rays: the *P-ray* is drawn parallel to the principal axis, the *F-ray* (or its extension) intersects the principal axis at the focal point, and the *C-ray* (or its extension) intersects the principal axis at the center of curvature.

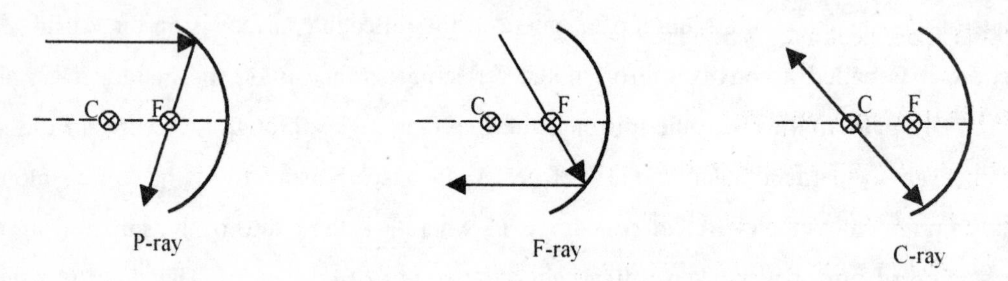

As you can see in the above figure for a concave mirror, the reflected P-ray intersects the principal axis at the focal point, the reflected F-ray is parallel to the principal axis, and the reflected C-ray hits the mirror at a 90 degree angle and travels back along the same path as the incident C-ray. When all three rays, coming from a given point on the object, are drawn, the intersection of the three reflected rays (or the intersection of their extensions) locates the image of that point.

In a version of the preceding diagram for a convex mirror the incident P-, F-, and C-rays would be drawn similarly except that the reflected rays would not intersect. Instead, the extensions of the reflected rays behind the mirror would intersect to form what is called a **virtual image** (see Figure 26-15 in the text). When the actual reflected light rays themselves intersect (not just their extensions) they form what is called a **real image**.

A geometrical analysis of a properly drawn ray diagram can be used to derive an equation that relates the image and object distances for reflection from a spherical mirror. The result of this analysis is the mirror equation

$$\frac{1}{d_o} + \frac{1}{d_i} = \frac{1}{f} = \frac{2}{R}$$

Similarly, an expression for the **magnification** of the image can be derived. The magnification is defined to be the ratio of the height of the image (h_i) to the height of the object (h_o). The result is

$$m = \frac{h_i}{h_o} = -\frac{d_i}{d_o}$$

In a the ray-diagram approach it will be clear from the diagram whether or not the image is real or virtual, upright or inverted. Algebraically, these attributes are distinguished by the sign that certain quantities have. Therefore, the mirror equation and the expression for the magnification are to be used in the context of certain *sign conventions* that distinguish among characteristics. The sign conventions for spherical mirrors are as follows:

Focal Length and Radius of Curvature

* f and R are positive for concave mirrors.
* f and R are negative for convex mirrors.

Magnification and Images

* m is positive for upright images.
* m is negative for inverted images.
* The image height is positive for upright images.
* The image height is negative for inverted images.

Image Distance

* d_i is positive for real images (those that are in front of the mirror).
* d_i is negative for virtual images (those that are behind the mirror).

Object Distance

* d_o is positive for real objects (those that are in front of the mirror).
* d_o is negative for virtual objects.

Example 26–1 A Convex Mirror A certain convex mirror has a radius of curvature of 75.0 cm. If a point source of light is placed 30.0 cm away from the mirror on the principal axis, where is the image of this object?

Picture the Problem The sketch shows a ray diagram for this situation. The diagram gives us some idea as to where the image should be.

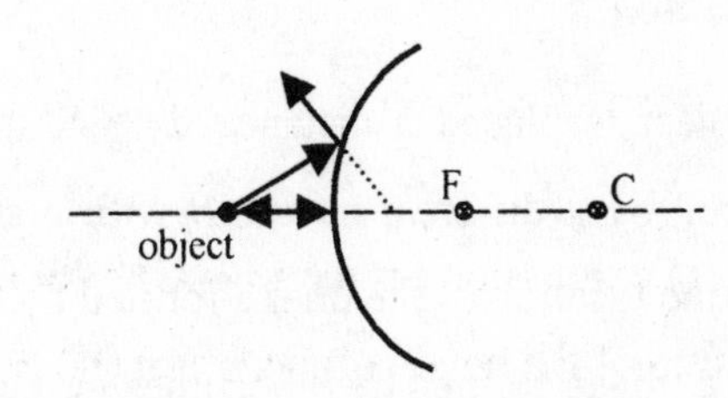

Strategy From the diagram we expect a virtual image inside of the focal point.

Solution

1. The values of R and d_o are: $R = -75.0\text{ cm},\ d_o = 30.0\text{ cm}$

2. The mirror equation gives: $\frac{1}{d_i} + \frac{1}{d_o} = \frac{2}{R} \Rightarrow d_i = \left[\frac{2}{R} - \frac{1}{d_o}\right]^{-1}$

3. The numerical result is: $d_i = \left[\frac{2}{-75.0\text{ cm}} - \frac{1}{30.0\text{ cm}}\right]^{-1} = -16.7\text{ cm}$

Insight The final result is consistent with our expectation based on the ray diagram. Take note that the radius of curvature was used as a negative number because it is a convex mirror even though the statement of the problem did not give a negative value. You must pay attention to the type of mirror being used. Because the object was a point source on the principal axis, the P-, C-, and F-rays were all the same, so I just drew an arbitrary ray as shown, it still works.

Exercise 26–2 An Extended Object If the object in Example 26–1 was an extended object of height 1.50 cm, what would be the height and orientation of the image?

Solution

From the solution to Example 26–1 we know both the object and image distances. From the definition of the magnification we can write

$$\frac{h_i}{h_o} = -\frac{d_i}{d_o} \Rightarrow h_i = -\frac{h_o d_i}{d_o} = -\frac{(1.50\text{ cm})(-16.67\text{ cm})}{(30.0\text{ cm})} = 0.834\text{ cm}$$

Since this height is positive, we know that it is also an upright image.

Practice Quiz

4. If for the reflection of a single object from a single spherical mirror we know that the image will be inverted, what else can we conclude about the image?

   (a) it is enlarged. **(b)** it is diminished. **(c)** it is real. **(d)** it is virtual. **(e)** none of the above

5. An object is placed a distance $d_o > R$ away from a concave spherical mirror *A*. If the mirror is replaced by a different mirror *B* with a smaller radius of curvature, which of the following is true about the location of the image formed by *B* compared to that formed by *A*?

   (a) The image formed by *B* becomes virtual, whereas the image formed by *A* was real.

   (b) The image formed by *B* becomes real, whereas the image formed by *A* was virtual.

(c) The image formed by B is closer to B than the image formed by A was to A.

(d) The image formed by B is farther from B than the image formed by A was to A.

(e) none of the above

6. A concave spherical mirror has a focal length of 1.20 m. If an object is placed 75.0 cm in front of the mirror, where is its image?

   (a) 2.00 m in front of the mirror

   (b) 2.67 m behind the mirror

   (c) 0.500 m behind the mirror

   (d) 0.450 m in front of the mirror

   (e) none of the above

26–5 The Refraction of Light

When light crosses the boundary between two different media, such as air and water, the light that penetrates into the second medium will, in general, travel in a different direction than the incident light; this phenomenon is known as **refraction.** Also, the speed of light in the second medium will generally be different from its speed in the first medium. The ratio of the speed of light in vacuum to its speed in a certain medium is called the **index of refraction** (n) of the medium:

$$n = \frac{c}{v}$$

Being the ratio of two speeds, the index of refraction is a dimensionless quantity. Note also that since c is the fastest known speed, $n \geq 1$. (See Table 26–2 in the text for values of n for different substances.)

Since the direction of the light ray changes upon transmission across the boundary, we require a way to determine the new direction of the ray. Let us call the angle that the incident ray makes with the normal line to the surface (in the first medium) θ_1 and the index of refraction of this medium n_1. The relationship between the angle of incidence and the angle, θ_2, that the refracted ray makes with the normal in the second medium (with index of refraction n_2) is

$$n_1 \sin\theta_1 = n_2 \sin\theta_2$$

This equation is called **Snell's law** (also known as the *law of refraction*). Some of the properties of refraction are illustrated in the following example.

Example 26–3 Refraction Consider a smooth boundary between air and water. What is the angle of refraction if, **(a)** light is incident from the air with a 35° angle of incidence, **(b)** light is incident from the air perpendicular to the surface, and **(c)** light is incident from the water with a 35° angle of incidence?

Picture the Problem The diagram is for part (a) with the incident light ray in air and the refracted ray in water.

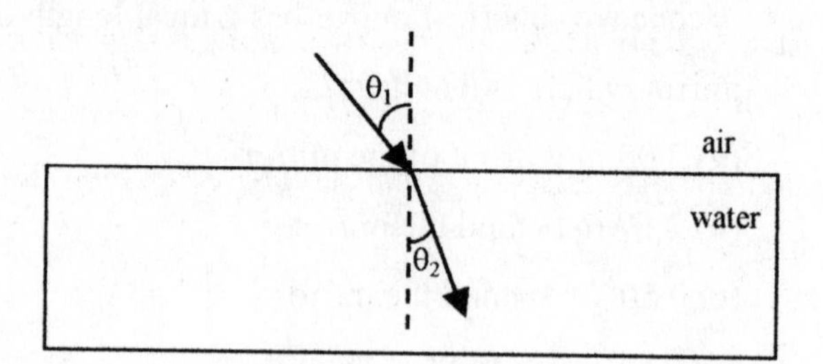

Strategy We can make direct use of Snell's law in each of the three cases.

Solution

Part (a)

1. Medium 1 is air and medium 2 is water. The values of n are from Table 26–2 in the text: $n_1 \sin\theta_1 = n_2 \sin\theta_2 \Rightarrow \theta_2 = \sin^{-1}\left[\frac{n_1 \sin\theta_1}{n_2}\right]$

2. The result is: $\theta_2 = \sin^{-1}\left[\frac{(1.00)\sin 35^\circ}{1.33}\right] = 26^\circ$

Part (b)

3. The angle of incidence is: $\theta_1 = 0.0^\circ$

4. Snell's law gives: $\theta_2 = \sin^{-1}\left[\frac{(1.00)\sin 0.0^\circ}{1.33}\right] = 0.0^\circ$

Part (c)

5. Medium 1 is water and medium 2 is air. Snell's law gives: $\theta_2 = \sin^{-1}\left[\frac{(1.33)\sin 35^\circ}{1.00}\right] = 50^\circ$

Insight Notice that in part (a) the refracted angle is smaller than 35°, and in part (c) it is bigger. These results reflect the general behavior that rays are refracted toward the normal if $n_2 > n_1$ and away from the normal if $n_2 < n_1$.

In general, when light is incident on a boundary there is both a reflected and a refracted ray. For a given pair of media, when $n_2 < n_1$, there is an angle of incidence beyond which no light is transmitted into the second medium; this angle is called the **critical angle** for **total internal reflection**, θ_c. The critical angle is reached when $\theta_2 = 90°$; therefore,

$$\sin\theta_c = \frac{n_2}{n_1}$$

Another effect that can be related to Snell's law is the total polarization of light that is reflected from a boundary. As mentioned in the previous chapter of the text, reflected light is generally partially polarized. At a certain angle of incidence, called **Brewster's angle**, θ_B (sometimes also called the *polarization angle*), the reflected light is completely polarized parallel to the reflecting surface. Brewster's angle is the angle of incidence such that the angle between the reflected and refracted rays is 90°. Using this fact in Snell's law gives

$$\tan\theta_B = \frac{n_2}{n_1}$$

Practice Quiz

7. What is the speed of light in ice (in m/s)?

   (a) 3.00×10^8 **(b)** 1.81×10^8 **(c)** 2.00×10^8 **(d)** 2.26×10^8 **(e)** 2.29×10^8

8. For a light ray incident on a boundary from air (n = 1.00) to crown glass (n = 1.52) the refracted ray makes an angle of 27° with the normal. What is the angle of incidence?

   (a) 63° **(b)** 57° **(c)** 33° **(d)** 44° **(e)** 17°

9. Which of the following cases cannot produce total internal reflection? A light ray goes from

   (a) air to water

   (b) water to ice

   (c) flint glass to crown glass

   (d) diamond to water

   (e) none of the above

26–6 Ray Tracing for Lenses

The light rays bend upon transmission into a different medium, we can also achieve image formation by refraction. This way of forming images uses a **lens** instead of a mirror. Lenses come in many different types. Here, we consider a basic *converging lens* (double convex) and a basic *diverging lens* (double concave). The principal axis of the lens is the line passing through the center of the lens making right angles with the surfaces. Incident rays that are parallel to the principal axis will converge to a focus (for a converging lens) at a focal point F on the axis and at a local length f from the center of the lens.

As with mirrors, we can characterize the image formed by refraction with ray tracing diagrams. Again, there are three commonly used rays: the *P-ray* is drawn parallel to the principal axis, the *F-ray* (or its extension) intersects the principal axis at the focal point, and the *M-ray* passes through the middle of the lens.

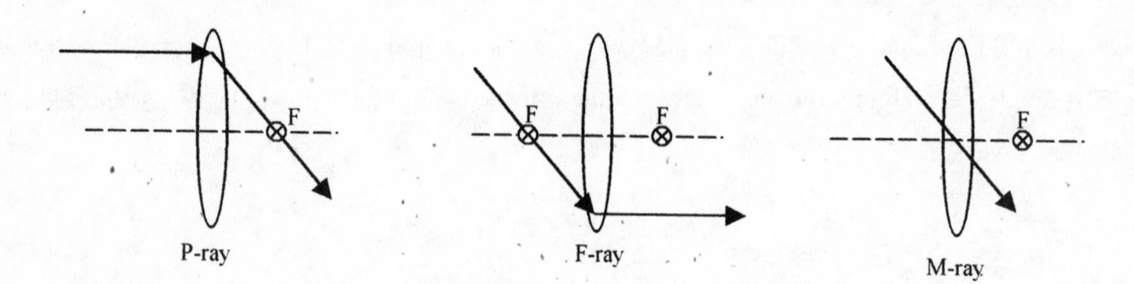

As you can see in the above figure for a (double) convex lens, the refracted P-ray intersects the principal axis at the focal point, the refracted F-ray is parallel to the principal axis, and the refracted M-ray passes through the center of the lens undeflected (which is a good approximation if the lens is thin). When all three rays, coming from a given point on the object, are drawn, the intersection of the three refracted rays locates the image of that point.

In a version of the above diagram for a concave lens the incident P-, F-, and M-rays would be drawn similarly except that the refracted rays would not intersect. Instead, the extensions of the refracted rays back in front of the lens would intersect to form a **virtual image** (see Figure 26–34 in the text).

26–7 The Thin-Lens Equation

The geometry of a ray diagram for lenses can be used to derive a relationship between the image and object distances, as well as the magnifications. The results look just like those for spherical mirrors. The relationship between d_i and d_o is called the **thin-lens equation**:

$$\frac{1}{d_o} + \frac{1}{d_i} = \frac{1}{f}$$

The magnification of the image also obeys the same relationship that it does for mirrors:

$$m = \frac{h_i}{h_o} = -\frac{d_i}{d_o}$$

The sign conventions for using the thin-lens equation and the magnification are as follows:

Focal Length

* f is positive for converging lenses
* f is negative for diverging lenses

Magnification

* m is positive for upright images
* m is negative for inverted images

Image Distance

* d_i is positive for real images (those that are on the opposite side of the lens from the object)
* d_i is negative for virtual images (those that are on the same side of the lens as the object)

Object Distance

* d_o is positive for real objects (those from which light diverges)
* d_o is negative for virtual objects (those toward which light converges)

Example 26–4 The Image of a Concave Lens A concave lens has a focal length of 65.0 cm. The image of a certain object is virtual and has a magnification of 0.250. Determine both the image and object distances.

Picture the Problem The sketch is of a ray diagram for this problem.

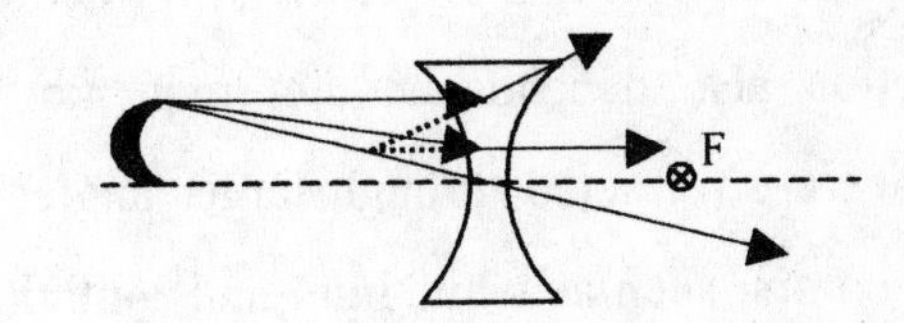

Strategy We have two equations that involve both unknowns, d_o and d_i, so we will solve them by substitution.

Solution

1. The magnification equation gives:

$$m = -\frac{d_i}{d_o} \Rightarrow d_i = -md_o$$

2. Substitute this into the thin-lens equation:

$$\frac{1}{d_o} + \frac{1}{d_i} = \frac{1}{f} \Rightarrow \frac{1}{d_o} - \frac{1}{md_o} = \frac{1}{f} = \frac{m-1}{md_o}$$

3. Solving for d_o gives:

$$d_o = \frac{f(m-1)}{m} = \frac{(-65.0\text{ cm})(0.250-1)}{0.250} = 195\text{ cm}$$

4. Use the magnification to get d_i:

$$d_i = -md_o = -(0.250)(195\text{ cm}) = -48.8\text{ cm}$$

Insight Notice again, as in Example 26–1 that we had to know to make the focal length negative because we were working with a diverging lens. If we had forgotten to do that, we would have gotten a positive image distance, indicating that the image was on the opposite side from what is indicated by the ray diagram. This discrepancy would have tipped us off to take a closer look at the problem. This consistency check will only work if you draw the ray diagrams, so do it when you solve these problems.

---

10. An object is 1.25 m away from a converging lens of focal length 95 cm. Which of the following correctly characterizes the image?

(a) virtual, enlarged, and inverted

(b) real, enlarged, and upright

(c) virtual, diminished, and upright

(d) real, enlarged, and inverted

(e) real, diminished, and inverted

26–8 Dispersion and the Rainbow

In addition to the fact that the index of refraction depends on the medium, for a given medium the index of refraction also depends on the frequency of the light that is being refracted. Generally, higher frequencies are refracted through larger angles. This means that light that is made up of a mixture of different colors, such as white light and sunlight, will be separated out, or dispersed, into those different colors on refraction by a medium in which the variation of n with frequency is wide enough that the effect can be noticed. Water and glass are examples of such media. This phenomenon is known as **dispersion,** and it is responsible for the **rainbow** that we often see when sunlight passes through water droplets or a glass prism.

Reference Tools and Resources

I. Key Terms and Phrases

geometrical optics the study of the reflection and refraction of light under the ray approximation

light ray a representation of the straight-line propagation of light

law of reflection the fact that the angle of incidence equals the angle of reflection for light reflecting from a smooth surface

object the source of the light from which an image will be formed by a mirror or a lens

image the representation of an object resulting from the convergence, or apparent convergence, of light rays coming from the object by reflection or refraction

upright image an image that is right-side-up with respect to the object

inverted image an image that is upside-down with respect to the object

magnification ratio of the image height to the object height

focal point the point on the principal axis at which incident rays that are parallel to the principal axis (or their extensions) will intersect after reflection or refraction

focal length the distance of the focal point from the mirror or lens

real image an image formed by the intersection of light rays

virtual image an image formed by the intersection of extensions of light rays into a region where no light is actually present

refraction the phenomenon that the direction of propagation of light generally changes when it crosses the boundary between two media

index of refraction the ratio of the speed of light in vacuum to the speed of light in a medium

Snell's law relates the angle of incidence to the angle of refraction

total internal reflection a phenomenon that occurs at incident angles greater than that for which the angle of refraction is 90°

Brewster's angle the angle at which reflected light is completely polarized

dispersion the spreading of light of different frequencies due to the fact that the index of refraction is frequency dependent

II. Important Equations

Name/Topic	Equation	Explanation
reflection	$\theta_i = \theta_r$	The law of reflection.
spherical mirrors	$\frac{1}{d_o} + \frac{1}{d_i} = \frac{1}{f} = \frac{2}{R}$	The mirror equation.

magnification	$m = \frac{h_i}{h_o} = -\frac{d_i}{d_o}$	This expression for the magnification applies to both mirrors and lens.
refraction	$n = \frac{c}{v}$	The definition of the index of refraction.
refraction	$n_1 \sin\theta_1 = n_2 \sin\theta_2$	Snell's law, which describes the refraction of light.
thin lenses	$\frac{1}{d_o} + \frac{1}{d_i} = \frac{1}{f}$	The thin-lens equation.

III. Know Your Units

Quantity	Dimension	SI Unit
magnification (m)	dimensionless	—
index of refraction (n)	dimensionless	—

Practice Problems

1. A light ray strikes a horizontal mirror and is reflected onto a vertical mirror. If θ = 17 degrees, and d = 1.35 meters, to the nearest degree what is ϕ?

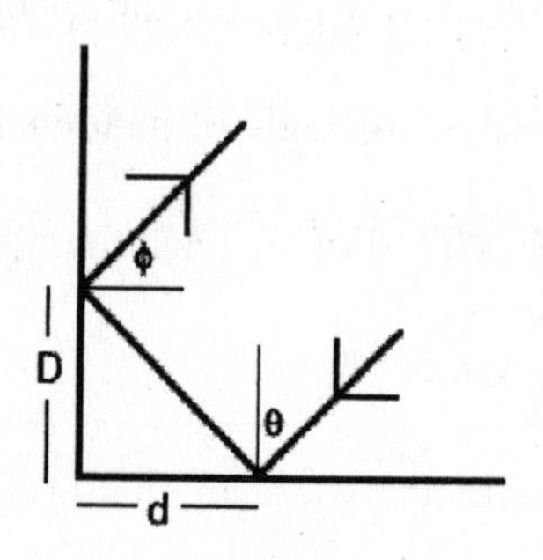

2. In problem 1 what is D, to the nearest hundredth of a meter?

3. If the object distance, d_o, is 12.06 cm, and the focal length, f, is 7.44 cm, to the nearest tenth of a cm, what is the image distance?

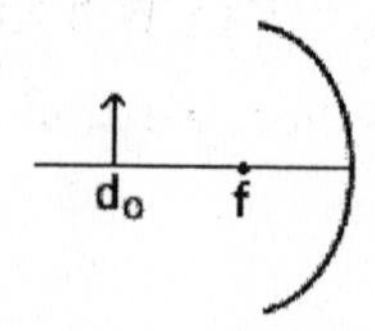

4. If the mirror in problem 3 was a convex mirror, what would be the answer?

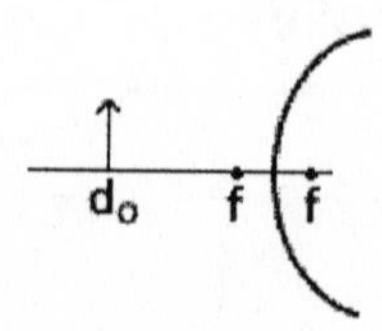

5. In the figure θ = 60 degrees and the index of refraction of the glass is 1.47. To the nearest hundredth of a degree what is ϕ?

6. In problem 5 if d = 3.59 cm, what is y, to the nearest tenth of a mm?

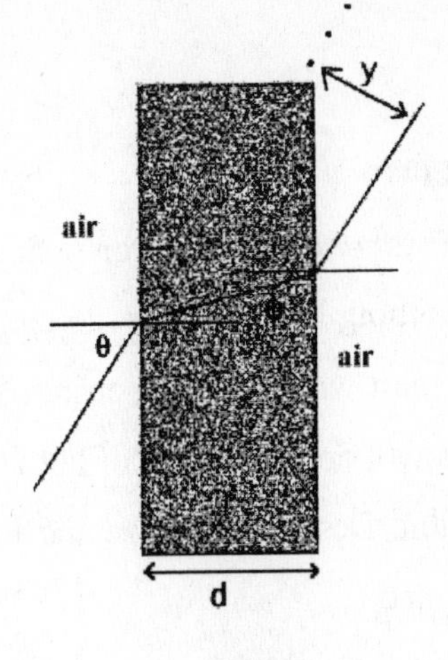

7. If d_o= 4.9 cm and f = 5.2 cm, what is the image distance, to the nearest tenth of a cm?

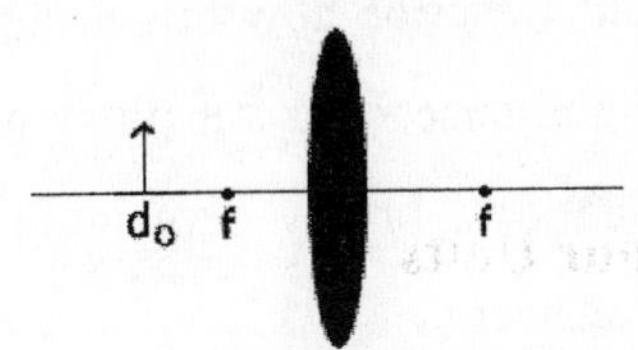

8. In problem 7 what is the answer if the lens is a negative lens, as shown here?

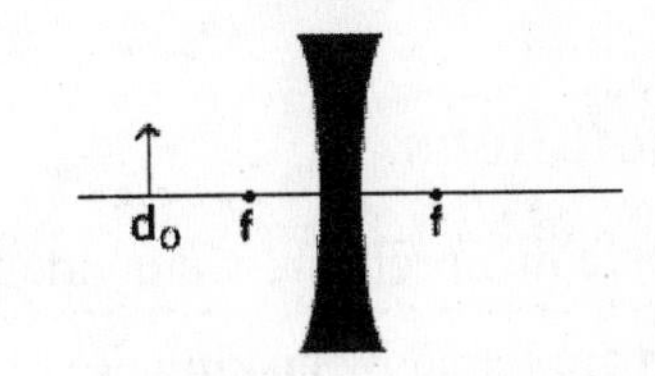

9. In the figure, total internal reflection occurs as shown. If θ_1 = 61° and θ_2 =77° what is the index of refraction of the glass prism, to two decimal places?

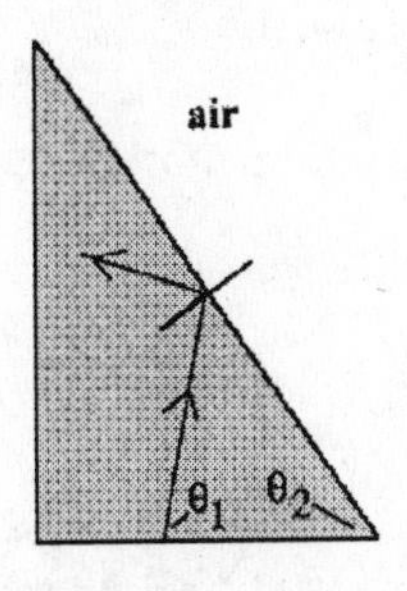

10. If n_1 = 1.046 and n_2 = 1.46, to the nearest tenth of a degree, what is the angle of incidence?

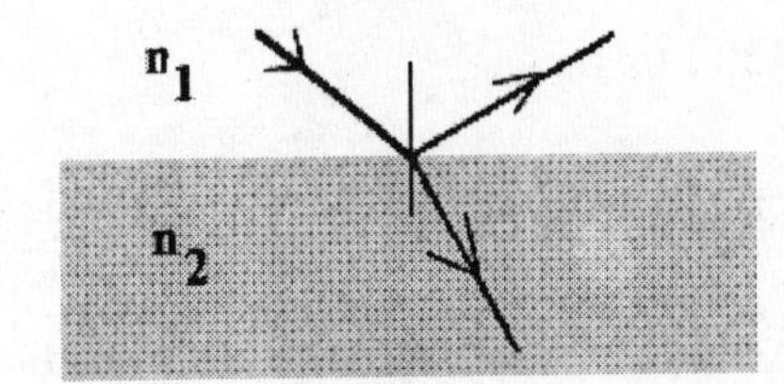

Puzzle

PHYSICS LITE

The figure shows a concave mirror with $f_M = 12.5$ cm and a converging lens with $f_L = 25$ cm. They are placed 50 cm apart with an object centered between them. The lens will create images of both the object and its reflection. Describe the size and location of the images formed.

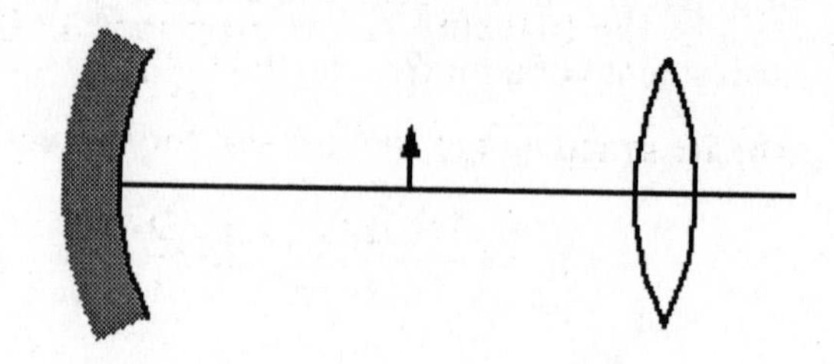

Bonus question: Describe how this setup could be useful in a lighthouse. Answer this question in words, not equations, briefly explaining how you obtained your answer.

Selected Solutions

13. (a) To solve this problem, we divide the part of the building that can be seen into three parts, two of which are found by making use of similar triangles.

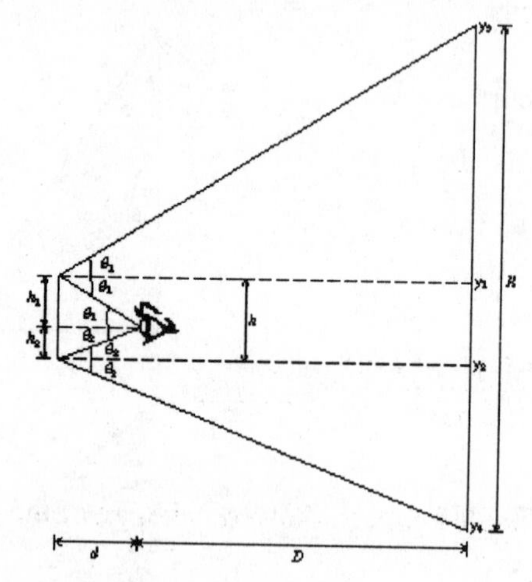

From the above diagram we can see that

$$\frac{h_1}{d} = \tan\theta_1 = \frac{y_3 - y_1}{D+d}, \quad \text{and} \quad \frac{h_2}{d} = \tan\theta_2 = \frac{y_2 - y_4}{D+d}$$

Since we know the total height of the mirror, h, we add these two equations to yield

$$\tan\theta_1 + \tan\theta_2 = \frac{h_1 + h_2}{d} = \frac{y_3 - y_4 - (y_1 - y_2)}{D+d}$$

and we use the facts that $h = h_1 + h_2 = y_1 - y_2$ and $H = y_3 - y_4$. Putting these results in the above equation gives

$$\frac{h}{d}=\frac{H-h}{D+d} \Rightarrow H=h\left(1+\frac{D+d}{d}\right)$$

We can now determine the value of H to be

$$H=(6.0\text{ cm})\left(1+\frac{95\text{ m}+0.50\text{ m}}{0.50\text{ m}}\right)=\boxed{12\text{ m}}$$

(b) Rearranging the above result for H gives

$$H=h\left(1+\frac{D+d}{d}\right)=h\left(2+\frac{D}{d}\right)$$

We can see that decreasing d increases the ratio D/d and thus, H, so, if the mirror is moved closer, the answer to part (a) will $\boxed{\text{increase}}$.

31. (a) Since the image is real, we know that the image distance is positive; the magnification must then be negative. Therefore,

$$d_i=-md_o=-(-3)(22\text{ cm})=\boxed{66\text{ cm}}$$

(b) Using the mirror equation gives

$$f=\left(\frac{1}{d_o}+\frac{1}{d_i}\right)^{-1}=\left(\frac{1}{22\text{ cm}}+\frac{1}{66\text{ cm}}\right)^{-1}=\boxed{17\text{ cm}}$$

45. Light travels at a constant speed v; therefore,

$$x=vt=\frac{c}{n}t \Rightarrow t=\frac{nx}{c}$$

The total time then is given by

$$t_{\text{total}}=\frac{n_1x_1}{c}+\frac{n_2x_2}{c}=\frac{1.33(3.31\text{ m})+1.51(1.51\text{ m})}{3.00\times10^8\text{ m/s}}=\boxed{22.3\text{ ns}}$$

47. For the empty glass, with θ_i being the angle between the vertical edge of the glass and the light ray, we have

$$\sin\theta_i=\frac{W}{\sqrt{W^2+H^2}}$$

and for the glass filled with water,

$$\sin\theta_{\text{refr}}=\frac{W/2}{\sqrt{(W/2)^2+H^2}}=\frac{W}{\sqrt{W^2+4H^2}}$$

Making use of Snell's law, $n_{\text{air}}\sin\theta_i=n_w\sin\theta_{\text{refr}}$, we can say

$$n_{air}\frac{W}{\sqrt{W^2+H^2}} = n_w\frac{W}{\sqrt{W^2+4H^2}} \Rightarrow \frac{n_{air}^2}{W^2+H^2} = \frac{n_w^2}{W^2+4H^2}$$

Further rearranging to solve for H gives

$$n_{air}^2(W^2+4H^2) = n_w^2(W^2+H^2) \Rightarrow H^2(4n_{air}^2 - n_w^2) = W^2(n_w^2 - n_{air}^2)$$

$$\therefore \quad H = W\sqrt{\frac{n_w^2 - n_{air}^2}{4n_{air}^2 - n_w^2}}$$

So, we can now calculate H to be

$$H = (6.2\text{ cm})\sqrt{\frac{(1.33)^2-(1.00)^2}{4(1.00)^2-(1.33)^2}} = \boxed{3.6\text{ cm}}$$

67. (a) To project a real image onto the wall, a convex lens should be used. Since convex lenses have positive focal lengths, real images are formed behind the lens; therefore, $\boxed{\text{the lens with focal length } f_1 \text{ should be used.}}$

(b) Here we need to find the image distance where $d_o + d_i = 3.0\text{ m}$; therefore,

$$d_i = -md_o = -(-2)(3.0\text{ m} - d_i) = 6.0\text{ m} - 2d_i$$

which gives,

$$3d_i = 6.0\text{ m} \Rightarrow d_i = 2.0\text{ m}.$$

The lens should be placed $\boxed{2.0\text{ m}}$ from the wall.

Answers to Practice Quiz

1. (d) **2.** (c) **3.** (b) **4.** (c) **5.** (c) **6.** (e) **7.** (e) **8.** (d) **9.** (a) **10.** (d)

Answers to Practice Problems

1. 73°
2. 4.42 m
3. 19.4 cm
4. −4.6 cm
5. 36.10°
6. 18.0 mm
7. −84.9 cm
8. −2.5 cm
9. 1.35
10. 54.4°

CHAPTER 27

OPTICAL INSTRUMENTS

Chapter Objectives

After studying this chapter, you should:

1. know the basics of how the human eye works as an optical system.

2. understand the basics of how a camera works.

3. be able to characterize the image of multiple-lens systems.

4. understand the optics of correcting eyesight.

5. understand how magnifying glasses, microscopes, and telescopes produce enlarged images.

6. know what is meant by lens aberrations.

Warm-Ups

1. Nearsighted people can see objects clearly if they are close, but not if they are far away. Does a nearsighted eye have a focal length that is too long or too short?
2. Reading glasses are supposed to give a virtual image of an object held at arms' length farther out, where the farsighted person can see clearly. Estimate the focal length of a reading-glass lens such that a newspaper held at 25 cm from the eyes will image at 40 cm from the eye.
3. In a simple refracting telescope there are two lenses, the objective lens and the eyepiece. Explain what each of these lenses does.
4. You wish to make a telescope using lenses that you have around. Let's say you have a magnifying glass and a pair of reading glasses. Estimate the highest magnification you can get.

Chapter Review

The geometrical optics discussed in the last chapter has many important applications in science and everyday life. This chapter discusses several of those applications.

27–1 The Human Eye and the Camera

The Human Eye

The human eye is the most important optical instrument we have. The outer coating of the eye is called the **cornea**. Most of the refraction of light rays in the eye occurs at the cornea. After passing through the cornea, light enters the eye through the **pupil**. The amount of light that passes through the pupil depends on its size, which is controlled by the **iris**. The iris expands in the dark and contracts in bright light to adjust the amount of light entering the eye. After entering the eye, light passes through an adjustable **lens**. In a process known as **accommodation**, the shape of the lens is adjusted by the **ciliary muscles** to help focus light onto the **retina** at the back of the eye.

In characterizing a person's vision there are two important distances, the **near point** and the **far point**. The near point, N, is the closest distance to the eye that an object can be placed and still be in focus. For young people with good vision the near point is typically about 25 cm. Since light rays from nearby objects require greater refraction, the ciliary muscles contract the lens as much as possible for objects at the near point, and the eye is under maximum *strain*. The far point is the greatest distance from the eye that an object can be and still be in focus. For people with normal vision, the far point is effectively infinity. The light rays of distant objects require little refraction, so the ciliary muscles don't need to contract the lens for objects at the far point, and the eye is *relaxed* in this case.

The Camera

A camera and the human eye operate similarly. Light enters a camera through its **aperture**. The amount of light that passes through the aperture depends on its area, which is controlled by the **f-number** setting. The f-number equals the ratio of the focal length of the lens, f, to the diameter of the aperture, D,

$$\text{f-number} = \frac{f}{D}$$

Since the aperture is circular, its area, and therefore the amount of light that passes through it, is proportional to the square of the diameter, D^2.

Another feature of a camera that controls the amount of light that enters it is a setting called the **shutter speed**. This setting refers to the amount of time that the shutter (which blocks light from entering the camera) is open. A shutter speed setting of 1000, for example, means that shutter speed is 1/1000th of a second, which is how long the shutter is open. Thus, the total amount of light that enters a camera depends on the combination of the settings for the f-number and the shutter speed.

Exercise 27–1 Camera Settings Suppose that for photograph 1 a camera is set at f/2.8 and a shutter speed setting of 500, and for photograph 2 the settings are f/5.6 and a shutter speed setting of 250. Which settings let in more light to the camera and by what percentage?

Solution

Let f_1 represent the f-number for photograph 1 (with D_1 as the associated aperture diameter) and f_2 for photograph 2 (with D_2 defined similarly to D_1). Since f-number = f/D, then the ratio of the f-numbers is

$$\frac{f_2}{f_1} = \frac{f/D_2}{f/D_1} = \frac{D_1}{D_2} = \frac{5.6}{2.8} = 2$$

Thus, D_1 is 2 times the size of D_2, and since the amount of light passing through the aperture depends on the area (the square of the diameter) the f-number setting for photograph 1 lets in 4 times the amount of light as the f-number setting for photograph 2.

Now, considering the shutter speed, the speed for photograph 1 is 1/500 sec and that for photograph 2 is 1/250 sec; therefore, the shutter speed setting for photograph 1 lets in half the light as that for photograph 2. Combining these two results, we see that the settings for photograph 1 let in 4/2 = 2 times more light than the settings for photograph 2. Hence, the settings are such that photograph 1 is taken with 100% more light than photograph 2 is.

The important thing to remember with camera settings is that you must consider both the f-number and shutter speed settings when trying to asses the amount of light used for a photograph.

Practice Quiz

1. Which of the following is not part of the human eye?

   (a) cornea (b) pupil (c) iris (d) near point (e) lens

2. Which of the following f-number settings represents the largest aperture diameter?

   (a) 2.8 (b) 4 (c) 5.6 (d) 11 (e) 16

3. Which of the following shutter speed settings represents the shortest amount of time for the shutter to be open?

   (a) 60 (b) 125 (c) 250 (d) 500 (e) 1000

27–2 Lenses in Combination and Corrective Optics

Optical systems involving more than one lens can be understood by noting two basic optical principles:.

* The image produced by one lens acts as the object for the next lens in the system

* The total magnification produced by a system of lenses equals the product of the magnifications produced by each individual lens.

These principles apply to many optical systems; in this section we use them to describe the corrective optics for eyesight.

Nearsightedness

The condition known as **nearsightedness**, or *myopia*, refers to the sight of a person who is able to focus nearby objects, but cannot focus very distant objects. For such a person the far point is a finite distance from the eye (instead of infinity, as for normal vision). With nearsightedness, light rays are refracted too much, causing the image to form in front of the retina. This condition is corrected by placing a diverging lens in front of the eye. When a diverging lens with the correct **refractive power** is used, it forms a virtual image of very distant objects ($d_o = \infty$) at the person's far point, which enables him/her to focus this object.

The refractive power of a lens is the inverse of its focal length:

$$\text{refractive power} = \frac{1}{f}$$

When the focal length is measured in meters, we get the SI unit of refractive power, called a **diopter**, where 1 diopter = 1 m^{-1}.

Farsightedness

The condition known as **farsightedness**, or *hyperopia*, refers to the sight of a person who is able to focus very distant objects, but cannot focus nearby objects. For such a person the near point is farther away from the eye as compared with normal vision. With farsightedness, light rays are not refracted enough, causing the image to form behind the retina. This condition is corrected by placing a converging lens in front of the eye. When a converging lens with the correct refractive power is used, it forms a virtual image of objects closer than the person's near point at locations beyond this near point which means that the person's eye can now focus these objects. Typically, the corrective lens is chosen such that the virtual image of objects placed at the normal near point of 25 cm from the eye will be located at the person's actual near point.

Example 27–2 A Two-lens System An object is placed 75.0 cm in front of a converging lens of focal length $f_1 = 15.0$ cm. A second lens of focal length 10.0 cm is located 12.0 cm from the first lens on the opposite side from the object. Determine the location and magnification of the image produced by this system.

Picture the Problem The sketch shows the object as an arrow in front of the system of two lenses.

object lens 1 lens 2

Strategy We follow the basic strategy of making the image formed by the first lens the object for the second lens.

Solution

1. The object distance, d_{o1}, for the first lens is 75.0 cm; use the thin-lens equation to get the image distance d_{i1}:

$$\frac{1}{d_{i1}}+\frac{1}{d_{o1}}=\frac{1}{f_1} \Rightarrow d_{i1}=\left[\frac{1}{f_1}-\frac{1}{d_{o1}}\right]^{-1} \therefore$$

$$d_{i1}=\left[\frac{1}{15.0\text{ cm}}-\frac{1}{75.0\text{ cm}}\right]^{-1}=\frac{75.0\text{ cm}}{4}=18.75\text{ cm}$$

2. Since the image from lens 1 is beyond lens 2, it is a virtual object ($d_{o2} < 0$) for lens 2. The object distance is:

$$d_{o2}=-(18.75\text{ cm}-12.0\text{ cm})=-6.75\text{ cm}$$

3. Using the thin-lens equation for the second lens, we can get the image distance d_{i2}:

$$d_{i2}=\left[\frac{1}{f_2}-\frac{1}{d_{o2}}\right]^{-1}=\left[\frac{1}{10.0\text{ cm}}-\frac{1}{-6.75\text{ cm}}\right]^{-1}=4.03\text{ cm}$$

4. The location of the image is:

4.03 cm behind lens 2

5. The magnification is the product of the individual magnifications:

$$M=\left(-\frac{d_{i1}}{d_{o1}}\right)\left(-\frac{d_{i2}}{d_{o2}}\right)=\frac{d_{i1}d_{i2}}{d_{o1}d_{o2}}$$

$$=\frac{(18.75\text{ cm})(4.0299\text{ cm})}{(75.0\text{ cm})(-6.75\text{ cm})}=-0.149$$

Insight The two things to be careful of in these situations are (a) to make sure that you account for the separation between the lenses and (b) to watch for virtual objects and to use their object distances as negative values. In the problem, notice that the final image is real and inverted.

Example 27–3 A Nearsighted Correction The left eye of a person has a far point of 3.50 m. Assuming that the lens of her glasses sits 2.00 cm from her eye, what refractive power should be used for a corrective lens?

Picture the Problem The sketch shows a human eye with a corrective diverging lens in front of it.

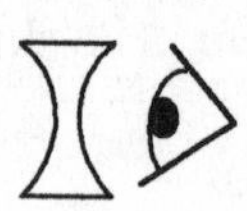

Strategy To correct nearsightedness, the diverging lens forms a virtual image of objects infinitely far away.

Solution

1. The object distance of the objects we need to consider is infinite: $d_o = \infty \Rightarrow \frac{1}{d_o} = 0$

2. The image distance from the lens is: $d_i = 3.50\text{ m} - 2.00\text{ cm} = 3.48\text{ m}$

3. Using the thin-lens equation gives: $\frac{1}{f} = \frac{1}{d_i} + 0 \Rightarrow \frac{1}{f} = \frac{1}{d_i} = \frac{1}{-3.48\text{ m}} = -0.287\text{ diopter}$

Insight In finding the object distance, it didn't matter that the lens was 2.00 cm from the eye because we were considering objects at infinity. It was needed, however, to account for this distance with regard to the image distance and to remember that the image is virtual, so the image distance must be negative.

Practice Quiz

4. If the refractive power of a lens is 0.250 diopter, what is the focal length of a lens with twice the refractive power?

   (a) 0.500 m **(b)** 2.00 m **(c)** 4.00 m **(d)** 8.00 m **(e)** 0.125 m

5. A person whose near point is 28 cm and whose far point is 85 cm would most likely be considered

   (a) farsighted **(b)** hyperopic **(c)** nearsighted **(d)** all of the above **(e)** none of the above

27–3 The Magnifying Glass

A **magnifying glass** is a converging lens which is used in much the same way as for the correction of farsightedness. The magnifying glass allows you to focus objects that are closer to your eye than your near point by forming a virtual image of the object at infinity (if you place the magnifying glass one focal length away from the object). Because the object is now closer to your eye it appears larger to you. Without the magnifying glass, the focused object would appear largest to you, having an angular size θ, if placed at your near point N. The **angular magnification**, M, of the object is the ratio of the angular size of the magnified object, θ′, to θ. It can be shown that this ratio also equals the ratio of the near point, N, to the focal length of the lens f,

$$M = \frac{\theta'}{\theta} = \frac{N}{f}$$

Practice Quiz

6. A magnifying glass of focal length f produces an angular magnification of M. All else being equal, what would be the angular magnification if the lens was replaced by one of focal length 3f?

   (a) M **(b)** 3M **(c)** 6M **(d)** M/3 **(e)** M/6

27–4 The Compound Microscope

A **microscope** is an optical instrument designed to magnify small objects. The simplest microscope, called a *compound microscope*, consists of just two lenses. One lens, called the **objective**, is placed such that the object is just beyond its focal point. The objective forms a real, enlarged image at the focal point of the second lens, called the **eyepiece**. Essentially, the eyepiece acts as a magnifying glass that forms a virtual image at infinity. To a good approximation, the angular magnification of the compound microscope is given by

$$M = -\frac{d_i N}{f_o f_e}$$

where d_i is the image distance for the image formed by the objective, N is the near point of the observer, f_o is the focal length of the objective, and f_e is the focal length of the eyepiece.

Practice Quiz

7. If the objective of a microscope is moved farther away from the object, what effect does this have on the size of the image?

   (a) It is enlarged.

   (b) It is diminished.

   (c) It becomes real from having been virtual.

   (d) It becomes virtual from having been real.

   (e) none of the above

27–5 Telescopes

A telescope is an optical instrument designed to magnify distant objects. As in a microscope, the main parts of a telescope are its objective and eyepiece. Since the objects viewed through telescopes are very far away — essentially infinitely far — the light from the object is focused at the focal point of the objective. The telescope is designed so that this point is also the focal point of the eyepiece. With this information, it can be shown that the angular magnification of the telescope is given by

$$M = \frac{f_o}{f_e}$$

where f_o is the focal length of the objective and f_e is that of the eyepiece.

Practice Quiz

8. What effect is produced if the eyepiece of a telescope is replaced with one of shorter focal length?
   (a) The image is enlarged.
   (b) The image is diminished.
   (c) The image becomes inverted.
   (d) The focal length of the objective increases.
   (e) The image is unaffected by changing the eyepiece.

9. The objective of a telescope has a focal length of 1.8 m. What is the focal length of an eyepiece that will produce a magnification of 500×?
   (a) 2.0 mm **(b)** 278 m **(c)** 900 m **(d)** 3.6 mm **(e)** none of the above

27–6 Lens Aberrations

Only ideal lenses bring all light rays to a perfect focus. Real lenses tend to form blurred images by focusing different rays at different points. The inability of lenses to focus rays precisely is called **aberration**. Among the more commonly encountered forms of aberration is **spherical aberration**, which is due to the shape of lenses ground to spherical sections. **Chromatic aberration** is also commonly seen and is due to the dispersion of light in the lens. In many optical systems chromatic aberration is reduced by using combinations of lenses.

Reference Tools and Resources

I. Key Terms and Phrases

accommodation the adjustment of the shape of the lens in a human eye to help focus objects

near point the closest distance to the eye at which an object can be placed and still be in focus

far point the greatest distance from the eye that an object can be and still be in focus

aperture the opening through which light enters a camera

f-number a camera setting that determines the diameter of the aperture

shutter speed the amount of time that the shutter of a camera remains open

nearsightedness the ability to focus nearby objects but not distant ones

refractive power the inverse of the focal length of a lens

farsightedness the inability to focus nearby objects

angular magnification the ratio of an object's angular size seen with an optical device to its angular size when viewed at the near point with an unaided eye

objective the first lens through which light from the object comes into a telescope or a microscope

eyepiece a magnifying lens through which the observer views images in a telescope or a microscope

aberration the inability of real lenses to sharply focus light rays

II. Important Equations

Name/Topic	Equation	Explanation
cameras	$\text{f-number} = \frac{f}{D}$	The f-number setting of a camera.
corrective optics	$\text{refractive power} = \frac{1}{f}$	The refractive power of a lens.
magnifying glass	$M = \frac{\theta'}{\theta} = \frac{N}{f}$	The angular magnification of a magnifying glass for a relaxed eye.
compound microscope	$M = -\frac{d_i N}{f_o f_e}$	The angular magnification of a compound microscope.
telescope	$M = \frac{f_o}{f_e}$	The angular magnification of a refracting telescope.

III. Know Your Units

Quantity	Dimension	SI Unit
refractive power	$[L^{-1}]$	diopter
angular magnification (M)	dimensionless	—

Practice Problems

1. The eye focuses by changing the focal length of its lens. What is the focal length, to the nearest hundredth of a cm, if the object distance is 59 cm?

2.5 cm

2. A normal eye like the one in problem 1 can focus objects at infinity on the back of the eye 2.5 cm away; therefore, in the relaxed state its focal length is 2.5 cm. Suppose that the eye is actually 7×10^{-1} mm longer than it should be what is the new far point, to the nearest cm?
3. To the nearest cm what is the focal length of a lens placed 2 cm in front of the eye that will correct the condition in problem 2?
4. As in problem 2 suppose the eye is defective, but this time by being 5×10^{-1} mm shorter than normal. What, then, is the near point, to the nearest tenth of a cm?
5. What is the power of the lens (2 cm in front of the eye) needed to correct the eye in problem 4, in diopters to two decimal places.
6. How far (out is positive, in is negative) to the nearest hundredth of a mm does the lens have to move to change the focus point from 3.43 m to 21 m?
7. A person with a near-point distance of 81 cm views an insect with a magnifying glass with a focal length of 2.61 cm. Assuming that he views it with a relaxed eye, what is the magnification to 2 decimal places?
8. A microscope with an objective focal length of 1 cm and an eyepiece of 3.06 cm is used to view an object 0.011 cm in front of the focal point of the objective lens. What is the magnification, to 1 decimal place (assume a near-point distance of 25 cm)?
9. A telescope has a magnification of 44 and a length of 1069. To the nearest mm, what is the focal length of the eyepiece?
10. In problem 9 what is the focal length of the objective?

Puzzle

PHISHICS

A person with good vision finds that she cannot focus on anything underwater. However, plastic goggles with "lenses" that are *flat* plastic disks allow her to see fish clearly. Explain how this can be. (Hint: This has nothing to do with salt or chlorine in the water.) Answer this question in words, not equations, briefly explaining how you obtained your answer.

Selected Solutions

7. $D = \dfrac{f}{\text{f-number}}$. Therefore, the smallest f-number has the largest aperture and *vice versa*.

Lens	Focal Length (mm)	*f*-number	Diameter (mm)	Rank
A	100	f/1.2	80	1
B	100	f/5.6	20	3
C	30	f/1.2	30	2
D	30	f/5.6	5	4

Ranking from largest to smallest: $\boxed{\text{A, C, B, D}}$.

21. (a) Let the object be located 24 cm to the left of the leftmost lens (1). From the thin-lens equation

$$\frac{1}{d_{1i}} = \frac{1}{f} - \frac{1}{d_{1o}}$$

$$d_{1i} = \left(\frac{1}{f} - \frac{1}{d_{1o}}\right)^{-1} = \left(\frac{1}{-12\text{ cm}} - \frac{1}{24\text{ cm}}\right)^{-1} = -8.00\text{ cm}$$

The image due to lens 1 is located 8.00 cm to the left of lens 1, so the object distance for the right lens (2) is $8.00\text{ cm} + 6.0\text{ cm} = 14.0\text{ cm}$. Using the thin-lens equation again gives

$$d_{2i} = \left(\frac{1}{f} - \frac{1}{d_{2o}}\right)^{-1} = \left(\frac{1}{-12\text{ cm}} - \frac{1}{14.0\text{ cm}}\right)^{-1} = -6.46\text{ cm}$$

So the final image is 6.46 cm in front of lens (2). Since lens (1) is only 6.0 cm in front of lens (2), this image is located $\boxed{\text{0.46 cm, or 4.6 mm, in front of lens (1)}}$ which is the lens closest to the object.

(b) The total magnification is the product of the two individual magnifications:

$$m_1 = -\frac{d_{1i}}{d_{1o}} = -\frac{-8.00\text{ cm}}{24\text{ cm}} = 0.33$$

$$m_2 = -\frac{d_{2i}}{d_{2o}} = -\frac{-6.46\text{ cm}}{14.0\text{ cm}} = 0.461$$

Therefore, $m = m_1 m_2 = \left(\frac{8.00}{24}\right)\left(\frac{6.46}{14.0}\right) = \boxed{0.15}$

29. We must subtract 2.0 cm from the distances relative to the eye to get the distances relative to the lens. Using the thin-lens equation and the fact that the lens forms a virtual image gives

$$\frac{1}{f} = \frac{1}{d_o} + \frac{1}{d_i} \Rightarrow f = \left(\frac{1}{d_o} + \frac{1}{d_i}\right)^{-1} = \left[\frac{1}{25\text{ cm} - 2.0\text{ cm}} + \frac{1}{-(19\text{ cm} - 2.0\text{ cm})}\right]^{-1} = \boxed{-65\text{ cm}}$$

Therefore,

$$\text{refractive power} = \frac{1}{f} = \frac{1}{0.23\text{ m}} + \frac{1}{-0.17\text{ m}} = \boxed{-1.5\text{ diopters}}$$

43. (a) Since a smaller focal point results in a larger magnification, $\boxed{f_1}$ can produce the greater magnification.

(b) $M_1 = 1 + \frac{N}{f} = 1 + \frac{25\text{ cm}}{5.0\text{ cm}} = \boxed{6.0}$

$$M_2 = 1 + \frac{25\text{ cm}}{13\text{ cm}} = \boxed{2.9}$$

51. (a) We use the thin-lens equation to find the object distance:

$$\frac{1}{d_o} = \frac{1}{f_o} - \frac{1}{d_i} = \frac{1}{f_o} + \frac{1}{mf_o}$$

$$d_o = f_o\left(1 + \frac{1}{m}\right)^{-1} = (4.00\text{ mm})\left(1 + \frac{1}{-40.0}\right)^{-1} = \boxed{4.10\text{ mm}}$$

(b) For a normal near point we get

$$M_{total} = m_o M_e = m_o \frac{N}{f_e}$$

Therefore,

$$f_e = \frac{m_o N}{M_{total}} = \frac{(-40.0)(25\text{ cm})}{125} = \boxed{-8.0\text{ cm}}$$

Answers to Practice Quiz

1. (d), **2**. (a), **3**. (e), **4**. (b), **5**. (c), **6**. (d), **7**. (b), **8**. (a), **9**. (d)

Answers to Practice Problems

1. 2.40 cm	**2**. 92 cm
3. –90 cm	**4**. 31.4 cm
5. 0.95 diopter	**6**. –0.62 mm
7. 31.03	**8**. –750.9
9. 24 mm	**10**. 1045 mm

CHAPTER 28

PHYSICAL OPTICS: INTERFERENCE AND DIFFRACTION

Chapter Objectives

After studying this chapter, you should:

1. understand Young's two-slit experiment in terms of wave interference for light.
2. be able to determine the locations of bright and dark fringes in the two-slit experiment.
3. be able to determine the conditions for constructive and destructive interference of reflected waves in air wedges and thin films.
4. be able to determine the locations of the dark fringes in single-slit diffraction.
5. understand and be able to use Rayleigh's criterion for the resolution of objects.
6. have a basic understanding of diffraction gratings.

Warm-Ups

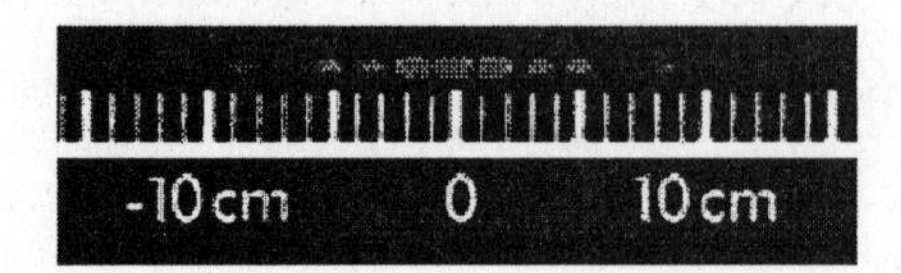

1. The accompanying picture shows light from a narrow laser beam as it appears on a screen after passing through two narrow slits. Examine the pattern and answer the following questions. What is the distance between two successiva minima due to interference between the slits? What is the distance between two successiva minima due to diffraction at the slits themselves? What would change if the separation between the slits was reduced? What would change if each slit was made wider?
2. The picture on the right shows what happens to a narrow laser beam after it passes through a small, round pinhole. The same happens to light passing through your pupil. In your textbook, find the criterion for resolving two light sources. Use the criterion to estimate at what minimum distance your eye is able to resolve the two headlights of an approaching car. (Estimate the diameter of your pupil to be 2 mm.) (Hint: Look at Example 24–3)

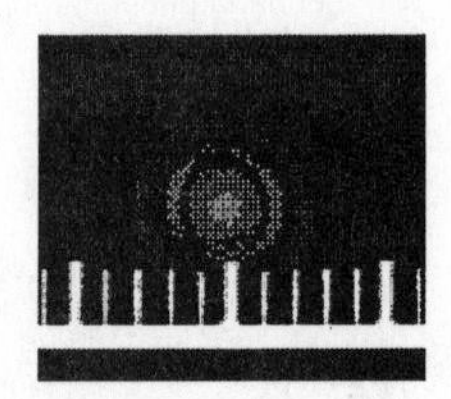

3. Why does a soap bubble reflect virtually no light just before it bursts?
4. Rock specimen slices are often 30 micrometers thick. Approximately how many wavelengths of visible light does that represent? What additional information would you need to be able to answer this question exactly?

Chapter Review

With geometrical optics most of the situations we considered were such that the wave properties of light were of little direct consequence. In this chapter we discuss aspects of the behavior of light that can only be understood in terms of its wave nature. This branch of optics is sometimes called *physical optics*.

28–1 Superposition and Interference

One of the key signatures of wave behavior is that of **superposition**, also called **interference**. This wave behavior for light is similar to what it is for sound waves, as discussed in Section 14–7. With sound the superposition of waves occurs when different waves coexist in a region such that the net displacement at a point is the sum of the displacements of the individual waves. For light, the electric and magnetic fields play the role of the displacement. When the net fields resulting from the combination of waves have larger magnitudes than the fields from the individual waves we call this **constructive interference**; when the combination results in fields of reduced magnitudes we call this **destructive interference**.

Interference effects are noticeable when the different light waves are of the same frequency, or **monochromatic**, and have a constant phase relationship, or are **coherent**. When the phase difference between waves is 0°, or some multiple of 360° (corresponding to path differences that are multiples of the wavelength) the waves are *in phase*. When the phase difference is 180°, or some odd multiple of it (corresponding to path differences of odd multiples of a half wavelength) the waves are said to be *completely out of phase*. When the phase relationship between the different waves varies randomly, the waves are said to be **incoherent**.

Practice Quiz

1. If two light waves have equal wavelengths, but one travels a quarter of a wavelength farther in distance, what is the phase difference between these waves?

   (a) 0° **(b)** 25° **(c)** 30° **(d)** 45° **(e)** 90°

28–2 Young's Two-Slit Experiment

Young's two-slit experiment is a classic experiment that demonstrates the interference properties of waves. In this experiment, monochromatic light is passed through a single slit to produce a small source of light. This light then shines on a setup containing two slits which act as independent coherent sources. By Huygens' principle we can treat each slit as a source of light spreading out in all forward directions. The light from these two slits then shines on a screen. On the screen there is a pattern of bright and dark *fringes* that are the result of constructive and destructive interference between the waves from the two slits.

In a typical setup the distance from the slits to the screen is much larger than the separation between the slits. This assumption simplifies the analysis. There will be constructive interference on the screen, where there are bright fringes, when the path difference, $\Delta\ell$, between the light waves from the two slits equals an integral multiple of the wavelength, λ. Under these assumptions, this is equivalent to the condition

$$\sin\theta = m\frac{\lambda}{d}, \quad m = 0, \pm1, \pm2, \ldots \qquad \text{[bright fringes]}$$

where θ is the angle from either slit to the relevant point on the screen (which are approximately equal under the current assumptions), and d is the separation between the slits. The dark fringes result from destructive interference that occurs for path differences that are odd multiples of $\lambda/2$, which leads to the condition

$$\sin\theta = \left(m - \tfrac{1}{2}\right)\frac{\lambda}{d}, \quad m = 0, \pm1, \pm2, \ldots \qquad \text{[dark fringes]}$$

At the center of the interference pattern on the screen is a bright fringe. The position of a given fringe is determined by its linear distance on the screen from the center of this central bright fringe. This linear distance is given by

$$y = L\tan\theta$$

where L is the distance between the screen and the plane of the two slits, and θ is the angle measured from the spot exactly between the slits to the fringe being located on the screen; under the assumption that $L \gg d$, this angle is approximately equal to the angle found from the previous two conditions for the bright and dark fringes.

Example 28–1 A Two-Slit Experiment Light of wavelength 455 nm is incident on a two-slit apparatus with a slit separation distance of 0.125 mm. What is the distance on a screen, 2.00 m away, between the first and fifth dark fringes?

Picture the Problem The diagram shows a two-slit setup and indicates the distance y between the first and fifth dark fringes.

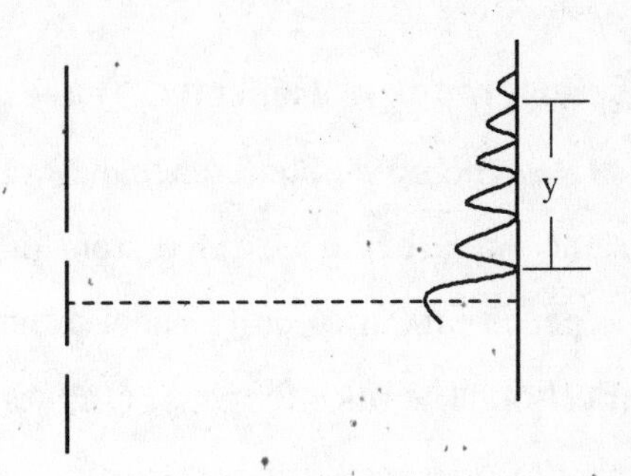

Strategy Since we know which dark fringes we are considering, we can determine their order numbers and their angles. We can then calculate the difference in their distances on the screen.

Solution

1. The first dark fringe has m = +1, so its angle is: $\sin\theta_1 = \left(1-\frac{1}{2}\right)\frac{\lambda}{d} \Rightarrow \theta_1 = \sin^{-1}\left(\frac{\lambda}{2d}\right) = 0.10428^\circ$

2. The fifth dark fringe has m = +5, so its angle is: $\sin\theta_5 = \left(5-\frac{1}{2}\right)\frac{\lambda}{d} \Rightarrow \theta_5 = \sin^{-1}\left(\frac{9\lambda}{2d}\right) = 0.93855^\circ$

3. The linear distance is determined by: $y = y_5 - y_1 = L\left[\tan\theta_5 - \tan\theta_1\right]$

4. Solving for the numerical result gives: $y = (2.00\text{ m})\left[\tan\left(0.93855^\circ\right) - \tan\left(0.10428^\circ\right)\right] = 0.0291\text{ m}$

Insight This problem also could have be done using m = 0 and m = –4.

Practice Quiz

2. Monochromatic light of wavelength λ incident on a two-slit apparatus produces an angular separation ϕ between bright fringes. If a light wave of wavelength λ/2 is used, which of the following statements is true about the angular separation between bright fringes?

   (a) It doubles to 2ϕ.

   (b) It increases to greater than ϕ, but not necessarily 2ϕ.

   (c) It's cut in half to ϕ/2.

   (d) It decreases to less than ϕ, but not necessarily ϕ/2.

   (e) none of the above

3. Light of wavelength 650 nm is incident of a two-slit apparatus with slit separation of 0.15 mm. At what angle will you find the first dark fringe?

   (a) 0.15° **(b)** 65 **(c)** 0.12° **(d)** 0.0022° **(e)** none of the above

28–3 Interference in Reflected Waves

In many cases the wave interference that we observe is not due to waves from different sources but rather to the same wave being reflected from different surfaces or different locations. The interference that results depends primarily on the path difference between the waves and phase changes that may occur upon reflection. The rules for phase changes upon reflection are the following:

* There is no phase change when light reflects from the boundary of a region with a lower index of refraction than that for the incident light.

* There is a 180° (half-wavelength) phase change when light reflects from the boundary of a region with a higher index of refraction than that for the incident light.

Air Wedge

One example where the interference between reflected waves occurs is found with an air wedge, which results when two plates of glass touch at one end and have a small separation at the other end. This setup creates a wedge-shaped air pocket between the plates. For this case we consider the dominant interference effect, which is between light reflected from the bottom surface of the top plate and light reflected from the top surface of the lower plate. The bottom of the upper plate is a glass-to-air boundary, so there is no phase change upon reflection ($n_{glass} > n_{air}$) for these rays. The top of the lower surface is an air-to-glass boundary, so there is a 180° phase change upon reflection from this surface. This phase change contributes a half wavelength to the *effective* path difference between the rays. The rest of the effective path difference comes from the fact that, for normal incidence, one of the rays travels an extra distance of 2d, where d is the thickness of the air gap. Thus, the effective path difference is

$$\Delta\ell_{eff} = \frac{\lambda}{2} + 2d$$

If this effective path difference is an integral number of wavelengths, there will be constructive interference:

$$\frac{1}{2} + \frac{2d}{\lambda} = m, \quad m = 1, 2, 3, \ldots \qquad \text{[constructive interference]}$$

If the effective path difference equals an odd multiple of a half wavelength, there will be destructive interference:

$$\frac{1}{2} + \frac{2d}{\lambda} = m + \tfrac{1}{2}, \quad m = 0, 1, 2, \ldots \qquad \text{[destructive interference]}$$

Thin Films

The interference of light due to thin films is very commonly seen and widely used in optics; almost every lens in any optical system is coated with a film in order to take advantage of these effects. The conditions for constructive and destructive interference are found by similar reasoning as with an air wedge. For thin films we consider the interference between a ray reflected at the top surface of the film, ray 1, and a ray reflected at the bottom surface of the film, ray 2.

The top surface of the film is an air-to-film boundary, so there is a 180° phase shift for ray 1. Whether or not there will be a phase shift at the bottom surface depends on the whether or not the index of refraction of the film is less than or greater than that of the *substrate* on which the film sits. If there is a phase shift for ray 2, it effectively cancels out the phase shift for ray 1. If there is no phase shift for ray 2, then there is a contribution of a half wavelength to the effective path difference. Once we know the phase difference due to the phase shifts (either 0 or 180°) we must consider the contribution resulting from the actual path difference, since ray 2 travels an extra distance of approximately twice the thickness of the film, t (for nearly normal incidence).

For ray 2 we must keep in mind that in the film the wavelength of light will be different from that in air (or vacuum) according to

$$\lambda_{\text{film}} = \frac{\lambda_{\text{vac}}}{n_{\text{film}}}$$

Therefore, the conditions for constructive or destructive interference occur when the distance that ray 2 travels in the film, 2t, is an integral multiple of λ_{film} or an odd multiple of $\lambda_{\text{film}}/2$. The actual conditions are determined on a case-by-case basis.

Example 28–2 Thin-Film Coatings You need to reduce the glare from a lens in an optical system. The index of refraction of the lens is 1.50. To reduce the glare from visible light you decide to coat the lens with a substance whose index of refraction is 1.35. Focusing on the wavelength of light at the middle of the visible part of the electromagnetic spectrum, 550 nm, with what minimum thickness should you coat the lens?

Picture the Problem The sketch shows the incident and reflected rays from the top and bottom boundaries of the thin film.

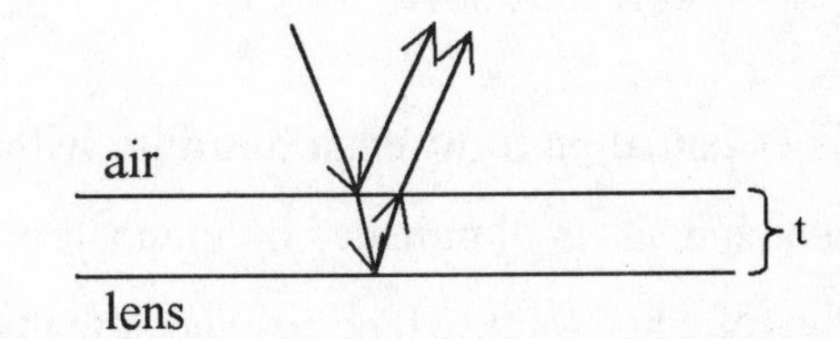

Strategy We want destructive interference of the

reflected waves. First, we consider the possible phase shifts upon reflection and then the needed path difference.

Solution

1. At the top boundary $n_{film} > n_{air}$, so: There is a 180° phase shift.

2. At the bottom boundary $n_{film} < n_{lens}$ so: There is a 180° phase shift.

3. The condition for the minimum thickness due to the path difference is: $2t_{min} = \frac{\lambda_{film}}{2} = \frac{\lambda_{vac}}{2n_{film}}$

4. Solving for the thickness gives: $t_{min} = \frac{\lambda_{vac}}{4n_{film}} = \frac{550\text{ nm}}{4(1.35)} = 102\text{ nm}$

Insight There are many other values of t that would produce destructive interference. See if you can find one more.

Practice Quiz

4. A thin film with an index of refraction of 1.66 covers a region of water (n = 1.33), with air being above the film. Which of the following statements is true about phase shifts upon reflection?
   (a) There is a 180° phase shift for light reflected off the top surface of the film and for light reflecting off the film-water boundary.
   (b) There is no phase shift for light reflected neither off the top surface of the film nor off the film-water boundary.
   (c) There is no phase shift for light reflected off the top surface of the film, but there is a 180° phase shift for light reflected off the film-water boundary.
   (d) There is a 180° phase shift for light reflected off the top surface of the film, but there is no phase shift for lighted reflected off the film-water boundary.
   (e) none of the above

5. For a situation in which a thin film with an index of refraction greater than that of the substance on which it sits is illuminated by monochromatic light incident from the air, constructive interference for the reflected light will occur when the thickness of the film equals
   (a) $\lambda_{film}/4$ **(b)** $\lambda_{film}/2$ **(c)** $\lambda_{air}/4$ **(d)** $\lambda_{air}/2$ **(e)** $3\lambda_{air}/4$

6. For a situation in which a thin film with an index of refraction less than that of the substance on which it sits is illuminated by monochromatic light incident from the air, destructive interference for the reflected light will occur when the thickness of the film equals

   (a) $\lambda_{film}/4$ **(b)** $\lambda_{film}/2$ **(c)** λ_{film} **(d)** $\lambda_{air}/4$ **(e)** $\lambda_{air}/2$

28–4 Diffraction

Besides interference, another characteristic of wave behavior is that waves bend when they pass by barriers or through openings; this phenomenon is known as **diffraction**. In this section we consider the diffraction of monochromatic light as it passes through a single small slit or opening of width W. If this diffracted light shines on a distant screen, we see an interference pattern on the screen. This pattern can be understood in terms of Huygens' principle in which every point within the slit can be treated as a separate source of light waves in all forward directions. The interference pattern is then a result of the superposition of all these waves. Taking $\theta = 0$ to be the direction directly along a line passing through slit and perpendicular to the distant screen, the angular locations of the dark fringes of the interference pattern are determined by

$$\sin\theta = m\frac{\lambda}{W}, \quad m = \pm 1, \pm 2, \pm 3, \ldots \qquad \text{[dark fringes]}$$

The numbers m take both positive and negative values to account for the fact that the diffraction pattern is symmetric, so that there are dark fringes on opposite sides of a central bright fringe.

Exercise 28–3 Single-Slit Diffraction Light of wavelength 675 nm is incident on a single slit of width 3.15 μm. What is the angular width of the bright fringe that is closest to the central maximum?

Solution

The width, $\Delta\theta$, of the first bright fringe (or first-order bright fringe) can be determined by the locations of the dark fringes on either side of it. The first- and second-order dark fringes correspond to m = 1 and m = 2, and they are located at an angles

$$\theta_1 = \sin^{-1}\left(\frac{\lambda}{W}\right) \quad \text{and} \quad \theta_2 = \sin^{-1}\left(\frac{2\lambda}{W}\right)$$

Therefore,

$$\Delta\theta = \theta_2 - \theta_1 = \sin^{-1}\left(\frac{2\lambda}{W}\right) - \sin^{-1}\left(\frac{\lambda}{W}\right) = \sin^{-1}\left(\frac{2(675\times10^{-9}\,\text{m})}{3.15\times10^{-6}\,\text{m}}\right) - \sin^{-1}\left(\frac{675\times10^{-9}\,\text{m}}{3.15\times10^{-6}\,\text{m}}\right) = 13.0^\circ$$

Practice Quiz

7. The effects of diffraction in a single slit will become more pronounced if

   (a) the width of the slit decreases.

   (b) the width of the slit increases.

   (c) the wavelength of the light is decreased.

   (d) The width of the slit has no effect.

   (e) none of the above

28–5 Resolution

Light enters optical instruments, such as the human eye, cameras, and telescopes, through apertures (such as the pupil of a human eye). This light, therefore, experiences diffraction, and this diffraction limits our ability to resolve images. The ability to visually distinguish closely spaced objects is called **resolution**. For a circular aperture of diameter D, the condition for the first dark fringe is given by

$$\sin\theta = 1.22\frac{\lambda}{D}$$

Rayleigh's criterion for the ability to resolve two objects is that the angular separation of their central bright spots must be greater than that of their first dark fringes. Since for small angles, $\sin\theta \approx \theta$, we write this criterion for limiting resolution as

$$\theta_{min} = 1.22\frac{\lambda}{D}$$

where θ_{min} is in radians.

Practice Quiz

8. The ability to resolve two objects will improve if

   (a) the diameter of the aperture decreases.

   (b) the diameter of the aperture increases.

   (c) the diameter of the aperture equals the wavelength of the light from the objects.

   (d) the wavelength of the light is longer.

   (e) none of the above

28–6 Diffraction Gratings

The diffraction of light plays an important role in Young's double-slit experiment. Increasing the number of slits creates an interference pattern (or diffraction pattern) that contains sharper and more widely spaced principal maxima. A system with a large number of slits for the purpose of analyzing light sources is called a **diffraction grating**. The angles at which these principal maxima occur for a diffraction grating is given by

$$\sin\theta = m\frac{\lambda}{d}, \quad m = 0, \pm1, \pm2, \ldots$$

where d is the separation between adjacent slits of the grating, and λ is the wavelength of the incident monochromatic light. Typically, diffraction grating are described in term of the number of lines (slits) per unit length, N. From this value the slit separation can be found as d = 1/N.

Example 28–4 A Diffraction Grating Your company is under contract to manufacture a diffraction grating for which the first-order (m = 1) maximum, for light of wavelength 475 nm, occurs at an angle of 25° away from the central maximum. How many lines per centimeter should your grating have?

Picture the Problem The sketch shows the diffraction grating and the first-order principal maximum.

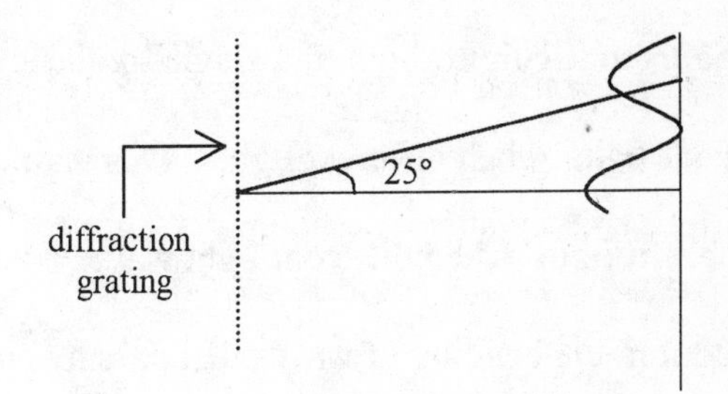

Strategy We can use the expression for the principal maxima to determine the slit spacing, from which we can calculate N.

Solution

1. The slit spacing is given by: $\sin\theta = m\frac{\lambda}{d} \Rightarrow d = \frac{m\lambda}{\sin\theta}$

2. The number of lines per unit length is: $N = \frac{1}{d} \Rightarrow N = \frac{\sin\theta}{m\lambda}$

3. Obtain the numerical result: $N = \frac{\sin\theta}{m\lambda} = \frac{\sin(25^\circ)}{475\times10^{-7}\text{ cm}} = 8900 \text{ lines/cm}$

Insight The main thing to be careful about here is the units in which N is to be expressed. We need to make sure that if we want lines/cm we don't actually calculate lines/m.

Practice Quiz

9. A diffraction grating has 3310 lines/cm. What is the distance between its slits?

   (a) 3.02 mm (b) 3.02 cm (c) 0.302 cm (d) 3.02 m (e) none of the above

10. Light of wavelength 650 nm is incident on a diffraction grating with 6666 lines/m. At what angle will you find a first order principal maximum?

    (a) 0.15° (b) 26° (c) 5.6° (d) 0. 25° (e) none of the above

Reference Tools and Resources

I. Key Terms and Phrases

superposition/interference the combination of waves such that the net "displacement" at a point is the sum of the displacements of the individual waves

constructive interference when waves superimpose such that the net displacements are increased

destructive interference when waves superimpose such that the net displacements are diminished

monochromatic light light of a single frequency (or wavelength)

coherent light when different light waves maintain a constant phase relationship

incoherent light when different light waves have a random phase relationship

diffraction the bending of waves that pass by obstacles or through openings

resolution a measure of the ability to visually distinguish closely spaced objects

diffraction grating a system with a large number of slits for the purpose of analyzing light sources

II. Important Equations

Name/Topic	Equation	Explanation
Young's two-slit experiment	$\sin\theta = m\frac{\lambda}{d}, \quad m = 0, \pm1, \pm2, \ldots$	The condition for bright fringes.
	$\sin\theta = (m - \frac{1}{2})\frac{\lambda}{d}, \quad m = 0, \pm1, \pm2, \ldots$	The condition for dark fringes.
single-slit diffraction	$\sin\theta = m\frac{\lambda}{W}, \quad m = \pm1, \pm2, \pm3, \ldots$	The condition for dark fringes.

resolution	$\theta_{min} = 1.22\dfrac{\lambda}{D}$	Rayleigh's criterion for when two objects can be barely resolved.
diffraction gratings	$\sin\theta = m\dfrac{\lambda}{d}, \quad m = 0, \pm 1, \pm 2, \ldots$	The condition for the principal maxima of a diffraction grating.

Practice Problems

1. Two radio towers are broadcasting on the same frequency. The signal is strong at A, and B is the first signal minimum. If d = 7.5 km, L = 13.7 km, and y = 2.47 km, what is the wavelength of the radio waves, to the nearest meter?

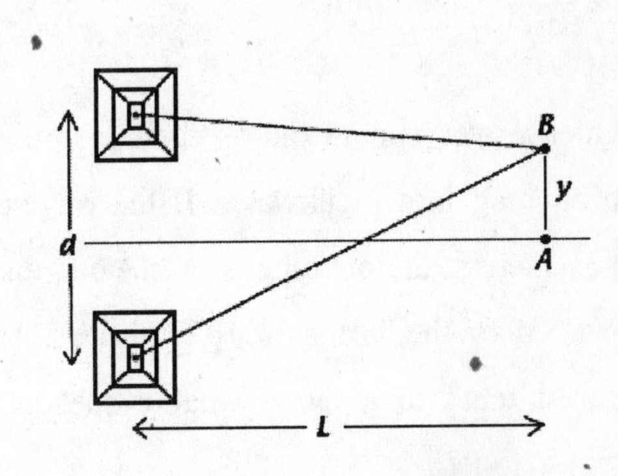

2. Water waves of wavelength 5.5 meters are incident upon a breakwater with two narrow openings separated by a distance of 271 meters. To the nearest thousandth of a degree, what is the angle corresponding to the first wave fringe maximum?

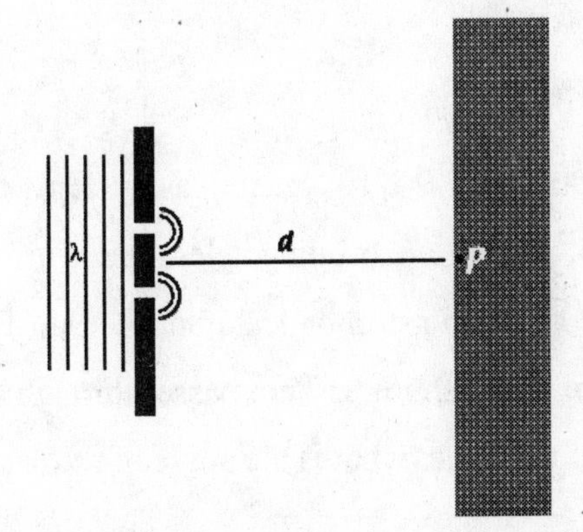

3. In problem 2 if the distance to the shore, d, is 1297 m, to the nearest tenth of a meter, what is the vertical distance above P to the first wave maximum?

4. An air wedge is created by putting a thread at one end. When viewed with 766 nm light there are 125 bright fringes. What is the thickness of the thread to the nearest tenth of a mm?

5. When viewed with a different color of light the air wedge in problem 4 displays 271 bright fringes. To the nearest nm what is the wavelength of the light?

6. A soap bubble is illuminated by a combination of red light (λ = 740 nm) and blue light (λ = 555 nm). What is the minimum thickness of the soap bubble film, to the nearest nm, if the red light is strongly reflected, and the blue light is not reflected?

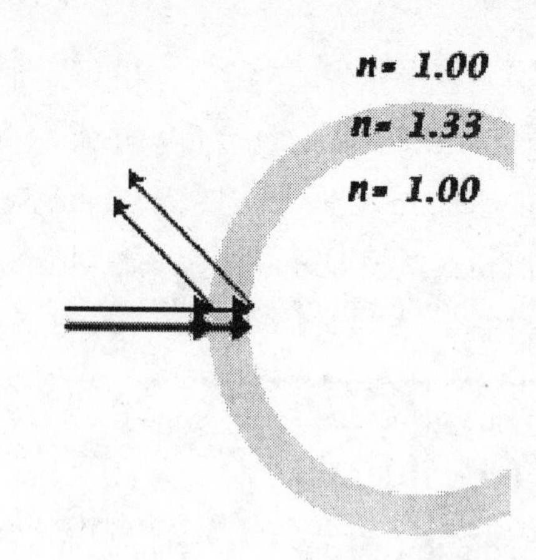

7. How many half wavelengths of the blue light fit into the soap bubble film?

8. Water waves spread out as they pass through an opening in a breakwater. If the wavelength is 35 meters, the opening is 60 meters, and the distance to the beach is 1070 meters, to the nearest tenth of a meter what is the distance from Y to P?

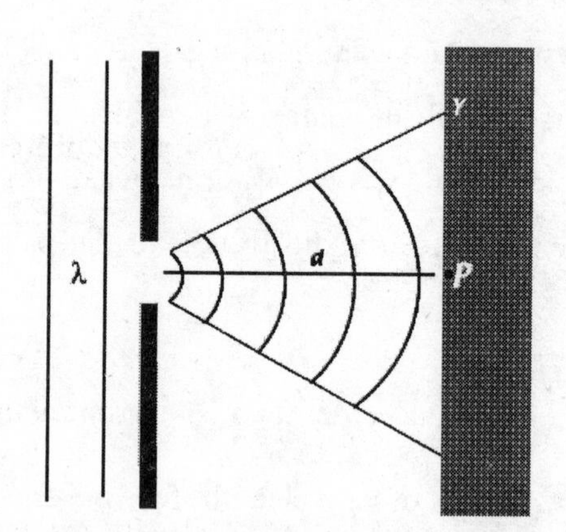

9. Two stars are 9.9 x 10^{16} meters away. How far apart must they be if they are to be resolved with light of 645 nm by an orbiting telescope with an aperture of 2.4 meters (the Hubble) to the nearest hundredth of an astronomical unit (the Sun-Earth distance: 149.6 x 10^9 m)?

10. Light of wavelength 612 nm passes through a diffraction grating having 2500 lines/cm. To the nearest tenth of a degree, what is the angle of the second-order maximum?

Puzzle

ONE POLARIZER, TWO POLARIZER

Two crossed polarizing filters (oriented with their axes at right angles to each other) almost completely annihilate a light beam (initially unpolarized.) If a third polarizer is placed between the two, the light intensity usually increases. Why is that? How would you orient the middle polarizer to maximize the intensity of transmitted light? What percentage of the original unpolarized beam would pass through such a setup?

Selected Solutions

7. The path differences equal to one and two wavelengths correspond to the lowest and second lowest frequencies, respectively. The path difference is given by

[distance from farthest speaker] – [distance from nearest speaker]

Therefore,

$$\left(2.00\text{ m}+\frac{0.525\text{ m}}{2}\right)-\left(2.00\text{ m}-\frac{0.525\text{ m}}{2}\right)=0.525\text{ m}$$

The lowest frequency is for path difference = λ:

$$f=\frac{v}{\lambda}=\frac{343\ \frac{m}{s}}{0.525\text{ m}}=\boxed{653\text{ Hz}}$$

The next lowest frequency is for path difference = 2λ:

$$f=\frac{v}{\lambda}=\frac{343\ \frac{m}{s}}{\frac{0.525\text{ m}}{2}}=\boxed{1310\text{ Hz}}$$

13. **(a)** For a second-order (m = 2) maximum the condition is: $\sin\theta=m\frac{\lambda}{d}=(2)\frac{\lambda}{d}=\frac{2\lambda}{d}$. Thus,

$$\lambda=\frac{1}{2}d\sin\theta=\frac{1}{2}(48.0\times10^{-5}\text{ m})\sin0.0990^\circ=\boxed{415\text{ nm}}$$

(b) Since the wavelength is directly proportional to the separation, the wavelength will $\boxed{\text{increase}}$ if the separation is increased.

(c) For this larger slit separation we get

$$\lambda=\frac{1}{2}(68.0\times10^{-5}\text{ m})\sin0.0990^\circ=\boxed{587\text{ nm}}$$

which is indeed larger as stated in part (b).

31. **(a)** Since $n_{glass}<n_{coating}$, there is no phase change at the coating-lens boundary. The condition for destructive interference is

$$\lambda_{vacuum}=\frac{2nt}{m}=\frac{2(1.480)(340.0\text{ nm})}{m}$$

The first few results give

m	1	2	3	4
λ (nm)	1006	503.2	335.5	251.6

The only wavelength that is within the visible range is $\lambda = \boxed{503.2\text{ nm}}$ which is the one that will be absent.

(b) Since $n_{glass} > n_{coating}$, there is a phase change at the coating-lens boundary. Adding 1/2 to the left side of the equation for destructive interference will account for the reflection at the coating-lens interface:

$$\frac{2nt}{\lambda_{vacuum}} + \frac{1}{2} = m \Rightarrow \lambda_{vacuum} = \frac{2nt}{m - \frac{1}{2}} = \frac{2(1.480)(340.0\text{ nm})}{m - \frac{1}{2}}$$

m	1	2	3	4
λ (nm)	2013	670.9	402.6	287.5

$\lambda = \boxed{670.9\text{ nm and }402.6\text{ nm}}$ will be absent.

39. The last dark fringes theoretically occur where θ approaches $\pm 90°$:

$$\sin(\pm 90°) = \pm 1 = m\frac{\lambda}{W}$$

$$m = \pm\frac{W}{\lambda} = \pm\frac{8.00 \times 10^{-6}\text{ m}}{553 \times 10^{-9}\text{ m}} = \pm 14.5$$

Thus, there are $\boxed{14}$ dark fringes produced on either side of the central maximum.

75. The smallest diameter corresponds to the minimum angle of resolution. This angle of resolution is subtended by the radius of the pupil; therefore, diameter $= 2y = 2L\tan\theta_{min}$. The angle can be determined by

$$\sin\theta_{min} = 1.22\frac{\lambda}{D} \Rightarrow \theta_{min} = \sin^{-1}\frac{1.22\lambda}{D} = \sin^{-1}\frac{1.22\lambda_{vacuum}}{nD}$$

The diameter, then, is given from

$$\text{diameter} = 2L\tan\left(\sin^{-1}\frac{1.22\lambda_{vacuum}}{nD}\right) = 2(0.0245\text{m})\tan\left[\sin^{-1}\frac{1.22(550\times 10^{-9}\text{m})}{(1.336)(0.00300\text{m})}\right] = \boxed{8.5\mu\text{m}}$$

Answers to Practice Quiz

1. (e) **2.** (d) **3.** (c) **4.** (d) **5.** (a) **6.** (a) **7.** (a) **8.** (b) **9.** (e) **10.** (d)

Answers to Practice Problems

1. 2572 m **2.** 1.163°

3. 26.3 m	**4.** 47.7 mm
5. 353 nm	**6.** 1252 nm
7. 6	**8.** 768.5 m
9. 0.22 AU	**10.** 17.8°

CHAPTER 29
RELATIVITY

Chapter Objectives

After studying this chapter, you should:

1. be able to calculate the time interval between events relative to the proper time.
2. be able to calculate the length of an object relative to the proper length.
3. be able to use the relativistic addition of velocities for one-dimensional motion.
4. be familiar with the expressions for relativistic momentum and energy.

Warm-Ups

1. Einstein's theory of special relativity predicts that the rate at which a moving clock runs will be slower than the rate of the same clock when stationary. Suppose you are shut in a rocket module which is moving through space at half the speed of light. Would you be able to observe the "slowing down" of the clock you brought with you from Earth?
2. Suppose you had a car engine that could convert the rest energy of ordinary water into kinetic energy of the car. Estimate the amount of water needed to power an average car for the lifetime of the car. (Use reasonable estimates for the number of useful years and for average yearly mileage.)
3. According to the theory of relativity the momentum formula p = mv has to be modified with the factor gamma. Which of the sketches of p vs. v/c best agrees with the relativistic momentum formula? Why?

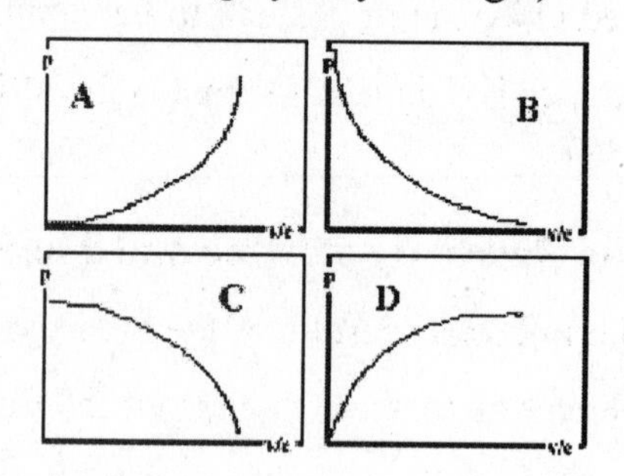

4. Estimate how fast you would have to run to have a relativistic mass of 1000 kg.

Chapter Review

It turns out that Newton's laws, despite their incredible success at describing physical behavior, are fundamentally incorrect. One of the corrections is contained in Einstein's theory of relativity. This theory is divided into two categories known as *special relativity* and *general relativity*. Special relativity is the main topic of this chapter, while general relativity is only briefly discussed. (The other major correction to Newton's laws is quantum physics, which is the topic of the next chapter.)

29–1 The Postulates of Special Relativity

The theory of special relativity is based on the following two postulates:

Equivalence of Physical Laws

The laws of physics are the same in all inertial frames of reference.

Constancy of the Speed of Light

The speed of light in vacuum, $c = 3.00 \times 10^8$ m/s, is the same in all inertial frames of reference, independent of the motion of the source or the receiver.

Recall that an inertial reference frame is one in which Newton's laws of motion are obeyed. Inertial frames of reference move with constant velocity relative to each other. The first postulates basically says that all the laws of physics are independent of the inertial reference frame in which they are being investigated.

The second postulate expresses the astounding fact that the speed of light in vacuum will have the same value, c, in all inertial frames of reference. As we will see, the consequences of this postulate are that the behavior of space and time is very different from what our everyday experience suggests. The reason for this difference is that for the speeds at which we move around in everyday life these unexpected effects are too small to be commonly noticed. Another important result from these postulates is that the speed of light in vacuum is the ultimate speed at which any object can move.

29–2 The Relativity of Time and Time Dilation

One of the important results of the theory of relativity is our knowledge that time is relative. That is, the rate at which time passes depends on the frame of reference. The basic result, known as **time dilation**, is that *moving clocks run slowly.* In relativity, we think of space and time in terms of **events**, where an event is just a particular point in space at a particular time. (If you prefer, you can think of something happening at that point and at that time, but it isn't necessary.) The amount of time between two events that are at the same point in space is called the **proper time**, Δt_0. The expression for how the time interval between events as measured in an arbitrary inertial frame, Δt, relates to the proper time is

$$\Delta t = \frac{\Delta t_0}{\sqrt{1 - v^2 / c^2}}$$

where v is the relative velocity between the two inertial frames. For objects moving at everyday speeds the smallness of the time dilation effect is such that most calculators will just give $\Delta t = \Delta t_0$. To handle these cases it is better to use the approximate equation

$$\Delta t = \Delta t_0 \left(1 + \frac{1}{2}\frac{v^2}{c^2} \cdots\right)$$

An additional consequence of the relativity of time is that simultaneity is relative; that is, two events that occur at the same time in one frame of reference will not generally be simultaneous in another frame of reference.

Example 29–1 A Moving Clock If exactly 1.0 hour has passed according to the clock on your wall, how much time will have passed according to someone moving past you at 93% of the speed of light?

Picture the Problem The diagram shows your clock and the observer moving by at speed v.

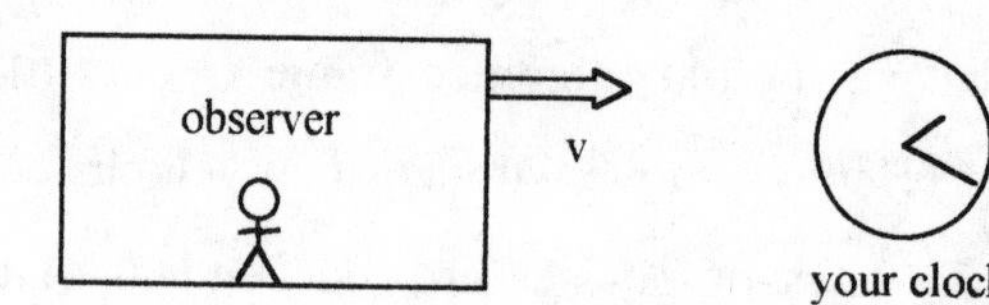

Strategy The time you measure on your clock is the proper time, Δt_0.

Solution

1. According to the time dilation equation:

$$\Delta t = \frac{\Delta t_0}{\sqrt{1 - v^2/c^2}} = \frac{\Delta t_0}{\sqrt{1 - (0.93c)^2/c^2}}$$

$$= \frac{1.0\ \text{h}}{\sqrt{1 - (0.93)^2}} = 2.7\ \text{h}$$

Insight Notice that to the observer it is *your* clock that is moving, so she sees your clock as being slow because 2.7 h has passed in her reference frame.

Practice Quiz

1. The statement that moving clocks run slowly implies that
   (a) people moving relative to you age more slowly than you do
   (b) people moving relative to you age more rapidly than you do
   (c) people moving relative to you age at the same rate as you do
   (d) This statement doesn't apply to the aging of people.
   (e) none of the above

2. For what relative velocity would moving clocks run at half the rate of stationary clocks?

(a) 0.50c **(b)** 0.87c **(c)** 0.25c **(d)** 1.7c **(e)** 0.75c

29–3 The Relativity of Length and Length Contraction

As with time, space is also relative. By this we mean that spatial distance, or length, depends on the frame of reference of the observer. This phenomenon is called **length contraction**. It has this name because the length of a moving object is shorter, along the direction of relative motion, than it would be if it were stationary. The length of an object as measured in a frame of reference at rest with respect to the object (the *proper frame*) is called its **proper length**, L_0. The length of this object as measured in an inertial frame of reference moving with speed v relative to the proper frame is given by

$$L = L_0\sqrt{1-\frac{v^2}{c^2}}$$

This length contraction effect only occurs along the direction of relative motion; lengths that are perpendicular to this direction are not affected.

Example 29–2 A Moving Rod Suppose, through sophisticated techniques, you measure the length of a rod that is moving by you at 93.0% of the speed of light to be 3.30 m. What length would you measure for this same rod if it were in your frame of reference?

Picture the Problem The diagram shows the moving rod in its reference frame and you.

Strategy If the rod were in your reference frame, you would measure the proper length, L_0.

Solution

1. Solve the length contraction equation for the proper length: $L = L_0\sqrt{1-\frac{v^2}{c^2}} \Rightarrow L_0 = L\left(1-\frac{v^2}{c^2}\right)^{-1/2}$

2. Obtain the numerical result: $L_0 = L\left(1-\frac{v^2}{c^2}\right)^{-1/2} = (3.30\text{ m})\left(1-[0.930]^2\right)^{-1/2}$

$$= 8.98\text{ m}$$

Insight You would measure the moving rod as shorter, so it makes sense that its proper length is longer.

Practice Quiz

3. The fact that the length of a moving object is shortened implies that

   (a) the proper length is always the shortest length you can measure

   (b) the proper length is always the longest length you can measure

   (c) you can't measure a moving length

   (d) the length of a moving object only appears to be shorter, but it is not actually shorter

   (e) none of the above

4. If the proper length of a rod that is laid out along the x direction is measured to be 1.0 m, how long will it be as measured in a frame moving at 0.90c relative to the proper frame along the y direction?

   (a) 1.0 m **(b)** 2.3 m **(c)** 0.9 m **(d)** 1.8 m **(e)** 3.2 m

29–4 The Relativistic Addition of Velocities

As stated above, relativity tells us that no object can travel faster than the speed of light in vacuum. This implies that the rules for velocity addition that we used previously (see Section 3–6) must be corrected by relativity theory. For simplicity, we will only consider the case for which all velocities are along the same line of motion. We consider two inertial frames of reference A and B where B, moves relative to A with a velocity v_{BA}. Now, let observers in A and B observe the motion of a point P. The velocity of point P as measured in A, v_{PA}, in terms of the velocity of P as measured in B, v_{PB}, is given by

$$v_{PA} = \frac{v_{PB} + v_{BA}}{1 + \frac{v_{PB} v_{BA}}{c^2}}$$

Keep in mind that the velocities in the above expression are one-dimensional vectors (not vector magnitudes), so that they can have negative values. Also recall that $v_{AB} = -v_{BA}$. The numerator in the above equation is the same as we used in Chapter three; the effect of the denominator is that no velocity will become greater than the speed of light.

Exercise 29–3 Velocity Addition Rocket A moves directly away from you at a speed of 0.87c. If rocket B is moving directly toward rocket A along the same line of motion at 0.75c relative to A, how fast, and in what direction is rocket B moving relative to you?

Solution

Let the subscript Y represent you. Then, the velocity addition equation gives

$$v_{BY} = \frac{v_{BA} + v_{AY}}{1 + \frac{v_{BA} v_{AY}}{c^2}}$$

If we take v_{AY} to be in the positive direction, then v_{BA} is in the negative direction. Thus,

$$v_{BY} = \frac{-0.75c + 0.87c}{1 + \frac{(-0.75c)(0.87c)}{c^2}} = \frac{0.12c}{0.3475} = 0.35c$$

Therefore, rocket B is moving away from you at 35% of the speed of light.

Practice Quiz

5. An object moving away from you at 0.80 times the speed of light shines a light in the direction of its forward motion. How fast is the beam of light moving away from you?

   (a) 1.8c **(b)** 0.80c **(c)** c **(d)** 0.60c **(e)** 1.7c

6. An object moving toward you at 0.80 times the speed of light shines a light in the direction away from you. How fast does the beam of light move away from you?

   (a) 1.2c **(b)** 0.20c **(c)** c **(d)** 0.60c **(e)** 0.80c

29–5 Relativistic Momentum and Mass

In order for the law of the conservation of momentum to be the same in all inertial frames of reference, we must generalize the expression for the momentum of an object to an expression that is consistent with the theory of relativity. We call this result the **relativistic momentum** of a particle

$$p = \frac{mv}{\sqrt{1 - v^2/c^2}}$$

Note that in this equation v is the speed of the particle and not the relative velocity between two frames of reference. You can see that for a speed small compared with c the denominator is nearly equal to 1, so that this relativistic momentum is completely consistent with the $p = mv$ that we normally use at low speeds. In fact, the expression $p = mv$ can be retained if we reinterpret the mass of an object to be

$$m = \frac{m_0}{\sqrt{1 - v^2/c^2}}$$

where m_0 is called the **rest mass**, and m is called the **relativistic mass**.

Practice Quiz

7. For a particle with a rest mass of 2.6 mg, what will be its relativistic mass if it moves at 70% of the speed of light?

   (a) 1.4 mg **(b)** 2.6 mg **(c)** 4.7 mg **(d)** 3.6 mg **(e)** 3.1 mg

29–6 Relativistic Energy and E = mc²

One of the most well known results of the theory of relativity is that mass is a form of energy. The equivalence between mass and energy is represented by the famous equation

$$E = mc^2$$

where m is the relativistic mass of the object. Note that the above equation includes both the energy of motion (its kinetic energy) and the energy content associated only with its rest mass, called the **rest energy**,

$$E_0 = m_0c^2$$

The kinetic energy can be found by taking the difference between the total energy E and the rest energy E_0. This relativistic kinetic energy is given by

$$KE = \frac{m_0c^2}{\sqrt{1 - v^2/c^2}} - m_0c^2$$

It can be shown that at low speeds this expression for the kinetic energy is completely consistent with the more familiar $KE = \frac{1}{2}mv^2$.

Example 29–4 Relativistic Energy and Momentum What is the linear momentum and kinetic energy of an electron that is moving at 65% of the speed of light?

Solution

The mass of an electron is 9.11×10^{-31} kg. The relativistic momentum is given by

$$p = \frac{mv}{\sqrt{1 - v^2/c^2}} = \frac{(9.11 \times 10^{-31}\ \text{kg})(0.65c)}{\sqrt{1 - (0.65)^2}} = 7.79 \times 10^{-31}\,\text{kg} \cdot \text{c}$$

$$= 7.79 \times 10^{-31}\,\text{kg} \cdot (3.00 \times 10^8\ \text{m/s}) = 2.3 \times 10^{-22}\ \text{kg} \cdot \text{m/s}$$

The kinetic energy is

$$KE = \frac{m_0c^2}{\sqrt{1-v^2/c^2}} - m_0c^2 = m_0c^2\left(\frac{1}{\sqrt{1-v^2/c^2}} - 1\right)$$

$$= \left(9.11\times10^{-31}\text{ kg}\right)\left(3.00\times10^{8}\text{ m/s}\right)^2\left[\frac{1}{\sqrt{1-(0.65)^2}} - 1\right] = 2.6\times10^{-14}\text{ J}$$

Notice that for the momentum calculation we called the mass "m" and for the kinetic energy calculation we called the same mass "m_0." The issue with the momentum equation is whether the factor $1/\sqrt{1-v^2/c^2}$ is written explicitly. If it is, then m is, in fact, the rest mass m_0; if it is not, then m is the relativistic mass. Thus, you need to pay attention to how you, and others, write the relativistic momentum. Both ways of doing it are in common use.

Practice Quiz

8. What is the energy content of 1.0 kg of mass?

   (a) 3.0×10^8 J **(b)** 9.0×10^{16} J **(c)** 0 J **(d)** 1.0 J **(e)** 1.0×10^{16} J

29–7 – 29–8 The Relativistic Universe and General Relativity

The theory of relativity now affects everyday life for most of us. Relativity is needed for many important medical applications; it is used to understand processes that lead to alternative (to coal) forms of energy, and it has become important for the pinpoint accuracy of navigation on Earth using the Global Positioning System (GPS).

In the discussion of special relativity we limited ourselves to frames of reference that have constant relative velocities. In formulating his **general theory of relativity**, Einstein argued that, with regard to the laws of physics, observers in accelerated frames of reference are identical to observers in gravitational fields. This apparent fact is called the **principle of equivalence**. Therefore, the general theory of relativity is equivalent to a new theory of gravity that replaces Newton's law of universal gravitation.

Among the many very interesting phenomena that studying the theory of general relativity has brought forth are the affect of gravity on light, and the ability to study the physics of **black holes**, and the prediction of **gravity waves**. A black hole is an object of such high density, and therefore high gravitational field, that the escape velocity from it exceeds the speed of light. Therefore, no object can escape from a black hole. For a spherical object of mass M to be a black hole its radius must be at most

$$R = \frac{2GM}{c^2}$$

Reference Tools and Resources

I. Key Terms and Phrases

inertial frame of reference a frame of reference (coordinate system) in which Newton's laws of motion are obeyed

event a specific point in space at a specific time

time dilation the principle of relativity that moving clocks run slowly

proper time the time interval between events that occur at the same point is space

length contraction the principle of relativity that moving lengths are shortened along the direction of relative motion

proper length the length of an object as measured in the rest frame of the object

relativistic momentum the expression for the momentum of a particle that is consistent with the conservation of momentum and the first postulate of special relativity

rest energy the energy content of the mass of an object as measured in an inertial frame at rest with respect to the object

II. Important Equations

Name/Topic	Equation	Explanation
time dilation	$\Delta t = \dfrac{\Delta t_0}{\sqrt{1 - v^2/c^2}}$	The dependence of time interval on the relative velocity between inertial frames of reference.
length contraction	$L = L_0\sqrt{1 - \dfrac{v^2}{c^2}}$	The dependence of length, along the direction of relative motion, on the relative velocity between inertial frames of reference.
addition of velocities	$v_{PA} = \dfrac{v_{PB} + v_{BA}}{1 + \frac{v_{PB}v_{BA}}{c^2}}$	The relativistic velocity addition equation for one-dimensional motion.
relativistic momentum	$p = \dfrac{mv}{\sqrt{1 - v^2/c^2}}$	The expression for the momentum of a particle that is consistent with the theory of relativity.

relativistic energy	$E = mc^2$	The expression of the equivalence between mass and energy.

Practice Problems

1. A flying saucer passes Earth with a relative speed of 0.51c. If 3.4 hours go by on the spaceship as seen from Earth, to the nearest hundredth of an hour how much time goes by on Earth.
2. During the time period on the spaceship in problem 1 how much time does the spaceship observer see go by on Earth?
3. An advanced spaceship is traveling from Earth to a star $x = 59.4$ ly away at a speed relative to the Earth-star system of 0.9887c. How long, to the nearest thousandth of a year does it take the ship to make the trip as seen from Earth?

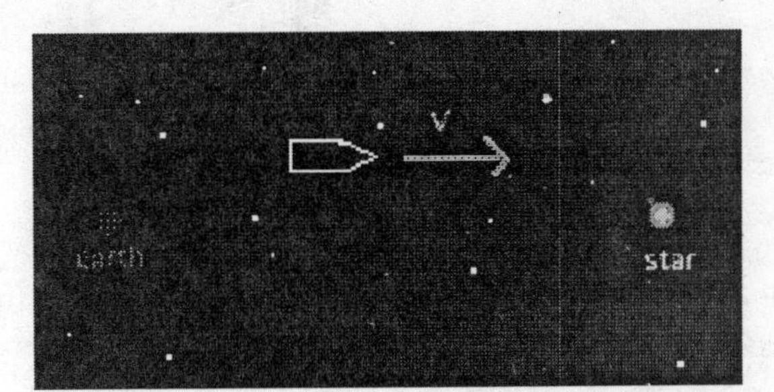

4. To the nearest tenth of a ly, what is the distance between Earth and the star as measured by the spaceship?
5. An observer sees a rocket moving away at $v = 0.69c$ while the rocket observer views a meteor moving away from him at 0.91c. What fraction of the velocity of light, to 3 decimal places, is the meteor traveling as viewed by the first observer?
6. What would be the answer to problem 5 if the meteor was moving toward the rocket with the given speed?
7. After a satellite is blown apart the first part has speed $v_1 = 0.69c$ and the second one has a speed $v_2 = 0.92c$. If the mass of the first is 150 kg, to the nearest tenth of a kg, what is the mass of the second piece?
8. A piece of space junk with a mass of 7.8 kg is traveling by an observer at 0.51c. To three decimal places, what is the ratio of its total energy to its rest energy?
9. In problem 8 what is the ratio of the relativistic kinetic energy to the classical kinetic energy, to two decimal places?
10. Calculate the result of the formula shown, to two decimal places, if E is the total relativistic energy, and p is the relativistic momentum for a particle of $m = 5.95$ kg and $v = 0.73c$.

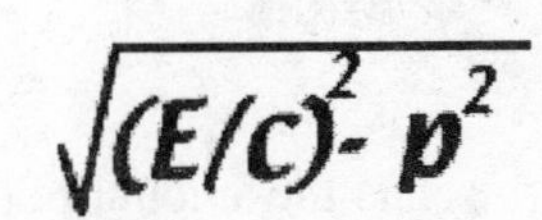

Puzzle

ETHER,... OR

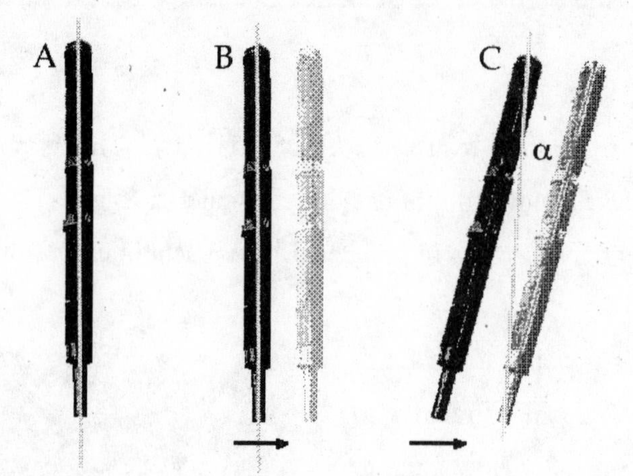

In an attempt to explain the null result of the Michelson-Morley experiment, a proposal was put forward that Earth is dragging ether, the medium that is carrying light waves, along with it. This explanation did not agree with the observation that light from distant stars comes in at an angle which depends on the velocity of Earth and that changes throughout the year. This phenomenon is called the *aberration of starlight*. See the figures: A (stationary telescope), B (moving telescope pointed at the star), and C (moving telescope adjusted for the aberration).

Give a rough estimate of the magnitude of the aberration angle, α? If the telescope tube was filled with a transparent substance of a high index of refraction, would you expect the aberration angle to change? If so, would it increase or decrease? Answer this question in words, not equations, briefly explaining how you obtained your answer.

Selected Solutions

13. (a) Since the time between each heartbeat will appear longer to the Earth-based observer, the measured heart rate will be $\boxed{\text{less than}}$ 72 beats per minute.

(b) From time dilation we have

$$\Delta t = \frac{\Delta t_0}{\sqrt{1-\frac{v^2}{c^2}}}$$

If Δt represents the amount of time between beats, then

$$\frac{1\text{ beat}}{\Delta t} = \frac{\sqrt{1-\frac{v^2}{c^2}}}{\Delta t_0}(72\text{ beats}) = \frac{\sqrt{1-\left(\frac{0.65c}{c}\right)^2}}{1\text{ min}}(72\text{ beats}) = \boxed{55\text{ beats/min}}$$

25. Outgoing Trip

Length contraction gives

$$L_1 = L_0\sqrt{1-\frac{v_1^2}{c^2}} \Rightarrow L_0 = \frac{L_1}{\sqrt{1-\frac{v_1^2}{c^2}}}$$

Return Trip

Length contraction gives

$$L_2 = L_0\sqrt{1-\frac{v_2^2}{c^2}} = \frac{L_1}{\sqrt{1-\frac{v_1^2}{c^2}}}\sqrt{1-\frac{v_2^2}{c^2}} = L_1\sqrt{\frac{1-\frac{v_2^2}{c^2}}{1-\frac{v_1^2}{c^2}}} = (7.5\text{ ly})\sqrt{\frac{1-\left(\frac{0.88c}{c}\right)^2}{1-\left(\frac{0.45c}{c}\right)^2}} = \boxed{4.0\text{ ly}}$$

35. Let us adopt the following definitions:

$v_1 = 0.60c$

$v_2 =$ the speed of asteroid 1 relative to asteroid 2

$v = 0.80c$

From the velocity addition equation we have that

$$v = \frac{v_1 + v_2}{1+\frac{v_1v_2}{c^2}}$$

We can solve this equation for v_2

$$v + \frac{vv_1}{c^2}v_2 = v_1 + v_2 \Rightarrow v_2 = \frac{v - v_1}{1-\frac{vv_1}{c^2}}$$

The final result is

$$v_2 = \frac{0.80c - 0.60c}{1-\frac{(0.80c)(0.60c)}{c^2}} = \boxed{0.38c}$$

51. $U_{spring} = \frac{1}{2}kx^2 = \Delta E = \Delta mc^2$

$$\Delta m = \frac{kx^2}{2c^2} = \frac{(544\ \frac{N}{m})(0.38\text{ m})^2}{2\left(3.00\times 10^8\ \frac{m}{s}\right)^2} = \boxed{4.4\times 10^{-16}\text{ kg}}$$

73. **(a)** Length contraction gives

$$L = L_0\sqrt{1-\frac{v^2}{c^2}}$$

Since the length is only contracted in the direction of motion, we have

$$L = \ell\cos\theta\sqrt{1-\frac{v^2}{c^2}} = (2.5\text{ m})\cos 45°\sqrt{1-\left(\frac{0.95c}{c}\right)^2} = \boxed{0.6\text{ m}}$$

(b) The direction perpendicular to the motion, y, is not affected by length contraction, while that along the direction of motion, x, is affected as in part (a) above. Therefore,

$$\tan\theta = \frac{y}{x} \quad\Rightarrow\quad \theta = \tan^{-1}\frac{y}{x}$$

Making use of the length contraction for x gives

$$\theta = \tan^{-1}\frac{\ell\sin\theta}{\ell\cos\theta\sqrt{1-\frac{v^2}{c^2}}} = \tan^{-1}\frac{\tan\theta}{\sqrt{1-\frac{v^2}{c^2}}} = \tan^{-1}\frac{\tan 45°}{\sqrt{1-\left(\frac{0.95c}{c}\right)^2}} = \boxed{70°}$$

Answers to Practice Quiz

1. (a) **2**. (b) **3**. (b) **4**. (a) **5**. (c) **6**. (c) **7**. (d) **8**. (b)

Answers to Practice Problems

1. 3.95 h **2**. 2.92 h

3. 60.079 yr **4**. 8.9 ly

5. 0.983 **6**. −0.591

7. 60.9 kg **8**. 1.163

9. 1.25 **10**. 5.95 kg·m/s

CHAPTER 30
QUANTUM PHYSICS

Chapter Objectives

After studying this chapter, you should:

1. understand how the photon hypothesis explains blackbody radiation, the photoelectric effect, and Compton scattering.
2. be able to calculate the energy and momentum of a photon.
3. be able to determine the de Broglie wavelength of a particle.
4. know the results for the uncertainty principle for position and momentum and for energy and time.

Warm-Ups

1. In your own words, explain what happens in the photoelectric effect?
2. A certain radio signal has a wavelength as long as your arm. Estimate the energy of one photon of this radiation.
3. Estimate your de Broglie wavelength when running as fast as you can.
4. Estimate the uncertainty in the momentum of a hydrogen nucleus given that we know it is in a diatomic hydrogen molecule of length 0.075 nm.

Chapter Review

In the previous chapter we discussed the theory of relativity as one of the principal corrections to Newton's laws that become most important for high speeds and strong gravitational fields. In this chapter we discuss the other main correction, which becomes important at small size scales; we refer to these results as quantum physics.

30–1 Blackbody Radiation and Planck's Hypothesis of Quantized Energy

One of the first indications of the need for a theory of quantum physics came from the study of the electromagnetic radiation given off by a **blackbody**. An ideal blackbody is an object that absorbs all of the electromagnetic radiation that is incident upon its surface. This fact means that a blackbody is a perfect absorber of radiation, and in order to maintain thermal equilibrium, a blackbody is also a perfect radiator of electromagnetic energy. One of the most useful properties of blackbody radiation is that the distribution of energy given off only depends on the temperature of the object. In fact, the frequency at which a blackbody gives off the maximum intensity of electromagnetic radiation is directly proportional to the absolute temperature T

$$f_{peak} = \left(1.04 \times 10^{11}\,s^{-1} \cdot K^{-1}\right)T$$

This expression is known as **Wien's displacement law.**

The dependence of the intensity of blackbody radiation on frequency could not be accurately explained by classical physics. In order to reproduce the experimental result Max Planck hypothesized that the radiant energy in a blackbody must be quantized in integral multiples of the constant h times the frequency:

$$E_n = nhf, \quad n = 1, 2, 3, \ldots$$

where the constant h is called **Planck's constant** and has the value

$$h = 6.63 \times 10^{-34}\,J \cdot s$$

Exercise 30–1 Wien's Displacement By how much is the frequency of the most intense light from a blackbody displaced if its temperature increases from 300 K to 380 K?

Solution

Using Wien's displacement law we can see that

$$\begin{aligned} f_2 - f_1 &= \left(1.04 \times 10^{11}\,s^{-1} \cdot K^{-1}\right)[T_2 - T_1] \\ &= \left(1.04 \times 10^{11}\,s^{-1} \cdot K^{-1}\right)[80\text{ K}] = 8.3 \times 10^{12}\text{ Hz} \end{aligned}$$

At first sight this may look like an unreasonably large shift, but consider that in this temperature range the frequency of peak intensity is already on the order of 10^{13} Hz.

Practice Quiz

1. At what wavelength does a 450 K blackbody radiate energy with the most intensity?

   (a) 4.38×10^{13} m

   (b) 2.14×10^{-14} m

   (c) 6.41×10^{-6} m

   (d) 6.67×10^{5} m

   (e) none of the above

30–2 Photons and the Photoelectric Effect

Planck's idea that electromagnetic radiation is quantized was considered to be a good explanation for blackbody radiation but not a general principle. It was Einstein who proposed that light, in general, comes in bundles of energy, called photons, that obey Planck's energy quantization result. Einstein argued that the energy of a photon is given by

$$E = hf$$

In addition to providing a more natural explanation of blackbody radiation, Einstein used his photon hypothesis to explain another phenomenon called the **photoelectric effect**. This effect occurs when light strikes the surface of a metal and ejects electrons. The minimum amount of energy needed to remove an electron from the surface of a metal is called the **work function**, W_0, for that metal. It was observed that no electrons are ejected from a metal unless the frequency of the incident light exceeds a certain **cutoff frequency**, f_0, given by

$$f_0 = \frac{W_0}{h}$$

It was further found that the maximum kinetic energy of the ejected electrons was independent of the intensity of the incident light and was only a function of its frequency:

$$KE_{max} = hf - W_0$$

Neither of these observations could be explained using classical physics. Einstein's photon hypothesis, however, provided a complete and accurate explanation of these results.

Example 30–2 Photons from the Sun The power output from the sun, called its luminosity, is 3.826×10^{26} W. Estimate the number of photons per second that are leaving the surface of the Sun.

Picture the Problem The sketch shows the Sun with light emanating from its surface.

Strategy Since every photon carries a certain amount of energy, we need to determine how many photons will produce an amount of energy consistent with the Sun's luminosity.

Solution

1. The amount of energy coming from the Sun in 1 second is: $E = Pt = (3.826 \times 10^{26}\,\text{J/s})(1.00\,\text{s}) = 3.826 \times 10^{26}\,\text{J}$

2. Most of the Sun's energy is radiated near the middle of the electromagnetic spectrum, giving an effective frequency of:

$$f_{eff} = 5.45 \times 10^{12}\ \text{Hz}$$

3. The number, n, of photons at the above frequency that give the calculated energy is:

$$E = nhf_{eff} \quad \therefore$$

$$n = \frac{E}{hf_{eff}} = \frac{3.826 \times 10^{26}\ \text{J}}{(6.63 \times 10^{-34}\ \text{J}\cdot\text{s})(5.45 \times 10^{12}\ \text{Hz})}$$

$$= 1.06 \times 10^{47}$$

Insight As you might have expected, a large number of photons are given off in just 1 second.

Practice Quiz

2. If the wavelength of photon-1 is 20% longer than the wavelength of photon-2, then

   (a) the energy of photon-1 is 20% greater than the energy of photon-2

   (b) the energy of photon-1 is 20% less than the energy of photon-2

   (c) the energy of photon-2 is 20% greater than the energy of photon-1

   (d) the energy of photon-2 is 20% less than the energy of photon-1

   (e) both photons have the same energy

3. What affect would increasing only the frequency of the light incident on a given metal have on the maximum kinetic energy of the electrons ejected from its surface?

   (a) It would have no effect on the maximum kinetic energy of the electrons.

   (b) It would decrease the maximum kinetic energy of the electrons by an amount directly proportional to the increase in frequency.

   (c) It would increase the work function of the metal, causing the maximum kinetic energy to decrease.

   (d) It would decrease the work function of the metal, causing the maximum kinetic energy to increase.

   (e) None of the above

30–3 The Mass and Momentum of a Photon

Although photons are like particles in that they have energy and linear momentum, one way in which photons are not like ordinary particles is that they are massless. In fact, using the results of relativity in the previous chapter it can be shown that any particle which travels at the speed of light must have zero rest mass.

However, as mentioned above, photons do carry linear momentum, and using the results of relativity, it can be shown that the momentum of a photon is given by

$$p = \frac{hf}{c} = \frac{h}{\lambda}$$

While the momentum of a typical photon is quite small, when large numbers of photons are incident on an object they can collectively provide a substantial *radiation pressure.*

Practice Quiz

4. If photon-1 has a longer wavelength than photon-2, then

   (a) the momentum of photon-1 is greater than the momentum of photon-2

   (b) photon-1 has more energy than photon-2

   (c) the momentum of photon-1 is less than the momentum of photon-2

   (d) both photons have the same momentum

   (e) none of the above

30–4 Photon Scattering and the Compton Effect

Another important verification of the photon idea was its ability to explain the **Compton effect**. The Compton effect is the result of how the direction and energy of light changes when scattered off electrons that are initially at rest. Using the photon idea, and the conservation of energy and momentum, an accurate explanation of the Compton effect emerged. Particularly, the shift in the wavelength of the scattered light is found to be

$$\Delta\lambda = \lambda' - \lambda = \frac{h}{m_e c}(1 - \cos\theta)$$

where λ' is the wavelength of the scattered photon, λ is the wavelength of the incident photon, and θ is the angle at which the photon scatters relative to its incident direction (see Figure 30–7 in the text). The quantity $h/m_e c$ is called the Compton wavelength of an electron.

Practice Quiz

5. For the Compton scattering of a photon, does the greater shift in wavelength occur if the photon is scattered through 0° or 180°?

   (a) 0°

(b) 180°

(c) They produce the same shift in wavelength.

(d) Neither angle produces any shift in wavelength.

(e) None of the above

30–5 The de Broglie Hypothesis and Wave-Particle Duality

That light, which was traditionally understood to be a wave, has both wavelike and particle-like properties led Louis de Broglie to propose that matter traditionally understood to be particles may also exhibit wavelike behavior. He postulated that the wavelength associated with a particle, such as an electron, would have the same relationship to its momentum as the momentum of a photon has to its wavelength

$$\lambda = \frac{h}{p}$$

The value of λ given by this expression is called the **de Broglie wavelength**.

Verification of de Broglie's radical idea is found in the fact that beams of particles exhibit interference and diffraction patterns indicative of waves. Like X-rays, electrons and other particles scattered from crystalline materials produce interference patterns consistent with the constructive and destructive interference of waves scattering from different layers of the crystal. If d is the distance between adjacent layers of the crystal structure, and θ is the angle between the surface of the layer and the incident direction of the beam, then the condition for constructive interference is

$$2d\sin\theta = m\lambda, \quad m = 1, 2, 3, \ldots$$

Now we see that all matter and energy, in general, have both wavelike and particle-like properties. We call this phenomenon the **wave-particle duality**.

Practice Quiz

6. What is the de Broglie wavelength of an electron moving at 90% of the speed of light?

   (a) 1.17×10^{-12} m **(b)** 5.64×10^{-22} m **(c)** 2.46×10^{-22} m **(d)** 2.29 m **(e)** 2.70×10^{-12} m

7. What is the de Broglie wavelength of a 0.50-kg ball thrown at 50.0 mi/h?

   (a) 2.7×10^{-35} m **(b)** 25 m **(c)** 1.3×10^{-35} m **(d)** 5.9×10^{-35} m **(e)** 11 m

30–6 The Heisenberg Uncertainty Principle

One of the most important lessons of quantum physics is that there is a fundamental uncertainty in nature that is related to the wave nature of matter; this fact is known as the **Heisenberg uncertainty principle**.

One form of this principle states that both the position and momentum of a particle cannot be known with arbitrary precision at the same time. We can write this result in the following way:

$$\Delta p_y \Delta y \geq \frac{h}{2\pi}$$

where Δp_y is the uncertainty in the y component of momentum, and Δy is the uncertainty in the y component of position.

Another important expression of the uncertainty principle relates to energy and time:

$$\Delta E \Delta t \geq \frac{h}{2\pi}$$

In the above expression, ΔE is the uncertainty in the energy of a system, and Δt is the amount of time for some process to occur within the system.

Example 30–3 Uncertainty If the kinetic energy of an electron is measured to be 250 eV, and the uncertainty in its momentum is ±5.0%, what is the minimum uncertainty in its position?

Picture the Problem The sketch shows an electron moving in a certain direction that we'll call the y direction.

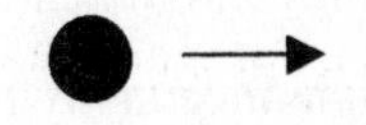

Strategy We need to determine its momentum so that we can find the uncertainty in it. From that information we can use the uncertainty principle to get the minimum uncertainty in the position.

Solution

1. From the kinetic energy we can determine the momentum:

$$KE = \frac{p^2}{2m_e} \quad \therefore \quad p = \sqrt{2m_e KE}$$

$$p = \sqrt{2\left(9.11\times10^{-31}\,\text{kg}\right)\left(250\,\text{eV}\right)\left(1.602\times10^{-19}\,\tfrac{\text{J}}{\text{eV}}\right)}$$

$$= 8.542\times10^{-24}\,\text{kg}\cdot\text{m/s}$$

2. The uncertainty in the momentum is from 5.0% less than the above result to 5.0% more:

$$\Delta p_y = 2(0.05)p$$

$$= 2(0.05)\left(8.542\times10^{-24}\,\text{kg}\cdot\text{m/s}\right)$$

$$= 8.542\times10^{-25}\,\text{kg}\cdot\text{m/s}$$

3. The minimum uncertainty in the position is given by the uncertainty principle:

$$\Delta p_y \Delta y_{min} = \frac{h}{2\pi} \quad \therefore$$

$$\Delta y_{min} = \frac{h}{2\pi \Delta p_y} = \frac{(6.63\times10^{-34}\ \text{J}\cdot\text{s})}{2\pi(8.542\times10^{-25}\ \text{kg}\cdot\text{m/s})}$$

$$= 1.24\times10^{-10}\ \text{m}$$

Insight Notice that when we used the uncertainty principle we used an equality instead on an inequality because we were seeking only the minimum uncertainty in position.

Practice Quiz

8. If the minimum uncertainty in an object's position is decreased by half, what can we say about the uncertainty in its momentum?
   (a) The uncertainty in momentum is at most half of what it was before the change.
   (b) The uncertainty in momentum is at least twice what is was before the change.
   (c) The uncertainty in momentum does not change.
   (d) The minimum uncertainty in momentum is precisely half of what it was before the change.
   (e) none of the above

30–7 Quantum Tunneling

Still another fascinating phenomenon that was discovered as a result of quantum physics is that of quantum mechanical **tunneling**. This phenomenon is the fact that particles can be observed to emerge from across regions where they are forbidden to go according to classical physics. This effect is readily observed with light and has practical applications in devices such as the Scanning Tunneling Microscope (STM).

Reference Tools and Resources

I. Key Terms and Phrases

blackbody an object that absorbs all of the electromagnetic radiation incident upon it

Wien's displacement law the direct proportionality between the frequency at which a blackbody emits energy with maximum intensity and its absolute temperature

Planck's constant the fundamental constant of quantum physics

photoelectric effect the ejection of electrons by light incident upon the surface of a metal

work function the minimum amount of energy needed to remove an electron from the surface of a metal

Compton effect the scattering of a photon that collides with an electron at rest

de Broglie wavelength the wavelength associated with a particle of matter

wave-particle duality the phenomenon that matter and energy have both wavelike and particle-like properties

Heisenberg uncertainty principle the result from quantum physics that there is a fundamental uncertainty in nature

tunneling the result from quantum physics that particles can emerge from across regions where they are forbidden to go according to classical physics

II. Important Equations

Name/Topic	Equation	Explanation
Wien's displacement law	$f_{peak} = \left(1.04 \times 10^{11} s^{-1} \cdot K^{-1}\right)T$	The frequency at which a blackbody radiates with maximum intensity is directly proportional to the absolute temperature.
photons	$E = hf$	The energy of a photon.
	$p = \frac{hf}{c} = \frac{h}{\lambda}$	The momentum of a photon.
de Broglie wavelength	$\lambda = \frac{h}{p}$	The wavelength of a particle of matter.
uncertainty principle	$\Delta p_y \Delta y \geq \frac{h}{2\pi}$	The Heisenberg uncertainty principle for position and momentum.
	$\Delta E \Delta t \geq \frac{h}{2\pi}$	The Heisenberg uncertainty principle for energy and time.

III. Know Your Units

Quantity	Dimension	SI Unit
Planck's constant (h)	$[M][L^2][T^{-1}]$	J·s

Practice Problems

1. The radiation peak of a star is at 13.9×10^{14} Hz. To the nearest kelvin, what is its temperature?
2. A certain molecule has a dissociation energy of 4.14 eV. What is the minimum frequency photon that can dissociate it, in PHz (peta Hz, 10^{15} Hz), to two decimal places.
3. A certain metal has a work function of 3.01 eV. What is its cutoff frequency in PHz to two decimal places?

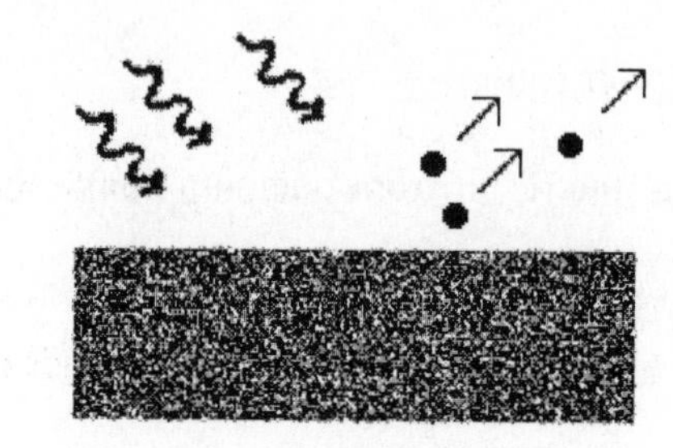

4. If light of frequency 2.95×10^{15} is incident on the metal in problem 3, what is the maximum kinetic energy of the ejected electrons in eV, to two decimal places?
5. An X-ray photon with a wavelength of 0.58 nm scatters off an electron at 137 degrees. What is the wavelength of the scattered X-ray in nm, to four decimal places?
6. What is the kinetic energy of the electron in problem 5 in eV, to one decimal place?

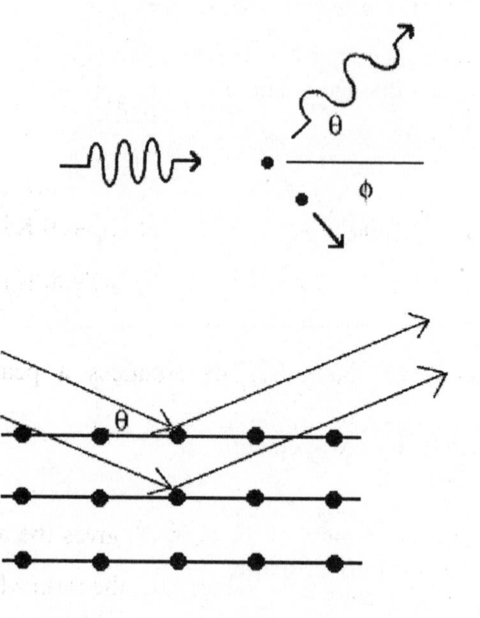

7. Neutrons with a speed of 1456 m/s are incident on a crystal with interatomic spacing of 0.249 nm. In nm, to four decimal places, what is the de Broglie wavelength of the neutrons?
8. To the nearest tenth of a degree, what is the angle of the first interference maximum in problem 7?
9. At a certain point an electron traveling horizontally at 42 m/s has a 1.51% uncertainty in its momentum. To the nearest tenth of a mm, what is its uncertainty in position?
10. If a neutron is confined to a region of size 3.9×10^{-15} m (roughly the size of a nucleus), what is its energy as estimated by the uncertainty principle, to the nearest hundredth of a MeV?

Puzzle

UNCERTAINTY

An electron passes through your apparatus at $t = 0$. You measure its position with accuracy x_0. How accurately can you calculate its position later on, at time t_1? (Hint: The uncertainty in the electron's velocity is related to the uncertainty in its momentum.)

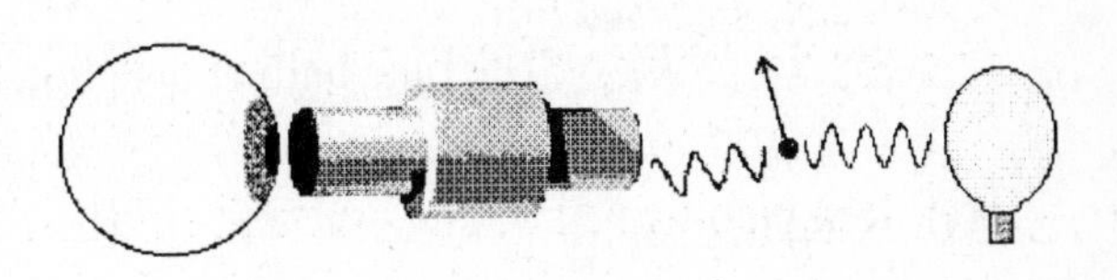

Selected Solutions

7. **(a)** Since $f_{peak} \propto T$, the halogen bulb has the $\boxed{\text{higher}}$ peak frequency.

(b) Since $\lambda \propto \frac{1}{f}$, the standard incandescent bulb has the $\boxed{\text{longer}}$ peak wavelength.

(c) $$\frac{f_{hal}}{f_{std}} = \frac{T_{hal}}{T_{std}} = \frac{3400\text{ K}}{2900\text{ K}} = \boxed{1.2}$$

(d) $$\frac{\lambda_{hal}}{\lambda_{std}} = \frac{T_{std}}{T_{hal}} = \frac{2900\text{ K}}{3400\text{ K}} = \boxed{0.85}$$

(e) $$f_{hal} = (1.04 \times 10^{11}\ \text{s}^{-1} \cdot \text{K}^{-1})(3400\text{ K}) = \boxed{3.5 \times 10^{14}\text{ Hz}}$$
$$f_{std} = (1.04 \times 10^{11}\ \text{s}^{-1} \cdot \text{K}^{-1})(2900\text{ K}) = \boxed{3.0 \times 10^{14}\text{ Hz}}$$

The halogen bulb produces a peak frequency closer to 5.5×10^{14} Hz than the standard incandescent bulb.

21. **(a)** The power of the bulb, P, gives the amount of energy per second, and the energy per photon is given by hc/λ. Therefore, the ratio $\lambda P/hc$ equals the number of photons per second:

$$\frac{\left(\frac{P}{E}\right)_{red}}{\left(\frac{P}{E}\right)_{blue}} = \frac{\frac{\lambda_{red}P_{red}}{hc}}{\frac{\lambda_{blue}P_{blue}}{hc}} = \frac{\lambda_{red}P_{red}}{\lambda_{blue}P_{blue}} = \frac{(650\text{ nm})(150\text{ W})}{(460\text{ nm})(25\text{ W})} = 8.5 > 1$$

So, $\left(\frac{P}{E}\right)_{red} > \left(\frac{P}{E}\right)_{blue}$. $\boxed{\text{The red bulb emits more photons per second than the blue bulb}}$.

(b) $\dfrac{E_{red}}{E_{blue}} = \dfrac{\frac{hc}{\lambda_{red}}}{\frac{hc}{\lambda_{blue}}} = \dfrac{\lambda_{blue}}{\lambda_{red}} = \dfrac{460\text{ nm}}{650\text{ nm}} = 0.71 < 1$

So, $E_{red} < E_{blue}$. The blue bulb emits photons of higher energy than the red bulb.

(c) Red bulb

$$\frac{P}{E} = \frac{\lambda P}{hc} = \frac{(650 \times 10^{-9}\text{ m})(150\text{ W})}{(6.63 \times 10^{-34}\text{ J}\cdot\text{s})\left(3.00 \times 10^{8}\ \frac{\text{m}}{\text{s}}\right)} = \boxed{4.9 \times 10^{20}\text{ photons/s}}$$

Blue bulb

$$\frac{(460 \times 10^{-9}\text{ m})(25\text{ W})}{(6.63 \times 10^{-34}\text{ J}\cdot\text{s})\left(3.00 \times 10^{8}\ \frac{\text{m}}{\text{s}}\right)} = \boxed{5.8 \times 10^{19}\text{ photons/s}}$$

41. For elastic scattering, the kinetic energy is conserved, so the kinetic energy of the recoiling electron equals the kinetic energy lost by the X-ray photon:

$$KE_{electron} = -\Delta KE_{photon} = K_i - K_f = 36\text{ keV} - 27\text{ keV} = \boxed{9\text{ keV}}$$

57. (a) $KE = \dfrac{1}{2}mv^2 = \dfrac{p^2}{2m} = \dfrac{h^2}{2m\lambda^2}$

Since $KE = \dfrac{h^2}{2m\lambda^2}$ and $m_e < m_p$, for identical de Broglie wavelengths, an electron has a greater kinetic energy than a proton.

(b) $\dfrac{KE_e}{KE_p} = \dfrac{\frac{h^2}{2m_e\lambda^2}}{\frac{h^2}{2m_p\lambda^2}} = \dfrac{m_p}{m_e} = \dfrac{1.673 \times 10^{-27}\text{ kg}}{9.11 \times 10^{-31}\text{ kg}} = \boxed{1840}$

67. (a)

$$\Delta p \Delta x \geq \frac{h}{2\pi} \Rightarrow \Delta p \geq \frac{h}{2\pi \Delta x}$$

$$\Delta p \geq \frac{6.63 \times 10^{-34}\text{ J}\cdot\text{s}}{2\pi(0.15 \times 10^{-9}\text{ m})} \quad \therefore \quad \Delta p \geq \boxed{7.0 \times 10^{-25}\text{ kg}\cdot\text{m/s}}$$

(b) $KE = \frac{1}{2}mv^2 = \frac{p^2}{2m} = \frac{\Delta p^2}{2m} = \frac{\left(7.0 \times 10^{-25}\ kg \cdot \frac{m}{s}\right)^2}{2(9.11 \times 10^{-31}\ kg)} = \boxed{2.7 \times 10^{-19}\ J}$

Answers to Practice Quiz

1. (c) **2**. (c) **3**. (e) **4**. (c) **5**. (b) **6**. (a) **7**. (d) **8**. (b)

Answers to Practice Problems

1.	13365 K	**2**.	1.00 PHz
3.	0.73 PHz	**4**.	9.21 eV
5.	0.5842 nm	**6**.	15.4 eV
7.	0.2727 nm	**8**.	33.2°
9.	182.6 mm	**10**.	1.37 MeV

CHAPTER 31
ATOMIC PHYSICS

Chapter Objectives

After studying this chapter, you should:

1. be familiar with some of the early model of the atom.
2. know how to determine which photons are in the emission spectrum of hydrogen.
3. understand Bohr's model of the hydrogen atom and its relation to de Broglie waves.
4. be able to specify the states of multi-electron atoms in terms of the four quantum numbers.
5. be able to write out the electronic structure of multi-electron atoms using the Pauli exclusion principle.
6. know the basic mechanisms for production of X-rays, laser light, and fluorescent and phosphorescent light.

Warm-Ups

1. What experiment suggested that atoms are not solid but that they may have a hard nucleus encircled by negative electrons?
2. A 5-mW laser pointer beam has a diameter of 3 mm. The wavelength of laser light is 680 nanometers. Estimate the number of photons per second that hit the laser spot on the wall.
3. Niels Bohr suggested that an atom emits light as an electron, orbiting the atom, moves closer to the nucleus. How did Bohr explain the fact that spectra of chemical elements (such as hydrogen) consists of discrete colors rather than the whole continuous spectrum?
4. The human retina can respond to the light energy of a single photon! The wavelength of visible light is of the order of 500 nm. Compare the energy of a photon of visible light to that of a 100-g rifle bullet traveling at the speed of sound (340 m/s).

Chapter Review

In this chapter we discuss the ideas that make up our modern understanding of the atoms. We focus primarily on the simplest atom, hydrogen. The ideas discussed in this chapter are the foundation of what is called *quantum physics*. The development of quantum physics (or *quantum mechanics*) revolutionized physics (and all physical sciences) in the 1900s.

31–1 Early Models of the Atom

The concept of the atom was originally proposed as the indivisible, fundamental quantity out of which things are made. The 1897 discovery of the electron, by J. J. Thomson, changed this belief by revealing that atoms must have internal structure. Thomson proposed what is called the "plum-pudding model" of the atom in which negatively charged electrons are embedded in a nearly uniform distribution of positively charged matter.

A new picture of atomic structure emerged after Ernest Rutherford scattered positively charged *alpha particles* from a thin gold foil. The results of Rutherford's experiments suggested that the structure of atoms is more like a miniature solar system rather than a plum pudding. Rutherford's model placed all the positive charge in an atom, and most of its mass, at a small, central location, called the **nucleus**, with the negatively charged electrons moving around the nucleus in orbits, similar to the planets orbiting the Sun. While Rutherford's model seemed more reasonable based on the scattering experiments, it was not consistent with experiments on the light given off by atoms, nor was it a stable structure according to Maxwell's electromagnetic theory.

31–2 The Spectrum of Atomic Hydrogen

By applying a large potential difference across a tube containing a low-pressure atomic gas causes the atoms of the gas to give off light. Passing this light through a diffraction grating separates the light into different wavelengths, producing a **line spectrum**. The process just described produces an *emission spectrum*. However, when light consisting of different wavelengths is passed through a gas, some of the wavelengths from this light will be absorbed by the gas, and the resulting light can then produce a line spectrum that is an *absorption spectrum*. Each atom has its own unique emission/absorption spectrum.

For hydrogen the wavelengths that make up its emission spectrum are given by the formula

$$\frac{1}{\lambda} = R\left(\frac{1}{n'^2} - \frac{1}{n^2}\right) \qquad \begin{array}{l} n' = 1, 2, 3, \ldots \\ n = n'+1, n'+2, n'+3, \ldots \end{array}$$

The constant R, called the *Rydberg constant*, is $R = 1.097 \times 10^7\ m^{-1}$. The best known wavelengths of this spectrum are those from the *Lyman series* (n′ = 1), the *Balmer series* (n′ = 2), and the *Paschen series* (n′ = 3); however, there are an infinite number of series because there is no upper limit on n′.

Exercise 31–1 The Paschen Series Determine the range of wavelengths that make up the Paschen series of the hydrogen spectrum.

Solution

The Paschen series has $n' = 3$. Therefore, the values of n range from 4 to ∞. For $n = 4$ we have

$$\frac{1}{\lambda_4} = R\left(\frac{1}{3^2} - \frac{1}{4^2}\right) = \left(1.097 \times 10^7 \text{ m}^{-1}\right)\left(\frac{1}{3^2} - \frac{1}{4^2}\right) = 5.333 \times 10^5 \text{ m}^{-1}$$

$$\Rightarrow \lambda_4 = 1.875 \times 10^{-6} \text{ m}$$

For $n = \infty$ we have $1/n = 0$. Therefore,

$$\frac{1}{\lambda_\infty} = R\left(\frac{1}{3^2} - \frac{1}{\infty}\right) = \left(1.097 \times 10^7 \text{ m}^{-1}\right)\left(\frac{1}{3^2} - 0\right) = 1.219 \times 10^6 \text{ m}^{-1}$$

$$\Rightarrow \lambda_\infty = 8.203 \times 10^{-7} \text{ m}$$

Can you determine to which parts of the electromagnetic spectrum these wavelengths belong?

Practice Quiz

1. Which of the following is not a wavelength in the Balmer series?

   (a) 656.3 nm (b) 486.2 nm (c) 364.6 nm (d) 434.1 nm (e) 541.4 nm

31–3 – 31–4 Bohr's Model of the Hydrogen Atom and de Broglie Waves

To deal with the inconsistency between Rutherford's model and the experiments on the spectrum of hydrogen, Bohr refined Rutherford's model with four assumptions:

(1) The electron in a hydrogen atom moves in a circular orbit around the nucleus.

(2) The only circular orbits that are allowed are those for which the angular momentum of the electron is given by $L_n = n\frac{h}{2\pi}$, for n = 1, 2, 3, ..., where h is Planck's constant.

(3) Electrons do not give off electromagnetic radiation while in an allowed orbit.

(4) Electromagnetic radiation is given off, or absorbed, by an atom only when an electron changes from one allowed orbit to another. The frequency of the single photon that is emitted or absorbed is $\Delta E = hf$, where ΔE is the energy difference between the two allowed orbits.

Use of the centripetal force for the circular motion from Coulomb's law leads to an expression for the radii of the allowed orbits in a hydrogen atom, often called **Bohr orbits**,

$$r_n = \left(\frac{h^2}{4\pi^2 mke^2}\right)n^2, \quad n = 1,2,3,...$$

where m is the mass of an electron, e is the elementary charge, and $k = \frac{1}{4\pi\varepsilon_0}$ from Coulomb's law. The speed with which an electron moves in a Bohr orbit is given by

$$v_n = \frac{2\pi ke^2}{nh}, \quad n = 1,2,3,...$$

The above results can be extended to any single-electron atom, such as singly ionized helium and doubly ionized lithium, by accounting for the charge of the additional protons in the nucleus. In general, we take a nucleus to contain Z protons giving it a charge of +Ze; the quantity Z is called the **atomic number** of the nucleus. The above results for r_n and v_n then become valid for single-electron atoms with heavier nuclei if we replace e^2 by Ze^2.

Each Bohr orbit has a certain amount of energy given by the sum of the kinetic and electric potential energies of the electron. For single-electron atoms the total energies of the orbits are given by

$$E_n = -\left(\frac{2\pi^2 mk^2e^4}{h^2}\right)\frac{Z^2}{n^2} = -(13.6\,\text{eV})\frac{Z^2}{n^2}, \qquad n = 1,2,3...$$

The lowest possible energy of the atom corresponds to n = 1 and is called the **ground state** of the atom. Orbits with energies higher than the ground state (n > 1) are called **excited states**. Using the above energy states, Bohr's model predicts an emission spectrum for hydrogen (Z = 1) that is described by the equation

$$\frac{1}{\lambda} = \left(\frac{2\pi^2 mk^2e^4}{h^3c}\right)\left(\frac{1}{n_f^2} - \frac{1}{n_i^2}\right)$$

where c is the speed of light. This result is consistent with the empirical result quoted previously, and in addition, it tells us from what quantities the Rydberg constant R is derived.

In 1923 deBroglie provided some physical insight into Bohr's model by showing that Bohr's condition on the allowed angular momenta of the orbiting electrons was equivalent to a standing wave condition for the electrons' matter waves: $n\lambda = 2\pi r_n$. This success helped the idea of matter waves to be taken more seriously. The properties of matter waves are determined by what is called **Schrodinger's equation**; this equation is the fundamental equation of quantum mechanics. Today, the most common view of matter waves is that the amplitude of a particle's matter wave, at a given location and at a given time, is related to the probability that the particle is located at that point at that time.

Example 31–2 Ionized Helium What is the energy of the 1st excited state of singly ionized helium?

Picture the Problem The sketch is of the first two Bohr orbits around the nucleus of a singly ionized helium atom (not to scale).

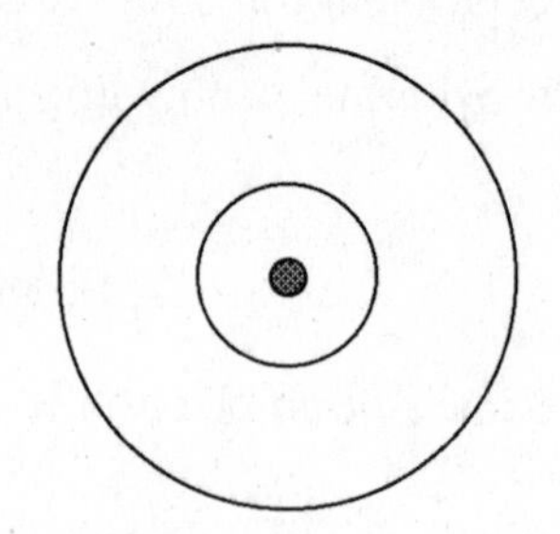

Strategy We use the more general expression for the allowed energies of single-electron atoms.

Solution

1. The atomic number of helium is: $Z = 2$

2. The 1st excited state is: $n = 2$

3. The energy then is: $E_2 = -(13.6\,\text{eV})\dfrac{Z^2}{n^2} = -(13.6\,\text{eV})\dfrac{2^2}{2^2} = -13.6\,\text{eV}$

Insight This is the same energy as the ground state of hydrogen, but for a very different situation. The energy is negative because, as with planets orbiting the Sun, the electron and proton form a bound system, so that the (negative) potential energy "overpowers" the kinetic energy.

Practice Quiz

2. According to Bohr's model of hydrogen, what is the energy of an electron in the n = 3 excited state?
   (a) –6.04 eV **(b)** –1.51 eV **(c)** –13.6 eV **(d)** –0.85 eV **(e)** –4.53 eV

3. According to Bohr's model of hydrogen, what is the radius of the orbit of an electron in the n = 3 excited state?
   (a) 5.29×10^{-11} m **(b)** 1.59×10^{-10} m **(c)** 4.76×10^{-10} m **(d)** 1.06×10^{-10} m **(e)** 2.38×10^{-10} m

31–5 The Quantum Mechanical Hydrogen Atom

Aside from relativistic effects, the results from solving Schrodinger's equation for the hydrogen atom produce the best model of this atom that we have. From this analysis we now describe the hydrogen atom using four parameters called *quantum numbers*.

The principal quantum number, n

The principal quantum number takes on all integer values, n = 1, 2, 3, ..., and determines the total energy of a given state. This total energy is given by, approximately, the same result produced by Bohr's model:

$$E_n = -(13.6\,\text{eV})\frac{Z^2}{n^2}$$

The orbital angular momentum quantum number, ℓ

For an electron in a state of principal quantum number n, the orbital angular momentum of the electron can have only certain values as determined by the orbital angular momentum quantum number, ℓ, which takes on the values

$$\ell = 0, 1, 2, \ldots, (n-1)$$

The magnitude of the angular momentum of an electron with a given value of ℓ is

$$L = \sqrt{\ell(\ell+1)}\frac{h}{2\pi}$$

Notice that an orbiting electron is allowed to have an angular momentum of zero.

The magnetic quantum number, m_ℓ

When a hydrogen atom is in an external magnetic field, the allowed energy values of the electron, with specific values of n and ℓ, depend on an additional quantum number, m_ℓ. The range of values for this quantum number is from $-\ell$ to $+\ell$ in increments of 1:

$$m_\ell = -\ell, -\ell+1, -\ell+2, \ldots -1, 0, 1, \ldots, \ell-2, \ell-1, \ell$$

This magnetic quantum number specifies the component of the orbital angular momentum along a single direction; this direction is usually chosen to be the z-axis. The result is

$$L_z = m_\ell \frac{h}{2\pi}$$

One of the new consequences of the complete quantum theory is that only one component of the angular momentum can be known precisely at a given time.

The electron spin quantum number, m_s

This last quantum number results from the fact that electrons contain intrinsic angular momentum that is a property of the electron itself (like its mass and charge). This type of angular momentum is called the spin angular momentum of the electron. The quantum number m_s takes on two possible values:

$$m_s = -\frac{1}{2},\ +\frac{1}{2}$$

When the values of all four quantum numbers are specified we call that a particular **state** of the hydrogen atom. Notice that several different states can have the same energy. The solution of the Schrodinger equation suggests that when the atom is in a particular state we should think of the matter wave of the electron in terms of a three-dimensional **probability cloud** instead of a specific path as suggested by Bohr's model. The probability cloud means that at any given time the electron has some finite probability of being anywhere around the nucleus.

Example 31–3 The Ground State of Hydrogen For an electron in the ground state of hydrogen find **(a)** the total energy, and **(b)** the magnitude of the orbital angular momentum. **(c)** Since it is possible to specify the value of only one component of the orbital angular momentum, what is that possible value for this atom?

Picture the Problem The sketch shows the Bohr orbit around the nucleus of the electron in a hydrogen atom.

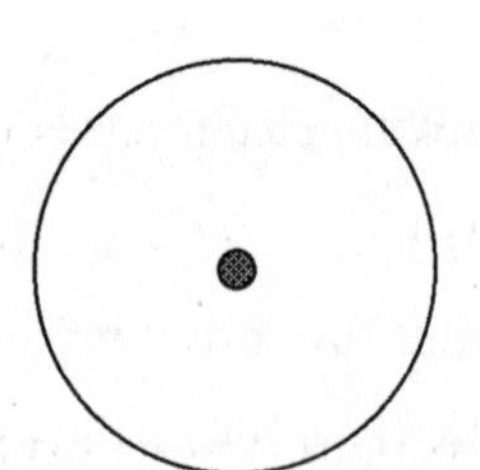

Strategy We must deduce the values of the relevant quantum numbers and use the above expressions to calculate the desired quantities.

Solution

Part (a)

1. The ground state of hydrogen has Z = n = 1, so the total energy is: $E_1 = -(13.6\,\text{eV})\frac{1^2}{1^2} = -13.6\,\text{eV}$

Part (b)

2. The range of the orbital angular momentum quantum number is: $\ell = 0\ldots(n-1) \Rightarrow \ell = 0$

3. The magnitude of the orbital angular momentum then is: $L = \sqrt{\ell(\ell+1)}\frac{h}{2\pi} = 0 \cdot \frac{h}{2\pi} = 0$

Part (c)

4. The range of the magnetic quantum number is: $m_\ell = -\ell, \ldots, +\ell \Rightarrow m_\ell = 0$

5. The "z-component" of the orbital angular momentum is: $L_z = m_\ell \frac{h}{2\pi} = 0 \cdot \frac{h}{2\pi} = 0$

Insight In part (b), L = 0 may seem odd for an orbiting electron, but keep in mind that the electron spends time in all directions at all distances. In part (c) the term "z component" is in quotes because it could really be any component. Since the atom is not in a magnetic field, no one component that we can call "z" is picked out.

Practice Quiz

4. How many possible values are there for ℓ when n = 4?

(a) 0 (b) 1 (c) 2 (d) 3 (e) 4

5. How many possible values are there for m_s when n = 4?

(a) 0 (b) 1 (c) 2 (d) 3 (e) 4

31–6 Multi-Electron Atoms and the Periodic Table

Unlike with hydrogen, the energies of the states of multi-electron atoms depend on both n and ℓ. Electrons that have the same value of n are said to be in the same **shell**. The case of n = 1 is called the K shell, n = 2 is called the L shell, and so on, in alphabetical order. Electrons with the same value of ℓ are in the same subshell. The case $\ell = 0$ is called the s subshell, $\ell = 1$ is the p subshell, $\ell = 2$ is the d subshell, and $\ell = 3$ is the f subshell. After the f subshell the names continue in alphabetical order.

Atoms naturally exist in their ground states unless some interactions are present to excite them. Finding the ground-state structure of multi-electron atoms requires what is known as the **Pauli exclusion principle**. This principle states that

only one electron can occupy a given state of an atom.

This means that no two electrons can have the same four quantum numbers. Because of this principle, the ground state of an atom is obtained by filling up the allowed energy states, starting with the lowest (n = 1), until all the electrons are accounted for (see Example 31–4 below).

The **electronic configuration** of an atom is the specification of how many electrons exist in certain subshells. A particular notation for this specification lists the value of n, followed by the name of the subshell with the number of electrons in that subshell as a superscript. For example, the ground state of helium consists of two electrons each with $n = 1$, $\ell = 0$, and $m_\ell = 0$, and one electron with $m_s = 1/2$, and the other with $m_s = -1/2$. The electronic configuration of this atom is $1s^2$. The electronic configuration of atoms leads to a better understanding of the **periodic table** of the elements. This table is a grouping of the chemical elements according to their properties. It is now known that elements with similar chemical properties correspond to elements with similar outer electron configurations.

Example 31–4 The Ground State of Boron The boron atom has an atomic number of 5; write out the electronic configuration of its ground state.

Solution

A neutral boron atom will have the same number of electrons as protons. Since its nucleus has an atomic number of 5, it also has 5 electrons surrounding the nucleus. For the n = 1 shell, ℓ can only be zero, corresponding to the s subshell. For the electrons in this 1s subshell, one can have $m_s = -1/2$ and the other can have $m_s = +1/2$ which places two electrons in this subshell; so the configuration starts with $1s^2$.

For the n = 2 shell, ℓ can only be either 0 or 1. There will be another two electrons in the s-subshell ($\ell = 0$) for the same reason as before; so we write $2s^2$ for the configuration of these electrons. This leaves one electron unaccounted for. This last electron goes into the p subshell ($\ell = 1$). Therefore, the electronic configuration of the ground state of boron is

$$1s^2 2s^2 p^1$$

Notice that the principal quantum number, n = 2, is only indicated once for both the s and p subshells. (However, you should be aware that some authors would write this as $1s^2 2s^2 2p^1$).

Practice Quiz

6. How many electrons can be placed in the L shell?
   (a) 2 (b) 4 (c) 6 (d) 8 (e) 10

7. How many electrons can be placed in the d subshell?
   (a) 2 (b) 4 (c) 6 (d) 8 (e) 10

8. Which of the following is the electronic structure of the ground state of nitrogen?
   (a) $1s^2 2s^2 p^1$ (b) $1s^2 2s^2 p^2$ (c) $1s^2 2s^2 p^3$ (d) $1s^2 2s^2 p^4$ (e) $1s^2 2s^2 p^5$

31–7 Atomic Radiation

The spectra of multi-electron atoms are more complicated than the spectrum of hydrogen. Three types of atomic radiation that are of practical use are discussed in this section. X-rays are high-energy photons that are often produced in tubes, called X-ray tubes, in which electrons are accelerated to high speeds and collided into a target. Most of the X-rays, called **bremsstrahlung**, are generated by the rapid deceleration of the electrons. Some of the X-rays are produced because of transitions of orbital electrons in the target atoms which occur because the collision between the incident electrons and the target removes some of the lower-energy electrons in target atoms. When the states vacated by these electrons are filled, X-ray photons are given off.

The light given off by a **laser** is also due to interactions at the atomic level. With laser light, excited atoms are stimulated to emit radiation by incident light. The emitted photons then stimulate other excited atoms to emit photons, and so on. All these photons have the same energy and phase, and move in the same direction. This process is why they are called lasers, which is an acronym for **L**ight **A**mplification by the **S**timulated **E**mission of **R**adiation.

Fluorescence occurs when atoms are illuminated by light and the energy of that light excites the atoms into a higher state. Once the illumination ceases, the electrons in the atoms spontaneously fall back to lower energy levels giving off (fluorescent) light of lower frequency than the light used to illuminate the atoms. A related phenomenon is **phosphorescence** in which the material continues to glow long after the original illumination ceases.

Reference Tools and Resources

I. Key Terms and Phrases

nucleus the central region of an atom that contains most of its mass and all of its positive charge

line spectrum the dicrete spectrum of the wavelengths of light given off by atoms

Bohr orbit the circular orbits of electrons around the nucleus of an atom in Bohr's model

atomic number the number of protons in the nucleus of an atom

ground state the case when the electrons in an atom are in their lowest possible energy levels

excited state the case when one or more electrons in an atom are in energy levels above the ground state

principal quantum number an integer value that determines the total energy of a state of hydrogen and sets the boundaries for the values of other quantum numbers

orbital angular momentum quantum number an integer value that determines the orbital angular momentum of an electron in an atom and sets the boundaries for the magnetic quantum number

magnetic quantum number a quantity of integer value that determines the possible values of a component of the orbital angular momentum of an electron in an atom

electron spin quantum number a value that determines the intrinsic angular momentum of an electron

state of an atom the specification of the quantum numbers for every electron in an atom

probability cloud the interpretation of matter waves for the electrons around the nucleus of an atom

shell electrons with the same the principal quantum number are in the same shell

subshell specification of the orbital angular momentum quantum number

Pauli exclusion principle the principle that only one electron can occupy a given state in an atom

electronic configuration specification of n, ℓ, and the number of electrons in each subshell

periodic table a table of the elements that organizes them by their chemical properties

laser light amplification by the stimulated emission of radiation

II. Important Equations

Name/Topic	Equation	Explanation
Bohr's model	$r_n = \left(\frac{h^2}{4\pi^2 mkZe^2}\right)n^2$ $n = 1,2,3,\ldots$	The radii of the Bohr orbits of single-electron atoms.
	$E_n = -(13.6\,\text{eV})\frac{Z^2}{n^2}$ $n = 1,2,3\ldots$	The allowed energy levels of single-electron atoms in Bohr's model.
hydrogen spectrum	$\frac{1}{\lambda} = \left(\frac{2\pi^2 mk^2e^4}{h^3c}\right)\left(\frac{1}{n_f^2} - \frac{1}{n_i^2}\right)$	The wavelengths in the spectrum of hydrogen.
orbital angular momentum	$L = \sqrt{\ell(\ell+1)}\frac{h}{2\pi}$	The magnitude of the orbital angular momentum from the associated quantum number.

magnetic quantum number	$L_z = m_\ell \frac{h}{2\pi}$	The magnetic quantum number specifies one component of the orbital angular momentum.

III. Know Your Units

Quantity	Dimension	SI Unit
Rydberg constant (R)	$[L^{-1}]$	m^{-1}

IV. Tips

The radius of the smallest Bohr orbit, for n = 1, is called the Bohr radius; it is typically denoted as a_0 and has a value of

$$a_0 = 5.29 \times 10^{-11} \text{ m}$$

It is often convenient to write the expressions for the radii of the Bohr orbits, and their energies, in terms of a_0. These expressions become

$$r_n = n^2 \frac{a_0}{Z}$$

$$E_n = -\frac{ke^2}{2a_0} \frac{Z^2}{n^2}$$

where, again, n = 1, 2, 3, … .

Practice Problems

1. In the Lyman series what is the wavelength of the n = 9 spectral line, to the nearest hundredth of a nanometer?
2. What is the ratio of v/c to 5 decimal places, of an electron in the n = 8 Bohr orbit?
3. The element Z = 5 has all but one of its electrons removed. To the nearest eV, what is the energy of the n = 6 state of this electron?
4. What is the energy of the photon emitted in the n = 5 to the n = 3 transition in hydrogen, to the nearest hundreth of an eV?
5. What is the wavelength of the photon in problem 4, to the nearest hundredth of a nanometer?
6. What is the frequency of the photon necessary to raise a hydrogen electron from the n = 2 to the n = 5 level, to the nearest THz (10^{12} Hz)?
7. What is the de Broglie wavelength of an electron traveling at 2.6 x 10^6 m/s, to the nearest hundredth of a nanometer?

8. What is the maximum angular momentum of a hydrogen electron in the n = 8 state in units of 10^{-34} J·s, to two decimal places?
9. What is the maximum value for the z component of the angular momentum in problem 8?
10. What is the energy, to the nearest eV, of a K_α X-ray for element with atomic number Z = 27?

Puzzle

ISN'T THIS BACKWARD?

According to classical mechanics, as an electron shifts from an outer orbit to an inner orbit its speed has to increase. Thus, the kinetic energy of the electron has to increase. Yet, according to Bohr, during such a process a photon will take away the excess energy. Where does the excess energy come from?

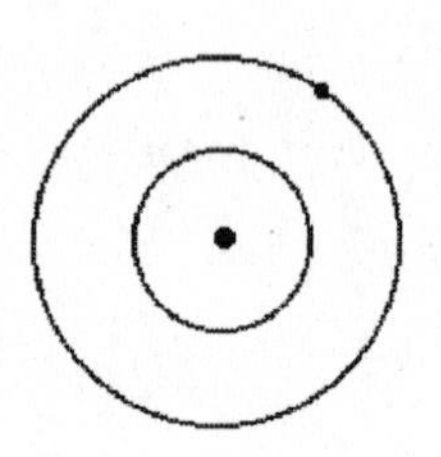
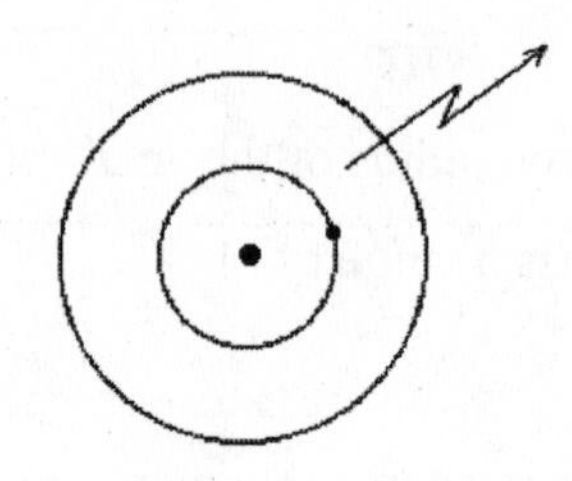

Selected Solutions

7. For the Lyman series we have: $\frac{1}{\lambda} = R\left(\frac{1}{1^2} - \frac{1}{n^2}\right) \Rightarrow \lambda = \frac{1}{R\left(1 - \frac{1}{n^2}\right)}$

The last form of the equation makes clear that the longest wavelength corresponds to the smallest n. Since $R = 1.097 \times 10^7\ \text{m}^{-1}$, we have

$$\lambda_2 = \frac{1}{R\left(1-\frac{1}{2^2}\right)} = \boxed{121.5\ \text{nm}}$$

$$\lambda_3 = \frac{1}{R\left(1-\frac{1}{3^2}\right)} = \boxed{102.6\ \text{nm}}$$

$$\lambda_4 = \frac{1}{R\left(1-\frac{1}{4^2}\right)} = \boxed{97.23\ \text{nm}}$$

19. **(a)** The photon's energy must equal the energy difference between the two states. Since the energy of a photon is given by $E = h\nu = \frac{hc}{\lambda}$, we can use the expression for the wavelengths to get

$$|\Delta E| = \left|Rhc\left(\frac{1}{n_f^2} - \frac{1}{n_i^2}\right)\right| = (1.097\times10^7\ \text{m}^{-1})(6.626\times10^{-34}\ \text{J}\cdot\text{s})\left(3.00\times10^8\ \frac{\text{m}}{\text{s}}\right)\left(\frac{1}{3^2} - \frac{1}{5^2}\right) = \boxed{1.55\times10^{-19}\ \text{J}}$$

(b) The energy of the photon will be $\boxed{\text{less than}}$ that found in part (a), because the absolute values of the lower n energy-state differences are greater than those of the higher n energy-state differences, and in both cases $\Delta n = 2$.

(c) $|\Delta E| = (1.097 \times 10^{7}\ \text{m}^{-1})(6.626 \times 10^{-34}\ \text{J}\cdot\text{s})\left(3.00 \times 10^{8}\ \frac{\text{m}}{\text{s}}\right)\left(\frac{1}{5^2} - \frac{1}{7^2}\right) = \boxed{4.27 \times 10^{-20}\ \text{J}}$

33. In the 3s subshell, $n = 3$, $\ell = 0$, $m_\ell = 0$, and $m_s = \pm\frac{1}{2}$.

n	ℓ	m_ℓ	m_s
3	0	0	$-\frac{1}{2}$
3	0	0	$\frac{1}{2}$

41. For each ℓ, there are $2(2\ell + 1)$ states.

(a) $n = 2 \;\therefore\; \ell = 0, 1$

total number of states $= 2[2(0) + 1] + 2[2(1) + 1] = 2 + 6 = \boxed{8}$

(b) $n = 3 \;\therefore\; \ell = 0, 1, 2$

total number of states $= 2[2(0) + 1] + 2[2(1) + 1] + 2[2(2) + 1] = 2 + 6 + 10 = \boxed{18}$

(c) $n = 4 \;\therefore\; \ell = 0, 1, 2, 3$

total number of states $= 18 + 2[2(3) + 1] = 18 + 14 = \boxed{32}$

59. (a) Using the expression for the radius of the orbit, we get

$$r_6 = \frac{36h^2}{4\pi^2 mkZe^2} \quad\Rightarrow\quad Z = \frac{36h^2}{4\pi^2 mke^2 r_6}$$

Therefore,

$$Z = \frac{36(6.626 \times 10^{-34}\ \text{J}\cdot\text{s})^2}{4\pi^2(9.11 \times 10^{-31}\ \text{kg})\left(8.99 \times 10^{9}\ \frac{\text{N}\cdot\text{m}^2}{\text{C}^2}\right)(1.602 \times 10^{-19}\ \text{C})^2(2.72 \times 10^{-10}\ \text{m})} = \boxed{7.00}$$

(b) $E_3 = -(13.6\ \text{eV})\frac{Z^2}{3^2} = -(13.6\ \text{eV})\frac{(7.00)^2}{9} = \boxed{-74.0\ \text{eV}}$

Answers to Practice Quiz

1. (e) 2. (b) 3. (c) 4. (e) 5. (c) 6. (d) 7. (e) 8. (c)

Answers to Practice Problems

1. 92.30 nm	**2**. 0.00091
3. −9 eV	**4**. 0.97 eV
5. 1285.41 nm	**6**. 689 THz
7. 0.28 nm	**8**. 7.90
9. 8.44	**10**. 6895 eV

CHAPTER 32

NUCLEAR PHYSICS AND NUCLEAR RADIATION

Chapter Objectives

After studying this chapter, you should:

1. know the basic constituents of the nucleus.
2. be familiar with the three processes of radioactive decay.
3. understand the exponential behavior of radioactive samples and know how radioactive dating works.
4. understand the effects of nuclear binding energy and why it leads to nuclear fission and fusion as energy sources.
5. know the difference between nuclear fission and nuclear fusion.
6. be aware of several practical applications of nuclear physics.

Warm-Ups

1. What aspect of short-lived radioactive substances make them a health hazard? What aspect of long-lived radioactive substances make them a health hazard?
2. The disintegration constant of technetium-99, an important medical tracer, is 10^{-13} per second. Estimate the half-life of technetium-99.
3. The activity of a radioactive sample depends on
    a. the property of the radioactive material and
    b. the size of the sample.

   The material is characterized by the disintegration (or decay) constant, which states what fraction of a sample will undergo the process per unit time. The disintegration constant of radioactive carbon-14 is equal to 3.84×10^{-12} per second. Based on that information, how many counts per minute would you expect for a mole of carbon-14? Explain your answer. (*Reminder*: A mole consists of 6×10^{23} items.)
4. The half-life of carbon-14 is 5730 years. Living organisms, such as plants, continuously replenish carbon-14 in their system from the atmosphere. After they die, the carbon-14 content no longer is maintained. You are given a piece of wood whose carbon-14 activity is 25% of what you observe in

living wood. How long has the wood been dead? (Explain in plain language how you obtained your answer by simple reasoning.)

Chapter Review

This final chapter primarily deals with nuclear and subnuclear physics. In this chapter we touch on topics at the very forefront of human understanding. Nuclear physics is very important in everyday life from the standpoint of energy generation and even medical applications. The chapter concludes with a brief discussion of gravity waves.

32–1 The Constituents and Structure of Nuclei

The nuclei of atoms consist of **protons** and **neutrons**; collectively these particles are called **nucleons**. The number of protons in a nucleus is called the **atomic number**, Z; the number of neutrons is called the **neutron number**, N. The total number of nucleons is called the **mass number**, A. Clearly,

$$A = Z + N$$

The chemical element to which a nucleus belongs is determined by the value of Z. Both Appendices D and E in your textbook list the atomic number of the chemical elements.

The notation used to specify the composition of the nucleus of a chemical element X is

$${}^{A}_{Z}X$$

Sometimes Z is omitted because the value of Z is specified by the chemical element X. Even though Z is the same for every nucleus of a certain chemical element, the number of neutrons may not be the same. Nuclei having the same Z but different values of N are said to be different **isotopes** of the same nucleus. For example, ${}^{1}_{1}H$ and ${}^{3}_{1}H$ are different isotopes of the hydrogen nucleus with 0 neutrons (A = 1) and 2 neutrons (A = 3), respectively.

The masses of nuclei are often quoted in terms of the **atomic mass unit**, u. By definition,

$$1\ u = 1.660540 \times 10^{-27}\ kg$$

The mass of an atom, quoted in this unit, is often called its atomic mass. Appendix E in your textbook gives the atomic masses of many common isotopes. Because of the equivalence between mass and energy, $E = mc^2$, the mass of a nucleus is sometimes given in units of E/c^2: $1\ u = 931.494\ MeV/c^2$. The size of a nucleus can be estimated by the empirical relationship

$$r = \left(1.2 \times 10^{-15}\ m\right) A^{1/3}$$

where A is the mass number. As this relationship shows, the radius of a nucleus is typically on the order of 10^{-15} m; this distance is called the **fermi** (fm).

A nucleus may contain many protons a very small distance apart. This situation leads to large electrostatic repulsion among the protons. Holding the nucleus together, against this repulsive force, is the **strong nuclear force**. This force is a fundamental force of nature that (a) is short range, acting only over a distance of a couple of fermi and (b) is attractive, acting with nearly equal strength among all nucleons. Since neutrons experience the strong nuclear force but do not experience electrostatic repulsion (being electrically neutral), their presence in a nucleus helps to stabilize the nucleus (hold it together). The most stable nuclei are those with nearly equal numbers of protons and neutrons ($N \approx Z$). The more protons in a nucleus, the less stable it is; no nucleus with more than $Z = 83$ protons is stable.

Exercise 32–1 A Lithium Isotope Estimate the mass, in u, and the radius, in fm, of the isotope ${}^{7}_{3}\text{Li}$.

Solution

Appendix E in the textbook gives the atomic mass of a neutral lithium-7 atom to be 7.016005 u. To estimate the mass of the nucleus we must subtract the mass of the $Z = 3$ electrons surrounding it. The mass of an electron is

$$m_e = 9.1094 \times 10^{-31}\ \text{kg}\left(\frac{1\ \text{u}}{1.660540 \times 10^{-27}\ \text{kg}}\right) = 5.4858 \times 10^{-4}\ \text{u}$$

Therefore,

$$M_{\text{Li-7}} = 7.016005\ \text{u} - 3\left(5.4858 \times 10^{-4}\ \text{u}\right) = 7.0144\ \text{u}$$

The given lithium isotope has $A = 7$. Therefore,

$$r_{\text{Li-7}} = \left(1.2 \times 10^{-15}\ \text{m}\right)A^{1/3} = \left(1.2 \times 10^{-15}\ \text{m}\right)7^{1/3} = 2.3 \times 10^{-15}\ \text{m} = 2.3\ \text{fm}$$

Practice Quiz

1. A particular isotope of carbon has 13 nucleons. How many neutrons are in the nucleus?

   (a) 5 (b) 6 (c) 7 (d) 8 (e) none of the above

2. Approximately by what factor is the radius of ${}^{16}_{8}\text{O}$ greater than that of ${}^{8}_{4}\text{Be}$?

   (a) 2 (b) 8 (c) 1.26 (d) 1.59 (e) 1.41

32–2 Radioactivity

An unstable nucleus will disintegrate into a different nucleus; when it does it will emit one or more particles. Also, a nucleus in an excited state can make a transition to a lower energy state and emit a high-energy photon. Both of these processes are referred to as *nuclear decay,* and the emission that occurs is called **radioactivity**; therefore, nuclear decay is often called **radioactive decay**.

The main types of particles emitted during radioactive decay are:

* *alpha particles* (α), are helium nuclei ^4_2He ; these emitted particles are referred to as α-rays.
* *electrons*, called β^--rays when emitted by nuclei.
* *positrons*, which are anti-electrons having the same mass as an electron but opposite charge, called β^+-rays when emitted by nuclei.
* *gamma rays* (γ), which are high-energy photons emitted when an excited nucleus decays to a lower energy state.

Radioactive decay that emits an alpha particle is called **alpha decay**. The initial unstable nucleus is called the **parent nucleus**, and the final nucleus is called the **daughter nucleus**. The daughter nucleus will have two fewer protons and two fewer neutrons than the parent nucleus. If the X represents the unstable parent, and Y is the daughter, we can write this process as

$$^A_ZX \rightarrow {}^{A-4}_{Z-2}Y + {}^4_2He$$

Notice that the atomic and mass numbers on the left-hand side equals the sum of the corresponding atomic and mass numbers on the right-hand side.

Radioactive decay that emits a β-ray (either β^+ or β^-) is called **beta decay**. In the beta decay emission of an electron, the basic process is that a neutron decays into a proton and an electron. Hence, the mass number remains the same, but the atomic number changes (increases) by 1,

$$^A_ZX \rightarrow {}^{A}_{Z+1}Y + e^-$$

The process of β^+ decay is more complicated to explain, but a similar result applies except that the atomic number decreases by 1,

$$^A_ZX \rightarrow {}^{A}_{Z-1}Y + e^+$$

When radioactive decay occurs, the total mass of the decay products (the final nucleus + emitted particles) is less than the mass of the initial nucleus. The difference in mass, Δm, results in a release of energy in the amount

$$E = |\Delta m| c^2$$

This fact can be used to predict how much kinetic energy the electron or positron should have as a result of beta decay. When the measured kinetic energies of the decay products in beta decay did not satisfy the conservation of energy, it was determined that another particle must be given off. This particle is called a **neutrino**. Neutrinos are very weakly interacting particles. A neutrino (ν) is given off in β^+ decay, and an antineutrion ($\bar{\nu}$) is given off during β^- decay.

Radioactive decay that emits a gamma ray photon (γ) is called **gamma decay**. This process occurs when a nucleus in an excited state decays to a lower energy state. An excited nucleus is indicated by placing an asterisk as a superscript on the symbol. Thus, we have

$$^{A}_{Z}X^{*} \rightarrow {}^{A}_{Z}X + \gamma$$

Both A and Z remain the same during gamma decay.

The rate at which nuclear decay takes place is called the **activity**, R. A common unit of measure for activity is the **curie** (Ci) which is defined as

$$1\ \text{Ci} = 3.7 \times 10^{10}\ \text{decays/s}$$

The SI unit of activity is the **becquerel** (Bq), which is defined as 1 Bq = 1 decay/s. Most commonly, activities are measured in millicuries (mCi) and microcuries (μCi).

Exercise 32–2 The Beta Decay of Carbon: Estimate the energy released when carbon-14 undergoes β^- decay.

Picture the Problem In the diagram, the completely filled circles represent neutrons and the others are protons. The diagram shows the beta decay of a carbon-14 isotope.

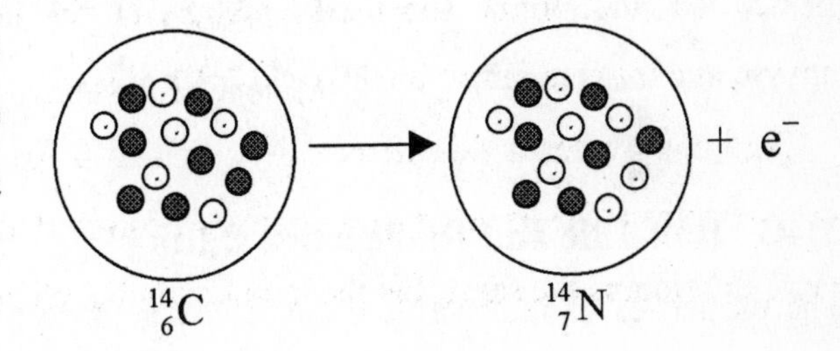

Strategy By comparing the masses of carbon-14 to that of nitrogen-14 + an electron we can estimate the energy released from the mass difference.

Solution

1. We can get the mass of a carbon-14 nucleus from Appendix E in the text: $M_C = 14.003242\ \text{u} - 6(5.4858 \times 10^{-4}\ \text{u}) = 13.999951\ \text{u}$

2. We can get the mass of a nitrogen-14 nucleus from Appendix E in the text: $M_N = 14.003074\ \text{u} - 7(5.4858 \times 10^{-4}\ \text{u}) = 13.999234\ \text{u}$

3. The mass difference is:

$$|\Delta M| = M_C - (M_N + m_e) = 13.99995\,u - (13.999783\,u)$$
$$= 0.000168\,u\left(\frac{931.494\,MeV/c^2}{u}\right) = 0.15649\,MeV/c^2$$

4. The energy released is:

$$E = |\Delta M|c^2 = \left(0.15649\frac{MeV}{c^2}\right)c^2 = 0.15649\ MeV$$

Insight Notice how many digits we had to keep in order to see the mass difference. Despite having to keep all these decimal places, the amount of energy is considerable. Another approach to doing this type of calculation is sometimes used. If you add the mass of 6 electrons to that of the carbon-14 nucleus, you have the mass of the carbon-14 atom. Also add the mass of 6 electrons to that of the nitrogen-14 nucleus, then, when you add the mass of the emitted electron, you get a total of 7 electrons, making the mass on the right-hand side that of the nitrogen-14 atom. Hence, the mass difference could be determined just from the differences in the atomic masses. This method allows you to do the problem in fewer calculations but the logic is less direct.

Practice Quiz

3. Which of the following represents a valid alpha decay process?

(a) ${}^{A}_{Z}X \rightarrow {}^{A}_{Z+1}Y + \alpha$ (b) ${}^{A}_{Z}X \rightarrow {}^{A}_{Z-1}Y + \alpha$ (c) ${}^{A}_{Z}X \rightarrow {}^{A-1}_{Z}X + \alpha$

(d) ${}^{A}_{Z}X \rightarrow {}^{A-4}_{Z-2}Y + \alpha$ (e) none of the above

31–3 Half-Life and Radioactive Dating

The properties of radioactive decay allow it to be used as a method of dating certain objects. This is because, for a given initial number of radioactive nuclei, N_0, the fraction of nuclei remaining at a given instant, N/N_0, depends, in a known way, on the amount of time t that has elapsed. This results in the well-known exponential decay equation

$$N = N_0 e^{-\lambda t}$$

where λ is called the **decay constant**. A large value of λ indicates a rapid decay process. A quantity that characterizes the speed of the decay process of a substance is its **half-life**. The half-life of a radioactive nucleus is the time interval required for the number of these nuclei to reduce by half. Using the above equation, we can determine that the half-life, $T_{½}$, is given by

$$T_{1/2} = \frac{\ln(2)}{\lambda}$$

The above behavior of radioactive nuclei results from the fact that the activity at a given time is directly proportional to the number of nuclei present at that time, with proportionality factor λ. Therefore, $R = \lambda N$. It then follows that

$$R = R_0 e^{-\lambda t}$$

If we measure the decay constant and present activity of a substance, then determine the initial activity of that substance by some means, we can approximately date the substance by solving the above equation for time:

$$t = -\frac{1}{\lambda}\ln\left(\frac{R}{R_0}\right)$$

In the case of carbon-14 dating of formerly living organisms, R_0 can be determined using the empirical fact that the ratio of carbon-14 to carbon-12 (which is not radioactive) tends to remain constant while organisms are alive. Therefore, we can measure the number of carbon-12 nuclei present today and assume it is the same as in the past. Knowing the ratio of carbon-14 to carbon-12 in living organisms, we can then find N_0 and set $R_0 = \lambda N_0$.

Example 32–3 Radioactive Decay A certain radioactive sample contains 5.00×10^{12} particles at a given instant. If its half-life is 12.2 minutes, how long will it take for 80% of the particles to decay?

Picture the Problem The sketch shows the sample at the initial instant, and after 80% has decayed.

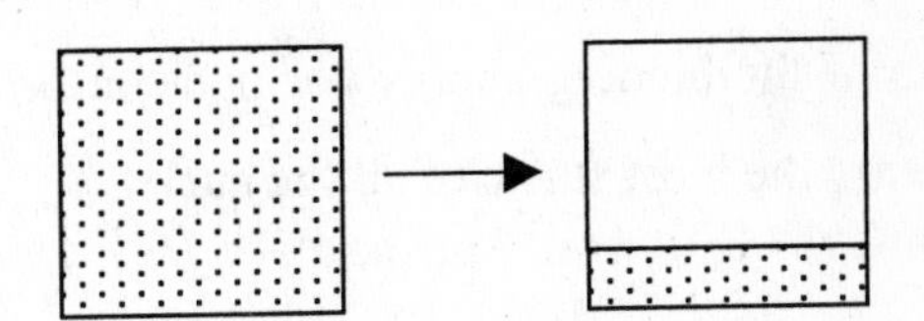

Strategy After 80% decays only 20% is left. We can also use the half-life to determine the decay constant, giving us enough information to solve the problem.

Solution

1. Using that $N = 0.2N_0$, we can solve for t: $\quad 0.2N_0 = N_0 e^{-\lambda t} \Rightarrow t = -\frac{\ln(0.2)}{\lambda}$

2. The decay constant can be written as: $\quad \lambda = \frac{\ln(2)}{T_{1/2}} = \frac{\ln(2)}{12.2\text{ min}} = 0.05682\text{ min}^{-1}$

3. The time for 80% to decay becomes: $\quad t = -\frac{\ln(0.2)}{0.05682\text{ min}^{-1}} = 28.3\text{ min}$

Insight Make sure you are able to solve for t to get the result in step 1.

Practice Quiz

4. A certain radioactive nucleus, A, has a decay constant λ and a half-life T. If a different radioactive nucleus, B, has a decay constant of λ/2, the half-life of B is
   (a) 2T (b) T (c) T/2 (d) 4T (e) none of the above

5. A certain radioactive nucleus, A, has a decay constant λ. A different radioactive nucleus, B, has a decay constant of λ/2. How do the activities compare at an instant when there are an equal number of nuclei of each type?
   (a) $R_B = 2R_A$ (b) $R_A = 2R_B$ (c) $R_A = R_B$ (d) $R_A = (2)^{1/3} R_B$ (e) none of the above

31–4 Nuclear Binding Energy

An interesting fact is that the mass of all stable nuclei with more than one nucleon is less than sum of the masses of the individual nucleons. This phenomenon occurs because of the **binding energy** of the nucleus. The binding energy is a measure of how tightly bound together the nucleons are; its magnitude equals the amount of energy required to break the nucleus apart into its constituent nucleons. This binding energy shows up as a reduction in mass, Δm, by the equivalence of mass and energy, $E = |\Delta m|c^2$. It is also useful to consider the binding energy per nucleon. An analysis of this quantity shows that the largest values of the binding energy per nucleon occur in the mass number range $50 < A < 75$. Nuclei in this range are the most stable of all the nuclei.

Exercise 32–4 Binding Energy Estimate the nuclear binding energy of ${}^{7}_{4}\text{Be}$.

Solution

The binding energy is the difference in mass (energy) between the sum of the masses of the individual nucleons and the mass of the given nucleus. Since Z = 4 and A = 7, the sum of the masses of the nuclei is

$$M_{sum} = 4m_p + 3m_n = 4(938.27\ \text{MeV/c}^2) + 3(939.57\ \text{MeV/c}^2) = 6571.79\ \text{MeV/c}^2$$

From Appendix E, the mass of ${}^{7}_{4}\text{Be}$ is

$$M_{nucleus} = M_{atom} - M_{electrons} = 7.016930\text{u} - 4(5.4858\times10^{-4}\,\text{u})$$

$$= 7.014736\text{u}\left(\frac{931.494\ \text{MeV/c}^2}{\text{u}}\right) = 6534.184\ \text{MeV/c}^2$$

The difference in mass is

$$|\Delta M| = 6571.79 \text{ MeV/c}^2 - 6534.184 \text{ MeV/c}^2 = 37.606 \text{ MeV/c}^2$$

Therefore, the binding energy, which is negative, is – 37.606 MeV.

Practice Quiz

6. Estimate the magnitude of the binding energy of deuterium in MeV/c^2.

   (a) 1.30 (b) 2.23 (c) 1.72 (d) 0.511 (e) none of the above

32–5 Nuclear Fission

Under certain conditions, nuclei can break apart into smaller nuclei in a process called **nuclear fission**. The energy released in nuclear fission is very large compared to the energy released in chemical reactions. The fission of one nucleus can indirectly initiate the fission of other nuclei causing a **chain reaction**. The ability to control the chain reaction of $^{235}_{92}\text{U}$ has lead to our modern use of nuclear power.

32–6 Nuclear Fusion

Given enough energy, light nuclei can combine to form a more massive nucleus in a process known as **nuclear fusion**. This larger nucleus has less mass than the individual nuclei that fused to form it. The mass difference is released as energy. As a potential energy source, nuclear fusion is even more powerful than fission. Because of the strong electric force of repulsion, or "Coulomb repulsion," among the nuclei when they are close together the nuclei must have a very high temperature. When fusion is initiated in a high-temperature gas containing nuclei the process is called a **thermonuclear fusion reaction**. It requires a temperature of about 10^7 K to ignite thermonuclear fusion of protons (hydrogen nuclei) into helium nuclei. This latter process is called the proton-proton cycle and is the main energy source of the Sun.

32–7 Practical Application of Nuclear Physics

Biological Effects of Radiation

The high-energy particles given off by radioactive decay can be very damaging to living tissue by ionizing its molecules and therefore changing its structure. Several quantities have been defined in order to quantify the biological effects of radiation. The first of these quantities is the **roentgen** (R), which measures the amount of ionization caused by X-rays and γ-rays. A dosage of one roentgen of X-rays or γ-rays is the amount that produces 2.58×10^{-4} C of charge, from ionized molecules, in 1 kg of dry air at STP. That is,

$$1 \text{ R} = 2.58 \times 10^{-4} \text{ C/kg}$$

Another quantity is the **R**adiation **A**bsorbed **D**ose, or **rad**. A dosage of 1 rad of any type of radiation delivered to a material is the amount of radiation that results in 0.01 J of absorbed energy by a 1-kg sample of material. So,

$$1 \text{ rad} = 0.01 \text{ J/kg}$$

Still another quantity is the **R**elative **B**iological **E**ffectiveness, or **RBE**. The *relative* in the RBE dose means relative to a 1-rad dose of 200-keV X-rays. A dose of 1 RBE of a particular type of radiation occurs when the ratio of the dose of 200 keV X-rays to the dose of the given type of radiation that would produce the same biological effect is 1. Thus, for a given biological effect,

$$\text{RBE} = \frac{\text{dose of 200 keV X-rays}}{\text{dose of given type of radiation}}$$

RBE is a dimensionless quantity. Another quantity that combines the rad and RBE is called the **biologically equivalent dose**. This dose is measured in a unit called **rem**, which stands for **R**oentgen **E**quivalent **M**an. The definition of dosage in rem is

$$\text{dose in rem} = \text{dose in rad} \times \text{RBE}$$

With this definition, 1 rem of any type of radiation produces the same biological damage.

Magnetic Resonance Imaging

The interaction of nuclei with magnetic fields is used in magnetic resonance imaging, or MRI. An important advantage of MRI is that the radiation produced is low-energy radiation. Because the photons have low energy, they cause very little tissue damage.

PET Scans

In a Positron-Emission Tomography scan, or PET scan for short, a patient is given a radioactive substance that undergoes $\beta+$ decay, giving off a positron. This positron almost immediately annihilates with an electron within the patient's body. The gamma rays given off by this radiation penetrate through and leave the body where they can be detected during the scan. The analysis of the results from such as scan can provide important biological information.

32–8 Elementary Particles

One of the many goals of physics is to understand the world at its most fundamental level. This endeavor has led to the study of **elementary particles** that are thought to be the "building blocks" of all matter. To the best of our current understanding elementary particles interact through the four fundamental forces of nature. These forces are called the strong nuclear force, the electromagnetic force, the **weak nuclear**

force, and gravity. You have seen the need for each of these except the weak nuclear force; an explanation of the weak force is beyond the scope of this text, but the weak force is active in the process of beta decay.

Particles that experience the weak nuclear force but not the strong nuclear force are called **leptons**. There are six known leptons, of which the electron is one. As far as we know, no lepton has internal structure, so these are truly elementary particles. Particles that experience both the weak and strong nuclear forces are called **hadrons**; protons and neutrons fall into this category. All hadrons are made up of elementary particles called **quarks**. Some hadrons consist of two quarks, these are called **mesons**, and some hadrons consist of three quarks, these are called **baryons**. There are six quarks; these quarks are grouped into pairs called *flavors*. The flavors of the quarks are *up* (u) and *down* (d), *charm* (c) and *strange* (s), and *top* (t) and *bottom* (b). Quarks carry electric charges that are fractions of e, either $\pm(2/3)e$ or $\pm(1/3)e$ depending on the quark. Quarks also carry another kind of charge called *color*. There are three different quark colors usually referred to as red, green, and blue. The theory that describes the interaction of quarks via the color charge is called **quantum chromodynamics** (QCD).

32–9 Unified Forces and Cosmology

It is widely believed that at the beginning of the universe, at the Big Bang, there was only one fundamental force of nature, called the **unified force**. As the universe evolved it is believed that the four forces we detect today separated from each other during processes that can be thought of as cosmological phase transitions. Today, scientists are trying to work backward to develop the theory of this unified force. One successful example is the **electroweak theory**, which has demonstrated that the electromagnetic and the weak nuclear forces are really just different aspects of the same basic force.

32–10 Gravity Waves

In the early 1900s Einstein developed a new theory of gravity called *general relativity*. This theory predicts the very interesting phenomenon of **gravity waves**. Basically, gravity waves are waves in space and time that travel at the speed of light. The direct detection of gravity waves has not yet occurred although there is strong indirect observational evidence for their existence. Today a major effort is underway to both detect gravity waves and use the information to gain deeper insight into the universe. This effort relies on the construction of several Laser Interferometer Gravitational Wave Observatories, or LIGOs. If they are successful, they will create an entirely new branch of science — gravitational wave astronomy.

Reference Tools and Resources

I. Key Terms and Phrases

nucleon either a proton or a neutron in the nucleus

atomic number the number of protons in a nucleus

neutron number the number of neutrons in a nucleus

mass number the number of nucleons in a nucleus

isotopes the nuclei with the same number of protons but different numbers of neutrons

atomic mass unit a unit of mass, u, commonly used for atoms and nuclei: $1\ \text{u} = 1.660540 \times 10^{-27}$ kg

fermi a unit of distance, fm, commonly used in atomic and nuclear physics: $1\ \text{fm} = 10^{-15}$ m

strong nuclear force one of the fundamental forces of nature that is responsible for holding nuclei together

radioactivity the high-energy emissions from nuclear decay due to a nucleus either being unstable or undergoing a transition from an excited state to a lower-energy state

activity the rate at which nuclear decay occurs

curie a common unit of measure for activity: $1\ \text{Ci} = 3.7 \times 10^{10}$ decays/s

half-life the time required for the number of radioactive nuclei to reduce by half

binding energy the amount of energy that must be added to a nucleus to break it apart into its constituent nucleons

nuclear fission when a nucleus breaks apart into smaller nuclei

nuclear fusion when light nuclei combine to form a heavier nucleus

elementary particles the fundamental particles out of which matter is made

weak nuclear force a fundamental force of nature important in radioactive decay

leptons elementary particles that experience the weak force but not the strong force

hadrons particles that experience both the weak and strong nuclear forces and are made up of quarks

quarks fundamental particles that make up the hadrons

unified force the single force of nature believed to be at work at the beginning of the universe

gravity waves waves of space and time predicted by Einstein's theory of gravity

II. Important Equations

Name/Topic	Equation	Explanation
nucleus	$r = \left(1.2 \times 10^{-15}\,\text{m}\right)A^{1/3}$	An empirical estimate of the radius of a nucleus.
radioactivity	$N = N_0 e^{-\lambda t}$	The time dependence of the number of radioactive nuclei.

III. Know Your Units

Quantity	Dimension	SI Unit
atomic number (Z)		
neutron number (N)	dimensionless	—
mass number (A)		
activity (R)	$[T^{-1}]$	Bq
half-life ($T_{½}$)	$[T]$	s
absorbed dose	$[L^2][T^{-2}]$	Gy
biologically equivalent dose	$[L^2][T^{-2}]$	Sv

IV. Tips

When it comes to the biological effects of radiation the SI units are less familiar than those mentioned in the textbook. Nevertheless, you should be aware of what they are. For the absorbed dose of radiation the SI unit is the gray (Gy): 1 Gy = 1 J/kg. For the biologically equivalent dose, the SI unit is the sievert (Sv): 1 Sv = 100 rem. These two units, the gray and the sievert, that are referred to in "Know Your Units" table.

Practice Problems

1. The nucleus $^{238}_{96}X$ has how many protons?
2. In problem 1 what is the number of neutrons?
3. To the nearest hundredth of a fermi, what is the radius of uranium-231?

4. A nucleus with an atomic mass of 238.05353 alpha decays into a nucleus of mass 234.03654. To the nearest hundredth of a MeV, how much energy is released?
5. A sample of 1.32 x 10^7 nuclei with a decay constant of 1.86 x 10^{-5} sits for 2 days. How many are left?
6. In problem 5 what was the initial activity, in decays/second (Bq)?
7. The carbon-14 activity of a 133-gram sample of ancient wood has an activity of 2.32 Bq. To the nearest year, what is its age?
8. A stable nucleus with 87 protons and 125 neutrons must have a mass of less than __ u (to the nearest ten thousanth of a u).
9. Uranium-236 fissions into ${}^{154}_{53}X$ and another nucleus with the release of three neutrons. What is the atomic mass number of the second nucleus?
10. In problem 9, what is the atomic number of the second nucleus?

Puzzle

A HOT HOUSE

Radium and radon are both naturally occurring radioactive substances found virtually everywhere. Why is radon considered a much more worrisome environmental problem?

Selected Solutions

7. (a) $K = \frac{1}{2}mv^2 \Rightarrow v = \sqrt{\frac{2K}{m}} = \sqrt{\frac{2K}{m_{He} - 2m_e}}$, where m = (mass of He atom) – (mass of 2 electrons).

Therefore,

$$v = \left[\frac{2(0.50\text{ MeV})}{(4.002603\text{ u})\left(\frac{931.494\text{ MeV/c}^2}{1\text{ u}}\right) - 2(0.511\text{ MeV/c}^2)}\right]^{1/2} = c\left[\frac{1.0\text{ MeV}}{3727.38\text{ MeV}}\right]^{1/2}$$

$$= \left(3.00\times10^8\ \tfrac{\text{m}}{\text{s}}\right)\sqrt{\frac{1.0}{3727.38}} = \boxed{4.9\times10^6\text{ m/s}}$$

(b) The closest approach occurs when all the kinetic energy is converted into electric potential energy:

$$\frac{k(Z_{gold}e)(Z_{helium}e)}{d} = K \Rightarrow d = \frac{k(Z_{gold})(Z_{helium})e^2}{K}$$

Therefore,

$$d = \frac{\left(8.99\times10^9\ \frac{\text{N}\cdot\text{m}^2}{\text{C}^2}\right)(79)(2)(1.602\times10^{-19}\text{C})^2}{(0.50\times10^6\text{ eV})\left(\frac{1.602\times10^{-19}\text{ J}}{1\text{ eV}}\right)} = \boxed{460\text{ fm}}$$

(c) Since $Z = 29$ for copper and $Z = 79$ for gold, the repulsive force the alpha particle experiences is much less when it approaches the copper nucleus. Thus, the distance of closest approach would be $\boxed{\text{less than}}$ that found in part (b).

15. (a) $\boxed{{}^{212}_{84}\text{Po} \rightarrow {}^{208}_{82}\text{Pb} + {}^{4}_{2}\text{He}}$

We can use the masses of the atoms, since the numbers of electrons will balance:

$$m_i = 211.988842\ \text{u}$$
$$m_f = 207.97664\ \text{u} + 4.002603\ \text{u} = 211.97924\ \text{u}$$
$$\Delta m = m_f - m_i = 211.97924\ \text{u} - 211.988842\ \text{u} = -0.00960\ \text{u}$$

Therefore,

$$E = |\Delta m| c^2 = (0.00960\ \text{u})\left(\frac{931.494\ \frac{\text{MeV}}{c^2}}{1\ \text{u}}\right)c^2 = \boxed{8.94\ \text{MeV}}$$

(b) $\boxed{{}^{239}_{94}\text{Pu} \rightarrow {}^{235}_{92}\text{U} + {}^{4}_{2}\text{He}}$

Again, the atomic masses are

$$m_i = 239.052158\ \text{u}$$
$$m_f = 235.043925\ \text{u} + 4.002603\ \text{u} = 239.046528\ \text{u}$$
$$\Delta m = m_f - m_i = 239.046528\ \text{u} - 239.052158\ \text{u} = -0.005630\ \text{u}$$

Therefore,

$$E = |\Delta m| c^2 = (0.005630\ \text{u})\left(\frac{931.494\ \frac{\text{MeV}}{c^2}}{1\ \text{u}}\right)c^2 = \boxed{5.244\ \text{MeV}}$$

23. First we need to find the decay constant from the half-life

$$T_{1/2} = \frac{\ln 2}{\lambda} \quad \Rightarrow \quad \lambda = \frac{\ln 2}{T_{1/2}} = \frac{\ln 2}{122\ \text{s}}$$

Thus, since $N = (10^{-4})N_0 = N_0 e^{-\lambda t}$, we have that

$$10^{-4} = e^{-\lambda t} \quad \Rightarrow \quad e^{\lambda t} = 10^4$$

$$\therefore \quad t = \frac{\ln 10^4}{\lambda} = (4 \ln 10)\left(\frac{122\ \text{s}}{\ln 2}\right) = \boxed{27.0\ \text{min}}$$

33. (a) ${}^{56}_{26}\text{Fe}$

Using the atomic masses,

$$m_i = 55.934939 \text{ u}$$
$$m_f = 26(1.007825 \text{ u}) + 30(1.008665 \text{ u}) = 56.463400 \text{ u}$$
$$\Delta m = 56.463400 \text{ u} - 55.934939 \text{ u} = 0.528461 \text{ u}$$

Therefore,

$$\frac{E}{56} = \frac{|\Delta m|c^2}{56} = \frac{(0.528461 \text{ u})\left(\frac{931.494 \frac{\text{MeV}}{c^2}}{1 \text{ u}}\right)c^2}{56} = \boxed{8.79033 \text{ MeV/nucleon}}$$

(b) ${}^{238}_{92}\text{U}$

Using the same procedure,

$$m_i = 238.050786 \text{ u}$$
$$m_f = 92(1.007825 \text{ u}) + 146(1.008665 \text{ u}) = 239.984990 \text{ u}$$
$$\Delta m = 239.984990 \text{ u} - 238.050786 \text{ u} = 1.934204 \text{ u}$$

Therefore,

$$\frac{E}{238} = \frac{|\Delta m|c^2}{238} = \frac{(1.934204 \text{ u})\left(\frac{931.494 \frac{\text{MeV}}{c^2}}{1 \text{ u}}\right)c^2}{238} = \boxed{7.57017 \text{ MeV/nucleon}}$$

51. **(a)** By definition of the rad:

$$E = (\text{dose in rad})m = (215 \text{ rad})(0.17 \text{ kg})\left(\frac{1 \text{ J}}{100 \times 0.01 \text{ J}}\right) = \boxed{0.37 \text{ J}}$$

(b) Using the expression for the amount of heat that produces a specific temperature change,

$$Q = mc_W\Delta T \Rightarrow \Delta T = \frac{Q}{mc_W} = \frac{E}{mc_W} = \frac{(215 \text{ rad})(0.17 \text{ kg})\left(\frac{1 \text{ J}}{100 \times 0.01 \text{ J}}\right)}{(0.17 \text{ kg})\left(4186 \frac{\text{J}}{\text{kg} \cdot \text{K}}\right)} = \boxed{0.51 \text{ mK}}$$

Answers to Practice Quiz

1. (c) 2. (c) 3. (d) 4. (a) 5. (b) 6. (b)

Answers to Practice Problems

1. 96
2. 142
3. 7.36 fm
4. 13.40 MeV
5. 530538
6. 246 Bq
7. 21325 yr
8. 213.7639 au
9. 79
10. 39